# Physiological Assessment of Human Fitness

## Second Edition

Peter J. Maud, PhD

New Mexico State University

Carl Foster, PhD

University of Wisconsin at La Crosse

*Editors*

**HUMAN
KINETICS**

**Library of Congress Cataloging-in-Publication Data**

Physiological assessment of human fitness / Peter J. Maud, Carl Foster, editors.-- 2nd ed.
    p. ; cm.
  Includes bibliographical references and index.
  ISBN 0-7360-4633-X (hard cover)
  1. Exercise--Physiological aspects. 2. Physical fitness--Testing. 3. Exercise tests.
  [DNLM: 1. Exercise--physiology. 2. Exercise Test. 3. Physical Fitness--physiology. QT 255 P5812 2006] I.
Maud, Peter J., 1938- II. Foster, Carl.
  QP301.P57 2006
  613.7'028'7--dc22

                              2005014253

ISBN: 0-7360-4633-X

The Web addresses cited in this text were current as of August 29, 2005, unless otherwise noted.

**Acquisitions Editor:** Michael S. Bahrke, PhD; **Developmental Editor:** Renee Thomas Pyrtel; **Assistant Editors:** Ann M. Augspurger, Bethany J. Bentley, and Kevin Matz; **Copyeditor:** Robert Replinger; **Proofreader:** Erin Cler; **Indexer:** Dan Connolly; **Permission Manager:** Dalene Reeder; **Graphic Designer:** Nancy Rasmus; **Graphic Artist:** Denise Lowry; **Photo Manager:** Sarah Ritz; **Cover Designer:** Jack W. Davis; **Photographer (cover):** © Getty Images; **Photographs (interior):** © Philippe Lopes (chapter 4), Everett Harman (chapter 7), Kenneth Rundell (chapter 10), Jill Kanaley (chapter 11), and Peter Maud (chapter 12); **Art Manager:** Kelly Hendren; **Illustrator:** Craig Newsom; **Printer:** Sheridan Books

Printed in the United States of America     10  9  8  7  6  5  4  3  2  1

**Human Kinetics**
Web site: www.HumanKinetics.com

*United States:* Human Kinetics, P.O. Box 5076, Champaign, IL 61825-5076
800-747-4457
e-mail: humank@hkusa.com

*Canada:* Human Kinetics, 475 Devonshire Road Unit 100, Windsor, ON N8Y 2L5
800-465-7301 (in Canada only)
e-mail: orders@hkcanada.com

*Europe:* Human Kinetics, 107 Bradford Road, Stanningley, Leeds LS28 6AT, United Kingdom
+44 (0) 113 255 5665
e-mail: hk@hkeurope.com

*Australia:* Human Kinetics, 57A Price Avenue, Lower Mitcham, South Australia 5062
08 8277 1555
e-mail: liaw@hkaustralia.com

*New Zealand:* Human Kinetics, Division of Sports Distributors NZ Ltd., P.O. Box 300 226 Albany, North Shore City, Auckland
0064 9 448 1207
e-mail: info@humankinetics.co.nz

# contents

# preface

Many changes have been made in this second edition of the text. Some chapters have been eliminated, new ones have been added, and in some cases material from two chapters has been combined into one. Author coverage of some chapters has also changed. Significant changes have been made in all chapters to provide up-to-date coverage of the topic being discussed.

One of the most often evaluated fitness parameters is maximum oxygen uptake, which has long been considered the gold standard for the evaluation of cardiorespiratory fitness. Direct assessment of this variable is preferred. Many tests, however, have been devised to estimate maximum oxygen uptake, or general cardiovascular condition, using indirect measures purported to correlate well with direct assessment.

Direct methods for determination of aerobic power are detailed in chapter 2 by Davis, who gives a historical overview of the open-circuit method of determination of maximum oxygen uptake and the added value of using one of the many automated breath-by-breath systems. He discusses test modalities, exercise-testing protocols, and criteria for the achievement of maximum values.

In chapter 3, Billat and Lopes describe indirect methods for the estimation of maximum oxygen uptake and provide detailed examples of specific protocols.

Lopes and White discuss heart rate variability in chapter 4. They provide the background information about heart rate sympathetic and parasympathetic control mechanisms and different methods of variability measurement. The effect of heart rate variability in response to exercise programs is discussed, as is the possible use of this measure to identify overtraining.

The so-called anaerobic and aerobic thresholds, identified primarily by changes in lactate accumulation and ventilatory and respiratory values, provide information about the capacity to perform exercise of an endurance nature. Foster and Cotter review these phenomena in chapter 5, discuss various terminologies used, and provide a rationale concerning appropriate methods of measurement.

Testing for anaerobic abilities, covered in chapter 6, is a somewhat controversial issue in that no labora- tory methods are universally accepted as appropriate. The two prime methods used are to measure either mechanical power output or oxygen deficit. Methods of measurement of these two variables are discussed, as are the limitations and problems of trying to quantify anaerobic energy contributions to exercise.

Muscular power is an important requisite for most athletic events. Methods of assessment range from field tests to laboratory techniques. In chapter 7 Harman describes the high-speed measurement of human mechanical power and gives a readily understandable review of the methodologies for recording mechanical power output. His discussion of specific technical requirements and techniques is of particular interest, especially to those with somewhat limited backgrounds in computer application.

Kraemer, Ratamess, Fry, and French provide a comprehensive review and an extensive bibliography of the strength-testing domain in chapter 8. Detailed methodologies for different modalities are provided with consideration given also to individual differences and specifics of performance method.

Evaluation of muscle tissue from muscle biopsy has been undertaken on numerous occasions (Gollnick & Matoba, 1984). Inclusion of a detailed chapter covering this area, however, may be controversial in that, as expounded by Gollnick and Matoba, routine evaluation of muscle tissue for prediction of athletic success should probably not occur. But this text is also designed for research purposes. Information obtained from muscle biopsy may contribute significantly to the study of chronic exercise effects. In chapter 9, McGuigan and Sharman provide an excellent introduction to, and comprehensive coverage of, muscle biopsy technique, histochemical preparation, fiber typing, and analysis of chemical composition of muscle tissue. These techniques allow complete analysis of muscle structure and function.

The use of near-infrared spectroscopy as a tool to measure tissue oxygen kinetics and blood flow dynamics noninvasively in the assessment of the athlete's response to exercise is described by Rundell and Im in chapter 10. This relatively new noninvasive technique is believed to hold great potential for evaluation of response to exercise training regimens.

Body composition is important when considering suitability for specific athletic events. Moreover, with national and world attention now being directed to the medical problems and associated potential astronomical costs of increased human obesity, body composition is important from a health fitness perspective. In chapter 11, Pollock, Kanaley, Garzarella, and Graves provide in-depth coverage of current body composition assessment techniques and complete bibliographical references.

Flexibility is also important for many athletes, whether they view performance in terms of time, distance, speed, or aesthetics. Many also suggest that flexibility is an important contributor to normal daily function from a health perspective. In chapter 12, Maud and Kerr discuss the limitations of certain measurement methods and provide details of preferred methods for assessment of joint range of motion and muscle length.

Chapter 13 is devoted to the collection of physiological data in the field, a topic that certainly warrants attention because it is too often neglected. As noted by authors Foster, Daniels, deKoning, and Cotter, athletic performance does not take place in the laboratory but rather in the field under a myriad of physical, environmental, sociological, and psychological conditions. Testing under these conditions results in more realistic exposure when considering athletic performance. The practical considerations for testing outside the laboratory are adequately covered.

## Exclusions

Although a rationalization could be given for the inclusion of more comprehensive evaluations of cardiovascular function by use of such means as ECG exercise tolerance testing, particularly when health is the prime concern, and assessment of visual abilities when considering either health or athletic performance, tests such as these are not included because of their specialized nature and probable necessity for some form of medical supervision. Similarly, the perceptual motor domain is not included because, as already discussed, the commonly held view is that general motor ability does not exist. Instead, most believe that several broad and independent perceptual motor abilities and physical proficiencies usually apply to specific athletic events.

# one

# Fitness Assessment Defined

Peter J. Maud, PhD

*New Mexico State University*

No definition of fitness appears to be widely accepted, nor is there agreement about what specific components should make up a fitness evaluation. Over 60 years ago Steinhaus (1936), evidently viewing fitness from the perspective of the physiologist, defined it as distance from death, a description somewhat like that of many in the medical profession, who tend to regard physical fitness as absence of disease. Willgoose (1961) has defined it as "a capacity for sustained physical activity" (p. 105). A more appropriate and universal definition of physical fitness, at least from a health perspective, may be the one found in *Mosby's Medical and Nursing Dictionary* (1986), which defines physical fitness as "the ability to carry out daily tasks with alertness and vigor, without undue fatigue, and with enough reserve to meet emergencies or to enjoy leisure time pursuits" (p. 880).

Differences in interpretation of the term probably depend on whether fitness is applied to health or relative to athletic competition. Higher levels of fitness are obviously necessary for success in athletics. Test items that are more specific are required for measurement of fitness attributes when successful, rewarding, and enjoyable participation in sport is desired. Here, where the primary intention is to discuss fitness from a physiological or physiologically related perspective, it is necessary to define fitness parameters and select those items appropriate for inclusion. Consideration must also be given to techniques that allow determination of changes that result from different types of training programs, whether information so gained is useful for planning health enhancement programs or for improving fitness for sport participation.

Fitness assessment may be viewed from several different perspectives, including determination of energy system utilization (which may be particularly important for sport participation), evaluation of fitness specifically for health enhancement purposes, or use of traditional component tests for determination of sport fitness and profiling. Evaluation of the perceptual motor domain and vision requirements for improved sport performance could also be included in fitness assessment. Additionally, study of the historical development of fitness testing is interesting and applicable to an understanding of current practices.

The inclusion of the term *physiological* in the title of the text may be somewhat debatable because several of the test areas assessed are not necessarily purely physiological. Some, for example, are more anatomical or morphological in nature. All of the testing areas included in the text, however, relate to attributes necessary for profiling human response to exercise.

## Historical Perspectives

This chapter includes a few examples of the types of tests that were used to assess fitness in earlier times. The tests described here are not necessarily those that were the most accepted, but they serve as examples of what was thought to be current and appropriate at that time. Three general areas of fitness—anthropometry; muscular strength, endurance, and power; and cardiovascular fitness—received prime consideration. An extensive review of the development of fitness tests and an excellent source for references is the text titled *Physical Fitness Appraisal and Guidance*, published in 1947 by a renowned pioneer of fitness testing, Thomas Kirk Cureton.

### Anthropometric Measurement

Some of the earliest tests used to evaluate or describe fitness were restricted to anthropometric measurement. An early pioneer of anthropometric assessment was Hitchcock of Amherst College, who measured such attributes as age, height, weight, chest girth, arm girth, forearm girth, and lung capacity, as described in the text that he coauthored with Seelye in 1893, and in which he described data collected approximately between the years 1861 and 1880. (Seaver, however, had published *Anthropometry in Physical Education* in 1890 and thus is perhaps the first modern author in this area.) Description of body type can be traced back to Hippocrates, who recognized two basic body types, and later to Rostan, who in the late 19th century proposed a classification system of three body types. Sheldon, in association with Stevens and Tucker (1940), made the first serious attempt to classify body types as having proportions that were a mixture of the three general classifications of ectomorph, mesomorph, and endomorph. The study of somatotype and anthropometry, and the study of their relationship to human motion are still important today, as illustrated by the formation of a new association, the International Society for the Advancement of Kinanthropometry in 1986, during the VIII Commonwealth and International Conference on Sport, Physical Education, Dance, Recreation, and Health held in Glasgow, Scotland. Beunen and Borms described the purpose and function of the new organization in 1990.

## Assessment of Muscular Strength, Endurance, and Power

In about 1880, Sargent probably initiated the move to include strength tests as a major component of fitness evaluation. In 1896, however, Kellogg described a "universal dynamometer" that made possible muscle strength testing that was more accurate. Sargent, in a 1921 journal article titled "The Physical Test of Man," described the vertical jump test—or Sargent jump, as it is now popularly known—one of the first tests of muscular power. Rogers (1927) developed the strength index, considered a general test of athletic ability, and a physical fitness index, derived by comparing the strength index with norms based on gender, weight, and age. He probably contributed more to the popularity of strength testing than any other individual did during the earlier part of the century. The test used seven measurements, consisting of left and right handgrip strength, back strength, leg strength, strength of the arms and shoulder girdle as measured in men by dips on the parallel bars and by pull-ups with overhand grip, and forced vital capacity. Others modified the original test. For example, MacCurdy (1933) eliminated the endurance factors

associated with the pull-up and dip tests by replacing them with static tests of arm strength using the back/leg dynamometer. McCloy (1939) later eliminated the lung capacity test, arguing that it was not a test of muscle strength.

Isometric tests evaluated by the cable tensiometer using methodology described by, for example, Clarke (1966), and the typical spring dynamometer are still in use, as are such field tests as dips, pull-ups, and push-ups. Still one of the more popular and applicable methods used to assess muscular strength is 1RM testing. Today, however, measurement of isometric and isokinetic strength by means of force transducers or load cells, and strain gauges, allows computer interface for data collection, thus facilitating test recording and evaluation.

## Evaluation of Cardiovascular Fitness

Many well-known tests of cardiovascular fitness date back to the early 20th century. One of the first compared heart rate and systolic blood pressure responses between the horizontal and standing positions (Crampton, 1913). The belief was that a fitter person would exhibit a maximal rise in blood pressure with no change in heart rate, "evidencing such a complete working of the splanchnic mechanism that the heart would not be called to help raise blood-pressure" (McCurdy & Larson, 1939, p. 267). The Barach cardiovascular test, described in 1919, was another test that used blood pressure and heart rate to evaluate cardiovascular function. This test, which was claimed to indicate the energy demands of the circulatory system, produced an index by adding systolic and diastolic pressures and then multiplying by heart rate.

One of the earlier tests used to investigate the response of heart rate to exercise and recovery was the Foster cardiovascular test published in 1914. Preexercise, immediate postexercise, and 45 s postexercise heart rates were obtained. These three heart rates comprised separate tests with norms established for all three conditions. The exercise component consisted of stepping up and down on the floor for 30 s at a rate of 180 steps per min, with heart rate measured manually for a 5 s period. Lack of standardization of the stepping exercise and the potential for error in measurement of heart rate tended to invalidate the test.

Schneider (1920) used a combination of pulse rate and blood pressure obtained in the horizontal and standing positions, and pulse rates taken immediately following 15 s of bench-stepping exercise and during recovery, to evaluate aviators during World War I. Tuttle described his well-known pulse ratio test, using a bench 13.5 in. (34.3 cm) high, in 1931. A 2 min pulse count obtained following standardized exercise was divided by a 1 min pretest resting rate.

McCurdy and Larson (1939, p. 274) described a test of "organic efficiency" and provided comprehensive normative tables for males aged 18 to 80 years. Test items included sitting diastolic pressure, breath-holding ability taken 20 s following 90 s of standardized bench-stepping exercise, the difference between preexercise standing pulse rate and pulse rate obtained 2 min following standardized exercise, standing pulse pressure, and vital capacity.

More recent tests have included the renowned Harvard step test developed by Brouha (1943) and other step tests that are still in use. Other exercise modalities, such as the treadmill and the bicycle ergometer, have replaced bench stepping in popularity, but heart rates taken during either exercise or recovery are still widely used to assess cardiovascular fitness.

## Contribution of Technology to Fitness Assessment

Some of the most significant advancements in fitness assessment have been made possible by the development of sophisticated equipment and techniques. For example, in the assessment of metabolic response, although the Douglas bag and associated chemical evaluation for gas composition using Scholander equipment may still be found in some laboratory settings, it is far more common to find elaborate, online computer equipment capable of providing instantaneous feedback using breath-by-breath analysis. Similarly, muscular strength, endurance, and power are commonly evaluated in the laboratory by isokinetic methods, with detailed computer analysis, rather than by spring dynamometer or cable tensiometer. Even heart rate measurement has been greatly facilitated and made more accurate by use of the ECG or small and simple, yet accurate, heart rate telemetry systems. In fact, the development of such systems, which now include the ability to make accurate measurement of beat-by-beat heart rate variability, is proving to be useful in the analysis of cardiovascular response to exercise and exercise training. Hydrostatic weighing, anthropometric measurement, and skinfold measurements are still used to assess body composition, but newer methods made possible by enhanced technologies now allow measurement by such means as air displace-ment plethysmography, bioelectric impedance, ultrasound, dual-energy projection, magnetic resonance imaging, and computed tomography. Advanced techniques in body tissue analysis and instruments like the electron microscope have greatly enhanced our ability to evaluate fitness changes at the cellular and ultracellular levels. From a historical perspective these advances have certainly changed the ways that fitness is assessed in the laboratory. Nevertheless, many techniques and methods used in earlier times are still appropriate, particularly the field tests used in many cases to evaluate sport fitness.

## Energy Systems Approach

One approach to fitness evaluation is to base all tests on the energy systems used during physical activity. The contribution to energy requirements by initial stores of adenosine triphosphate (ATP) present within the muscle and the subsequent restoration of these stores by creatine phosphate (CP), by the anaerobic breakdown of glycogen, or by aerobic utilization of glycogen, fat, and protein, would require assessment. Fox, Robinson, and Wiegman (1969) classified activity by the prime source of energy being used and proposed four time periods. Period 1 comprised activities that lasted less than 30 s. They suggested that these activities depended primarily on the contribution of ATP and CP. Period 2, from 30 to 90 s, used mainly the phosphagens and anaerobic glycolysis. Period 3, lasting from 90 to 180 s, depended mainly upon anaerobic glycolysis and aerobic metabolism, and period 4, consisting of exercise past 180 s, was reported to be primarily aerobic in nature. Shephard (1978) described another classification system using a similar approach but with five phases. Phase 1 was a single maximum contraction; phase 2 comprised very brief events that lasted less than 10 s; phase 3 was for events that lasted 10 to 60 s; phase 4 was for activity that lasted from 1 min to 1 h; and phase 5 was for prolonged events that lasted in excess of 1 h. Skinner and Morgan (1984) have proposed another four-period classification system based on more recent research, particularly in the areas of power output and lactate tolerance. They suggested that it consists of a peak anaerobic power phase lasting 1 to 10 s, in which initial stores of ATP and CP would be the main energy sources; a mean anaerobic power phase of 20 to 45 s, in which anaerobic glycolysis, in addition to ATP and CP stores, would be the prime energy contributor; a lactic acid tolerance phase lasting 1 to 8 min; and an aerobic phase of 10 or more min, in which aerobic metabolism would be the prime energy source.

Use of gas analysis during maximal work tests to derive maximum oxygen uptake is well documented as indicating the contribution of the aerobic system to fitness. Methods for assessing other systems' contributions are far more controversial. Numerous tests have been devised to measure mechanical power output during time periods thought to represent the energy contribution phase. A short, single contraction, as in the vertical jump test described by Sargent (1921), would, if considered only from the perspective of the energy requirements for the activity, indicate the contribution of ATP stores. The 2 to 4 s required to complete the Margaria stair-run test (Margaria, Aghemo, & Rovelli, 1966) or the 5 s Wingate test (Bar-Or, 1987; Bar-Or et al., 1980) would represent peak anaerobic power in which ATP and CP stores should be the main energy contributors. The 30 s Wingate test (Bar-Or et al., 1980) measures mechanical mean anaerobic power output during the anaerobic glycolytic phase. But there are many problems in assuming that the power output during specified time periods does in fact represent specific energy contributions. Muscle biopsy techniques may allow quantification of ATP and CP stores, but the size, strength, and speed of contraction of the muscle, the predominant fiber type, the structural arrangement of the individual muscle, and coordination all affect power output. Because it is easy to measure, blood lactate is commonly used to predict anaerobic glycolytic ability, but diffusion of lactate into the blood does not provide an accurate assessment of individual muscle contribution. Furthermore, motivation plays an important role in tests requiring maximal effort. In the measurement of maximum oxygen uptake, parameters that denote achievement of a true maximal effort have been identified, but measurable characteristics to identify maximal effort are less clearly defined for anaerobic testing.

## Health Fitness

Probably the most commonly accepted components of health fitness include cardiovascular endurance, body composition, flexibility, and muscular strength and endurance. These four components, however, are not necessarily accepted as the only ones that need to be assessed. Medically oriented health fitness evaluation centers offer far more comprehensive programs that may include, for example, extensive cardiovascular evaluations with an ECG exercise tolerance test, blood chemistry analysis and blood count, maximum oxygen uptake measurement, pulmonary function tests, and orthopedic assessments (Maud & Longmuir, 1983).

The American College of Sports Medicine (ACSM) (2000), for safety reasons, has recommended that subjects be screened before they undergo moderate to vigorous tests. These recommendations include using the PAR-Q questionnaire developed by the Canadian Society for Exercise Physiology (1994) as an initial screening procedure. The type of subsequent testing, submaximal or maximal, and whether medical supervision is required depends on the age of the subject and whether risk factors for cardiopulmonary or metabolic diseases are present. The ACSM guidelines should be reviewed and followed before conducting any fitness tests.

Although there may be debate as to whether or not extensive medical testing in the evaluation of fitness for the general population is beneficial, and the American College of Cardiology and the American Heart Association (1986) generally do not believe that diagnostic exercise testing is of value to apparently healthy people, such testing is readily available. Certainly such tests as blood chemistry, blood pressure screening, and ECG evaluations are invaluable when assessing health status and predisposition toward cardiovascular disease and diabetes (Smith, 1988).

## Fitness Evaluation for Athletic Participation

A plethora of tests has been developed to evaluate the fitness of athletes representing a variety of sports and activities. Reviewing all athletic group fitness profiles or individual tests administered would be a monumental task. Methods used to evaluate athletic fitness depend on the requirements of the individual sport or event, the reason for administering the tests, the availability of equipment and facilities, the practicality of assessment, and the personal perspectives of the researcher.[1]

To give the reader some idea about the variety of tests that have been employed to assess athletes, fol-

---

[1] A detailed description of test methods and protocols used by a variety of Australian sporting organizations may be found in the authoritative text published for the Australian Sports Commission (2000), edited by Christopher Gore and titled, *Physiological Tests for Elite Athletes.*

lowing is a brief outline obtained from 14 studies covering the period from 1976 to 2004, a total of 28 years. Athletes from the following team sports were studied: basketball (Parr, Hoover, Wilmore, Bachman, & Kerlan, 1978), football (Wilmore et al., 1976), soccer (Raven, Gettman, Pollock, & Cooper, 1976), rugby union (Maud & Schultz, 1984), rugby league (Gabbett, 2002), Australian football (Parkin, 1982), field hockey (Rate & Pyke, 1978), team handball (Rannou, Prioux, Zouhal, Gratas-Delamarche, & Delamarche, 2001), and lacrosse (Withers, 1978). Individual and dual sports included are racquetball (Salmoni, Guay, & Sidney, 1988), tennis (Carlson & Cera, 1984), downhill skiing (National Alpine Staff, 1990), middleweight boxing (Guidetti, Musulin, & Baldari, 2002), and cheerleading (Thomas, Seegmiller, Cook, & Young, 2004). (Although not universally recognized as a sport, cheerleading is included because it has become a complex physical activity requiring many of the same physical attributes as other sports do.) These studies indicate the commonalities and diversities of test items used. Note that little has changed relative to the test items used during the 28-year period. Probably the most significant changes have been in the development of new equipment, such as heart rate monitors and small portable oxygen uptake systems that allow collection of data during athletic endeavors. Also occurring during this time has been a great increase in the amount of research being conducted to study the acute and chronic responses to exercise in normal, athletic, and special populations, which may or may not be applicable to fitness assessment.

• Many investigators have evaluated cardiovascular fitness by direct determination of maximum oxygen uptake, in studies of rugby (Maud & Schultz, 1984), football (Wilmore et al., 1976), soccer (Raven et al., 1976), lacrosse (Withers, 1978), field hockey (Rate & Pyke, 1978), basketball (Parr et al., 1978), handball (Rannou et al., 2001), boxing (Guidetti et al., 2002), tennis (Carlson & Cera, 1984), and cheerleading (Thomas et al., 2004). Others have assessed this parameter by indirect means, using a timed 15 min run for Australian football players (Parkin, 1982), a 2 mi (3.2 km) (McCurdy & Larson, 1939) or 1 mi (1.6 km) (National Alpine Staff, 1990) run for downhill skiers, a 15 min run for racquetball players (Salmoni et al., 1988), and by a multistage shuttle run in one of the rugby studies (Gabbett, 2002).

• Anaerobic capacities have been evaluated by a 440 yd (402 m) (National Alpine Staff, 1990) run for downhill skiers; by the Wingate test for rugby (Maud

& Schultz, 1984), handball (Rannou et al., 2001), and racquetball (Salmoni et al., 1988) athletes; by the Margaria stair-run test for lacrosse (Withers, 1978), field hockey (Rate & Pyke, 1978), and tennis (Carlson & Cera, 1984) players; and by postexercise blood lactate levels in lacrosse players (Withers, 1978) and team handball players (Rannou et al., 2001).

• Grip strength has traditionally been used as a general measure of muscular strength, as in the studies of rugby (Maud & Schultz, 1984), soccer (Raven et al., 1976), Australian football (Parkin, 1982), tennis (Carlson & Cera, 1984), boxing (Guidetti et al., 2002), and racquetball (Salmoni et al., 1988). Leg strength has also been evaluated (Parkin, 1982; Salmoni et al., 1988), as has arm and shoulder strength by the bench press in soccer (Raven et al., 1976), football (Wilmore et al., 1976), and cheerleading (Thomas et al., 2004). Several studies (Carlson & Cera; Parr et al., 1978; Thomas et al., 2004) used isokinetic evaluation to assess strength.

• Testing for muscular endurance has commonly been accomplished by use of field tests. Five such tests were used in the Australian football study (Parkin, 1982). Two studies (Carlson & Cera, 1984; Parr et al., 1978) used isokinetic endurance evaluation.

• Testing for muscular power, one of the most important attributes for successful performance in many games and sports, has frequently used the vertical jump (Maud & Schultz, 1984; National Alpine Staff, 1990; Parkin, 1982; Raven et al., 1976; Salmoni et al., 1988; Gabbett, 2002). Besides being used to assess muscular strength and endurance, isokinetic evaluation was used for power evaluation of tennis (Carlson & Cera, 1984) and basketball players (Parr et al., 1978).

• The sit-and-reach test has been the most widely used measure of flexibility, despite its controversial nature. The only other flexibility assessments used in the studies being examined were wrist and shoulder flexibility in racquetball players (Salmoni et al., 1988) and back hyperextension in basketball players (Parr et al., 1978).

• Skinfold measurement was the most prevalent method for estimating body composition (Carlson & Cera, 1984; Maud & Schultz, 1984; Parkin, 1982; Rate & Pyke, 1978; Raven et al., 1976; Wilmore et al., 1976; Withers, 1978; Parr et al., 1978; Thomas et al., 2004). Four of the studies (Carlson & Cera, 1984; Wilmore et al., 1976; Parr et al., 1978; Thomas et al., 2004) also used the underwater weighing technique. One study (Carlson & Cera, 1984) also described skeletal widths and circumferences, and another described somatotype (Wilmore et al., 1976). In the boxing study (Guidetti et al., 2002) cross-sectional area of the arm and forearm was also described.

- Other data collected to describe athletic attributes have included measurement of speed by timing a 40 yd (36.6 m) dash (National Alpine Staff, 1990; Rate & Pyke, 1978; Salmoni et al., 1988; Gabbett, 2002) or, in the case of the Australian football study (Parkin, 1982), by using 15, 40, and 55 m run times and a 40 m run following a 15 m running start. Investigators measured agility by timing shuttle runs (National Alpine Staff, 1990) or agility runs (National Alpine Staff, 1990; Rate & Pyke, 1978; Raven et al., 1976; Salmoni et al., 1988; Gabbett, 2002).

These examples illustrate the diversity of sports evaluated and assessment methods used. They also illustrate the intermixing of two different types of tests, the so-called field tests and those that require special equipment or are conducted in the laboratory setting.

# Field Tests

For the purpose of this chapter, field tests are defined as those tests that may be completed outside the laboratory environment and do not require specialized equipment for data collection or recording. This definition excludes tests that may be conducted in the field using specialized equipment varying from the relatively simple, such as skinfold calipers, to the sophisticated, such as the equipment used to determine oxygen uptake of athletic activities in the simulated competitive environment. Most tests described in this text are not field tests, although some are included in the chapters dealing with the indirect determination of aerobic power and measurement of muscular strength. Both of these types of tests have a place in the evaluation of health fitness and for the evaluation of athletes. Table 1.1 gives examples of both types of commonly used tests.

**Table 1.1  Classification of Test Types: Field Tests and Tests That Require a Laboratory or Specialized Equipment**

| Fitness parameter | Examples of field tests | Examples of tests either conducted in a laboratory or requiring specialized equipment |
|---|---|---|
| **Aerobic power** | | |
| (a) Maximal tests | 1. Time to cover a specific distance<br>2. Distance covered in a specific period<br>3. Time taken, or distance covered, in a shuttle run to exhaustion, with incremental speed increases at specified time intervals | 1. Continuous tests<br>Maximum oxygen uptake obtained during a continuous progressively increased workload test using a specific exercise modality[†]<br>a. Ramp test with workload continuously increasing<br>b. Test with specific workload increase at specified time intervals<br>2. Discontinuous tests<br>Like 1b except that specific recovery periods are interspersed between exercise stages |
| (b) Submaximal tests | 1. Recovery heart rate following specific-height bench stepping at a specified rate for a specific period | 1. Steady state heart rate response to a specified workload during a specific period[†] |
| **Anaerobic power** | | |
| (a) Peak anaerobic power | 1. Vertical jump height<br>2. Standing broad jump distance<br>3. Timed, short, specific-distance sprints in which time to completion is usually in the 5 s to 10 s range[‡] | 1. Peak power output, usually recorded in W or W/kg, either per 1 s or 5 s, obtained during all-out exercise lasting 5 s to 10 s[††] |
| (b) Mean anaerobic power | 1. Specific-distance sprints in which time to completion is usually within the 30 s to 60 s range | 1. Mean power output, usually recorded in W or W/kg, obtained during all-out exercise over a 20 s to 60 s period[††]<br>2. Oxygen deficit achieved during all-out exercise over a 20 s to 30 s period |

| Fitness parameter | Examples of field tests | Examples of tests either conducted in a laboratory or requiring specialized equipment |
|---|---|---|
| Body composition | 1. Estimates of percent body fat from circumference measurements<br>2. Body height, weight, and frame size | 1. Skinfold measurement<br>2. Underwater weighing<br>3. Air displacement plethysmography<br>4. Bioelectrical impedance<br>5. Ultrasound<br>6. Magnetic resonance imaging and computed tomography |
| Flexibility | 1. Linear measurement from one body segment or specific identifiable site to another or to an external object | 1. Use of goniometer and inclinometer to measure range of motion in degrees |
| Muscular strength | 1. 1RM measurement<br>2. Use of muscular strength to overcome gravitational resistance of body or body part[‡‡] | 1. Isometric strength measurement using a cable tensiometer or dynamometer<br>2. Isometric, isotonic, and isokinetic measurement of strength using force transducers or load cells, and strain gauges |
| Muscular endurance | 1. Number of repetitions completed using a specific weight resistance or a percentage (e.g., 70%) of 1RM<br>2. Time to maintain a specific weight or percentage of 1RM in a set position<br>3. Number of repetitions that can be completed against gravitational pull on body or body part | 1. Measurement of isokinetic endurance by measurement of number of maximum effort contractions that can be made before the maximum force drops below a specified percentage (e.g., 70%) of maximum<br>2. Time that a specific muscle group can maintain a joint at a specific angle using a percentage of maximum (e.g., 70%) force as the load |

[†]Examples of types of equipment that may be used for exercise modality include the treadmill, cycle ergometer, arm ergometer, kayak ergometer, rowing ergometer, cross-country ski ergometer, and swimming flume.

[‡]Although running is usually the mode of activity tested, other activities such as bicycling, swimming, or rowing could be used.

[††]The two most commonly used modes of activity are all-out pedaling on the cycle ergometer or continuous, maximum effort vertical jumping.

[‡‡]Use of typical tests such as the pull-up are complicated by the fact that the greater the number of repetitions achieved, the greater the reliance on muscular endurance rather than muscular strength.

# Overtraining

One area of research that certainly has important implications relative to elite performance is that of overreaching, which results in a short-term loss of performance capacity, or overtraining, which results in a relatively long-term negative effect on performance (Kreider, Fry, & O'Toole, 1998). Although numerous physiological, biochemical, psychological, and immunological signs and symptoms have been described as being present with overreaching or overtraining, finding tests that can identify markers that precede the ultimate drop in performance associated with these phenomena has been difficult. Obviously, tests that could predict the onset of overreaching or overtraining, probably specific to the individual, would be invaluable to the coach and sport physiologist when training athletes. Such tests have yet to be identified. See the comprehensive text by Kreider et al. (1998) for in-depth coverage of this area.

# Perceptual Motor Domain

The performance of complex skills depends on neuromuscular coordination produced in response to sensory feedback and its subsequent processing. Testing within this domain is fraught with problems, mainly stemming from lack of agreement about the specific parameters that define the area.

## Motor Ability

During the 1920s it was hypothesized that ability to perform motor tasks was an inherent characteristic much like intelligence. Researchers therefore believed that they could develop tests similar to those used to measure IQ to predict ability to perform the motor tasks involved in sport and other complex movement patterns. Brace, in 1927, was one of the earliest researchers to develop such a test battery, comprising

20 different stunts designed to evaluate "inherent motor skill" ability. In 1929 Cozens published a test that purported to identify "general athletic ability." This was followed by Johnson's test in 1932 used to evaluate "native neuromuscular skill capacity." McCloy then published a modification of the Brace test in 1937 in an attempt to evaluate "motor educability." Subsequently came the realization that general motor ability does not exist but that there may be a number of rather broad, yet relatively independent, motor abilities, as described by Fleishman (1964). He used a two-classification system to describe motor abilities, one consisting of perceptual motor abilities and the other of physical proficiencies. His tests of perceptual motor ability consisted of 11 items: "control precision, multi-limb coordination, response orientation, reaction time, speed of arm movement, rate control (timing), manual dexterity, finger dexterity, arm-hand steadiness, wrist-finger speed, [and] aiming." The physical proficiency battery included "extent flexibility, dynamic flexibility, static strength, dynamic strength, explosive strength, trunk strength, gross body coordination, equilibrium, [and] stamina" (Fleishman, 1975, p. 1132). This brief discussion of motor abilities indicates that many traits may contribute to fitness for athletic performance, particularly reaction time, balance, movement speed, agility, and coordination. Applicable to specific sports, these abilities need to be evaluated, and they form a significant part of many fitness-testing batteries.

## Vision Testing

Whether vision testing should be separate from other sensory tests or from the psychomotor domain is debatable. Vision can certainly affect athletic performance, and vision testing has been part of the assessment of athletes in the sports medicine program at the United States Olympic Committee training center in Colorado Springs. In sports in which aiming is a crucial skill component, such as archery and shooting, visual abilities are paramount, but vision testing in other sports also has been undertaken. Tests have been designed to evaluate such traits as visual acuity (the sharpness and clarity of vision), dynamic visual acuity (the ability to see moving objects clearly), vision pursuit (the ability to follow the pathway of moving objects), depth perception (the ability to judge distance and speed), and eye–hand–body coordination. If tests such as these differentiate performance ability, then one can argue that they should be part of the process used in the evaluation of the athlete, particularly if training can remedy any deficiencies. Further discussion of this issue, however, is beyond the scope of this text.

## Rationale for Text Test Items

Several authorities have suggested items that should be included in a typical fitness evaluation. The typical test items usually covered include aerobic fitness, peak and mean anaerobic power, anthropometry and body composition, flexibility, and muscular strength and endurance, all of which are used for both health fitness evaluation and assessment of athletic potential and ability. This text includes all of these items despite the argument put forward by Åstrand and Rodahl (1986) that many such test items, "including evaluation of flexibility, skill, strength, etc., are related to special gymnastic or athletic performance" and "are not really suitable for an analysis of basic physiological functions" (p. 355). Many of these items, however, may have a profound effect on physiological performance and, therefore, these tests should be included. Inclusion of such items in tests of health status or athletic ability should provoke little argument because they are crucial to both areas. Several new methods for assessment of physiological response to exercise such as measurement and applicability of heart rate variability, near-infrared spectrophotometry and its use in athletic assessment, and measurement of muscle structure and function are also included.

## Summary

Undoubtedly, athletic competition requires fitness beyond that necessary for optimal health. But the value of specific fitness test items to athletes and coaches, and the use that can be made of data collected, have been much debated (Gollnick & Matoba, 1984; Noakes, 1988). Ultimately, physiological fitness professionals are responsible for assessing the value of the different areas that they might evaluate and the specific methods that they might use for those evaluations. Obviously, they will take different approaches depending on whether the goal is to evaluate health fitness, to assess fitness for successful athletic participation, or to research the response of the human body to varied exercise intensities and regimes.

# two

# Direct Determination of Aerobic Power

## James A. Davis, PhD

*California State University at Long Beach*

Measurement of aerobic power or maximal oxygen uptake ($\dot{V}O_2$max) is one of the oldest and most common measurements in exercise physiology. This chapter outlines the historical development of the equipment used to make the measurement, the exercise test modes and protocols, and the criteria used to determine achievement of $\dot{V}O_2$max. The last section of the chapter deals with reference values for $\dot{V}O_2$max.

The parameters of aerobic function are aerobic power or maximal oxygen uptake ($\dot{V}O_2$max), work efficiency, time constant for $\dot{V}O_2$ kinetics, and the lactate threshold (Whipp, Davis, Torres, & Wasserman, 1981). Maximal $\dot{V}O_2$ is an important parameter because it represents the upper limit of aerobic exercise tolerance. Endurance activities are performed at some fraction of $\dot{V}O_2$max. If $\dot{V}O_2$max is low, then the level of endurance performance is necessarily constrained.

## Measurement of $\dot{V}O_2$max

Whole-body oxygen uptake can be determined from cardiovascular measurements or from respiratory measurements, using equations based on the Fick principle. From cardiovascular measurements,

$$\dot{V}O_2 = C.O. (CaO_2 - C\bar{v}O_2),$$

where C.O. is cardiac output, $CaO_2$ is content of $O_2$ in arterial blood, and $C\bar{v}O_2$ is content of $O_2$ in mixed venous blood. From respiratory measurements,

$$\dot{V}O_2 = \dot{V}_A (F_IO_2 - F_{\bar{A}}O_2)$$

where $\dot{V}_A$ is alveolar ventilation, $F_IO_2$ is the fraction of $O_2$ in inspired gas, and $F_{\bar{A}}O_2$ is the fraction of $O_2$ in the mean alveolar gas; this equation is not rigorous because it does not account for the normally small differences between inspired and expired alveolar volumes. Thus, the determinants of $\dot{V}O_2$ are the heart (C.O.), the mechanical properties of the lungs and the chest wall ($\dot{V}_A$), diffusion of $O_2$ from the alveolus into the pulmonary capillary blood ($F_{\bar{A}}O_2$ and $CaO_2$), and the extraction of $O_2$ from the capillary blood by the muscle cell ($C\bar{v}O_2$). A person with a high $\dot{V}O_2$max necessarily has good function in each of these determinants. Conversely, a sedentary person has relatively poor function for each determinant, which results in a low $\dot{V}O_2$max. If a person has pathology associated with one of these determinants (e.g., coronary artery disease that would limit the cardiac output during exercise), then the $\dot{V}O_2$max will be very low. Hence, one of the important reasons for measuring $\dot{V}O_2$ during graded exercise testing (GXT) is to establish whether the $\dot{V}O_2$max is normal.

Two issues are unresolved regarding the measurement of $\dot{V}O_2$max. The first is the criteria used to establish whether $\dot{V}O_2$max has been achieved during GXT. The most widely accepted criterion is that the $\dot{V}O_2$ plateaus during the later stages of the GXT as the work rate continues to increase. Many subjects, however, clearly reach their limit of tolerance during GXT without demonstrating a plateau in $\dot{V}O_2$. The second unresolved issue is the recommendation that if a subject does not demonstrate a plateau in $\dot{V}O_2$, then the term should be $\dot{V}O_2$peak, not $\dot{V}O_2$max. A later section of this chapter more fully discusses both of these unresolved issues.

## Equations

First, it is necessary to present the measurements that need to be made. The basic equation is as follows:

$$\dot{V}O_2 = \dot{V}_I F_I O_2 - \dot{V}_E F_{\bar{E}} O_2 \qquad (1)$$

where $\dot{V}_I$ is inspired ventilation, $\dot{V}_E$ is expired ventilation, $F_I O_2$ is the fraction of $O_2$ in the inspired gas, and $F_{\bar{E}} O_2$ is the fraction of $O_2$ in the mixed expired gas. However, it is generally assumed that inspired nitrogen ($\dot{V}_I F_I N_2$) equals expired nitrogen ($\dot{V}_E F_{\bar{E}} N_2$). Thus,

$$\dot{V}_I F_I N_2 = \dot{V}_E F_{\bar{E}} N_2 \text{ or}$$

$$\dot{V}_I = \frac{\dot{V}_E F_{\bar{E}} N_2}{F_I N_2} \text{ where } F_{\bar{E}} N_2 = 1 - \left( F_{\bar{E}} O_2 + F_{\bar{E}} CO_2 \right) \qquad (2)$$

Replacement of $\dot{V}_I$ with $\dfrac{\dot{V}_E F_{\bar{E}} N_2}{F_I N_2}$ in equation $(1)$ yields

$$\dot{V}O_2 = \left( \frac{\dot{V}_E F_{\bar{E}} N_2}{F_I N_2} \cdot F_I O_2 \right) - \dot{V}_E F_{\bar{E}} O_2 \qquad (3)$$

$$\dot{V}O_2 = \dot{V}_E \left( \frac{F_{\bar{E}} N_2 \cdot F_I O_2}{F_I N_2} - F_{\bar{E}} O_2 \right) \qquad (4)$$

Under normoxic (room air) conditions, dry $F_I O_2$ equals 0.2093 and dry $F_I N_2$ equals 0.7904. The ratio of dry inspired $O_2$ to dry inspired $N_2$ is 0.265. Hence, equation (4) can be simplified to

$$\dot{V}O_2 = \dot{V}_E [F_{\bar{E}} N_2 (0.265) - F_{\bar{E}} O_2] \qquad (5)$$

Therefore, the quantities to measure for determination of $\dot{V}O_2$ at the mouth are $\dot{V}_E$, $F_{\bar{E}} O_2$, and $F_{\bar{E}} CO_2$.

By convention, $\dot{V}O_2$ is expressed under standard temperature and pressure dry (STPD) gas conditions. Because $\dot{V}_E$ is typically measured under atmospheric temperature and pressure saturated (ATPS) gas conditions, it must be converted to STPD gas conditions. The STPD correction factor can be calculated using the following equation:

$$\text{STPD correction factor} = \frac{273\ ^\circ K}{\left(273\ ^\circ K + T_A\right)} \times \frac{\left(P_B - P_{H2O}\right)}{760} \qquad (6)$$

where $^\circ K$ is degrees Kelvin, $T_A$ is the gas temperature at the time of measurement, $P_B$ is the barometric pressure, and $P_{H2O}$ is the water vapor pressure of saturated gas at $T_A$. Values of $P_{H2O}$ at various values of $T_A$ can be found in many exercise physiology textbooks.

## Historical Development of $\dot{V}O_2$ Measurement

One of the first open-circuit equipment configurations used to measure $\dot{V}O_2$ during exercise was the Tissot spirometer–volumetric gas analyzer system. The Tissot spirometer measured expired ventilation. An aliquot sample of the mixed expired gas was drawn from the spirometer and analyzed by a volumetric gas analyzer (e.g., Haldane or Scholander) for fractional concentrations of $O_2$ and $CO_2$.

A slight variation of the Tissot spirometer–volumetric gas analyzer system was the Douglas bag–volumetric gas analyzer system. In later years, lightweight meteorological balloons replaced the heavy, bulky Douglas bags and a gas meter replaced the Tissot spirometer for measurement of gas volume. Over time, electronic gas analyzers replaced the volumetric gas analyzers. This system is called the meteorological balloon–electronic gas analyzer system.

The next development in equipment configuration was the semiautomated system (Wilmore & Costill, 1974). In this system, the subject's expired air first went into a mixing chamber and then into a gas meter. A pump pulled approximately 500 ml of gas per minute from the mixing chamber and sent it to a 2 L latex bag; a second pump pulled the gas from the bag and sent it to the electronic analyzers for determination of $F_{\bar{E}} O_2$ and $F_{\bar{E}} CO_2$. The system used three bags. While the gas analyzers were sampling the contents of one bag, a second bag was being filled with gas from the mixing chamber and a third bag was being evacuated. The bags were part of a spinner device that was manually rotated 120° each minute so that the evacuated bag would move to a new position and be filled with gas from the mixing chamber. The other bags would likewise move to new positions. A problem with this mixing-chamber system is that the measurement of ventilation and the mixed expired gas fractions occur at different times. Ventilation is measured without any delay, but measurement of the mixed expired gases is delayed. This delay has two components. One is the transport delay between the subject's mouth and the analyzers. The other delay is due to the response time of the analyzers. Unless the mixed expired gas fractions are not changing, the calculated $\dot{V}O_2$ will be in error for this system. The magnitude of the error depends on the degree of temporal misalignment between the measurements of ventilation and the mixed expired gas fractions. Fortunately, at heavy exercise, the ventilation is usually large enough to cause the temporal misalignment to be quite small, resulting in little error in the calculated $\dot{V}O_2$max value.

Following the semiautomated system was the automated system developed by Beckman Instruments (Wilmore, Davis, & Norton, 1976) called the metabolic measuring cart (MMC). This system measured ventilation with a turbine volume transducer (Davis & Lamarra, 1984), had a mixing chamber, and measured the mixed expired gas fractions with $O_2$ and $CO_2$ electronic gas analyzers. The MMC did not have the latex bag arrangement of the semiautomated system. The gas drawn from the mixing chamber was sent directly to the $O_2$ and $CO_2$ gas analyzers. A small computer was an attractive feature of the MMC. The computer sampled the analog signals from the turbine volume transducer and the electronic gas analyzers over a duration of typically 1 min. The computer would then calculate the standard variables of interest (e.g., $\dot{V}O_2$, $\dot{V}CO_2$, and R, the respiratory exchange ratio) and output them via a printer a few seconds after the end of the sampling interval.

The MMC had many advantages over its predecessors. It eliminated hand calculation of the variables of interest and manual turning of the spinner device after each sampling interval, and it provided nearly online analysis of $\dot{V}O_2$ and $\dot{V}CO_2$. A limitation of the MMC was that it made no attempt to align ventilation and the mixed expired gas fractions temporally. However, the second generation of the MMC, called the MMC Horizon, was programmed to time align these signals so that the accuracy of the computed $\dot{V}O_2$ and $\dot{V}CO_2$ would be independent of the ventilation magnitude (Jones, 1984).

The equipment configurations just described collected or sampled many breaths over some time interval, typically 1 min. The calculated $\dot{V}O_2$ represented the average $\dot{V}O_2$ during the sampling interval. These types of equipment configurations are sufficient for the measurement of $\dot{V}O_2$max. But other parameters of aerobic function, most notably the time constant for $\dot{V}O_2$ kinetics, require a greater density of data. Hence, an equipment configuration was designed to provide online, breath-by-breath measurement of ventilation and the gas concentrations (Beaver, Wasserman, & Whipp, 1973). Now, more than 12 companies manufacture over 20 automated systems. In a comprehensive review of automated systems, Macfarlane (2001) provides a table showing the details of these systems. Of the 22 listed as laboratory-based systems, 12 are breath-by-breath systems, 7 are mixing-chamber systems, and 3 can provide either breath-by-breath or mixing-chamber measurements.

The manufacturers of current automated systems use a variety of technologies to measure $\dot{V}_E$, $F_{\bar{E}}O_2$, and $F_{\bar{E}}CO_2$. Measurement of $\dot{V}_E$ is made with a turbine device, a pneumotachometer, Pitot tubes, or a hot wire sensor. Technically, these four devices measure flow. Volume is found by integrating the area under the flow curve. A few manufacturers use a mass spectrometer to measure both $F_{\bar{E}}O_2$ and $F_{\bar{E}}CO_2$, but most of them use separate analyzers to measure the fractional concentrations of each gas. Most manufacturers use infrared radiation absorption technology for the $F_{\bar{E}}CO_2$ measurement and a polargraphic electrode, a paramagnetic analyzer, or a fuel cell for the $F_{\bar{E}}O_2$ measurement.

Several issues are specific to automated systems. One is the temporal alignment of the ventilation and gas concentration measurements. If misalignment is present, especially in breath-by-breath systems, large errors (as much as 30% at high breathing frequencies) can occur in the calculation of $\dot{V}O_2$ (Proctor & Beck, 1996). Some of the sensors for the $\dot{V}_E$ measurement and the $F_{\bar{E}}O_2$ and $F_{\bar{E}}CO_2$ measurements generate nonlinear outputs given linear inputs. Automated systems typically solve these nonlinear hardware problems with software corrections. Knowing the temperature and humidity of gas collected in a meteorological balloon is easy, but it is not so easy to know this information during each minute of a GXT for the flow and gas concentration sensors of an automated system. Hughson, Northey, Xing, Dietrich, and Cochrane (1991) have shown that errors in the gas temperature measurement, or the assumed value of that measurement, can result in errors in the computed $\dot{V}O_2$, for example, a 3.6% error in $\dot{V}O_2$ for a 2 °C error in the temperature measurement or the assumed value of that measurement. A number of independent studies have been published dealing with the validity and reliability of automated systems. A summary of these studies presented by Macfarlane (2001) provides evidence that several of these systems are both valid and reliable. Finally, the cost of a typical automated system—$20,000 to $25,000—is a significant issue. Given the complexities of these systems, some investigators purchase maintenance contracts with their systems, which further increase the cost.

Of the equipment configurations described, only the automated systems and the meteorological balloon–electronic gas analyzer system are used routinely today. Fay, Londeree, LaFontaine, and Volek (1989) used the latter system to measure $\dot{V}O_2$max in female distance runners. Expired gas was directed through 3.175 cm ID smooth rubber tubing, corrugated plastic tubing, and a three-way valve into meteorological balloons. The gas was collected continuously in 30 s intervals after the subject indicated that she could only run 90 s longer. An aliquot sample was taken from each meteorological balloon and measured for $\bar{F}_EO_2$ with a paramagnetic gas analyzer and $\bar{F}_ECO_2$ with an infrared gas analyzer. Reference gases for calibration of the electronic gas analyzers were verified with the Scholander apparatus. Expired gas volumes were measured with a gas meter. Oxygen uptake was calculated using equation (5), presented earlier in this chapter.

McArdle, Katch, and Pechar (1973) analyzed the test–retest reliability data for the $\dot{V}O_2max$ measurement by the meteorological balloon–electronic gas analyzer system. They performed duplicate cycle ergometer GXTs on 15 college-age males. The mean ±SD $\dot{V}O_2max$ values for the first and second tests were 4.157 ± 0.445 and 4.146 ± 0.480 $L \cdot min^{-1}$, respectively. The standard error of estimate was low (0.094 $L \cdot min^{-1}$), and the correlation coefficient was high (.959). Hence, the meteorological balloon–electronic gas analyzer system can provide excellent test–retest reliability.

Whether one uses a meteorological balloon–electronic gas analyzer system or an automated system to measure $\dot{V}O_2max$, several general fea-tures of any system are desirable. First, the system should not have any leaks. Second, the resistance to inspiration or expiration caused by the system should be less than 5 cm $H_2O$ pressure at any ventilation. Third, the device used to measure gas volume should provide measurements to within 3% of the true value. Fourth, the device or devices used to measure fractional gas concentrations should be able to measure both $O_2$ and $CO_2$ to within 0.0003 of their true values.

## Exercise Test Modes

The most popular modes for graded exercise testing are the cycle ergometer and the motor-driven tread-mill. Both modes can be controlled manually or by computer. For cycle ergometers, the braking mecha-nism is done either by friction or electronically. The electronic cycle ergometers are more expensive and more difficult to calibrate. Maximal $\dot{V}O_2$ values in typical people are approximately 10% higher when measured during treadmill running compared with cycle ergometry (see figure 2.1) (Davis & Kasch, 1975).

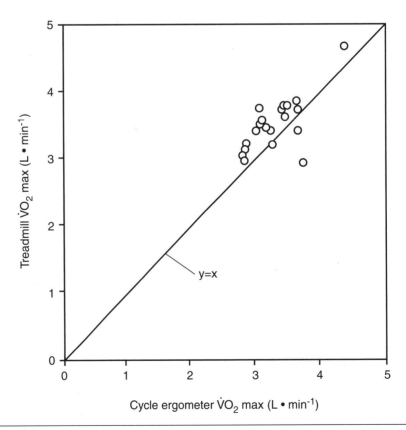

**Figure 2.1** Maximal $\dot{V}O_2$ measured during treadmill running compared with cycle ergometry.

Data from Davis and Kasch, 1975.

## Exercise Test Protocols

The most widely used exercise testing protocols are continuous and graded. The study of Buchfuhrer et al. (1983) is worthy of review before specific GXT protocols are discussed. Buchfuhrer et al. examined the possibility that the $\dot{V}O_2max$ measurement was protocol dependent. They found that "fast" protocols (i.e., those with large work rate increments per minute) and "slow" protocols (i.e., those with small work rate increments per minute) caused underestimations of the true $\dot{V}O_2max$ value, which was found using "intermediate speed" protocols. Buchfuhrer et al. (1983) suggested that the fast protocols caused subjects to terminate the GXT early because they had insufficient muscle strength to accommodate the large work rate increases during the final stages of the test. Two reasons probably explain why Buchfuhrer et al. (1983) found low $\dot{V}O_2max$ values for the slow protocols. First, these protocols, which lasted an average of 18 min for cycle ergometry and 26 min for treadmill exercise, would likely result in a significant increase in core temperature. This increase, in turn, would result in a redistribution of the cardiac output so that less blood (and, therefore, less $O_2$) would be going to the exercising musculature and more blood would be going to the cutaneous circulation in an effort to dissipate heat. Less blood flow (and therefore less $O_2$ delivery) to the working muscles at maximal work rates would explain the lower $\dot{V}O_2max$ values found for the slow protocols. A second plausible explanation for this finding is subject motivation. The slow protocols are particularly exhausting, requiring high motivation on the part of the subject to deal with the high levels of lactate and heat associated with heavy, prolonged exercise. Buchfuhrer et al. (1983) found the highest $\dot{V}O_2max$ values with GXT protocols that lasted 8 to 12 min.

Using the results of Buchfuhrer et al. (1983), GXT protocols should be designed in such a way to cause the test to end somewhere between 8 and 12 min. An example will demonstrate how the increment size can be found. Consider a subject with a predicted $\dot{V}O_2max$ of 3,000 $ml \cdot min^{-1}$ for a cycle ergometer GXT. According to Wasserman and Whipp (1975), the relationship between $\dot{V}O_2$ in $ml \cdot min^{-1}$ and work rate (WR) in watts (W) for cycle ergometry is given by the following linear regression equation:

$$\dot{V}O_2 = 10 \; ml \cdot min^{-1} \cdot WR + 500 \; ml \cdot min^{-1} \qquad (7)$$

Solving for the work rate that would result in a predicted $\dot{V}O_2$ of 3,000 $ml \cdot min^{-1}$ yields 250 W. The predicted increment size per minute that would produce a test duration of 8 min is 31 W (250 W/8 min). For a test duration of 12 min, the predicted increment size per minute would be 21 W (250 W/12 min). A prudent choice of the increment size per minute would be 25 W, which would be predicted to produce a test duration of approximately 10 min.

The most widely used cycle ergometer GXT protocols have a warm-up period of approximately 4 min. The work rate during the warm-up period is typically unloaded cycling or a light work rate such as 15 W. Immediately after 4 min of warm-up, the work rate is incremented by $x$ W each minute until the subject reaches his or her limit of tolerance; $x$ is the increment size that is predicted to produce a test duration somewhere between 8 and 12 min from the time when the work rate increments begin. These cycle ergometer protocols can be used to test the entire spectrum of subjects, from elite athletes to patients with cardiopulmonary disease. Only the increment size needs to be adjusted. Regarding the pedal frequency during cycle ergometry GXT, Hermansen and Saltin (1969) found that 60 rpm gave higher $\dot{V}O_2max$ values than did 50, 70, or 80 rpm. Hence, the pedal frequency for cycle ergometry GXT should be 60 rpm.

The Balke test (Balke & Ware, 1959) is a widely used treadmill GXT protocol. It is basically a treadmill walking test; the speed is 3.3 mph (5.3 kph) until a grade of 25% is reached. Thereafter, the grade is constant and the speed is increased 0.2 mph (0.32 kph) each minute. The grade is 0% for the first minute. The grade is raised to 2% at the end of the first minute and increased 1% per minute thereafter until it reaches 25%. For low-fit subjects, this protocol elicits valid $\dot{V}O_2max$ values. For fitter subjects, however, the test duration is very long. Also, these subjects are required to walk at grades above 20% during the last few minutes of the test and, according to McArdle et al. (1973), complain of severe local discomfort in the lower back and calf muscles, which may limit their ability to achieve maximal work rates. McArdle et al. (1973) compared $\dot{V}O_2max$ values measured using the Balke treadmill test to those measured using a continuous running treadmill test in reasonably fit male subjects. They found that the Balke test yielded $\dot{V}O_2max$ values that were approximately 4% lower than those obtained on the continuous running test. The Balke test could be modified to have a faster walking speed, which would likely result in increased $\dot{V}O_2max$ values for fitter subjects.

Maksud and Coutts (1971) provide a typical running treadmill protocol. It begins with the subject running at 6 mph (9.7 kph) on the level (0% grade) for 2 min. Thereafter, the grade is increased by 2.5% each 2 min; the speed is constant throughout the test.

Each of the protocols just described is continuous. Some GXT protocols are discontinuous. An example is the treadmill test of Mitchell, Sproule, and Chapman (1958). The test begins with the subject walking at 3 mph (4.8 kph) for 10 min at 10% grade. The subject then rests for 10 min. Next, the subject runs for 2.5 min at 6 mph (9.7 kph) up a 2.5% grade. After another 10 min rest period, the subject runs for 2.5 min at 6 mph (9.7 kph) up a 5.0% grade. This procedure (rest followed by a 2.5% grade increase) continues until the subject reaches his or her limit of tolerance.

The $\dot{V}O_2$max obtained using a continuous protocol is the same as that obtained using a discontinuous protocol. McArdle et al. (1973) compared the continuous Maksud and Coutts treadmill protocol with the discontinuous Mitchell et al. (1973) treadmill protocol in 15 college-aged males. The mean $\dot{V}O_2$max values for the continuous and discontinuous protocols were similar, 4.109 and 4.145 $L \cdot min^{-1}$, respectively. What differed markedly was the test duration:

12.3 min for the continuous test and 67.3 min for the discontinuous test. Both the subjects and the investigators preferred the continuous test to the discontinuous test because of the long test duration of the discontinuous test.

Over the years, cardiologists have designed a number of clinical treadmill protocols to test patients with heart disease. Two of these protocols are those of Bruce (Bruce, Kusumi, &. Hosmer, 1973) and Ellestad (1986)—see figures 2.2 and 2.3, respectively. The Bruce protocol is the most widely used clinical treadmill protocol.

## Criteria for Achievement of $\dot{V}O_2$max

The most widely accepted criterion for the achievement of $\dot{V}O_2$max during GXT is a plateau in $\dot{V}O_2$ as the work rate continues to increase. Typically, however, less than 50% of subjects tested demonstrate a plateau. Cumming and Borysyk (1972) administered GXT to 65 men aged 40 to 65 years. Only 43% of

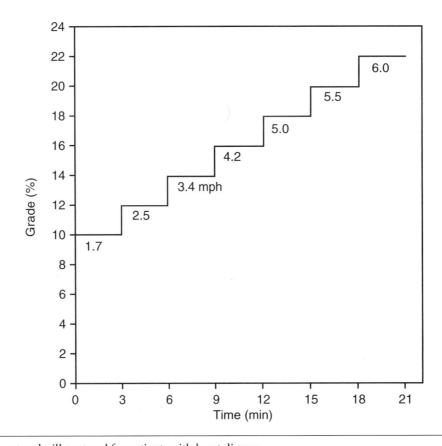

**Figure 2.2** Bruce treadmill protocol for patients with heart disease.

Adapted, by permission, from the American College of Sports Medicine, 1986, *ACSM's guidelines for exercise testing and prescription*, 3rd ed. (Lippincott, Williams, and Wilkins), 20.

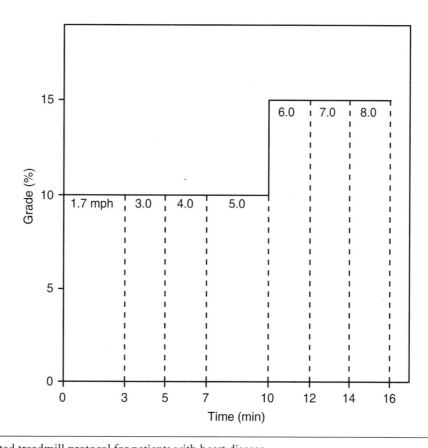

**Figure 2.3**    Ellestad treadmill protocol for patients with heart disease.

Adapted, by permission, from the American College of Sports Medicine, 1986, *ACSM's guidelines for exercise testing and prescription*, 3rd ed. (Lippincott, Williams, and Wilkins), 20.

these men met the plateau requirement. Freedson et al. (1986) found that less than 40% of 301 adults undergoing GXT demonstrated a plateau. Cumming and Friesen (1967); Cunningham, Van Waterschoot, Paterson, Lefcoe, and Sangal (1977); and Åstrand (1952) found that less than 50% of young boys who underwent GXT demonstrated a plateau. Indeed, Noakes (1988) has pointed out that the original investigators who developed the plateau criterion failed to find a true plateau in $\dot{V}O_2$ as the work rate continued to increase.

Three other criteria often used to defend the achievement of $\dot{V}O_2$max are (a) blood lactate concentration in the first 5 min of recovery >8 mmol/L, (b) respiratory exchange ratio at test termination >1.00, and (c) heart rate at test termination >90% of age-predicted maximum (220 – age). The third criterion is the least rigorous because of the well-known large variation in maximal heart rate at any given age.

It has been suggested that if a subject fails to demonstrate a plateau in $\dot{V}O_2$ as the work rate continues to increase, he or she then reached a $\dot{V}O_2$peak, not a $\dot{V}O_2$max. Given the concerns raised regarding the pla-

teau criterion for achievement of $\dot{V}O_2$max (Noakes, 1988), however, many investigators accept that the subject has achieved $\dot{V}O_2$max if he or she meets the plateau criterion or two of the three secondary criteria.

For patients, especially those with heart disease, a special term has been developed, namely, symptom-limited $\dot{V}O_2$max. This quantity is simply the highest $\dot{V}O_2$ measured before the patient had to stop exercising because of symptoms like severe angina.

Thoden (1991) has developed a novel solution to identifying the "true" $\dot{V}O_2$max value. He uses a protocol with athletes that has two phases (see figure 2.4) preceded by a 5 min warm-up. The progressive phase is a GXT that is initiated at about 30% of the athlete's predicted $\dot{V}O_2$max value and progresses at about 10 to 15% of that value for each work rate increment. The duration of each work rate increment is 2 min. The typical duration of the progressive phase is 8 to 12 min. Following the progressive phase, the athlete recovers at low exercise intensity until his or her heart rate returns to about 100 beats · min⁻¹, which typically takes 5 to 15 min. At this point, the verification phase begins. This phase is a constant-load (square-wave) test at a work rate one increment higher than that achieved during the progression phase. The typical duration is 3 to 5 min. An increase of <2% compared with that obtained in the progressive phase indicates that $\dot{V}O_2$max has been achieved. If the highest $\dot{V}O_2$ during the verification phase is $\geq 2\%$ above that found during the progressive phase, then the verification test is repeated at the next work rate prescribed by the protocol. This procedure continues until the increase is <2%. The "true" $\dot{V}O_2$max is taken as the highest $\dot{V}O_2$ found in either the progressive phase or the verification phase. The $\dot{V}O_2$ response to the progressive phase will be linear up to the maximum after a short delay at the start of the phase. For the square-wave forcing function of the verification phase, the $\dot{V}O_2$ response will be exponential up to the maximum. Thoden (1991) reports excellent test–retest reliability for this protocol. For 15 subjects with $\dot{V}O_2$max ranging from 46 to 76 ml $\cdot$ min$^{-1}$ $\cdot$ kg$^{-1}$, no significant difference occurred from test to retest. The correlation coefficient was .95 between the two tests.

## Reference Values

Many factors are known to influence $\dot{V}O_2$max. Bed rest causes it to go down, whereas endurance exercise training increases it (Saltin et al., 1968). When $\dot{V}O_2$max is expressed in the units of L $\cdot$ min$^{-1}$, large people (those with increased height or weight) typically have higher values than small people do. Women typically have lower $\dot{V}O_2$max values than men do. Old adults generally have lower $\dot{V}O_2$max values than young adults do. These factors (i.e., extent of physical activity in leisure time, height, weight, gender, and age) can be used to predict $\dot{V}O_2$max with some degree of confidence. Jones, Makrides, Hitchcock, Chypchar, and McCartney (1985) performed cycle ergometer GXTs on 50 male and 50 female subjects of various fitness levels who ranged in age from 15 to 71 years. From the data collected, they developed the following multiple linear regression equation:

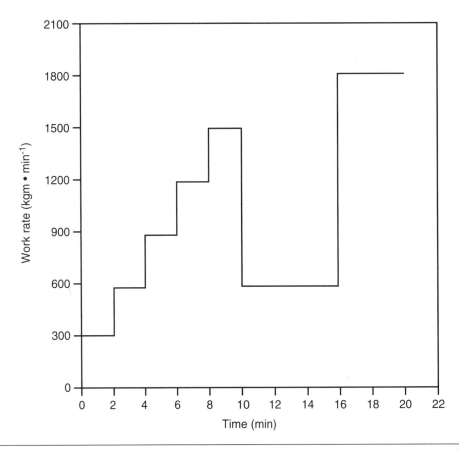

**Figure 2.4**  An example of a cycle ergometer protocol used by Thoden (1991) to determine "true" $\dot{V}O_2$max in athletes. The protocol is preceded by a 5 min warm-up (not shown). The protocol has three parts, namely, a progressive phase, a recovery period, and the verification phase. The protocol shown in this figure is specific to athletes with predicted $\dot{V}O_2$max values between 2 and 4 L $\cdot$ min$^{-1}$.

$$\dot{V}O_2\text{max (L·min}^{-1}) = 0.025 \, (Ht) - 0.023 \, (Age) -$$
$$0.542 \, (Gender) + 0.019 \, (Wt) + 0.15 \, (Lei) - \qquad (8)$$
$$2.32 \, \text{L·min}^{-1}$$

where *Ht* is standing height in cm, *Age* is in years, and *Wt* is body weight in kg. For male subjects, the gender code is 0. For female subjects, the gender code is 1. *Lei* is leisure time spent per week in physical activity. The four grades of leisure activity are as follows: grade 1 for <1 h per wk, grade 2 for 1 to 3 h per wk, grade 3 for 3 to 6 h per wk, and grade 4 for >6 h per wk. An example illustrates how this equation can be used to predict $\dot{V}O_2$max. Assume that a 44-year-old male subject is 183 cm tall, that he weighs 70 kg, and that he jogs 2.5 h per week (hence, his leisure activity grade is 2). Plugging these numbers into the above equation yields a predicted $\dot{V}O_2$max of 2.87 L·min$^{-1}$. The multiple correlation coefficient for this equation is .892 and the standard error of estimate is 0.415 L·min$^{-1}$. This latter statistic can be used to compute the upper and lower 95% confidence limits of the predicted value. In the above example, the predicted $\dot{V}O_2$max value of 2.87 L·min$^{-1}$ would have lower and upper 95% confidence limits of 2.05 and 3.69 L·min$^{-1}$, respectively.

Recently, Davis, Storer, Caiozzo, and Pham (2002) developed multiple linear regression equations that allow computation of the lower reference limit for $\dot{V}O_2$max. The unique feature of these equations is that they were developed using sedentary people. Hence, if the measured $\dot{V}O_2$max value of a subject falls below the lower reference limit, it cannot be argued that the low measured value was due to sedentary living. For each gender, three equations with different predictor variables were developed. Shown below are the prediction equations for men including the squared multiple correlation coefficient ($R^2$), the standard error of estimate (SEE), and the one-sided 95% confidence interval (95% CI) for each equation. The units for the predictor variables of age, height (Ht), mass, and fat-free mass (FFM) are years, cm, kg, and kg, respectively.

$$\dot{V}O_2\text{max (L·min}^{-1}) = -0.0282 \, (Age) + 0.0205 \, (Ht) +$$
$$0.3200 \, \text{L·min}^{-1};$$

$$R^2 = .580; \, SEE = 0.398 \, \text{L·min}^{-1};$$
$$\text{and } 95\% \, CI = 0.660 \, \text{L·min}^{-1} \qquad (9)$$

$$\dot{V}O_2\text{max (L·min}^{-1}) = -0.0296 \, (Age) +$$
$$0.0167 \, (Mass) + 2.6538 \, \text{L·min}^{-1};$$

$$R^2 = .640; \, SEE = 0.369 \, \text{L·min}^{-1};$$
$$\text{and } 95\% \, CI = 0.612 \, \text{L·min}^{-1} \qquad (10)$$

$$\dot{V}O_2\text{max (L·min}^{-1}) = -0.0262 \, (Age) +$$
$$0.0266 \, (FFM) + 2.1154 \, \text{L·min}^{-1};$$

$$R^2 = .657; \, SEE = 0.359 \, \text{L·min}^{-1};$$
$$\text{and } 95\% \, CI = 0.595 \, \text{L·min}^{-1} \qquad (11)$$

The corresponding prediction equations for women are shown below.

$$\dot{V}O_2\text{max (L·min}^{-1}) = -0.0171 \, (Age) +$$
$$0.0160 \, (Ht) - 0.2740 \, \text{L·min}^{-1};$$

$$R^2 = .500; \, SEE = 0.286 \, \text{L·min}^{-1};$$
$$\text{and } 95\% \, CI = 0.474 \, \text{L·min}^{-1} \qquad (12)$$

$$\dot{V}O_2\text{max (L·min}^{-1}) = -0.0199 \, (Age) +$$
$$0.0135 \, (Mass) + 1.6267 \, \text{L·min}^{-1};$$

$$R^2 = .557; \, SEE = 0.269 \, \text{L·min}^{-1};$$
$$95\% \, CI = 0.446 \, \text{L·min}^{-1} \qquad (13)$$

$$\dot{V}O_2\text{max (L·min}^{-1}) = -0.0183 \, (Age) +$$
$$0.0230 \, (FFM) + 1.3405 \, \text{L·min}^{-1};$$

$$R^2 = .554; \, SEE = 0.270 \, \text{L·min}^{-1};$$
$$95\% \, CI = 0.448 \, \text{L·min}^{-1} \qquad (14)$$

These equations were developed on a relatively large sample (115 men and 115 women) that ranged in age from 20 to 70 years. The sample included about 23 subjects in each age decade. The GXT mode was the cycle ergometer. Thus, the predicted $\dot{V}O_2$max values are specific to cycle ergometer GXT. The equations can be used to predict $\dot{V}O_2$max for treadmill GXT simply by multiplying the predicted value by 1.10 to raise that value by 10%.

An example will illustrate how the lower reference limit can be calculated. The predicted $\dot{V}O_2$max for a 40-year-old male who is 175 cm tall is 2.780 L·min$^{-1}$ using the equation for men with the predictor variables of age and height. The lower 95% confidence limit (lower reference limit) is calculated as the predicted $\dot{V}O_2$max minus the one-sided 95% confidence interval (0.660 L·min$^{-1}$). Thus, the lower reference limit for this subject is 2.120 L·min$^{-1}$ (2.780 – 0.660 L·min$^{-1}$).

## Summary

If the only aerobic function parameter of interest is $\dot{V}O_2$max, then the meteorological balloon–electronic gas analyzer system is the simplest and most inexpensive of the systems currently used to make the measurement. But if the user wants to measure all the parameters of aerobic function, from rest to maximal exercise tolerance, then an automated system is the system of choice. Independent studies have shown that some of the commercially available automated systems provide valid and reliable measurements of ventilation and $\dot{V}O_2$ (Macfarlane, 2001). Should readers desire information concerning the $\dot{V}O_2$max measurement in high-performance athletes, I suggest that they read the excellent chapter by Thoden (1991).

# Indirect Methods for Estimation of Aerobic Power

Véronique Billat, PhD
*University of Evry Val d'Essonne, France*

Philippe Lopes, PhD
*University of Evry Val d'Essonne, France*

Maximal oxygen uptake ($\dot{V}O_2max$) is defined as the point at which oxygen consumption reaches a peak and plateaus, or only increases slightly, in response to an increased work rate (Wasserman, Hanson, Sue, Whipp, & Casaburi, 1999). Criteria for establishment of $\dot{V}O_2max$ include an increase in $\dot{V}O_2$ of less than 2.1 ml $\cdot$ kg$^{-1}$ $\cdot$ min$^{-1}$ for an increase in speed equal to 1 km $\cdot$ h$^{-1}$ (Taylor, Buskirk, & Henschel, 1955), a blood lactate concentration of 8 to 12 mM, a respiratory exchange ratio equal to or greater than 1.1, and a heart rate equal to at least 90% of the theoretical maximal heart rate in beats $\cdot$ min$^{-1}$ as derived from the equation 220 – age in years (Åstrand & Rodahl, 1986).

Because of ease, cost effectiveness, and safety, submaximal testing is widely used as a substitute for the direct measurement of $\dot{V}O_2max$. Indirect methods of estimating $\dot{V}O_2max$ have been developed for the assessment of fitness levels and for individualizing exercise prescriptions. Two fundamental relationships allow indirect predictions of $\dot{V}O_2max$:

- a linear relationship between $\dot{V}O_2$ and mechanical power output, and
- a linear relationship between $\dot{V}O_2$ and heart rate according to the Fick equation (see p. 23).

However, neither $\dot{V}O_2$ nor HR increases linearly at high power outputs during incremental exercise test in humans (Zoladz, Duda, & Majerczak, 1998a, 1998b; Zoladz, Szkutnik, Majerczak, & Duda, 1999), which limits the utility of the method.

These two fundamental relationships make it possible to link heart rate (HR) and power output. HR increases in proportion to power output until about 85% of the velocity at $\dot{V}O_2max$ (v$\dot{V}O_2max$). v$\dot{V}O_2max$ combines $\dot{V}O_2max$ and economy into a single factor to identify aerobic differences between athletes. These relationships are illustrated in this chapter by use of a hypothetical comparison between two runners: Christopher and Didier (the latter being the fitter of the two). The relationship between their HR and running speed is shown in figure 3.1. The deflection of HR when approaching v$\dot{V}O_2max$ (see figure 3.1 and figure 3.2) has been extensively reported (Bodner & Rhodes, 2000). In 1968, Brooke, Hamley, and Thomason reported that HR response to incremental testing does not always follow a strictly linear path. Their findings showed that some individuals among a group of racing cyclists demonstrated a HR/workload relationship that was sigmoidal (see figure 3.2). This HR/workload curve was characterized by three distinct sequential phases: an anticipatory phase (until 40% of p$\dot{V}O_2max$[1]), a linear phase (45 to 85% of p$\dot{V}O_2max$), and a curvilinear phase in which the slope of HR/workload decreased.

Figure 3.1 shows that Didier has a lower heart rate than Christopher does at each speed. This presumably indicates a larger stroke volume for Didier, because they had the same running economy (that is, the

---

[1]p$\dot{V}O_2$max refers to the power output at $\dot{V}O_2$max.

19

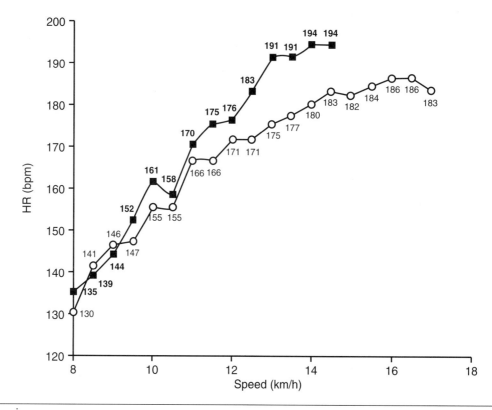

**Figure 3.1**   v$\dot{V}O_2$max and Hrmax determination on the track. The heart rate–speed relationship for Christopher is shown in black squares, and that for Didier, the fitter of the two, is shown in white circles.

same $\dot{V}O_2$) at each speed. As can be seen in figure 3.1, the heart rate versus speed relationship is not linear. A deflection in the relationship between HR and speed can be seen at 12.5 km · h$^{-1}$ for Christopher, which corresponds to 92% of v$\dot{V}O_2$max (14 km · h$^{-1}$), and 14 km · h$^{-1}$ for Didier, which corresponds to 87% of v$\dot{V}O_2$max (16 km · h$^{-1}$). Furthermore, when comparing HR curves, Didier exhibits a marked HR deflection that is due to the difference in physical fitness levels. Didier was previously a good marathon runner (2 h 40 min). From a physiological point of view, Didier was able to increase his stroke volume until almost the end of the exercise, thus redistributing a higher fraction of his cardiac output toward active muscle mass. This deflection may, however, be an artifact resulting from the test protocol in which increments in speed of 0.5 km · h$^{-1}$ were used every minute rather than a series of steady-state stages, which imposes the issue of HR kinetics on the heart rate–speed relationship.

There is a deflection for HR versus $\dot{V}O_2$ and for HR versus speed (power output) curves. This represents a disproportionate increase in $\dot{V}O_2$ and a decrease in HR versus power output (see figure 3.2). According to the Fick equation, the decrease in HR must be

compensated by an increase in the stroke volume or by an increase in arteriovenous oxygen difference to allow an increase of $\dot{V}O_2$ with the increase in work rate (Åstrand & Rodahl, 1986).

The absence of consistency of the slope of HR versus work rate and $\dot{V}O_2$ provides a reason why the indirect measurement of $\dot{V}O_2$max, based on the HR$\dot{V}O_2$ and HR/power output relationships may be inappropriate. Accordingly, indirect measurement of $\dot{V}O_2$max is not recommended for competitive athletes who need to follow changes in their $\dot{V}O_2$max precisely.

In the remainder of the chapter, the physiological basis for the indirect determination of $\dot{V}O_2$max will be examined to allow the reader to use appropriate methods rather than a pure cookbook approach. Estimation methods for the determination of $\dot{V}O_2$max using no exercise or tools will then be considered, as will the use of HR monitors and appropriate exercise modalities. Finally, estimation of $\dot{V}O_2$max by measurement of HR variability at rest will be discussed. A final table will review the methods that are available for healthy subjects, taking into account their fitness levels, age, and test objectives. (Remember, it is recommended that subjects, irrespective of age,

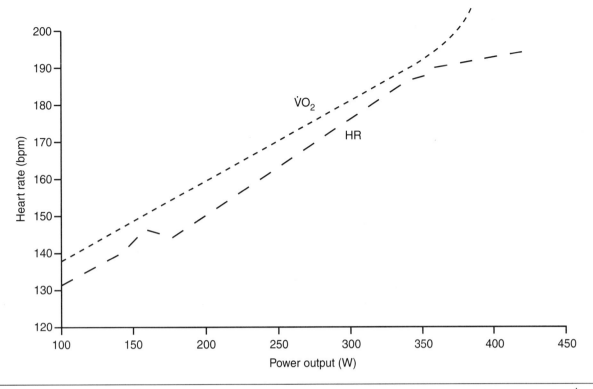

**Figure 3.2** Heart rate and $\dot{V}O_2$ curve during an incremental exhaustive on a bicycle. The maximal oxygen uptake ($\dot{V}O_2$) curve is depicted by - - - -. The HR response (—  —) shows a HR deflection point associated with an increase in power output at about 180 beats · min⁻¹, that is, at 92% of HRmax and 88% of the maximal heart rate reserve (HRmax – HR)/(HR – HR at rest). The relationship between HR and the power output is linear between 145 and 180 beats · min⁻¹, which corresponds to 40% and 80% of the power output associated with $\dot{V}O_2$max (p$\dot{V}O_2$max).

seek medical clearance should they have any risk factors that would put them at risk while undertaking an exercise test or participating in an exercise program.)

# Indirect Estimation of $\dot{V}O_2$max Using Power Output or Velocity

In 1984 Daniels, Scardina, Hayes, and Folet introduced the term *velocity at* $\dot{V}O_2$*max* (v$\dot{V}O_2$max) as a useful variable that combined $\dot{V}O_2$max and economy into a single factor with the potential to identify aerobic differences between individual runners or categories of runners. v$\dot{V}O_2$max explained individual differences in performance that $\dot{V}O_2$max or running economy alone did not. Daniels et al. found in high-level female runners that v$\dot{V}O_2$max was slightly faster than the average velocity over 3,000 m (about 9 min) and that v3,000 m was equal to 98% ± 2% of v $\dot{V}O_2$max (Billat, Renoux, Pinoteau, Petit, & Koralsztein,

1994). Morgan, Baldini, Martin, and Kohrt (1989) showed that variations in 10 km run times attributable to v $\dot{V}O_2$max were smaller than those associated with either $\dot{V}O_2$max or running economy. Daniels et al. calculated v $\dot{V}O_2$max by extrapolating from the regression curve the relationship of running velocity and $\dot{V}O_2$ to $\dot{V}O_2$max (see figure 3.3). Because the $\dot{V}O_2$max increase is mainly due to stroke volume increase, the decrease in HR at similar speeds may indicate a stroke volume increase and $\dot{V}O_2$max improvement. Use of a heart rate monitor can provide further indications of this increase in $\dot{V}O_2$max (see figure 3.4).

Figure 3.4 shows the direct link between oxygen uptake and power output. The increase in $\dot{V}O_2$ associated with 1 km · h⁻¹ speed increments is the basis for all the indirect determinants of $\dot{V}O_2$max in the field. From 8 to 20 km · h⁻¹ the aerodynamic component of the energy cost of running is negligible or below the accuracy of current $\dot{V}O_2$max determination methods, that is, 3% (Pugh, 1970).

This is why the maximal aerobic power, v$\dot{V}O_2$max, is often better correlated with performance than

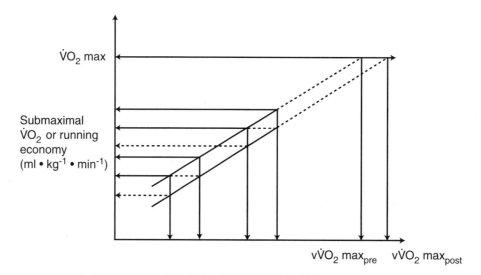

**Figure 3.3**   Relationship between $\dot{V}O_2$ and treadmill running velocity in female runners (n = 30), modified from Daniels et al. (1984). Also shown is the velocity that mathematically corresponds to $\dot{V}O_2$ ($v\dot{V}O_2$max). Submaximal $\dot{V}O_2$ was calculated from four 6 min runs at velocities of 230, 248, 268 and 283 m·min$^{-1}$, with a 4 to 7 min recovery between submaximal runs. $\dot{V}O_2$max was measured separately with an incremental test based on the 5,000 m race pace adding a 1% grade to the treadmill each minute until exhaustion. The highest $\dot{V}O_2$ reached during the max test was considered the $\dot{V}O_2$max. The running economy improvement can be seen with the dashed lines; that is, the decrease of submaximal $\dot{V}O_2$ at each speed allows an improvement of $v\dot{V}O_2$max with no improvement in $\dot{V}O_2$max.

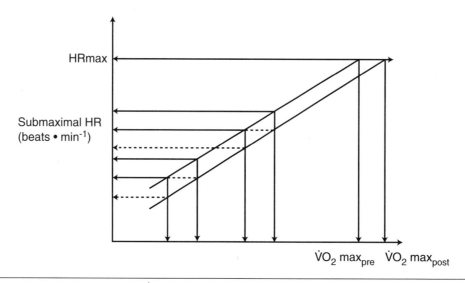

**Figure 3.4**   Relationship between HR and $\dot{V}O_2$max. The decrease in submaximal HR (dashed lines) at each speed allows an improvement of $\dot{V}O_2$max for the same HRmax after training.

is $\dot{V}O_2$max alone. Given that time over distance gives average speed, it is logical that speed may be better correlated with performance than is $\dot{V}O_2$max alone. One could conclude that it is not necessary to measure $\dot{V}O_2$max to predict performance and that the best predictor of performance is performance itself. Nevertheless, $\dot{V}O_2$max is the main determinant of performance even among elite runners (Billat et al., 1996; Billat, Renoux, Pinoteau, Petit, & Koralsztein,1994b).

# Determination of $\dot{V}O_2$max From Speed

Velocity or power output is directly related to $\dot{V}O_2$ according to equation (1) (Billat et al., 1994b):

$$v\dot{V}O_2\text{max} = \dot{V}O_2\text{max}/Cr \qquad (1)$$

where $v\dot{V}O_2$ is expressed in m·min$^{-1}$, $\dot{V}O_2$max is expressed in $ml \cdot kg^{-1} \cdot min^{-1}$, and Cr is the gross energy cost of the run in $ml \cdot kg^{-1} \cdot m^{-1}$.

At each speed ($m \cdot min^{-1}$) Cr was calculated from the ratio of the steady-state $\dot{V}O_2$ ($ml \cdot kg^{-1} \cdot min^{-1}$) above the preexercise resting level (determined with the subject standing on the treadmill), divided by the running speed ($m \cdot min^{-1}$). For instance, Christopher has a $v\dot{V}O_2max$ of 14 $km \cdot h^{-1}$ (233 $m \cdot min^{-1}$) performing an incremental test with increases in speed of 1 $km \cdot h^{-1}$ every 2 min until exhaustion (Billat, Bernard, Pinoteau, Petit, & Koralsztein, 1994; Billat, Sirvent, Py, Koralsztein, & Mercier, 2003). $\dot{V}O_2$ was measured directly. His $\dot{V}O_2$ increase above rest was 50 $ml \cdot kg^{-1} \cdot min^{-1}$. Therefore, his Cr was equal to $233/50 = 0.214$ $ml$ $O_2 \cdot kg^{-1} \cdot m^{-1}$ or the equivalent of 1.07 $cal \cdot kg^{-1} \cdot m^{-1}$.

On average, Cr is 0.210 $ml$ $O_2 \cdot kg^{-1} \cdot m^{-1}$. This means that each time the speed increases by 1 $km \cdot h^{-1}$ (i.e., 16.6 $m \cdot min^{-1}$), the additional Cr is 0.210 $ml$ $O_2 \cdot kg^{-1} \cdot m^{-1} \times 16.6 = 3.5$ $ml$ $O_2 \cdot kg^{-1} \cdot min^{-1}$. Therefore,

$$v\dot{V}O_2max = \dot{V}O_2max / Cr, \text{ or}$$

$$v\dot{V}O_2max \ (km \cdot h^{-1}) = \dot{V}\backslash O_2max / 3.5$$

Hence, when $v\dot{V}O_2max$ is known, $\dot{V}O_2max$ can be estimated by using this average Cr.

If $v\dot{V}O_2max = 14$ $km \cdot h^{-1}$, then

$$\dot{V}O_2max = 3.5 \ ml \ O_2 \cdot kg^{-1} \cdot min^{-1} \times 14$$

$$= 49 \ ml \ O_2 \cdot kg^{-1} \cdot min^{-1}$$

The same principles apply when using the cycle ergometer for indirect measurement of $\dot{V}O_2max$. Power output is substituted for speed, and the average cycling cost estimated to be ~10 to 12 $ml \cdot min^{-1} \cdot W^{-1}$ (Åstrand & Ryhming, 1954; Riley, Wasserman, Fu, & Cooper, 1996; Wasserman et al., 1999). For example, a power output equal to 300 W will provide the following estimate of $\dot{V}O_2max$:

$$\dot{V}O_2max = 300 \ W \times 12 \ ml \cdot min^{-1}$$

$= 3{,}600$ $ml \cdot min^{-1}$ (This is irrespective of body weight.)

If the simple evaluation of $v\dot{V}O_2max$ cannot be used as an estimation of the physiological basis of $\dot{V}O_2max$ improvement, then inspection of the relationship between HR and velocity can provide some additional information (see figure 3.1). Indeed, the slope of the HR to speed, or power output, gives a kind of "cardiac economy." This is expressed in meters run or walked per heartbeat. This concept is not currently used but may be useful in the future as greater use of HR monitors occurs.

The ratio between oxygen uptake and HR is oxygen pulse ($ml$ $O_2 \cdot beat^{-1}$). The Fick equation is the fundamental relationship to consider relative to the cardiorespiratory approach to physical fitness:

$$\dot{V}O_2 = Q \times (a - vO_2diff) \tag{2}$$

where $\dot{V}O_2$ is in $ml \cdot min^{-1}$, Q is the cardiac output in $ml \cdot min^{-1}$, and $a - vO_2diff$ is the difference in oxygen content between the arterial and venous blood (in ml of $O_2$ per 100 ml of blood).

The $a - vO_2diff$ reflects the ability of muscle to extract oxygen, whereas Q depends on cardiorespiratory function. Cardiac output is the product of stroke volume and heart rate. The stroke volume (SV) in $ml \cdot beat^{-1}$ is the quantity of blood ejected into the aorta from the left ventricle for each contraction of the heart. Therefore,

$$Q = HR \times SV \tag{3}$$

The Fick equation shows that the linear relationship between HR and oxygen uptake depends on a steady increase in cardiac stroke volume and arteriovenous oxygen difference. In the healthy young adult, departures from linearity of cardiac output and peripheral blood flow have generally been reported as very small (Shephard, 1984). However, a HR deflection is observed with short-stage protocols of 0.5 $km \cdot h^{-1}$ increments (see figure 3.1) (Javorka, Zila, Balharek, & Javorka, 2003). With the Conconi test (Conconi, Ferrari, Ziglio, Droghetti, & Codeca, 1982), which uses stages of 200 m performed on the track, the stage duration decreases with speed increase and thus explains most of the HR deflection with speed (Jeukendrup, Hesselink, Kuipers, & Keizer, 1996). This HR deflection with speed increases has also been associated with the onset of blood lactate accumulation (speed at a blood lactate concentration of 4 mM). For review, see Billat & Koralsztein (1996) and Billat et al. (2000). On the other hand, subjects who have a HR deflection even with long-stage protocols (3 min long) are capable of increasing their stroke volume until the end of the test. This increase in stroke volume also occurs at high power output (83 ± 12% of $\dot{V}O_2max$), particularly in triathletes with HR deflection appearing at 75 ± 7% of $\dot{V}O_2max$ (Bodner & Rhodes, 2000). As opposed to healthy young subjects (Hofmann et al., 1997), older adults, who may have difficulty in sustaining stroke volume at high work rates, have disproportionate increases of HR (Shephard & Sidney, 1978). Individual values of predicted $\dot{V}O_2max$ generally show a variability of ±10% of the mean error (Shephard, 1984). A recent study reported an overestimation of $\dot{V}O_2max$ by 3 $ml \cdot kg^{-1} \cdot min^{-1}$ (53 versus 50.3 $ml \cdot kg^{-1} \cdot min^{-1}$, +

6%) when using the Åstrand nomogram (Åstrand & Rodahl, 1986).

Indirect $\dot{V}O_2$ measurements are based on the assumption that the HR at a given speed or power output is linked to the oxygen pulse (i.e., the oxygen uptake for each heartbeat in ml $O_2 \cdot beat^{-1}$).

This model is expressed in equation (4):

$$speed = \dot{V}O_2/Cr \qquad (4)$$

where Cr represents the energy cost of running

$$speed = Q \times (a - vO_2 diff)/Cr$$

$$speed = HR \times SV \times (a - vO_2 diff)/Cr$$

$$speed/HR = SV \times (a - vO_2 diff)/Cr$$

Given that stroke volume is the main parameter modified by training, the cardiac economy (speed/HR) provides an excellent index for cardiac improvement. It has been reported that the power output at a given HR was unchanged with age (Gore, Booth, Bauman, & Owen, 1999). Indeed, the power output at a HR of 150 beats $\cdot$ min$^{-1}$ and 170 beats $\cdot$ min$^{-1}$ was unchanged with age between 18 and 78 years in a cross-sectional study of 1,043 subjects having the same self-reported activity. This ratio between power output (or speed) and heart rate is evidently characteristic of a person's level of physical conditioning. But the ratio (meters run per heartbeat) increases in response to training. This increase may be due to an increase in the stroke volume (SV) or better blood redistribution within the muscle mass (Boutcher, McLaren, Cotton, & Boutcher, 2003). With training, maximal heart rate remains unchanged (except in persons over age 40 in which training may result in increased maximum HR). Because maximal HR is unchanged, an increase in stroke volume appears to be the main reason for the increased cardiac output. An increase of 13% has been reported for 8 male students, aged 19 to 27 years, after 16 weeks of cycle ergometer interval training held three times a week (Ekblom & Hermansen, 1968). Therefore, even if indirect estimation of $\dot{V}O_2$max does not allow determination of a specific increase in stroke volume, the increase of the cardiac efficiency can provide an index of such adaptation in response to aerobic training.

This decrease in HR for a given speed is interesting because it allows estimation of the distance run for each beat of the heart, thus providing a kind of "cardiac economy."

An understanding of the physiological basis for the indirect estimation of $\dot{V}O_2$max provides a rationale for the development of the following test methods used to estimate $\dot{V}O_2$max. These methods include

- nonexercise measurements,
- performance over distances between 1,500 m and 10,000 m,
- a timed testing protocol performed on a track,
- a timed test performed on a track with additional use of a heart rate monitor, and
- heart rate variability at rest.

The aim of this chapter is not to draw up a catalogue of the numerous indirect tests that may be used to estimate $\dot{V}O_2$max, but rather to provide the reader with an update of the simplest, more accurate, more consistent, and more promising methods currently in use. A new method will be proposed, using HR monitors, for the estimation of $\dot{V}O_2$max during mountain climbing (running or walking). Extensive research has not yet validated this method, but a preliminary investigation carried out in 15 subjects of different gender and varying age provides sufficient accuracy to warrant inclusion in this text.

# Determination of $\dot{V}O_2$max Using Nonexercise Measurements

Philosophically, the estimation of $\dot{V}O_2$max should not be an aim in itself, but rather a way to assess physical fitness before starting a regular activity program for personal health or fitness development. In this example, Christopher is coming for a physiological assessment before developing an 8-month training program, designed to allow completion of a marathon in a time of 3 h 50 min.

The first methods to be examined for the estimation of $\dot{V}O_2$max use activity questionnaires and individual physical characteristics. Christopher is 35 years of age, is 174 cm in height, weighs 69 kg, has 16% body fat, and has a body mass index (BMI) (weight/height) of 22.8 kg $\cdot$ m$^{-2}$. He is married, has no children, works 50 h $\cdot$ week$^{-1}$, and is the sales director in a large company. He currently rides a mountain bike for 2 h every Sunday and runs once per week for 1 h at a speed of 10 km $\cdot$ h$^{-1}$. He had a physical activity rating (PA-R) of 6, derived from the Physical Activity Rating (PA-R) questionnaire developed by Jackson et al. (1990) (see figure 3.5).

The equation to predict $\dot{V}O_2$max using Christopher's personal data follows (Matthews, Heil, Freedson, & Pastides, 1999):

## NASA Code for Physical Activity

Use the number (0 to 7) that best describes your general activity level for the previous month.

Do not participate regularly in programmed recreation sport or heavy physical activity

0   Avoid walking or exertion, for example, always use elevator, drive whenever possible instead of walking

1   Walk for pleasure, routinely use stairs, occasionally exercise sufficiently to cause heavy breathing or perspiration

Participate regularly in recreation or work requiring modest physical activity, such as golf, horseback riding, calisthenics, gymnastics, table tennis, bowling, weightlifting, yard work

2   10 to 60 minutes per week

3   Over 60 minutes per week

Participate regularly in heavy physical exercise such as running or jogging, swimming, cycling, rowing, skipping rope, running in place or engaging in vigorous aerobic activity exercise such as tennis, basketball, or handball

4   Run less than 1 mile (1.6 kilometers) per week or spend less than 30 minutes per week in comparable physical activity

5   Run 1 to 5 miles (1.6 to 8 kilometers) per week or spend 30 to 60 minutes per week in comparable physical activity

6   Run 5 to 10 miles (8 to 16 kilometers) per week or spend 1 hour to 3 hours per week in comparable physical activity

7   Run over 10 miles (16 kilometers) per week or spend over 3 hours per week in comparable physical activity

**Figure 3.5**   Physical Activity Rating (PA-R) questionnaire used by Jackson et al. (1990) and Matthews, Heil, Freedson, and Pastides (1999) to assess Physical Activity Status (PAS) to estimate $\dot{V}O_2$peak.

Adapted, by permission, from Jackson et al., 1990, "Prediction of functional aerobic capacity without exercise testing," *Medicine and Science in Sports and Exercise* 22: 863-870; and Matthew, Heil, Freedson, and Pastides, 1990.

$$\text{predicted } \dot{V}O_2\text{max (ml·kg}^{-1}\text{·min}^{-1}) = (0.133 \times \text{age}) - (0.005 \times \text{age}^2) + (11.403 \times \text{gender}) + \tag{5}$$
$$(1.463 \times \text{PA-R score}) + (9.17 \times \text{height}) - (0.254 \times \text{body mass}) + 34.142$$

where age is in years, gender is represented by 1 for male and 0 for female, physical activity rating (PA-R) is obtained from figure 3.5, height is in meters, and body mass is in kilograms.

Using equation (5), the predicted $\dot{V}O_2$max (ml · kg$^{-1}$ · min$^{-1}$) for Christopher would be

$$\dot{V}O_2\text{max} = (0.133 \times 35) - (0.005 \times 1{,}225) +$$
$$(11.403 \times 1) + (1.463 \times 6) + (9.17 \times 1.74) -$$
$$(0.254 \times 69) + 34.142 \text{ or}$$

$$\dot{V}O_2\text{max} = 51.3 \text{ ml·kg}^{-1}\text{·min}^{-1}$$

Because Christopher is a real person, comparisons are going to be made between values derived from a directly measured $\dot{V}O_2$max test and results from the indirect methods as described in this chapter. Christopher had his $\dot{V}O_2$max (50.0 ml · kg$^{-1}$ · min$^{-1}$) directly measured during a running test on a track.

The predicted value derived from equation (5) is very close to that directly measured (51.3 versus 50.0 ml · kg$^{-1}$ · min$^{-1}$). This equation emphasizes body mass as compared with the PA-R score. If the same

physical characteristics are used with a maximal PA-R score, then the $\dot{V}O_2$max value becomes 52.7 ml · kg$^{-1}$ · min$^{-1}$. Each additional PA-R point, therefore, improves the $\dot{V}O_2$max by 1.46 ml · kg$^{-1}$ · min$^{-1}$.

Note that questionnaires are used primarily for epidemiologic purposes to classify cardiorespiratory fitness (Matthews et al., 1999). In large populations, valid methods of predicting cardiorespiratory fitness or quantifying peak functional capacity are needed (Ainsworth, Richardson, Jacobs, Leon, & Sternfeld, 1999; Williford et al., 1996). The self-reported physical activity status scoring method used by Matthews, as shown in figure 3.5, has been validated for adult males and females between the ages of 19 and 79 years (Matthews et al., 1999).

Jackson et al. (1990) used a similar approach to estimate $\dot{V}O_2$max. A number of important factors including age, gender, BMI or percent body fat (%BF), and a ranking of physical activity were identified. For their ranking, the physical activity rating questionnaire (PA-R) described in figure 3.5 was used. Using these different factors, two formulas that predicted

$\dot{V}O_2$max without an exercise test were developed. In the first formula, referred to as the N-Ex BF model, the score from the PA-R, measured percentage of body fat, age, and gender were used to predict $\dot{V}O_2$max. The second formula, referred to as the N-Ex BMI model, included the same variables except %BF, which was replaced by BMI. Both the N-Ex BMI model and the N-Ex BF model were valid predictors of $\dot{V}O_2$max. The formulas are as follows, with the first number in each formula being a constant that was developed based on the person's gender:

For the N-Ex BMI model,

Female: $\dot{V}O_2$max = 56.363 + 1.921 (PA-R score)
 − 0.381 (age) − 0.754 (BMI)   (6)

Male: $\dot{V}O_2$max = 67.350 + 1.921 (PA-R score)
 − 0.381 (age) − 0.754 (BMI)   (7)

For the N-Ex BF model,

Female: $\dot{V}O_2$max = 50.513 + 1.589 (PA-R score)
 − 0.289 (age) − 0.552 (%BF)   (8)

Male: $\dot{V}O_2$max = 56.376 + 1.589 (PA-R score)
 − 0.289 (age) − 0.552 (%BF)   (9)

Using Jackson's equation (1) (males) applied to Christopher, who is 35 years old, is 174 cm tall, weighs 69 kg, has a body fat percentage of 16%, has a BMI of 22.8 kg·m$^{-2}$, and has a physical activity rating (PA-R) of 6, then

$\dot{V}O_2$max = 67.350 + 1.921 (PA-R) − 0.381 (age)
 − 0.754 (BMI)

= 67.350 + 1.921 (6) − 0.381 (35) − 0.754 (22.8)

= 48.3 ml·kg$^{-1}$·min$^{-1}$

This value is close to the previous one and to the value from the direct method of measurement obtained from the track test.

If the N-Ex BF model is applied to Christopher,

$\dot{V}O_2$max = 56.376 + 1.589 (PA-R) − 0.289 (age)
 − 0.552 (%BF)

= 56.376 + 1.589 (6) − 0.289 (35) − 0.552 (16)

= 47 ml·kg$^{-1}$·min$^{-1}$

All these equations compare reasonably well with the direct $\dot{V}O_2$max measurement. The BMI and the percentage of body fat methods both depend on the balance between energy intake and expenditure of the subjects. Total energy expenditure alone can be used for the prediction of $\dot{V}O_2$max with mean habitual daily energy expenditure being reported as one of the best predictor of $\dot{V}O_2$max (r = .916, n = 120, P < .00001) (Berthouze et al. 1995).

# Estimation of $\dot{V}O_2$max Using Simple Calculations and the Critical Speed Concept

When athletes request advice on physical conditioning, some may have already competed in timed race events. An estimation of $\dot{V}O_2$max without any further testing can be determined using the distance and time achieved in these events. For instance, Christopher had previously run the 800 m in 2 min 40s, the 1,500 m in 6 min 10 s, the 3,000 m in 13 min 39 s, the 5,000 m in 22 min 19 s, the 10,000 m in 46 min 30 s. Determination of the critical velocity can be used to predict his $\dot{V}O_2$max. The critical velocity is the theoretical speed that a person can sustain for a very long (infinite, in theory) time, which is assumed to be based on the hyperbolic relationship between speed and time to exhaustion (Hill, 1927) (figure 3.6).

The equation derived from the hyperbola depicted in figure 3.6 is

$$tlim = AWC/(P − CP)   (10)$$

where tlim is the time to exhaustion in seconds at the power output P in watts, AWC is the anaerobic work capacity in joules (i.e., the energy that can be provided by anaerobic metabolism and desaturation of oxygen reserves) (0.5 L, i.e., 10 kJ), and CP is the critical power in watts that the subject can sustain for several hours provided that fluid and food intake during exercise is adequate.

Experimental studies have demonstrated that people can sustain critical velocity for 20 to 40 min, depending on fitness level. Running at critical velocity induces an increase of blood lactate and $\dot{V}O_2$ (Brickley, Doust, & Williams, 2002; Perrey et al., 2003; Pringle & Jones, 2002; Smith & Jones, 2002) and cannot be sustained for as long a time as postulated by Monod and Scherrer (1965). This nonlinear (hyperbola) relationship requires an appropriate regression equation. Until recently, equation (10) was thought to be the most accurate and appropriate when using the performance variables of distance and time. Morton et al. (1997), however, have demonstrated that there were no significant differences between critical power estimates or between anaerobic work capacity estimates from any model formulation with

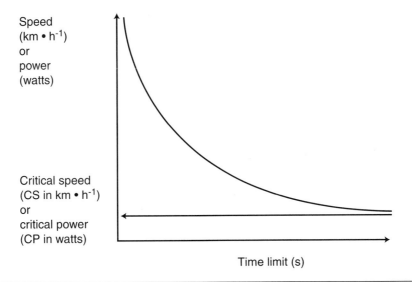

**Figure 3.6** The velocity (or power)/time relationship is a hyperbola: speed = 1/time limit. Time limit is the time to exhaustion at a given speed.

the critical power estimated either from constant load or from incremental exercise. This hyperbola can be transformed into a regression line if speed is replaced by race distance with the time of race (personal best) on the x-axis. Equation (10) is now transformed into equation (11)

$$\text{Distance limit} = \text{AWC} + \text{CP} \times \text{tlim} \qquad (11)$$

where the distance limit is the distance (in meters) covered during the time limit, AWC is the distance (in meters) that can be covered by anaerobic metabolism, CP is the critical power in watts, and tlim is the maximal duration of exercise in seconds.

The critical speed, which is the slope of the relationship between distance limit and time limit, allows the estimation of $v\dot{V}O_2$max by interpolation of the velocity during 5 min, which is the average time to exhaustion when running at $v\dot{V}O_2$max (Dabonneville, Berthon, Vaslin, & Fellmann, 2003; Tong, Fu, & Chow, 2001). Furthermore, the determination of critical velocity (CV) can also provide a good estimation of anaerobic work capacity in addition to the maximum aerobic power output. To estimate $\dot{V}O_2$max from the critical velocity or power (on the cycle ergometer), one can interpolate the velocity at 5 min (table 3.1). The time to exhaustion at $v\dot{V}O_2$max is reproducible within a subject, but there is great interindividual variability among subjects (Billat, Blondel, & Berthoin, 1999). From the critical velocity measurement, it is possible to interpolate and estimate the time to sustain an exercise at $v\dot{V}O_2$max. Some subjects can sustain $v\dot{V}O_2$max for longer than 5 min, which adds imprecision to the measurement.

For instance, exercise duration of 5 min could be the time limit at 105% $v\dot{V}O_2$max for a subject who might have a tlim of 8 min at 100% $v\dot{V}O_2$max (table 3.1). However, subjects who have higher $v\dot{V}O_2$max and $\dot{V}O_2$max are not necessarily the ones who have a longer tlim $v\dot{V}O_2$max. Using the critical velocity model, Billat, Richard, Lonsdorfer, and Lonsdorfer (2001) demonstrated that the time to exhaustion at $v\dot{V}O_2$max depends on the ratio between the anaerobic capacity and the difference in velocity between $v\dot{V}O_2$max and the lactate threshold (vLT) velocities, as in equation (12)

$$\text{tlim at } v\dot{V}O_2\text{max} = \text{AWC}/(v\dot{V}O_2\text{max} - \text{vLT}) \quad (12)$$

where tlim is in seconds, AWC is the distance (in meters) that can be covered by anaerobic metabolism, $v\dot{V}O_2$max is in m · s$^{-1}$, and vLT (velocity at the lactate threshold) is in m · s$^{-1}$.

Therefore, there are two ways to get a long time to exhaustion at $v\dot{V}O_2$max: either have a high anaerobic work capacity, or have good endurance as defined by a narrow difference of speed between $v\dot{V}O_2$max and CV or vLT. See equation (12). The critical speed model allows estimation of $v\dot{V}O_2$max by interpolation of the distance run during 5 min (Tong et al., 2001; Dabonneville et al., 2003). No specific exhaustive 5 min test is required. Alternatively, $v\dot{V}O_2$max evaluation using a single 5 min running field test has been found to be appropriate, irrespective of ability level, age, or sport specialty (Tong et al., 2001; Dabonneville et al., 2003). Some caution, however, is required when using this estimation method. When a subject is requested to run as fast as possible for 5 min, average velocity has been found reliable, but it explains only 35% of the variance in the measured $\dot{V}O_2$max (Tong et al., 2001). Familiarization sessions should be held to achieve best performances (Tong et al., 2001). The 5 min run test is similar to the Cooper's 12 min test but easier to administer. Both tests require subjects to cover maximum distance within the given period. Cooper (1968) has shown that the 12 min run test was highly correlated (r = .9) with $\dot{V}O_2$max. Cooper's study used military personnel who were likely to be highly motivated. Tong et al. interpreted the weak relationship that they found between the average 5 min velocity (v5min) and $\dot{V}O_2$max as resulting from the significant contribution of the lactate threshold and anaerobic capacity to energy requirements over the 5 min period. The 5 min run test is performed at a self-determined pace, and well-trained athletes are better able to select a pace that will reflect v$\dot{V}O_2$max (Billat, Flechet, Petit, Muriaux, & Koralsztein, 1999).

**Table 3.1   Time to Exhaustion (±SD) at the Velocity Associated With $\dot{V}O_2$max: Some Experimental Approaches in Chronological Order**

| Authors | Number of subjects $\dot{V}O_2$max (ml·kg$^{-1}$·min$^{-1}$) v$\dot{V}O_2$max (km·h$^{-1}$) | Ergometer protocol and environment | % $\dot{V}O_2$max and tlim (min) |
|---|---|---|---|
| Horvath and Michael (1970) | 14 female college students 39.5 ± 3.7 ml·kg$^{-1}$·min$^{-1}$ | Cycle ergometer | 100%: 3 ± 2 min |
| Higgs (1973) | 20 active college women 41.3 ml·kg$^{-1}$·min$^{-1}$ | Treadmill | 100%: 4.63 min |
| Volkov, Shirkovets, and Borilkevich (1975) | 4 recreational runners 60.8 ± 3.2 ml·kg$^{-1}$·min$^{-1}$ | Treadmill | 100%: 5.4 ± 3.25 min |
| Lavoie and Mercer (1987) | 5 international women rowers 61.4 ± 4.5 ml·kg$^{-1}$·min$^{-1}$ | Cycle ergometer | 100%: 3.83 ± 1.11 min |
| Camus, Juchmes, Thys, and Fossion (1988) | 7 male students 57.6 ± 9.0 ml·kg$^{-1}$·min$^{-1}$ | Cycle ergometer | 100%: 5.50 min |
| Billat et al. (1994a) | 8 male middle-distance runners 69.4 ± 3.7 ml·kg$^{-1}$·min$^{-1}$ 21.3 ± 0.9 km·h$^{-1}$ | Treadmill | 100%: 6.7 ± 1.9 min |
| Billat et al. (1995) | 16 male middle-distance runners 75.5 ± 5.3 ml·kg$^{-1}$·min$^{-1}$ 22.3 ± 1.1 km·h$^{-1}$ | Treadmill | 90%: 17.4 ± 5.25 min 100%: 5.25 ± 1.31 min 105%: 2.57 ± 0.43 min |
| Billat et al. (1996) | 14 long-distance runners 72 ± 4 ml·kg$^{-1}$·min$^{-1}$ 22.5 ± 0.8 km·h$^{-1}$ | Treadmill | 100%: 5.3 ± 1.3 min |
| Billat et al. (1999) | 6 physical education students 60.6 ± 5.9 ml·kg$^{-1}$·min$^{-1}$ 17.0 ± 2.0 km·h$^{-1}$ | Track | 90%: 12.2 ± 3.7 min 100%: 5.5 ± 1.5 min 120%: 2.1 ± 0.5 min 140%: 1.2 ± 0.3 min |
| Billat et al. (1999) | 8 endurance-trained runners 71 ± 3 ml·kg$^{-1}$·min$^{-1}$ 20.5 ± 0.8 km·h$^{-1}$ posttraining 74 ml·kg$^{-1}$·min$^{-1}$ 21.1 ± 0.8 km·h$^{-1}$ (P < .01) | Track | 100%: 5.0 ± 0.9 min after 4 weeks of normal training 100%: 4.7 ± 0.7 min (NS vs. pretraining) |
| Billat et al. (2001) | 11 male middle-distance runners, 41 ± 10 yr old 54.4 ± 6.0 ml·kg$^{-1}$·min$^{-1}$ 16.0 ± 2.0 km·h$^{-1}$ | Track | 90%: 10.6 ± 2.8 min 95%: 6.6 ± 1.5 min 100%: 4.0 ± 0.7 min 105%: 3.3 ± 0.7 min |

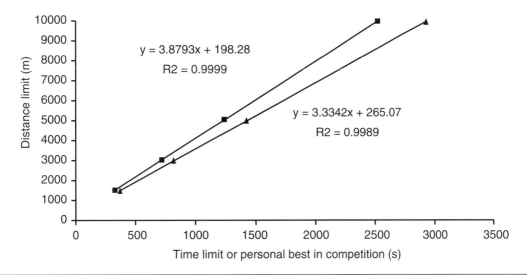

**Figure 3.7**   Christopher's critical velocity (CV) before and after training. CV (in m · s⁻¹) is the slope of the regression line between the distance run (m) and the time to exhaustion (s). The critical velocity equals 3.3342 m · s⁻¹ (or 12 km · h⁻¹) before training (▲) and 3.8793 m · s⁻¹, or 14 km · h⁻¹, after training (■). The slope is steeper after training.

Figure 3.7 provides the calculation of Christopher's critical speed (velocity), which is the slope (distance/time limit) of the regression line. Taking the equations of the regression line linking the distances in which he competed (1,500 m; 3,000 m; 5,000 m; and 10,000 m) with the associated personal best before and after training, we can estimate the $v\dot{V}O_2max$ enhancement.

## $v\dot{V}O_2max$ Before Training

According to equation (11), distance limit = AWC + CP × tlim. Therefore,

distance limit = 265.07 + 3.3342 × time limit (13)

where 265.07 m is the anaerobic work (distance) capacity in meters and 3.3342 is the critical speed in m · s⁻¹.

Therefore, if we are looking for the distance run in 5 min (300 s) to get $v\dot{V}O_2max$ we have

distance limit at $v\dot{V}O_2max$ = 265.07 + (3.3342 × 300)

= 1,265.3 m

Hence, $v\dot{V}O_2max$ = 1,265/300

= 4.21 m·s⁻¹

= 15.1 km·h⁻¹

Therefore, to estimate $\dot{V}O_2$ (in ml · kg⁻¹ · min⁻¹) from a speed (in km · h⁻¹), we simply multiply the velocity by 3.5 ml · kg⁻¹ · min⁻¹. The value of 3.5 ml · kg⁻¹ · min⁻¹ corresponds to the resting metabolic rate (or 1

MET). So $\dot{V}O_2max$ is estimated to be 15.1 × 3.5, or 53 ml · kg⁻¹ · min⁻¹.

Hence, at $v\dot{V}O_2max$ (15 km · h⁻¹), Christopher's energy expenditure is 15 METs. This so-called metabolic scope makes the difference between Christopher and a Kenyan female runner who can increase her energy expenditure up to 21 METS, that is, 75 ml · kg⁻¹ · min⁻¹ (Billat et al., 2003).

## $v\dot{V}O_2max$ After Training

According to equation (11),

distance limit = 3.8793 × time limit + 198.28

where 3.8793 is the critical speed in m · s⁻¹ and 198.28 is the anaerobic work (distance) capacity in meters.

Distance limit = (3.8793 × 300) + 198.28

= 1,362.0 m, and

$v\dot{V}O_2max$ = 1,362/300

= 4.54 m·s⁻¹

= 16.3 km·h⁻¹, hence,

$\dot{V}O_2max$ = 16.3 × 3.5

= 57.2 ml·kg⁻¹·min⁻¹

This gives an 11% improvement in $\dot{V}O_2max$ and $v\dot{V}O_2max$. This improvement is not difficult to achieve with an interval training program based on $v\dot{V}O_2max$ (Billat, 2001a; Billat, 2001b; Seiler & Kjerland, in press; Snyder et al., 1993). After training,

Christopher is now able to run the 5,000 m in 18 min 10 s instead of 22 min 19 s.

The estimation of $v\dot{V}O_2$max using only the performance of the subject in different races requires the following assumptions:

1. To plot the relationship between dlim and tlim, the distances covered in the running event must be between 1,500 m and 10,000 m for the following reasons:

   - The range of the times to exhaustion must be such that throughout the exercise the anaerobic stores are completely utilized (di Prampero, 1986). This is not the case for exercises lasting more than 45 min, in which the glycogen reserve is more crucial.

   - The efficiency of transformation of chemical energy into mechanical energy must be of constant intensity throughout the exercise. This is possible when running below 20 to $22 \text{ km} \cdot \text{h}^{-1}$ so that the aerodynamic component of the cost of running remains below 10% (Pugh, 1970).

2. The performances must have been carried out in the year preceding this estimation. The consistency of the performance can be checked by examining the coefficient of determination of the regression line ($R^2$), which must be over .98. If it is not, one of the race times is out of range and does not fit with the others. Therefore, it is more desirable to plot at least four race times over four distances or time limits (figure 3.7).

# Estimation of $\dot{V}O_2$max Using an Estimation of the Oxygen Cost of Running and Walking

Although $\dot{V}O_2$max has generally been accepted as the physiological variable that best described the capacities of the cardiovascular and respiratory systems, running velocity at maximal oxygen uptake ($v\dot{V}O_2$max), which gives a practical foundation of aerobic demands and capability during running performance, was not assessed until 50 years later. The reason for this temporal gap could be that elite runners did not use exercise physiology laboratories. In the 1980s, aerobic training and running created a social demand, encouraging grants to evaluate fitness (including $\dot{V}O_2$max) and provide scientific training advice. However, the procedure for directly measuring $\dot{V}O_2$max is time consuming and requires trained personnel and special equipment used under controlled conditions (Safrit, Glaucia Costa, Hooper, Patterson, & Ehlert, 1988).

The objective of field testing is to determine $v\dot{V}O_2$max and thus to predict $\dot{V}O_2$max. This approach provides an alternative to the all-out 12 min test of Cooper (1968). Balke (1963) suggested that the distance covered during 15 min of running or walking was a valid indicator of $\dot{V}O_2$max. Cooper, whose study intensified efforts to establish the concurrent validity of distance runs, reported a correlation of .90 between $\dot{V}O_2$max and the distance covered during a 12 min run or walk. Cooper's test is one of the most widely used up to this time, but it involves great motivation and appropriate pacing (Leger & Boucher, 1980). Earlier studies (Cooper, 1968; Leger & Boucher, 1980) of distance run tests yielded high validity coefficients. Cooper's test was based on the linear relationship between running velocity and $\dot{V}O_2$. The limitations and accuracy of the prediction depended on the interindividual variation of the energy cost of running. For less fit subjects, walking tests should be considered.

## Running

During running, Leger and Boucher (1980) showed the reliability of an indirect continuous multistage field test to predict $\dot{V}O_2$max, the University of Montreal Track Test (UMTT). The speed increased from $8.5 \text{ km} \cdot \text{h}^{-1}$ by $1 \text{ km} \cdot \text{h}^{-1}$ every 2 min until the subject could no longer maintain the pace. The so-called maximal aerobic speed (MAS) (Leger & Mercier, 1984) was used to predict $\dot{V}O_2$max. This method was based on Shephard's linear equation (Leger and Lambert, 1982) for the energy cost of running, corrected for the wind effect by Pugh's formula (Pugh, 1970), which yields the following equation:

$$\dot{V}O_2\text{max track (ml·kg}^{-1}\text{·min}^{-1}) = 0.0324 \text{ (MAS track)}^2$$
$$+ 2.143 \text{ MAS track} + 14.49 \qquad (14)$$

Four years later, Leger and Mercier (1984) reinvestigated the gross energy cost of horizontal treadmill and track running and showed the following linear equation:

$$\dot{V}O_2\text{max (ml·kg}^{-1}\text{·min}^{-1}) = 3.5 \text{ speed (km·h}^{-1}) \quad (15)$$

or

$$\dot{V}O_2\text{max (expressed in METs)} = \text{speed (km·h}^{-1}) \ (16)$$

Thus, there was little difference in the test score, which might be expressed as MAS in $\text{km} \cdot \text{h}^{-1}$ or as maximal aerobic power in METs (1 MET, or $\dot{V}O_2$max at rest = 3.5 ml $O_2 \cdot$ kg$^{-1} \cdot$ min$^{-1}$), with speeds between 8 and 20 km $\cdot$ h$^{-1}$ and for adults with average height and weight.

For running speeds above 20 km $\cdot$ h$^{-1}$, equation (15) can be further refined to take into account the aerodynamic energy cost of running. Therefore,

$$\dot{V}O_2max \ (ml \cdot kg^{-1} \cdot min^{-1}) = 2.209 + 3.163 \ \text{speed} \ (km \cdot h^{-1}) + 0.000525542 \ \text{speed}(km \cdot h^{-1})^3 \quad (17)$$

$$MAS \ (m \cdot s^{-1}) = 0.97 \ v1,500 \ (m \cdot s^{-1}) - 0.47 \quad (18)$$

Here MAS is the maximal aerobic speed, and v1,500 is the velocity over 1,500 m.

Kuipers, Verstappen, Keizer, Geurten, and van Kranenburg (1985) had previously proposed the following equation to adjust the maximal velocity (Vmax) based on the length of time it was maintained:

$$Vmax \ (km \cdot h^{-1}) = v + delta \ v \times (t/120) \quad (19)$$

where v is the last velocity completed for 120 s; t is the number of seconds of the final, not completed stage, in which velocity was sustained; delta v is the value of the increase in velocity in km $\cdot$ h$^{-1}$ from the last stage that was maintained; and 120 is the number of seconds in the stage.

For example, if the subject is able to sustain 22 km $\cdot$ h$^{-1}$ for just 1 min (instead of 2) at the last stage of a test during an incremental test of 1 km $\cdot$ h$^{-1}$ every 2 min, Kuipers' equation would calculate the Vmax (km $\cdot$ h$^{-1}$) as follows:

$$Vmax \ (km \cdot h^{-1}) = 22 + [1 \times (60/120)]$$

where v is the velocity maintained during the next to last stage, 1 is the value of the increase in velocity in km $\cdot$ h$^{-1}$ at the last stage maintained during the 110 s of the last stage.

Then

$$Vmax \ (km \cdot h^{-1}) = 22 + (1/2) = 22.5$$

In summary, we can consider the advantage of some of the methodologies that are dependent upon the intent of the investigator. From a practical point of view, Leger's protocol can be used on the track in which the last velocity sustained for at least 2 min is considered the velocity associated with maximal oxygen consumption. To check that the runner reaches his or her maximal oxygen consumption, it would be useful to measure the last heart rate values recorded in the final two stages. If these are separated by more than 5 beats $\cdot$ min$^{-1}$, it is advisable to check whether the runner is able to run faster at a higher heart rate.

## Walking

Two walking tests can be considered for use with untrained subjects: the Rockport Fitness Walking Test (RFWT) (Dolgener, Hensley, Marsh, & Fjelstul, 1994) and the UKK 2 km walking test (Kukkonen-Harjula et al., 1998; Laukkanen, Oja, Pasanen, & Vuori, 1992; LaVoie & Mercer, 1987).

The Rockport Fitness Walking Test, a 1 mi (1,609 m) walking test, is one of the two tests recommended. It has been validated for college males and females (Dolgener et al., 1994). The subject is required to walk as fast as possible for the duration of the test.

The formula for $\dot{V}O_2$max calculation is

$$\dot{V}O_2max = 132.853 - (0.0769 \times \text{weight}) - (0.3877 \times \text{age}) + (6.315 \times \text{gender}) - (3.2649 \times \text{time}) - (0.1565 \times HR) \quad (20)$$

where weight is in kg, age is in years, gender has the value of 0 for women and 1 for men, time is the time to complete the 1 mi (1,609 m) walk (in decimal minutes), and HR is in beats per minute (beats $\cdot$ min$^{-1}$) recorded immediately after stopping.

Christopher performed a 1 mi (1,609 m) walk in 12 min 4 s or 12.06 min (8 km $\cdot$ h$^{-1}$). He did this at a maximal speed technically possible for a non-competitive walker. His Rockport Fitness Walking Test $\dot{V}O_2$max is

$$\dot{V}O_2max = 132.853 - (0.0769 \times 69) - (0.3877 \times 35) + (6.315 \times 1) - (3.2649 \times 12.06) - (0.1565 \times 185)$$
$$= 132.853 - 5.3061 - 13.57 + 6.315 - 39.39 - 28.95$$
$$= 51.5 \ ml \cdot kg^{-1} \cdot min^{-1}$$

This value compares well with the data from the direct $\dot{V}O_2$ measurement, which was 50 ml $\cdot$ kg$^{-1}$ $\cdot$ min$^{-1}$. Furthermore, the HR, which is taken at the end of the exercise, tends to offset the tendency for the RFWT to be a relatively easy test for fit persons. Christopher is still not fit. If his HR had been 150 beats $\cdot$ min$^{-1}$ at the end of the test (35 beats $\cdot$ min$^{-1}$ less), he would have had an RFWT estimated $\dot{V}O_2$max of 57.4 ml $\cdot$ kg$^{-1}$ $\cdot$ min$^{-1}$. This number is not too high for a subject capable of walking at 8 km $\cdot$ h$^{-1}$. Subjects who are unable to run like this test, and it is a good test for people who are beginning a physical activity program. One of the goals for indirect measurement of $\dot{V}O_2$max tests, like the RFWT test, is to encourage participation in exercise programs.

The UKK Institute in Finland has validated a similar kind of walking test for adults (20 to 64 years old) and overweight adults—the UKK 2 km walk test (Kukkonen-Harjula et al., 1998; Laukkanen, et al., 1992; LaVoie & Mercer, 1987). The equation for men is

$$\dot{V}O_2max\ (ml\cdot kg^{-1}\cdot min^{-1}) = 184.0 - 4.65\ (time) - 0.22\ (HR) - 0.26\ (age) - 1.05\ (BMI) \quad (21)$$

The equation for women is

$$\dot{V}O_2max\ (ml\cdot kg^{-1}\cdot min^{-1}) = 116.2 - 2.98\ (time) - 0.11\ (HR) - 0.14\ (age) - 0.39\ (BMI) \quad (22)$$

Christopher performed this 2 km walking test at 7.8 km · h$^{-1}$. He achieved a time of 15 min 23 s (i.e., 15.38 min). He has a BMI of 22.8 kg · m$^{-2}$ and a heart rate of 148 beats · min$^{-1}$. His $\dot{V}O_2$max estimated from the 2 km walking test was

$$\dot{V}O_2max\ (ml\cdot kg^{-1}\cdot min^{-1}) = 184.0 - 4.65\ (15.38) - 0.22\ (148) - 0.26\ (35) - 1.05\ (22.8)$$

$$\dot{V}O_2max\ (ml\cdot kg^{-1}\cdot min^{-1}) = 46.9$$

This number is also close to the direct $\dot{V}O_2$max measured on the track. This 2 km test is sensitive to training and can be used as a reasonably accurate field test to predict changes in $\dot{V}O_2$max resulting from aerobic training in healthy nonathletic adults (Laukkanen et al., 1992). Furthermore, it can be programmed into HR monitor software with the results given automatically from the information provided by equation (21) or equation (22).

# Estimation of $\dot{V}O_2$max Using a Heart Rate Monitor

As previously noted, maximal performance tests allow the prediction of v$\dot{V}O_2$max and $\dot{V}O_2$max. However, from v$\dot{V}O_2$max, that is, the minimal speed that elicits maximal oxygen uptake, a $\dot{V}O_2$ plateau occurs despite the increase of speed (Billat et al., 1994a). Indeed, an athlete reaches and sustains $\dot{V}O_2$max during exhaustive exercise lasting more than 60 s according to the time constant of oxygen kinetics (Billat et al., 1994). For exercise intensity above that which corresponds to the velocity-time asymptote, or critical velocity (CV) (Morgan et al., 1989), it has been found that a slow component of the oxygen uptake ($\dot{V}O_2$) kinetics becomes manifest some 80 to 110 s following the onset of exercise (Gaesser & Poole, 1996; Whipp & Wasserman, 1972). One consequence of this slow component is that it creates a broad range of exercise intensities at which maximal oxygen uptake ($\dot{V}O_2$max) will occur if the exercise is continued to the point of fatigue. Therefore, the so-called velocity associated with $\dot{V}O_2$max, which has been defined

as the minimal velocity at which $\dot{V}O_2$max occurs in an incremental exercise protocol (v$\dot{V}O_2$max) (Billat et al., 1994a), would not be the sole velocity at which $\dot{V}O_2$max occurs. It has been shown that $\dot{V}O_2$max can be achieved during exercise at constant power during a range of intensities that are higher or lower than that at which $\dot{V}O_2$max occurs during incremental exercise (Whipp, 1994). But in that range of exercise intensities, one velocity allows the longest time for exercise at $\dot{V}O_2$max, which is defined as the asymptote in the time at $\dot{V}O_2$max velocity relationship. Indeed, during a 400 m race of 50 s an athlete sustains $\dot{V}O_2$max for about 30 s. Many speeds elicit $\dot{V}O_2$max, from 90% (the critical speed) to 150% of v$\dot{V}O_2$max. Hence, the relationship between time for sustaining $\dot{V}O_2$max and speed is a curve that can be modeled according to the oxygen kinetics and the critical speed model (Billat et al., 2003). Therefore, it is extremely difficult to determine v$\dot{V}O_2$max without knowing the HR because we cannot be sure that a runner, like Christopher, is not able to go beyond his true v$\dot{V}O_2$max.

The estimation of $\dot{V}O_2$max is much enhanced with HR monitoring, allowing the use of nonexhaustive tests similar to the Åstrand submaximal test, with its well-known nomogram (figure 3.8) (Åstrand & Ryhming, 1954). But problems inherent in submaximal testing, such as nonlinearity of $\dot{V}O_2$ and HR over the entire range of effort, changes in heart rate not related directly to work output, and population specificity, are well known. These limitations explain the fact that the nomogram of Åstrand is accurate at ±10%.

The protocol has been clearly explained in the previous edition of this book. Briefly, the Åstrand protocol is designed to determine maximal oxygen consumption by exercising the subject at a submaximal workload and measuring the steady state heart rate. The protocol defines several workloads at which the subject may be evaluated. The workload for male subjects spans from 300 kpm · min$^{-1}$ to 1500 kpm · min$^{-1}$ in 150 kpm · min$^{-1}$ steps. The workload for females covers the area of 300 kpm · min$^{-1}$ to 900 kpm · min$^{-1}$ in 75 kpm · min$^{-1}$ steps. The workload should be difficult enough to maintain a steady heart rate of at least 120 beats · min$^{-1}$. Subjects should be able to complete the minimum of six minutes and reach steady state conditions (i.e., until the heart rate varies no more than 4 beats · min$^{-1}$). The initial power selected should be relatively low and gradually increased until subjects younger than 50 years of age achieve a heart rate of around 140 beats · min$^{-1}$ and older subjects achieve a heart rate of approximately 120 beats · min$^{-1}$. In addition, the perceived exertion measure from the Borg scale should most likely be 13 or 14.

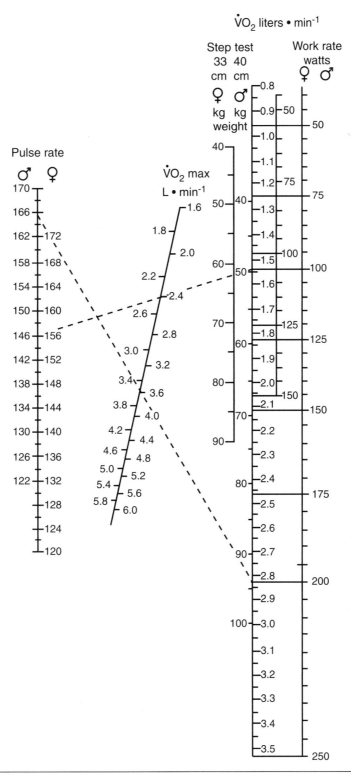

**Figure 3.8** The adjusted nomogram for the calculation of maximal oxygen uptake from submaximal pulse rate and $O_2$ uptake values (cycling, running, or walking). $O_2$ uptake can be estimated by reading horizontally from the workload scale (workload used in the cycle test) to the pulse rate scale (final HR achieved). The predicted maximal $O_2$ uptake ($\dot{V}O_2$max, in liters) can be read in the $\dot{V}O_2$max scale in the middle. In this test, Christopher had a HR of 150 beats · min$^{-1}$ at 150 W. Therefore, he had a predicted absolute value of 3,100 ml · min$^{-1}$ and a relative value per kilogram of body mass of 45 ml · kg$^{-1}$ · min$^{-1}$. This number is 8.8% off the directly measured $\dot{V}O_2$max.

Reprinted, by permission, from P.O. Åstrand, 1960, "Aerobic work capacity in men and women with special reference to age," *Acta Physiologica Scandinavica* 49(suppl 169): 67.

The steady-state HR is recorded. The maximal oxygen uptake is then predicted from HR and the power setting on the ergometer, using the Åstrand nomogram (figure 3.8).

It is possible to estimate the maximal aerobic speed (MAS) and $\dot{V}O_2$max from the heart rate (HR) at a given velocity of 10 km · h$^{-1}$ and 9 km · h$^{-1}$ from a nomogram (figure 3.9). For instance, Christopher has a MAS estimated at 13.7 km · h$^{-1}$ with the 10 km · h$^{-1}$ scale and 13.9 km · h$^{-1}$ with the 9 km · h$^{-1}$ scale with the associated increase in HR from 9 to 10 km · h$^{-1}$. Because Christopher wants to train at v$\dot{V}O_2$max once a week on a bike, his MAS on a bicycle can also be estimated. Christopher needs to ride for 10 minutes at $10 \times 2.5 = 25$ km · h$^{-1}$ and multiply the MAS obtained on the 10 km · h$^{-1}$ scale by 2.5, which becomes the 25 km · h$^{-1}$ scale on the bike. The factor 2.5 is the ratio between the energy cost of running and cycling (with no wind) per unit of distance.

It is possible to estimate cycling maximal aerobic speed from running. For instance, Christopher weighs 69 kg and has a $\dot{V}O_2$max of about 48.8 ml · kg$^{-1}$ · min$^{-1}$ estimated from the 9 km · h$^{-1}$ scale. This gives a gross $\dot{V}O_2$ of $69 \times 48.8 = 3,367$ ml · min$^{-1}$. Hence, considering the average $\dot{V}O_2$/power output value of 12 ml · W$^{-1}$ gives a maximal aerobic power (MAP) on a cycle ergometer of 3,367/12 = 280 W. The equation of Capelli et al. (1993) can be applied with this example:

$$AP\ (W) = av + bv^3 \qquad (23)$$

$$\text{Calculation for } a = (69 + 11)\ 9.81 \times 0.005$$
$$= 3.924$$

where 69 kg is body weight, 11 kg is the weight of the bike, 9.81 is the acceleration of gravity, and 0.005 is the average rolling resistance coefficient, and where v = velocity in m · s$^{-1}$ and b = 0.176, which represents the average coefficient of friction and aerodynamics using the low handlebar position (di Prampero, Cortili, Mognoni, & Saibene, 1979), and W = 280 watts, Christopher's maximal aerobic power.

$$AP\ (W) = av + bv^3$$
$$280 = 3.924v + 0.176v^3 \text{ or}$$
$$0 = 0.176v^3 + 3.924v - 280$$

Solving the equation[1] for v (maximal aerobic speed),

$$v = 11.0379 \text{ m·s}^{-1} \text{ or } 39.73 \text{ km·h}^{-1}$$

---

[1]This equation can be solved using the Mathcad Software (Mathsoft Inc., USA) or Microsoft Excel.

## Estimation of $\dot{V}O_2$max From the Running or Walking Ascent of a Mountain

It is also possible to obtain an indirect estimation of $\dot{V}O_2$max from the running or walking ascent of a mountain (figure 3.10). For example, Christopher spends his holidays in the mountains. He vacations at 1,093 m and in the morning climbs to the top of the local mountain, which is 1,648 m. Therefore, he ascends 555 m in 58 minutes or 3,480 s. His average HR during the climb was 128 beats · min$^{-1}$, and he weighs 69 kg. Power output may be determined from the following equation (Borg, 1982):

$$\text{power output} = (m \times g \times h)/t \qquad (24)$$

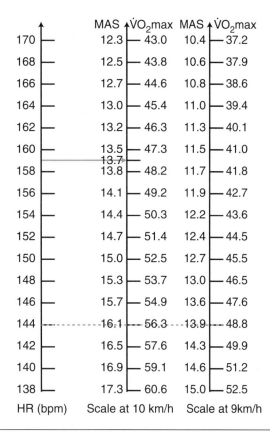

| HR (bpm) | MAS | $\dot{V}O_2$max | MAS | $\dot{V}O_2$max |
|---|---|---|---|---|
| 170 | 12.3 | 43.0 | 10.4 | 37.2 |
| 168 | 12.5 | 43.8 | 10.6 | 37.9 |
| 166 | 12.7 | 44.6 | 10.8 | 38.6 |
| 164 | 13.0 | 45.4 | 11.0 | 39.4 |
| 162 | 13.2 | 46.3 | 11.3 | 40.1 |
| 160 | 13.5 | 47.3 | 11.5 | 41.0 |
| | 13.7 | | | |
| 158 | 13.8 | 48.2 | 11.7 | 41.8 |
| 156 | 14.1 | 49.2 | 11.9 | 42.7 |
| 154 | 14.4 | 50.3 | 12.2 | 43.6 |
| 152 | 14.7 | 51.4 | 12.4 | 44.5 |
| 150 | 15.0 | 52.5 | 12.7 | 45.5 |
| 148 | 15.3 | 53.7 | 13.0 | 46.5 |
| 146 | 15.7 | 54.9 | 13.6 | 47.6 |
| 144 | 16.1 | 56.3 | 13.9 | 48.8 |
| 142 | 16.5 | 57.6 | 14.3 | 49.9 |
| 140 | 16.9 | 59.1 | 14.6 | 51.2 |
| 138 | 17.3 | 60.6 | 15.0 | 52.5 |
| HR (bpm) | Scale at 10 km/h | | Scale at 9km/h | |

**Figure 3.9** Estimation of the maximal aerobic speed (MAS) and $\dot{V}O_2$max from the heart rate (HR) at a given velocity of 9 and 10 km · h$^{-1}$ in males and females between 20 and 40 years old. For instance, Christopher has a MAS estimated at 13.7 km · h$^{-1}$ in the 10 km · h$^{-1}$ scale and 13.9 km · h$^{-1}$ in the 9 km · h$^{-1}$ scale. The HR increased from 144 to 159 beats · min$^{-1}$ with the speed increase from 9 to 10 km · h$^{-1}$.

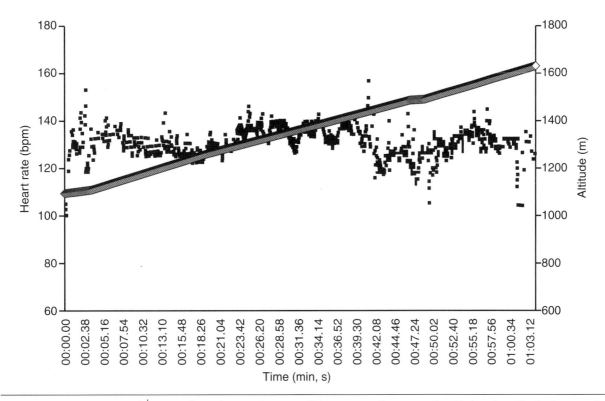

**Figure 3.10** Estimation of $\dot{V}O_2$max from the running or walking ascent of a mountain. Christopher climbed 555 m in 58 min with an average HR of 128 beats·min$^{-1}$.

where m is mass (body weight in kilograms); g is the acceleration of gravity, which is 9.81 m·s$^{-2}$; and h is the height of the ascent in meters.

Christopher's power output would be

$$\text{power output} = (m \times g \times h)/t$$
$$= 69 \times 9.81 \times 555/3{,}480$$
$$= 108 \text{ W}$$

Given that the efficiency of climbing is the same as that for cycle ergometry or for the step test, it may be possible to use the Åstrand nomogram (Åstrand and Ryhming, 1954) (figure 3.8 and figure 3.10) to estimate $\dot{V}O_2$max. By doing this, we find that Christopher has an estimated $\dot{V}O_2$max of 3.4 L·min$^{-1}$ or 49.2 ml·kg$^{-1}$·min$^{-1}$. If he climbs faster, we will find the same result because the work done is similar but the time taken is shorter, that is, 43 versus 58 min (see figure 3.11). The power output, therefore, is higher (146 versus 108 W) but with a higher HR (143 versus 128 beats·min$^{-1}$) (see figure 3.10). With this increased power output but concomitant increase in HR, we find the same value of estimated $\dot{V}O_2$max using the Åstrand nomogram (Åstrand and Ryhming, 1954) (see figure 3.8). This new method gives closer results to the directly measured $\dot{V}O_2$max than the supposedly more sophisticated indirect method. Christopher would have had the same opportunity for indirect $\dot{V}O_2$max measurement had he cycled up the mountain on his mountain bike. All he would need would be a HR monitor.

# Estimation of $\dot{V}O_2$max Using HR Variability at Rest

This section presents a new method for estimating $\dot{V}O_2$max from HR variability at rest developed by the Polar Electro Company and named the Polar Fitness Test. In the Polar Fitness Test, the subject gets a score—the Polar OwnIndex, which estimates maximal aerobic power. This test has been developed using artificial neural network calculation, which is a widely used information processing technique. The factors included in the estimation model of the Polar OwnIndex are resting heart rate, heart rate variability, gender, age, height, body weight, and self-assessment of the level of long-term physical activity. Heart rate variability is measured during 5 min in which at least 250 R-R intervals are recorded in a supine position. For reproducible HR and HR variability measurements, the tests should be conducted in standardized conditions. The tests should be performed in a

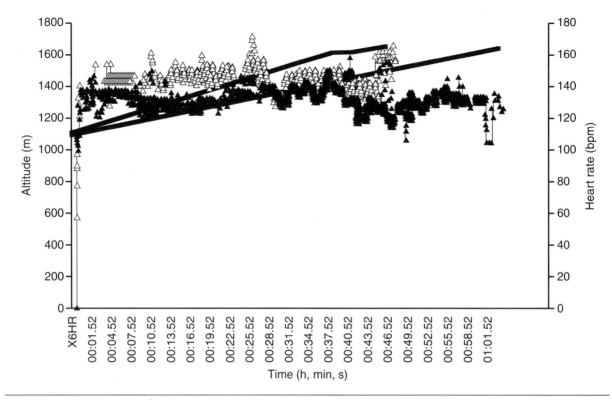

**Figure 3.11**   Estimation of $\dot{V}O_2$max from the running or walking ascent of a mountain. Christopher climbed 555 m in 43 min with an average heart rate of 143 beats · min$^{-1}$ in the second trial (upper straight line and white triangles for heart rate).

peaceful environment, because talking (even a cough), noisy music, and ringing telephones could change HR and HR variability values. The physical activity scale has been modified from NASA–JCS physical activity scale (Ross & Jackson, 1990). The physical activity scale provides four categories including low, middle, high, and top categories, which correspond to the value of 0-1, 2-4, 5-6, and 7 and over, in the NASA scale, respectively. In addition, the physical activity score should remain the same, providing that a person's regular exercise habits have not changed during the previous 6 months.

The index is given at the end of the 5 min period, and the value corresponds to $\dot{V}O_2$max. The correlation coefficient between the laboratory measured $\dot{V}O_2$max and the artificial neural network prediction was .97, with the mean error in the $\dot{V}O_2$max prediction being 6.5% in 25 randomly selected subjects (Vainams et al., 1996). Subjects should avoid heavy physical efforts, alcoholic beverages, and pharmacological stimulants on the test day and the day before.

Christopher had a Polar OwnIndex value of 48, which corresponds to a $\dot{V}O_2$max estimated to be 48 ml · kg$^{-1}$ · min$^{-1}$. This score is very close to his real $\dot{V}O_2$max as measured directly (50 ml · kg$^{-1}$ · min$^{-1}$).

## Summary

Whatever the methods chosen, generally according to the objectives of the subjects tested, one may appreciate the limits of each method and the physiological significance of the results provided. Table 3.2 provides a summary of the possible methods in relation to the population to be tested and the equipment available.

The main aim of indirect methods of estimating $\dot{V}O_2$max is to involve subjects in a project of personal development, including regular exercise, with the initial fitness evaluation being only a pretext for that purpose.

Table 3.2    **Choice of Test According to the Tools Available and the Populations Tested**

| Tools | Subjects | Testing method |
|---|---|---|
| Nothing but a pencil | S | Questionnaires |
| | RE | Questionnaires |
| | ET | MAS using CV concept<br>or just the best performance over 3 km or 10 km |
| Stopwatch and tape measure | S | Walking test for 1 mi (1.6 km) |
| | RE | Incremental test from 8 km·h$^{-1}$ to exhaustion with 1 min long stages and 0.5 km·h$^{-1}$ speed increment |
| | ET | - Incremental test from 10 km·h$^{-1}$ to exhaustion with 2 min long stages and 1 km·h$^{-1}$ speed increment<br>- 5 min all-out test |
| Stopwatch, tape measure, heart rate monitor (and cycle ergometer for S subjects) | S | - Åstrand nomogram on cycle ergometer |
| | RE | - Walking or running test for 1 mi or 2 km<br>- Incremental test from 8 km·h$^{-1}$ to exhaustion with 1 min long stages and 0.5 km· h$^{-1}$ of speed increment |
| | ET | - Incremental test from 10 km·h$^{-1}$ to exhaustion with 2 min long stages and 1 km·h$^{-1}$ of speed increment<br>- 5 min all-out test<br>- Running or walking test in the mountains using the Åstrand nomogram |
| Beat-by-beat heart rate monitor | S<br>RE | Heart rate variability test during a 5 min rest test |

S = sedentary subjects, RE = regular exercisers, ET = endurance trained subjects, CV = critical velocity.

# four

# Heart Rate Variability: Measurement Methods and Practical Implications

**Philippe Lopes, PhD**
*University of Evry Val d'Essonne, France*

**John White, PhD**
*University of Nottingham, England*

This chapter is composed of five main sections. The first section deals with heart control mechanisms and the specific anatomy of the conducting system. The second section provides insight into the methods used to analyze heart rate variability (HRV) and discusses the importance of respiration control. The third section deals with selected practical implications of HRV in exercise performance. The fourth section presents a practical example of data collection and analysis. The final section summarizes the potential importance of heart rate variability in coronary heart disease morbidity and mortality.

## Control Mechanisms and the Conducting System

The electrical activity of the heart has been observed since 1856, when two investigators applied the cut end of a frog's nerve muscle preparation to the left ventricular wall of an exposed heart (Van Capelle, 1987). They discovered that the electrical activity of the heart during beating stimulated a muscular twitch of the nerve muscle. Since this early work, electrophysiological knowledge was almost entirely derived in animal studies. Advanced electrical recording methods, however, have provided more information about

the human heart electrical conducting system and the generation of the electrocardiogram (ECG) signal.

Heart rate is derived from the ECG signal and is under the control of various mechanisms. Internal control within the heart maintains heart rhythm, whereas variations of the heart rate are under the control of the nervous system.

Electrical potentials exist across the membranes of all the cells of the body. Within the cardiac muscle, some cells, known as excitable cells, have the ability to conduct with an action potential, the magnitude of which varies in different types of cardiac tissues (Clark, 1978).

The internal mechanisms for the regulation of heartbeats consist of several specialized fibers. The heart requires an electrical stimulus to initiate its action. The normal heartbeat starts with the propagation of an electrical impulse in the sinoatrial node, the natural pacemaker of the heart, which is located at the top of the right atrium (figure 4.1). The regular electrical discharge from this node initiates the usual rhythmic contraction of the entire heart. The impulse spreads through the atrial muscle, causing atrial systole. This contraction is registered by the atrioventricular node, located at the bottom of the atrium, and conveyed by the Purkinje fibers to the ventricular muscles, causing ventricular systole. This series of contractions is preceded by a minute change

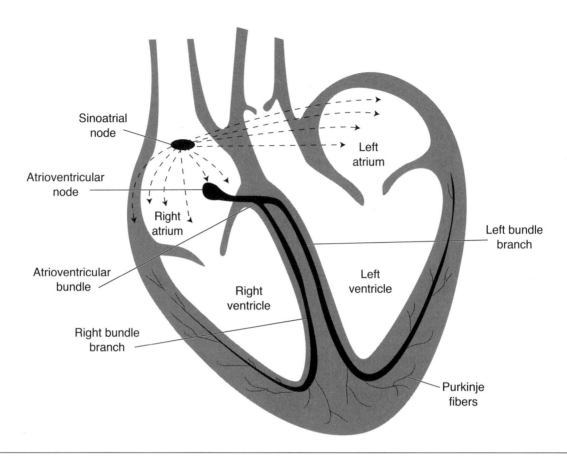

**Figure 4.1**   Electrical conduction system of the heart.

in electrical voltage (depolarization), which can be monitored at the body surface in the form of an electrocardiogram, or ECG signal.

To collect the ECG waveform, a series of electrodes need to be placed on the chest of a subject as illustrated in figure 4.2. The derived ECG consists of three major components, which include the P-wave, the QRS complex, and the T-wave (see figure 4.3). These components are associated with atrial activity, depolarization of the ventricles, and repolarization of the ventricles, respectively. Following contraction, the muscle relaxes to allow the heart to fill with blood again, and a refractory period occurs during which the muscle is unable to contract again. The rate of depolarization determines the length of the refractory period (Fozzard, 1979).

The sinoatrial node, which is under the influence of both the sympathetic and the parasympathetic nervous systems, sets a specific rate at around 120 beats · min$^{-1}$ at normal body temperature (Guyton, 1977). The vagus nerve inhibits the node, however, which keeps the actual resting heart rate down to somewhere in the range of 70 to 80 beats · min$^{-1}$. The parasympathetic nervous system exerts its influence through the vagus nerve by releasing the

neurotransmitter acetylcholine at the nerve endings, which results in a slowing of both the activity at the pacemaker and the cardiac impulse passing into the ventricles (Guyton, 1977). The sympathetic nervous system exerts its influence by releasing the neurotransmitter norepinephrine at the sympathetic

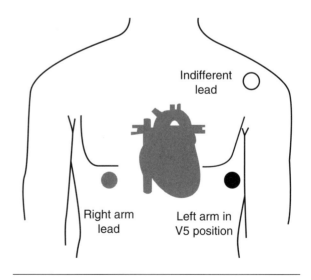

**Figure 4.2**   ECG electrodes position.

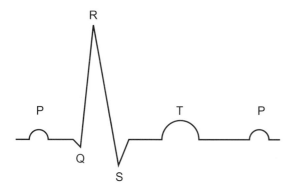

The P-wave represents atrial depolarization
The QRS complex represents ventricular depolarization
The T-wave represents ventricular repolarization

**Figure 4.3** Electrocardiogram (ECG).

nerve endings in the heart. This stimulation acts mainly at the pacemaker, increasing the rate of sinoatrial node discharge and accelerating atrioventricular conduction of impulses (Guyton, 1977).

Under normal circumstances, when a healthy individual is resting, parasympathetic tone is predominant (Vassalle, 1976). If the administration of the appropriate drug (e.g., atropine) blocks this parasympathetic influence, the heart speeds up considerably, demonstrating what is called a pronounced tachycardia. The fact that the parasympathetic tone is the dominant one under resting conditions can be shown by blocking the sympathetic innervation of the sinoatrial node (e.g., propanolol).[1] Thus, removing parasympathetic influence causes a big increase in heart rate, demonstrating that the vagus nerve

is exerting considerable control in keeping heart rate down. On the contrary, removing sympathetic influence causes only a small decrease in heart rate, showing that its normal excitatory influence under resting conditions is only minimal.

## Measurement Methods

Heart rate variability (HRV) refers to the amount of heart rate fluctuation in the mean heart rate and reflects the modulation of cardiac function by autonomic and other physiological regulatory systems (Akselrod et al., 1981). Respiration has an important influence on heart rate variation. This phenomenon, called respiratory sinus arrhythmia (RSA), is a rhythmical fluctuation in heartbeat intervals characterized by a shortening and lengthening of the intervals in a phased relationship with inspiration and expiration, respectively (Berntson, Caciopo, & Quigley, 1993). Fluctuations in heart rate are commonly assessed by RR time interval measurements[2] as shown in figure 4.4.

In a controlled environment in which respiration is monitored, RR intervals lengthen with expiration and shorten with inspiration, as illustrated in figure 4.5.

From the ECG waveform, the measured RR intervals representing the cardiac rhythm are displayed using an interbeat interval series (IBIS) tachogram (Janssen, Swenne, de Bie, Rompelman, & van Bemmel, 1993). As shown in figure 4.6, cardiac beat is displayed along the x-axis and each RR interval (measured in ms) is displayed along the y-axis.

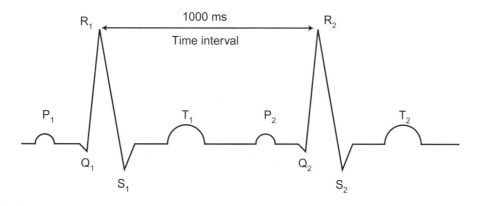

**Figure 4.4** Schematic representation of an RR time interval.

[1]Propanolol is a beta-adrenergic blocker that blocks the sympathetic system, and thus variation in heart period is assumed to be predominantly vagally mediated.

[2]RR intervals correspond to the time in ms between two R spikes of the ECG complex. In this example, the time interval $R_1R_2 = 1,000$ ms.

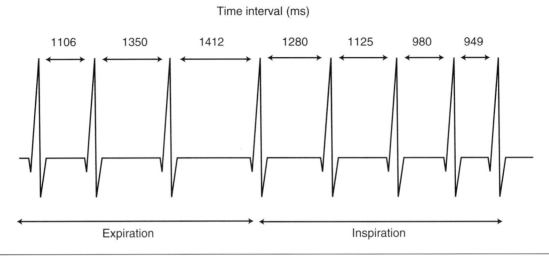

**Figure 4.5** Influence of respiration on RR intervals (ms).

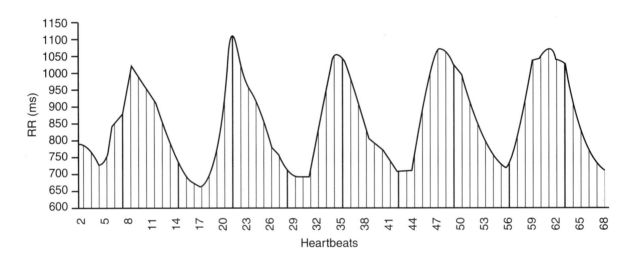

**Figure 4.6** Representation of RR intervals in the time domain.

## Heart Rate Variability Recording

Recent advances in computing have allowed for a variety of physiological signals to be recorded. When using commercially available ECG equipment, the sampling rate for the acquisition of RR intervals is usually 1,000 Hz, providing an accuracy of ±1 ms between RR intervals.

### Duration of Recording

The duration of the recording varies with RR intervals collected over periods ranging from as little as 30 RR intervals to as long as 24 or 48 h. A distinction can be made between "long term" HRV for records of 24 h and more and "short term" HRV for records collected between 2 min and a few hours. The Task Force of the European Society of Cardiology and the North American Society of Pacing and Electrophysiology (1996) advise a minimum of 5 min of recording.

### Artifact and Arrhythmias

Usually, the collection of ECG data and subsequent RR intervals contains errors or noise because of artifacts or arrhythmias. Careful subject preparation and maintenance of recording equipment should help reduce these inaccuracies. Because the analysis of HRV depends on the integrity of the data, eliminating the noise is necessary.

To optimize R-wave peak identification, various automated techniques including template matching or interpolation algorithms can be used. In the case of artifacts, techniques using smoothing or filtering of the digitized data are common (Rinoli & Porges, 1997; Task Force of the European Society of Cardiology & the North American Society of Pacing and Electrophysiology, 1996). Researchers often verify the integrity of the data manually, particularly with short-term recordings, but this process can be tedious for longer recordings. In the case of arrhythmias, HRV analysis is not possible, particularly with persistent atrial fibrillation.

# Time Domain Analysis of Heart Rate Variability (Statistical Measures)

Analysis of the beat-to-beat variation can be performed using time domain analysis methods. Time domain variables can be derived either directly from the measurement of RR intervals or from the difference between RR intervals (table 4.1).

Table 4.1   **Time Domain Analysis (Statistical Measures) in a Subject Breathing at 6 Breaths·Min$^{-1}$ and at 12 Breaths·Min$^{-1}$**

| Variable | Units | Value |
|---|---|---|
| **Statistical measures, at 6 breaths·min$^{-1}$** | | |
| Mean RR | (s) | 0.869 |
| STD | (s) | 0.076 |
| Mean HR | (1/min) | 69.62 |
| STD | (1/min) | 6.20 |
| rMSSD | (ms) | 45.2 |
| NN50 | (count) | 28 |
| pNN50 | (%) | 20.3 |
| **Statistical measures, at 12 breaths·min$^{-1}$** | | |
| Mean RR | (s) | 0.860 |
| STD | (s) | 0.026 |
| Mean HR | (1/min) | 69.96 |
| STD | (1/min) | 2.19 |
| rMSSD | (ms) | 22.5 |
| NN50 | (count) | 0 |
| pNN50 | (%) | 0.0 |

Variables calculated directly from the measurement of RR intervals generally include the following:

- Mean RR (ms or seconds): mean RR interval between all normal beats

$$Mean\ RR = \frac{\sum_{i=1}^{n} NN_i}{n}$$

where $NN_i$ represents normal-to-normal RR intervals (artifact-free RR intervals)

- RRmax – RRmin (ms or seconds): difference between the average maximum RR interval and the average minimum RR interval in the window of measurement or during a single or few breaths (Fouad, Tarazi, Ferrario, Fighaly, & Alicandri, 1984)
- SDNN (ms or seconds): standard deviation of all normal RR intervals

$$SDNN = \sqrt{\frac{\sum_{i=1}^{n}\left(NN_i - \bar{N}\bar{N}\right)^2}{n-1}}$$

- SDNN index (ms or seconds): mean of the standard deviations of all normal RR intervals for all 5 min segments of the entire recording period (long-term recording)

$$SDNN_{index} = \frac{\sum_{i=1}^{n} SDNN_i}{n}$$

- SDANN (ms or seconds): standard deviation of the averaged normal sinus RR intervals for all 5 min segments of the entire recording period (long-term recording)

Variables derived from the difference between RR intervals include the following:

- NN50: the number of adjacent RR intervals that varied by more than 50 ms
- pNN50 (%): the percentage of adjacent RR intervals that varied by more than 50 ms
- rMSSD (ms): the root mean square of the difference between the coupling intervals of adjacent RR intervals

$$rMSSD = \sqrt{\frac{\sum_{i=1}^{n}\left(NN_{i+1} - NN_i\right)^2}{n-1}}$$

The task force recommends SDNN, SDANN, and rMSSD for the HRV analysis in the time domain

(Task Force of the European Society of Cardiology & the North American Society of Pacing and Electrophysiology, 1996). Note that the mean RR or mean HR is nearly the same for the two breathing frequencies. Heart rate variability measures, however, are much greater for the 6 beats·min⁻¹ frequency.

## Geometric Methods of Heart Rate Variability Analysis

Geometric methods present RR intervals in geometric patterns. The triangular index is a measure in which the length of RR intervals serves as the x-axis of the plot and the number of each RR interval length serves as the y-axis. It is computed as the total number of NN intervals divided by the height of the histogram of all NN intervals using a discrete scale.

Another geometric method, the triangular interpolation of NN (TINN), provides the baseline width of the distribution measured as a base of a triangle, approximating the NN interval distribution HRV (see figure 4.7).

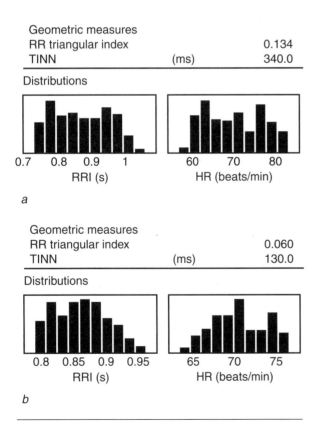

**Figure 4.7**   Time domain analysis (geometric measures) in a subject breathing *(a)* at 6 breaths·min⁻¹ and *(b)* at 12 breaths·min⁻¹.

## Frequency Domain Analysis of Heart Rate Variability (Spectral Analysis)

Classical statistical analysis of RR intervals (RRmax – RRmin, SDNN, rMSSD) has been used for several decades to provide a sensitive measure of cardiac vagal tone, but these measures are often influenced by the underlying mean heart rate (Billman, Schwarz, & Stone, 1982). Spectral analysis, on the other hand, is thought to provide an understanding on the effects of sympathetic and parasympathetic systems on heart rate variability (Akselrod et al., 1981; Akselrod, 1985; Pomeranz, Macauley, Caudill, et al.,1985). Furthermore, frequency domain measures are thought to provide an advantage over time domain measures, in that the measures offer more information about the relative contributions of sympathetic and parasympathetic function (Akselrod et al., 1981; Malliani, Pagani, Lombardi, & Cerutti, 1991; Pagani et al., 1986).

Early work pointed out a few bands in the spectrum of HRV that could be interpreted as markers of physiological relevance (Penaz, Roukenz, & Van Der Waal, 1968; Sayers, 1973). Investigators used Fast Fourier Transformation (FFT) for the analysis of heart rate time series and the associated mechanisms of thermoregulation, blood pressure, and respiratory effects on heart rate to these respective harmonic activities. Later work introduced a parametric method of spectral estimation, based on an autoregressive approach (AR), with an automatic decomposition of spectral components that could easily obtain parameters associated to the various frequency bands (Baselli et al., 1985; Hedelin, Wiklund, Bjerle, et al., 2000).

Figure 4.8 shows a typical RR interval spectrum calculated over 256 consecutive beats. Three main activities associated with the power spectrum can be recognized (Cerutti et al., 1994):

- A very low-frequency (VLF) component usually allocated at less than 0.04 Hz. The power output below 0.04 Hz is thought to represent fluctuations that obviously occur very slowly and are possibly due to circadian rhythms and peripheral vasomotor and thermoregulatory influences (Appel, Berger, Saul, Smith, & Cohen, 1989).

- A low-frequency (LF) component at around 0.1 Hz (frequency band between 0.04 and 0.15 Hz). The power output in the low frequency (LF) band is partly dependent on sympathetic

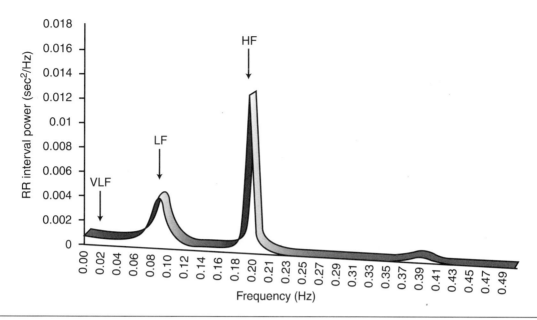

**Figure 4.8**   Power spectral densities (autoregressive method) with the spectral components identified in a subject breathing at 12 breaths · min⁻¹.

tone because of baroreceptor activity (Malliani et al., 1991).

- A high-frequency (HF) component synchronous with the respiration rate (inside a wide range of frequencies, generally from 0.2 Hz up to 0.5 Hz depending on the breathing frequency and considered an indicator of vagal activity) (Akselrod et al., 1981; Malliani et al.).

The power spectral components normally reported are presented in figure 4.8 and the associated power spectrum analysis in table 4.2. VLF, LF, and HF central frequencies bands (Hz) and associated power are presented in absolute values (s² or ms²) or in normalized units (nu), that is, percentage value of power in respect to the total power (TP) less the VLF component (Task Force of the European Society of Cardiology & the North American Society of Pacing and Electrophysiology, 1996) with $LF_{nu}$ and $HF_{nu}$ calculated as follows:

$$LF_{nu} = 100 \times \frac{LF}{\left(TP - VLF\right)}$$

and

$$HF_{nu} = 100 \times \frac{HF}{\left(TP - VLF\right)}$$

Strong evidence has been obtained that HF power might be considered a reliable marker of vagal control of heart rate (Akselrod et al., 1981) and that LF

Table 4.2   **Power Spectral Calculations in a Subject Breathing at 12 Breaths · Min⁻¹ (from figure 4.8)**

| Frequency band | Peak (Hz) | Power (ms²) | Power (%) | Power (nu) |
|---|---|---|---|---|
| VLF | 0.0000 | 25 | 8.2 | |
| LF | 0.0918 | 138 | 45.8 | 46.6 |
| HF | 0.1992 | 139 | 46.0 | 46.9 |
| LF/HF | | | 0.994 | |

provides a marker of sympathetic activity (Akselrod et al., 1981; Pagani et al., 1986). The LF component is, to some degree, an index of sympathetic activity because pharmacological interventions that increase or block the sympathetic drive (e.g., using propanolol) can alter it (Malliani et al., 1991). However, this LF component also represents some parasympathetic activity (Eckberg, 1997). The LF/HF ratio has also been accepted as an index of sympathovagal modulation of the sinoatrial node by the Task Force of the European Society of Cardiology and the North American Society of Pacing and Electrophysiology (1996). There has been some recent criticism, however, because of the controversy surrounding the physiological significance of the LF component (Eckberg, 1997).

## Neurophysiological Mechanisms of Heart Rate Variability Control

Neural regulation of circulatory functions is mainly achieved through the excitatory influence of the sympathetic and vagal outflows, which are tonically and phasically modulated by the interactions of at least three major factors: the central integration, peripheral inhibitory baroreceptor-vagal reflex mechanisms (with negative feedback characteristics), and peripheral excitatory sympathetic reflex mechanisms (with positive feedback characteristics) (Malliani, Pagani, & Lombardi, 1986; Malliani et al., 1991). As illustrated in figure 4.9, several investigators have tried to model the complexity of neural regulation and viewed the dynamic state of the sympathetic-vagal interaction as a push-pull, or reciprocal, relationship (Malliani et al.). The core hypothesis of these investigators is that this interaction can be broadly explored in the frequency domain with the following assumptions:

- Heart rate variability related to respiration defined as the high-frequency (HF) spectral component is a marker of vagal modulation.

- Heart rate and arterial pressure variability defined as the low-frequency (LF) spectral component is a marker of sympathetic modulation.

- A reciprocal relation exists between these two rhythms, a relationship similar to that characterizing the sympathetic-vagal balance.

The model proposes a schematic representation of the opposite feedback mechanisms that, in addition to central integration, subserves the neural control of the cardiovascular system. Baroreceptive and vagal afferent fibers from the cardiopulmonary region mediate negative feedback mechanisms (exciting the vagal and inhibiting the sympathetic outflow), and sympathetic afferent fibers mediate positive feedback mechanisms (exciting the sympathetic and inhibiting the vagal outflow).

## Validity and Reliability of Heart Rate Variability Measures

Interindividual variability was tested in 25 healthy subjects aged 20 to 82 on two occasions: 7:00 a.m. and 9:00 a.m. Heart rate and RSA amplitude showed relative stability (Hrushesky, 1991). Interindividual variability was also tested in "several individuals" every 2 h during 24 h, and it was demonstrated that RSA amplitude appeared to be most stable and nearest the 24 h mean value early in the day, usually between 6:00 a.m. and noon.

Other work has found time and frequency domain measures of heart rate variability to be highly reproducible with short-term ECG recordings (Kyrozi, Maounis, Chiladakis, Vassilikos, Manolis, & Cokkinos, 1995). Indeed, two consecutive 10 min ECG recordings were analyzed and showed high correlation coefficients between time and frequency domain, with pNN50 showing the highest correlation coefficient and SDNN the lowest (r =.96 and r =.63, p < .001, respectively).

This was further confirmed in a study that demonstrated similar reproducibility patterns, especially the high-frequency power during controlled respiration. Reproducibility was evaluated by means of the interclass correlation coefficient (ICC), comparing baseline values with measurements taken at the second week and the seventh month. All time domain measurements had an ICC ≥ .75 except for the standard deviation of RR intervals, which had an ICC of

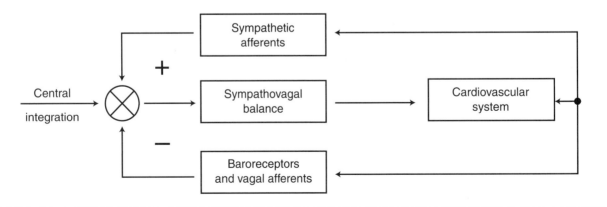

**Figure 4.9**   Cardiovascular neural regulation model.

Reprinted, by permission, from A. Malliani et. al., 1991, "Cardiovascular neural regulation explored in the frequency domain," *Circulation* 84: 482-492.

.57. The frequency domain parameters obtained by means of either FFT or AR showed similar reproducibility (Pitzalis et al., 1996).

## Nonlinear Analysis of Heart Rate Variability

Time and frequency domain measures require the RR interval data to be stationary. Nonstationary data included abrupt changes in RR interval variation often seen in physiological manipulations (e.g., exercise, posture change) or in pathological conditions (e.g., hypertension, panic attacks). Because changes in RR interval are slow, ignoring this violation is common and often reasonable.

In a variety of other physiological and environmental conditions in which rapid dynamic changes in autonomic control are likely to occur, investigators have used nonlinear methods to quantify the complexity of heart dynamics (Goldberger, 1996). The nonlinear techniques make a fundamentally different assumption about the data: that much of its behavior is aperiodic, but still not random. For example, approximate entropy (ApEn) quantifies the randomness or predictability of RR intervals (Pincus, 1991, 1995). Although a number of techniques are emerging, especially in the clinical setting (Goldberger, 1991; Lombardi, 2002; Makikallio, Tapanainen, Tulppo, & Huikuri, 2002), one in particular, the Poincaré plot, is receiving increasing recognition in the field of sport and exercise.

The Poincaré plot is a scattergram in which each RR interval is plotted as a function of the previous RR interval (see figure 4.10, a-b). It provides a graphical representation of cardiovascular dynamics resulting in an elliptical type of shape. The ellipse is fitted onto the so-called line of identity at 45° to the normal axis. The standard deviation of the points perpendicular to the line of identity, denoted by SD1, describes short-term RR variability due to the respiratory component of HRV. The standard deviation along the line of identity, denoted by SD2, describes long-term variability (Tulppo, Mäkikallio, Takala, Seppänen, & Huikuri, 1996). In figure 4.10, a and b, the different breathing frequencies (6 and 12 breaths · min⁻¹) produced two ellipses of different shape and magnitude with heart rate remaining stable.

A young subject would normally show, at rest, high heart rate variability associated with a big ellipse, as demonstrated in figure 4.11. An older subject exhibiting reduced heart rate variability, however, could produce a much reduced ellipse in which the RR points would gravitate around the mean RR and both lines of identity (see figure 4.12).

## Respiration Control in Heart Rate Variability Measurement

More recently, much attention has been given to the methodology used to analyze HRV, especially when inferring parasympathetic tone (Jennings & McKnight, 1994). Respiration has a profound influ-

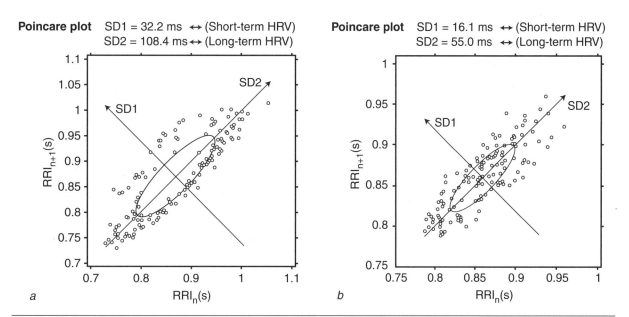

**Figure 4.10**   Quantitative beat to beat analysis of RR intervals (Poincaré plot) of a subject breathing at (a) 6 breaths · min⁻¹ and (b) 12 breaths · min⁻¹.

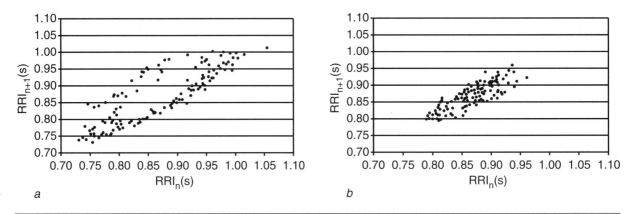

**Figure 4.11**   Poincaré plot of a young subject breathing at *(a)* 6 breaths · min[-1] and *(b)* 12 breaths · min[-1].

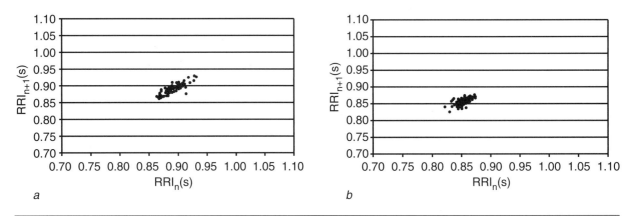

**Figure 4.12**   Poincaré plot of a 52-year-old man exhibiting several CHD risk factors and breathing at *(a)* 6 breaths · min[-1] and *(b)* 12 breaths · min[-1].

ence on heart rate, a phenomenon called respiratory sinus arrhythmia (RSA) (Davies & Neilson, 1967), and the frequency at which HF appears depends on the respiratory rate. Therefore, several authors have emphasized the importance of controlling respiratory rate and volume when using HRV measures as noninvasive estimation of cardiac vagal tone (Brown, Beightol, Koh, & Eckberg, 1993; Grossman, Karemaker, & Wieling, 1991; Hayano, Mukai, et al., 1994a; Hirsch & Bishop, 1981).

Many early studies that considered the effects of respiration on RR interval variation suggested that respiration evoked changes in human heart rate. As early as the 1930s, it was found that vagal excitation of the sinus node was virtually absent during inspiration (indicated by shorter intervals between heartbeats). In contrast, during expiration, intervals were found to be of longer duration between successive heartbeats, which led to a decrease in heart rate (Anrep, Pascual, & Rossler, 1936). A later study examining the separate effects of inspiration and

expiration on heart rate also found that the respiratory effect on heart rate is mainly brought about by inspiration and that cardiac intervals tend to increase as the respiratory intervals increase (Davies & Neilson, 1967). Other workers proposed that the changes in heart rate induced by breathing probably reflect fluctuations in vagal tone towards the sinus node and that abnormal results may reflect abnormalities in the withdrawal of vagal tone including inactivation of acetylcholine at the receptor site (Mehlsen, Pagh, Nielsen, Sestoft, & Nielson, 1987). It appears that RSA depends on both respiratory frequency and tidal volume (depth), with respiratory parameters of breathing frequency and tidal volume therefore exerting profound influences on the quantity, periodicity, and timing of vagal cardiac efferent activity in conscious humans (Eckberg, 1983).

A more recent study investigated the effects of a variety of breathing frequencies and tidal volumes on RR interval variation power spectra (Brown et al., 1993). The study also considered relevant published literature in this field to elucidate whether studies had considered the potential effect of respiration on RR interval power spectra. The study concluded that

- respiratory rate and tidal volume exert significant influence on respiratory power,
- breathing rate and tidal volume did not significantly affect average RR intervals, and
- most researchers ignored the influences of breathing frequency and tidal volume on the measurement of RR interval.

Figure 4.13 depicts the power spectral density of a subject at rest breathing spontaneously at respiratory rates of 6, 8, 10, 12, 14, and 16 breaths · min$^{-1}$ and at a nominal tidal volume of 30% of vital capacity. The power spectra show specific HF frequency peaks that are closely related to the respiratory frequency. Note that each respiratory frequency of 6, 8, 10, 12, 14, and 16 breaths · min$^{-1}$ has a corresponding HF frequency band centered on 0.10, 0.13, 0.16, 0.20, 0.23, and 0.26 Hz, respectively. In addition, the HF power spectrum declined significantly when the respiratory rate was increased from 6 to 16 breaths · min$^{-1}$.

In addition, the effects of breathing frequency on the relationship between spectral components of HRV and the mean cardiac vagal tone have been investigated (Hayano, Mukai, et al., 1994a). The

study showed increased HF amplitude and decreased LF/HF ratio without changes in mean RR interval under both the respiratory control and β-adrenergic blockade conditions.

In summary, evidence is accumulating that respiratory frequency and tidal volume exert profound influences on HRV measurement, in terms of altering the amplitude and magnitude of RR interval variation without altering mean RR intervals, and that changes in these respiratory parameters lead to a variation in autonomic heart activity.

# Practical Implications

This section discusses the importance of selected variables including the influences of age, aerobic fitness exercise, and overtraining on heart rate variability.

## Age and Heart Rate Variability

A decrease in HRV appears to occur with increasing age (Pomeranz et al., 1985). This decline of HRV has been attributed to a decline in efferent vagal cardiac tone and decreased β-adrenergic responsiveness (Molgaard, Sorensen, & Bjerregaard, 1991; O'Brien, O'Hare, & Corrall, 1986). More interestingly, the

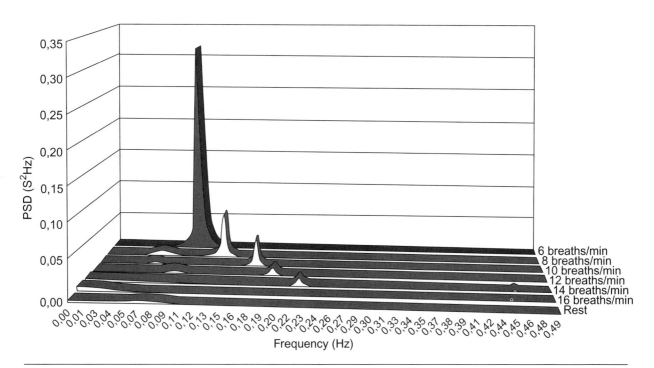

**Figure 4.13**   RR interval power spectral density of a subject breathing spontaneously (at rest) and at respiratory rates of 6, 8, 10, 12, 14, and 16 breaths · min$^{-1}$ and at a nominal tidal volume of 30% of vital capacity.

age-related loss of vagal cardiac activity has been partly attributed to the loss of arteriolar compliance (De Meersman, 1993a). These findings were based on short-term measurements of RR intervals at 6 breaths $\cdot$ min$^{-1}$ respiratory frequency and 30% of subjects' vital capacity.

Another study investigated the age and RSA amplitude relationships in a group of 326 healthy subjects, aged 10 to 75 years. Subjects were arbitrarily divided into 14 age-defined categories, each spanning 5 years. An obvious age-related fall in amplitude occurred (slope, 10% per decade, r = .65, p < .01). In absolute terms the RSA amplitude decreased from 8.5 beats per breathing cycle in the youngest group (10-15 years old) to just over 1.3 beats per breathing cycle in the oldest age group (70-80 years old) tested (Hrushesky, 1991).

## Aerobic Fitness and Heart Rate Variability Measures

The reduced levels of resting heart rates commonly observed in endurance athletes have been shown to be related to the cardiovascular effects of training (Park & Crawford, 1985). Furthermore, the interbeat variation in heart rate has been demonstrated to be a function of conditioning status in young subjects (White, Semple, & Norum, 1990). Also, the involvement of the autonomic nervous system during exercise and physical training is well documented (Green, 1990; Seals & Chase, 1989), and physically trained subjects have a generally increased parasympathetic activity and lower heart rates, particularly at rest, when compared with sedentary individuals (Dixon, Kamath, McCartney, & Fallen, 1992; Frick, Elovainio, & Somer, 1967).

Following the observation that well-trained athletes had more prominent RSA than would ordinarily be expected for age, it was suggested that RSA amplitude could be used to measure aerobic conditioning (Hrushesky, 1991). The study was designed to compare the behavior of RSA and the maximal oxygen consumption during either graded walking exercise or flexibility training over a 20-week test period. Sixteen sedentary adults (7 males and 9 females) between 22 and 35 years of age with no medical problems were stratified by sex, smoking, status, age (22-26 and 27-35), and baseline $\dot{V}O_2$max and randomized to perform 12 weeks of either graded walking or stationary flexibility exercises (four times a week for 50 min). RSA measurements were performed at the beginning, after 6 weeks, after 12 weeks, and 4 weeks after discontinuation of exercise

with subjects breathing at 10 cycles $\cdot$ min$^{-1}$. $\dot{V}O_2$max was measured on study entry and after 12 weeks during a Bruce multistage treadmill test. Overall, the changes in the RSA amplitude and the changes in $\dot{V}O_2$max between the beginning and completion of the 12-week exercise period were significantly correlated (r = .51; p = .04). The results demonstrated that $\dot{V}O_2$ max is more likely to be positively affected by walking than by flexibility exercises. The modest exercise program, however, did not significantly improve the average $\dot{V}O_2$ max in either study group. On the other hand, the longitudinal behavior of RSA amplitude clearly differentiated the two forms of exercise. The average RSA amplitude rose progressively during the 12-week exercise period by 80%. Four weeks following the discontinuation of the walking program, half of this exercise-induced RSA rise was lost. The study concluded that quantitative evaluation of the RSA provides information about aerobic conditioning that may be useful in a variety of health-related areas.

A longitudinal study was undertaken in an attempt to clarify the effects of chronic aerobic exercise on autonomic neural activity (De Meersman, 1992), specifically parasympathetic outflow, in nine young, healthy male athletes before and after an 8-week high-intensity aerobic training program. Subjects performed a maximal oxygen consumption test ($\dot{V}O_2$max) using an incremental cycle ergometer protocol. Measurements were taken 1 week before the beginning of the training season and within 1 week following the end of the training season. True $\dot{V}O_2$max criteria (plateau or drop in $\dot{V}O_2$max in spite of increased workload and RER > 1.1 and volitional indication of exhaustion) were obtained for all subjects. RSA, as measured by comparing RR intervals during inhalation and exhalation, was performed for 3 min with 5 s inhalation and 5 s exhalation (6 cycles $\cdot$ min$^{-1}$) with tidal volume synchronized with an oscilloscope and timer at 30% of the subject's vital capacity. The athletes trained 7 d $\cdot$ week$^{-1}$ for approximately 1.5 to 2 h daily, twice a week at a training intensity of about 85 to 90% of predicted maximal heart rate (MHR) and four to five sessions a week at 75% of MHR. The author found a 7.3% increase in $\dot{V}O_2$ max associated with a significant 23.1% increase in parasympathetic activity (RSA increase) between pre- and postprogram results (p < .05) and concluded that enhanced aerobic capacity increases efferent parasympathetic tone.

The relationships between RSA under resting conditions and maximal aerobic power derived from a standardized incremental exercise have also been investigated among subjects with various fitness levels (White et al., 1990). Twenty-three adult males undertook seated ECG measurements at a respiratory rate of 6 cycles $\cdot$ min $^{-1}$ for approximately 1 min, but tidal volume was not controlled. RSA (heart rate modulation index percentage) was calculated using the following formula: (mean longest RR – mean shortest RR)/mean RR. $\dot{V}O_2$max was determined using standard criteria on a cycle ergometer. A correlation coefficient of $r = .85$ ($p < .001$) was observed between HRMI (mean $= 30.2 \pm 1.7\%$) and $\dot{V}O_2$max (mean $= 59.3 \pm 16$ ml $\cdot$ kg$^{-1}$ $\cdot$ min$^{-1}$), suggesting that RSA could represent a useful passive estimate of cardiorespiratory entrainment resulting from aerobic conditioning.

A summary of studies that consider differences in parasympathetic activity between endurance and sedentary subjects together with the number and age of subjects assessed, length of recording, and the major conclusions is given in table 4.3. These studies used power spectral analysis of RR intervals but did not control respiratory parameters. Most studies involved young male athletes and sedentary individuals, and determined HF at rest in the supine position except for one 24 h Holter ECG study (Goldsmith, Bigger, Steinman, & Fleiss, 1992). The conclusions suggested that exercise endurance training increases parasympathetic activity (Goldsmith et al., 1992) and modifies heart rate control through neurocardiac mechanisms (Dixon et al., 1992), with higher amplitudes of LF and HF in athletes compared with sedentary subjects. Note, however, that only one study controlled breathing frequency (Puig et al., 1993).

It has been well recognized that highly fit athletes with high $\dot{V}O_2$max show resting bradycardia (De Meersman, 1993b; Kenney, 1985; Smith, Hudson, Graitzer, & Raven, 1989). Cardiac vagal tone seems to be enhanced at rest in long-term physically trained athletes (De Meersman, 1992; Dixon et al., 1992; Goldsmith et al., 1992; Puig et al., 1993), although more recent investigation showed persistent sympathetic activation in trained athletes (Furlan et al., 1993). Furthermore, habitual aerobic exercise appears to play a role in the maintenance of

**Table 4.3   A Summary of Studies That Considered Differences in Parasympathetic Activity Between Endurance Athletes and Sedentary Subjects Using Power Spectral Analysis**

| Studies | Number of subjects | Age (years) | Recording position | High frequency at rest — Athletes | High frequency at rest — Sedentary | Major conclusions |
|---|---|---|---|---|---|---|
| Puig et al.[†] (1993) | 33 athletes 33 sedentary | 23.4 ± 5.5 24.3 ± 7.6 | 15 min supine | 2,258 (± 2,349 ms$^2$) | 1,179** (± 1,542 ms$^2$) | Higher amplitude of athletes over sedentary in low and high frequencies because of increased vagal activity. |
| Sacknoff, Gleim, Stachenfeld, & Coplan (1994) | 12 athletes 18 sedentary | 26 ± 1.6 30 ± 0.3 | 15 min supine | 2,022 (± 240 ms$^2$) | 5,839$^{NS}$ (± 1,839 ms$^2$)[‡] | Frequency domain analysis of heart rate variability may not be an accurate indicator of cardiac vagal tone. |
| Goldsmith et al. (1992) | 8 athletes 8 sedentary | 24-38 24-38 | 24 h Holter ECG | 1,399 (± 776 ms$^2$) | 318*** (± 193 ms$^2$) | Exercise training may increase parasympathetic activity. |
| Dixon et al. (1992) | 10 athletes 14 sedentary | 28.4 ± 3.5 27.4 ± 2.6 | 45 min at rest 15 min supine | 62.2 (± 10.7) beats·min$^{-1}$·Hz$^{-1}$ | 43.7*** (± 22.4) beats·min$^{-1}$·Hz$^{-1}$ | Endurance training modifies heart rate control in whole or part through neurocardiac mechanisms. |

†Breathing frequency was controlled at 15 breaths·min$^{-1}$. The remaining studies controlled neither breathing frequency nor tidal volume.

‡ Standard deviation of all normal-to-normal RR intervals was significantly greater in athletes as compared with controls (92.3 ± 7.7 ms versus 73 ± 9.3 ms, $P < .05$).

* $P < .05$, ** $P < .01$, *** $P < .001$, NS = not significant

augmented HRV in active men when compared with age- and weight-matched sedentary subjects (Dixon et al., 1992). This phenomenon suggests the beneficial role of long-term aerobic exercise in mitigating the age-dependent loss of HRV in physically active persons.

In addition, a study that is not often cited has given further evidence of the importance of maintaining a regular pattern of physical activity (Hrushesky, 1991). It has been found that respiratory sinus arrhythmia rose by 80% after a 12-week walking program in previously sedentary adults, whereas 4 weeks after discontinuation of the walking program, half of the exercise-induced respiratory sinus arrhythmia was lost. If such observations are further substantiated, the increased heart rate variability and vagal activity that has been documented to be associated with exercise could, in part, explain the cardioprotective effect of regular physical activity.

Indeed, a recent study that investigated the effects of 6 months of endurance training in borderline hypertensive subjects on blood pressure and baroreflex sensitivity (phenylephrine method) showed a significant reduction in resting systolic and diastolic blood pressures by 9.7 and 6.8 mmHg, respectively, and baroreflex sensitivity was increased from $14.0 \pm 1.8 \text{ ms} \cdot \text{mmHg}^{-1}$ to $17.5 \pm 2.0 \text{ ms} \cdot \text{mmHg}^{-1}$ (Somers, Conway, & Sleight, 1986). This study was also the first longitudinal study to report an increase in baroreflex sensitivity with increased physical fitness in human hypertension. Interestingly, it seems that physical training may lower blood pressure in normotensive subjects as well (Somers, Conway, Johnston, & Sleight, 1991).

## Heart Rate Variability During Exercise

Relatively few studies have investigated the effects of exercise on HRV measures. A decrease in time (SDNN) and frequency domain variables (HF and LF), as expressed in absolute or normalized units (Casadei, Cochrane, Johnston, Conway, & Sleight, 1995; Kamath, Fallen, & McKelvie, 1991; Shin, Minamitani, Onishi, Yamazaki, & Lee, 1995; Yamamoto, Hughson, & Peterson, 1991), is seen from rest to exercise. The changes in HRV indices, however, are inconsistent when exercise intensity increases and after recovery.

The effects of steady-state exercise on the power spectrum of heart rate variability were studied in 19 healthy subjects (Kamath et al., 1991). Continuous ECG signals were recorded during (1) 15 min of rest in the supine state, (2) 10 min of standing, (3) 10 min of steady-state exercise at 50% of maximum predicted power output on a cycle ergometer, and (4) 15 min of postexercise recovery in the supine state. Steady-state exercise caused a significant suppression of both LF and HF components, with LF peak power rising to significantly high levels throughout 15 min of the postexercise recovery period. The authors suggested that neuroregulatory control of heart rate plays a major role in adaptive responses to exercise and in postexercise recovery, whereas humoral factors are probably more important in maintaining heart rate during steady-state exercise.

In another study, eight subjects completed submaximal exercise tests with work rates of 20 W, or at 30, 60, 90, 100, and 110% of the predetermined ventilatory threshold (Tvent) (Yamamoto et al., 1991). HF decreased dramatically ($p < .05$) when the subjects exercised compared with rest and continued to decrease until the intensity reached 60% Tvent. The LF/HF ratio remained unchanged up to 100% Tvent, whereas it increased abruptly ($p < .05$) at 110% Tvent. The authors suggested that a progressive decrease in vagal activity from rest to a work rate equivalent to 60% Tvent, and an increase in sympathetic activity only when exercise intensity exceeded Tvent.

The response of the autonomic nervous system to dynamic exercise was investigated in athletes and nonathletes with power spectral analysis of HRV (Shin et al., 1995). Thirteen healthy subjects (5 athletes and 8 nonathletes) performed continuous ECG recordings during (1) 15 min of rest in a sitting position on a bicycle ergometer, (2) the dynamic exercise test to the point of exhaustion, and (3) a 15 min postexercise period. In athletes and nonathletes, LF and HF powers gradually decreased with exercise. As recovery progressed, they continued to increase gradually but remained below resting level. During rest and postexercise, HF power in athletes was significantly ($p < .05$) higher than in nonathletes. In addition, the recovery of HR and HF powers during early recovery was more rapid in athletes than in nonathletes. The authors suggested that in athletes the lower HR during rest and the more rapid recovery of HR postexercise was due to a high level of HF power, indicating that vagal activity was enhanced by the adaptive changes in neural regulation produced by long-term physical training.

A recent study was performed in 11 healthy young men at rest and during incremental cycle ergometry (Casadei et al., 1995). The HRV indices were calculated at 43, 57, 72, and 86% $\dot{V}O_2$max. HF fell at the onset of exercise, consistent with a reduction in cardiac vagal activity. Conversely, $HF_{nu}$ increased with increasing work rates. $LF_{nu}$, reflecting cardiac sympathetic activity, was no longer detectable in severe exercise when the adrenergic drive is known to be elevated. The authors concluded that autoregressive spectral analysis of the RR interval variability does not adequately reflect the autonomic changes that occur during incremental exercise. In particular, the evidence indicated that as the cardiac vagal tone falls with increasing levels of exercise, a greater percentage of the residual power of the high-frequency component may be due to nonneural mechanisms.

## Overtraining and Heart Rate Variability

The overtraining syndrome, characterized mainly by reduced performance and pronounced fatigue, is a result of an excess training load together with low-quality recovery (Budgett, 1998; Uusitalo, 2001). The overtraining syndrome (particularly the parasympathetic type) is assumed to be a consequence of an imbalance between long-term inappropriate high training volume in endurance sports and too little time for regeneration (Lehmann et al., 1998). In addition, autonomic nervous system dysfunction or imbalance has been presented as one reason for the signs and symptoms of the overtraining syndrome (Armstrong & VanHeest, 2002). Only a few studies, however, have focused on the relationships between overtraining and HRV (Hedelin et al., 2000; Hedelin, Bjerle, & Henriksson-Larsén, 2001; Pichot et al., 2002; Pichot, Roche, Gaspoz, et al., 2000; Portier, Louisy, Laude, et al., 2001; Uusitalo, Uusitalo, & Rusko, 2000).

Autonomic nervous system activity using heart rate variability was investigated in seven middle-distance runners, aged 24.6 ± 4.8 years, during their usual training cycle composed of 3 weeks of heavy training periods followed by a relative resting week (Pichot et al., 2000). The spectral analysis results calculated during sleep demonstrated a significant and progressive decrease in parasympathetic indices (pNN50, HF, $HF_{nu}$) of up to –41% (p < .05) during the 3 weeks of heavy training, followed by a significant increase during the relative resting week of up to +46% (p < .05). The indices of sympathetic activity (LF, $LF_{nu}$) followed the opposite trend, first an increase up to +31% and then a decrease of –24% (p < .05), respectively. The results suggested that heavy training shifted the cardiac autonomic balance toward a predominance of the sympathetic over the parasympathetic drive. When recorded during the night, heart rate variability

appeared to be a better tool than resting heart rate to evaluate cumulated physical fatigue, because it magnified the induced changes in autonomic nervous system activity.

Another study examined the effects of intense endurance training on autonomic balance using spectral analysis of heart rate (HR) and systolic blood pressure (SBP) (Portier et al., 2001). Eight elite runners were tested twice: after a relative rest period (RRP) of 3 weeks and after an intense 12-week training period (ITP) for endurance. At the end of each phase, the subjects were tested by means of a $\dot{V}O_2$max test and a tilt-table test. LF was significantly lower in the supine position (p < .05) during ITP. Upright tilting was accompanied by a 22.6% reduction in HF values during the rest period, whereas in ITP the HF spectral power rose by 31.2% (p < .01) during tilt, characterizing a greater parasympathetic system control. Sympathetic control represented by the LF/HF ratio regressed markedly (p < .01) in response to the tilt test in ITP. The authors concluded that spectral analysis of SBP in the high frequencies shows that the changes in cardiac parameters were coupled with a decrease in sympathetic vasomotor control (–18%) and a reduction in diastolic pressure (–3.2%) in the response to the tilt test at the end of the ITP. It is suggested, therefore, that spectral analysis could provide a means of demonstrating impairment of autonomic balance for the purpose of detecting a state of fatigue that could result from overtraining.

More recently, a controlled laboratory study followed six sedentary subjects (32.7 ± 5.0 years) going successively through 2 months of intensive physical training and 1 month of overload training on a cycle ergometer followed by 2 weeks of recovery (Pichot et al., 2002). Maximal power output over 5 min, $\dot{V}O_2$, and standard indices of heart rate variability were monitored during the whole protocol. During the intensive training period, physical performance increased significantly ($\dot{V}O_2$ peak: +20.2%, p < .01; maximal power output over 5 min: +26.4%, p < .0001) as well as most of the indices of heart rate variability (mean RR, pNN50, SDNN, $SDNN_{index}$, rMSSD, and HF, all p < .05) with a significant shift in the autonomic nervous system toward a predominance of parasympathetic activity (LF/HF, $LF_{nu}$, $HF_{nu}$, p < .01). During the overload training period, a stagnation of the parasympathetic indices was associated with a progressive increase in sympathetic activity (LF/HF, p < .05). During the week of recovery, a sudden significant rebound of the parasympathetic activity occurred (mean RR, pNN50, rMSSD, and HF, all p < .05). After 7 weeks of recovery, all heart rate variability indices tended to return to the prestudy values. The authors concluded that the autonomic nervous system status depends on cumulated physical fatigue because of increased training loads. Therefore, heart rate variability analysis may be an appropriate tool to monitor the effects of physical training loads on performance and fitness and could eventually be used to prevent overtraining states.

## Practical Example

This section provides a practical example with the collection of RR interval data in a subject at rest and the treatment of the data before the heart rate variability analysis procedure.

## Data Collection

The RR interval data can be collected by a range of commercial instruments. For this particular example, the Polar S810 (Polar Electro Oy, Finland) transmitter and receiver have been used with the Polar Precision Performance software (version 4.00.025).

The recommended place for the transmitter is below the chest muscles against the bare skin (see figure 4.14). To ensure good skin contact and, consequently, accurate measurement results, it is important to perform the following:

1. Carefully wet the electrodes with water or gel the contact.

2. Place the electrode surfaces flat against the chest.

3. Tighten the transmitter sufficiently.

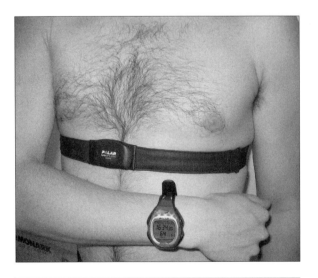

**Figure 4.14**   Heart rate transmitter position.

When recording RR intervals, attach the receiver on a waist belt to ensure perfect transmission of every heartbeat. Alternatively, fit the receiver at the wrist as recommended by the manufacturer.

The Polar S810 heart rate monitor does not store the ECG data in its memory. Therefore, it is usually impossible to know whether an extra beat or a measurement error caused certain extraordinary data.

## Sources of Interference and Error Correction

Erroneous data can also be collected because of other sources of interference including electromagnetic interference with high-voltage lines, televisions, cars, wireless bicycle computers, exercise equipment, and cellular phones. Interference could come from another heart rate monitor if the encoded heart rate transmission that eliminates interference from other heart rate monitor users fails. Similarly, many fitness devices with electronic components, such as an LED display, may send interfering signals that can cause erroneous readings. Usually, however, eliminating the source of interference allows clean data collection.

Figure 4.15 provides a 10 min RR interval recording of a subject at rest. Note that three beats are erroneous and needed to be corrected to perform subsequent time and frequency domain analysis.

To correct or clean the data, two options are available:

1. On the *Edit* menu, select *Error Correction*.

2. Right-click a heart rate curve and then select *Error Correction* in the shortcut menu that appears.

The *error correction* function appears together with the beats that are erroneous as shown in figure 4.16. The *error correction* function has two settings, namely *filter power* and *minimum protection zone*.

The *filter power* algorithm tries to find all possible measurement errors by using median and moving-average-based filtering methods. The algorithm influences the magnitude of the deviation in the heart rate curve that will be interpreted as erroneous and is corrected. Five correction levels are available: very high, high, moderate, low, and very low. As the filter power level increases, smaller differences from the trend curve are considered errors.

The *minimum protection zone* is a limit above which the *filter power* does not adjust heart rate readings. Its function is opposite that of the *filter power*. For example, if an error is detected with a high *filter power* that exceeds the conditions on correction, a

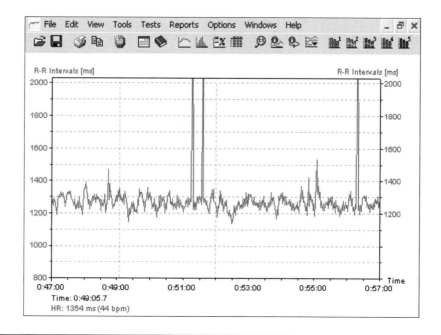

**Figure 4.15**    Ten-minute RR interval data recorded in a subject at rest.

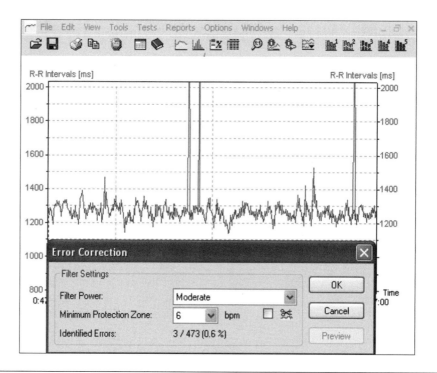

**Figure 4.16**    Error correction function of the Polar Precision performance software.

high *minimum protection zone* prevents the heart rate reading from being changed. The *minimum protection zone* values can be set from 1 to 20 beats · min⁻¹. In most cases, the default corrections of the software provide good results (*filter power* set at moderate and *minimum protection zone set at* 6 beats · min⁻¹). The algorithm then computes the appropriate RR interval to substitute for the detected errors according to the previous and the next normal RR intervals. One of the advantages of this algorithm is that it maintains the total time of the recording; that is, the sum of RR intervals is the same as the elapsed time.

The algorithm, however, may not remove supraventricular extra beats. To solve that problem, a checkbox is provided beside the scissors in the *error correction* window so that the algorithm can be forced to correct those sudden peaks regardless of the small effects in the total time.

## Heart Rate Variability Analysis

After the error corrections are done, that is, after the RR interval data are corrected, the time and frequency domain analysis can be performed. To perform an analysis on the data corrected in the previous steps, the data needs to be selected. To select the data, the mouse pointer is moved to the desired position below the time axis. The left mouse button is pressed and held down while the pointer is moved to the end of the part that is to be selected. The data selected needs to be exported as a text file (.txt) by using the *File* menu and clicking on *Export as text*. Figure 4.17 shows the RR interval data that have been selected for heart rate variability analysis.

Noncommercial heart rate variability analysis software for Windows (HRV analysis software v1.1) is also available from the Biomedical Signal Analysis Group from the University of Kuopio in Finland (http://it.uku.fi/biosignal/). The developers must grant permission before users can download the software program. After importing the RR data previously saved as a text file into the HRV analysis software program, the software provides an output analysis as shown in figure 4.18.

# Heart Rate Variability Measures in Coronary Heart Disease Morbidity and Mortality

The following section provides further insights about the role of autonomic dysfunction in cardiovascular disease and the possible role of heart rate variability in detecting patients at risk. This phenomenon was recognized long ago by the Chinese: "If the pattern of the heartbeat becomes as regular as the tapping of a woodpecker or the dripping of rain from the roof, the patient will be dead in four days" (Wang Su Ho, Chinese physician, 265-317 CE: A Treatise on Qualities of the Pulse).

The first study to link autonomic disturbances in CHD was performed in the Cardiac Department of the Royal Victoria Hospital in Belfast (Webb, Adgey, & Pantridge, 1972). Seventy-four patients were seen within 30 min of the onset of acute myocardial infarction, and 68 (92%) had signs of autonomic imbalance. Excessive vagal activity (patients with sinus bradycardia with heart rate ≤60, or atrioventricular block, or systolic blood pressure ≤100 mmHg in the absence of pronounced bradycardia) was evident in 41 (55%). Sympathetic overactivity (patients with sinus tachycardia with heart rate ≥100, with or without transient hypertension) was found in 27 (36%).

Later, a study of 176 patients admitted to a coronary care unit with acute myocardial infarction was con-

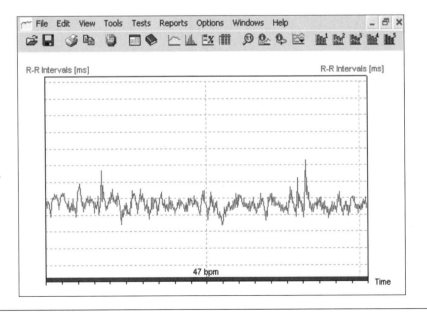

**Figure 4.17**   RR interval selected for heart rate variability analysis.

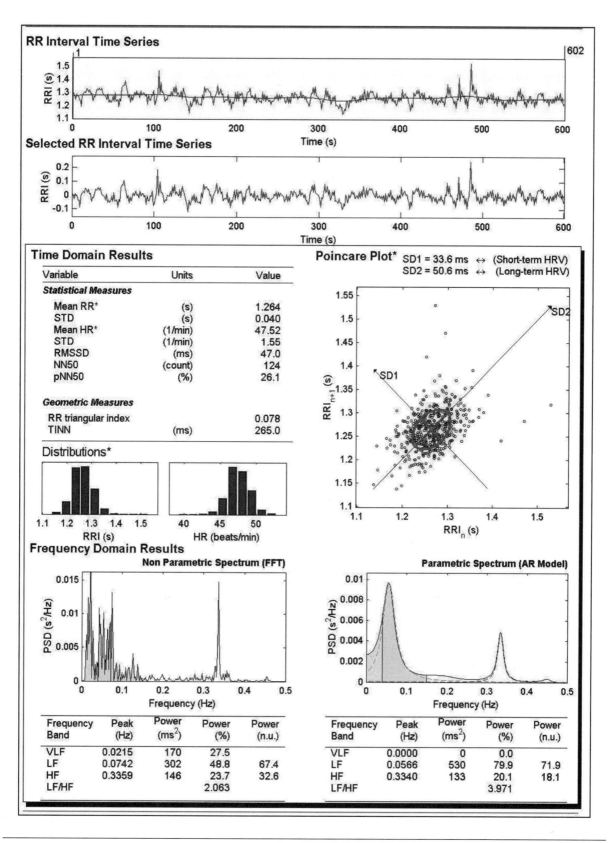

**Figure 4.18** RR interval time series, heart rate variability analysis in the time and frequency domains, and Poincaré plots of the RR interval data selected previously.

ducted to determine the incidence of sinus arrhythmia and to compare the associated findings and prognosis in patients with and without this arrhythmia (Wolf, Varigos, Hunt, & Sloman, 1978). A 60 s ECG recording was performed immediately after admission, and 30 consecutive RR intervals were used to determine the presence or absence of sinus arrhythmia by calculating the variance of these intervals. Those with a variance value over 1,000 ms$^2$ were considered to have sinus arrhythmia, and those with a variance of less than 1,000 ms$^2$ were considered not to have sinus arrhythmia. The former had lower hospital mortality (patients were discharged 9 to 14 d after admission)—4.1% compared to 15.5%—and they tended also to have a higher incidence of inferior infarction (46% with no deaths compared with 37% and four deaths) and a lower incidence of anterior infarction (38% with three deaths compared with 47% and nine deaths).

## Heart Rate Variability Measures in Uncomplicated CHD

Evaluation of parasympathetic nervous function in CHD has been conducted using standard noninvasive cardiovascular tests and with the findings related to the clinical and angiographic features of CHD (Airaksinen, Ikaheimo, Linnaluoto, Neimela, & Takkunen, 1987). The study involved 57 men and 6 women with CAD (mean age 50, range 33-59 years), 20 healthy controls (mean age 46, range 36-55 years), and 22 patients (mean age 47, range 35-59 years) with atypical chest pain. All patients underwent left-sided cardiac catheterization, and a stenosis causing reduction of >50% in the luminal diameter of a coronary artery was regarded as an important lesion. Heart rate was measured after 10 min of supine rest, and the patients were asked to take six deep breaths at no more than 40% of the vital capacity. The difference between the maximum and minimum heart rate during each breath was measured. Less variation in heart rate during deep breathing occurred in patients with CHD than occurred in healthy subjects (difference between means of 5 beats · min$^{-1}$, 95% confidence interval 1 to 10 beats · min$^{-1}$, p < .05), but not in patients with chest pain. Also, the number or location of affected vessels did not significantly affect the variation of heart rate during deep breathing. The patients with chest pain also had slightly diminished heart rate responses to deep breathing, which might suggest that reduced physical activity caused by disabling symptoms may contribute to the blunting of heart rate variation. The severity of symptoms, however, did not correlate with the heart variation in deep breathing.

More recently, the spectral components of RR interval variability under controlled respiration (15 breaths · min$^{-1}$) were analyzed in 56 patients (age range 35-73 years) referred for coronary angiography (Hayano et al., 1990). Fourteen patients had multivessel disease (group M), 21 had one-vessel disease (group S), and 21 had nonsignificant disease or normal coronary arteries (group N). There were 43 healthy controls (age range 36-71 years) (group C). The patients had no clinical evidence of heart failure, hypertension, diabetes mellitus, or acute stage of infarction and had taken no medication for 3 d. The magnitudes of LF and HF were represented by the coefficient of component variance (CCV), which provided the amplitude relative to the mean RR interval. The age- and sex-adjusted mean of CCV$_{HF}$ significantly decreased with advancing angiographic severity (1.64 ± 0.09%, 1.66 ± 0.12%, 1.22 ± 0.13%, and 0.81 ± 0.16% for groups C, N, S, and M, respectively) (p = .0001). The CCV$_{HF}$ was unrelated to left ventricular function, previous myocardial infarction, or stenosis of any specific artery including the sinoatrial and atrioventricular node arteries. The CCV$_{LF}$ decreased only in group M (p = .0462). These results indicate that coronary artery disease is associated with vagal dominant impairment in autonomic cardiac function and that reduction in the vagal cardiac function correlates with the angiographic severity.

In a subsequent study, the same authors investigated cardiac autonomic function by relating heart rate spectral components to clinical and angiographic findings in 80 patients who were undergoing coronary angiography (Hayano et al., 1991). The age- and sex-adjusted magnitude of the respiratory spectral component, which is an index of cardiac vagal tone, showed a significant negative correlation with the extent of coronary atheromatosis (r = −.43, p < .0001) and a less significant negative correlation with the severity of coronary stenosis (r = −.30, p = .007). These relationships were independent of previous myocardial infarction and of left ventricular function. Stepwise regression analysis showed that the respiratory spectral component contributed to atheromatosis independently of established coronary risk factors (partial R$^2$ = 9.4%, p = .002), but not to stenosis. The authors confirmed the hypothesis that decreased cardiac vagal activity is associated with an increased risk of coronary atherosclerosis.

More recently, the association of baseline cardiac autonomic activity (1987-1989) with incident CHD after 3 years (1990-1992) of follow-up of the Atherosclerosis Risk in Communities Study cohort selected was examined in 137 incident cases of CHD and a stratified random sample of 2,252 examinees free of

CHD at baseline (Liao et al., 1997). Baseline, supine, resting beat-to-beat heart rate data were collected. HF (0.16-0.35 Hz) and LF (0.025-0.15 Hz) spectral powers and high- and low-frequency ratio, estimated from spectral analysis, and standard deviation of all normal RR intervals, calculated from time domain analysis, were used as the conventional indices of cardiac parasympathetic, sympatho-parasympathetic, and their balance, respectively. Incident CHD was defined as hospitalized myocardial infarction, fatal CHD, or cardiac revascularization procedures during 3 years of follow-up. The age, race, gender, and other CHD risk factor-adjusted relative risks (and 95% confidence intervals) of incident CHD comparing the lowest quartile with the upper three quartiles of high-frequency power, low-frequency power, high- and low-frequency power ratio, and standard deviation of RR intervals were 1.72 (95% CI 1.17-2.51), 1.09 (95% CI 0.72-1.64), 1.25 (95% CI 0.84-1.86), and 1.39 (95% CI 0.94-2.04), respectively. The findings from this population-based, prospective study suggest that altered cardiac autonomic activity, especially lower parasympathetic activity, is associated with the risk of developing CHD.

## Heart Rate Variability Measures in the Prognosis of CHD and Mortality

The most important clinical application of HRV measurement is the risk stratification of patients after MI. To test the hypothesis that HR variability is a predictor of long-term survival after MI, the 24 h Holter tapes of 808 patients who survived MI were analyzed (Kleiger, Miller, Bigger, & Moss, 1987). HRV was defined as the standard deviation of all normal RR intervals in a 24 h continuous ECG recording made 11 ± 3 d after MI. In all patients demographic, clinical, and laboratory variables were measured at baseline. Mean follow-up time was 31 months. The association between HRV and mortality was examined by fitting multivariate proportional hazards models to adjust for potentially important covariates. The relative risk of mortality was 5.3 times higher in the group with HRV < 50 ms than in the group with HRV > 100 ms. HRV remained a significant predictor of mortality after adjusting for clinical, demographic, other Holter features, and ejection fraction ($\chi^2$ = 19.63, df = 1, p < .0005). The authors explained these findings, sug-

gesting that decreased HR variability correlates with increased sympathetic or decreased vagal tone, which may predispose to ventricular fibrillation.

A subsequent study examined the relation between HRV and other clinical, electrocardiographic, and angiographic variables in 100 patients undergoing elective coronary angiography using 24 h ECG (Rich et al., 1988). HRV was inversely correlated with HR (r = −.38, p = .001), diabetes mellitus (r = −.22, p = .025) and digoxin use (r = −.29, p = .004), but not with left ventricular ejection fraction, extent of coronary artery disease, or other clinical, electrocardiographic, or angiographic variables. All patients were followed for 1 year. Major clinical events after initial discharge occurred in 10 patients, including 6 deaths and 4 coronary bypass operations. Left ventricular ejection fraction was the only variable that correlated with the occurrence of a clinical event (p = .002). Decreased HRV and ejection fraction were the best predictors of mortality (both p < .01), and the contribution of HRV to mortality was independent of ejection fraction, extent of coronary artery disease, and other variables. Furthermore, 11 patients with HRV < 50 ms had an 18-fold increase in mortality compared with patients with HR variability > 50 ms (p = .001). Thus, the authors suggested that decreased HR variability is a potent independent predictor of mortality in the 12 months following elective coronary angiography in patients without recent myocardial infarction. Although the risk of overall mortality is increased in patients with reduced HRV, the predictive value of abnormal HRV seems to be relatively low in risk stratification. In combination with other risk factors, such as the ejection fraction and late potentials, the risk stratification increases sufficiently (Farrell et al., 1991).

To evaluate HRV early after MI and its relation to clinical and hemodynamic data, 54 patients (42 men and 12 women, mean age 60.4 ± 11 years) with evidence of MI were studied by analyzing the 24 h Holter tapes recorded on day 2 or 3 (Casolo et al., 1992). HRV was also measured in 15 patients with unstable angina and in 35 age-matched normal subjects. HRV was lower in MI than in unstable angina patients (57.6 ± 21.3 versus 92 ± 19 ms; p < .001) and controls (105 ± 12 ms; p < .001). No difference was found for infarct site. However, HR was significantly related to mean 24 h HR, peak creatine kinase-MB, and left ventricular ejection fraction (all p < .0001). Patients belonging to Killip class[4] >1 or who required the use of diuretics or digitalis had lower counts (p <

---

[4]Canadian classification of heart failure based on the activity level that initiates the onset of symptoms. The classification is made of six categories: Killip class I, II, IIa, IIb, III, IV, with Killip class I classification referring to patients with no symptoms, normal activities, and clear lungs and Killip class IV classification referring to patients in cardiogenic shock.

.004, p < .001, and p < .024, respectively). Six patients died within 20 days after admission to the hospital. In these patients, HRV was lower than in survivors (31.2 ± 12 ms versus 60.9 ± 20 ms, p < .001), and a value <50 ms was significantly associated with mortality (p < .025). The authors confirm that HRV is decreased during the early phase of MI and is significantly related to clinical and hemodynamic indexes of severity. The causes for the observed changes in HRV during MI may be reduced vagal or increased sympathetic outflow to the heart, and the authors suggested that early measurements of HR variability during MI may offer important clinical information and contribute to the early risk stratification of patients.

A subsequent study, determining whether spectral measures of heart period (RR) variability predict death when measured late after infarction, examined 331 patients (age = 59 ± 9 years) (Bigger, Fleiss, Rolnitzky, & Steinman, 1993). The 24 h power spectral density was computed from ECG recordings made 1 year after infarction using FFT. Four components of the heart period power spectrum were assessed as follows: ULF (<0.0033 Hz), VLF (0.0033-0.04 Hz), LF (0.04-0.15 Hz), HF (0.15-0.40 Hz), total power (TP) (0-0.40 Hz), and the ratio of LF/HF. These variables increase to steady-state values by 3 months after infarction, however, and the prognostic significance of recovery values is unknown. Each measure of RR variability had a strong and significant univariate association with mortality; the relative risks for these variables ranged from 2.5 to 5.6 (p < .01). After adjustment for age, New York Heart Association functional class, left ventricular ejection fraction, and ventricular arrhythmias, some measures of heart rate variability still had a strong and significant independent association with all-cause mortality. The authors concluded that spectral measures of heart period variability, measured late after infarction, predict death.

More recently, the relation between both time and frequency domain analyses of RR variability and mortality was examined in a series of 226 consecutive patients with acute myocardial infarction admitted to three district hospitals in London (Vaishnav et al., 1994). All patients underwent 24 h Holter monitoring early after infarction (mean 83 h, range 48 to 180), and time and frequency domain analyses of RR variability were performed using commercially available software. During an 8-month follow-up period (range 3 to 12 months), 19 cardiac deaths (8.4%) occurred. Time domain analysis confirmed reduced RR variability (SDNN, SDANN) among nonsurvivors compared with survivors. But there was no difference between the groups when the percentage of absolute differences between successive RR intervals > 50 ms (pNN50) and the root-mean-square of successive differences (rMSSD) were analyzed. Frequency domain analysis demonstrated a significant difference between those who died and the survivors when the LF and HF components were analyzed (p < .05). None of these measures of RR variability was related to infarct site or left ventricular ejection fraction. In conclusion, the authors confirmed the association between low RR variability and mortality after acute myocardial infarction. The mechanism did not appear to relate exclusively to decreased parasympathetic tone, however, and the increased risk of early mortality associated with reduced RR variability reflects an imbalance in sympathovagal function that is unrelated to left ventricular function.

## Cardiac Arrhythmia and Mortality

Sudden arrhythmic death is responsible for a significant number of deaths in the United Kingdom and the United States, particularly among patients with a history of ischemic heart disease, myocardial infarction (MI), or congestive heart failure (CHF) (Goldstein, 1989; Kannel, Plehn, & Cupples, 1988). Whereas it was once believed that dangerous rhythms such as ventricular tachycardias and fibrillation were manifestations of high electrical disorder (chaos) within the cardiovascular system (Goldberger & West, 1987), it is now clear that such arrhythmias appear more commonly among individuals demonstrating highly periodic (regular) cardiac electrical activity. More specifically, low heart rate variability (HRV) is associated with higher incidence of mortality during or after MI because of sudden arrhythmic death (Bigger et al., 1993; Casolo et al., 1992; Farrell et al., 1991; Kleiger et al., 1987; Odemuyiwa et al., 1991; Rich et al., 1988; Singer et al., 1988; Vaishnav et al., 1994).

## Reduced Cardiac Autonomic Control in Cardiovascular Disease

Recent studies have shown that reduced cardiac autonomic control correlates with the severity of cardiac disease and the risk of mortality (Billman et al., 1982; Myers et al., 1986; Wolf et al., 1978). Investigations into the relation between autonomic cardiac function and the clinical and angiographic features of coronary heart disease found that decreased cardiac vagal activity was associated with an increased risk of coronary arteriosclerosis (Hayano et al., 1990, 1991).

# Summary

The rhythm of the heart is under the influence of both the sympathetic and the parasympathetic nervous systems. The parasympathetic nervous system exerts its influence through the vagus nerve, which results in slowing the activity of the heart. The sympathetic nervous system exerts its influence by increasing the rate of sinoatrial node discharge and accelerating atrioventricular conduction of impulses. In addition, under resting conditions, parasympathetic tone is predominant in keeping the heart rate down because removing sympathetic influence results in only a small decrease in heart rate.

Heart rate variability refers to the amount of heart rate fluctuation in the mean heart rate and reflects the modulation of cardiac function by autonomic and other physiological regulatory systems. Beat-to-beat measurements are usually collected between 3 min and 24 h depending on the purpose of the investigation. Analysis of RR interval must be performed with clean and often stationary data. Several linear (time and frequency domain), geometric, and nonlinear techniques (approximate entropy, Poincaré plots) could be used to assess both branches of the autonomic nervous system noninvasively.

Respiration has a profound influence on heart rate, a phenomenon called respiratory sinus arrhythmia (RSA). Several authors have emphasized the importance of controlling respiratory rate and volume when using HRV measures as noninvasive estimation of cardiac vagal tone.

A decrease in HRV seems to occur with increasing age. This decline of HRV has been attributed to a decline in efferent vagal cardiac tone, decreased β-adrenergic responsiveness, and loss of arteriolar compliance.

Physically trained subjects have a generally increased parasympathetic activity and lower heart rates, particularly at rest, compared with sedentary individuals. In addition, habitual aerobic exercise appears to play a role in the maintenance of augmented HRV in active men when compared with age- and weight-matched sedentary subjects. In addition, exercise endurance training may increase parasympathetic activity and modify heart rate control through neurocardiac mechanisms.

On the other hand, the overtraining syndrome, a result of autonomic system dysfunction or imbalance, could be investigated using HRV, but little data are available. HRV could also be used to monitor the autonomic response to training intensity with the evaluation of cumulated physical fatigue to verify the time course recovery after intensive training periods and to prevent overtraining states.

From rest to exercise, a decrease in time and frequency domain variables (SDNN, HF, and LF) in absolute terms is seen until 60% of ventilatory threshold with LF/HF unchanged up to 100% but abruptly increased at 110% of ventilatory threshold. These, together with an abrupt increase in HF at near-maximal exercise, would indicate the influence of nonneural mechanisms at maximal intensities.

Several studies have suggested that altered cardiac autonomic activity, especially lower parasympathetic activity, was associated with the risk of developing CHD. Furthermore, the vagal impairment in autonomic cardiac function was associated with the severity of coronary artery disease assessed by angiographic methods. Interestingly, the most important clinical application of HRV measurement is the risk stratification of patients after myocardial infarction. In addition, some authors consider that decreased HRV is a potent independent predictor of mortality.

In conclusion, the following recommendations for further study are offered:

- Although a task force was set up to standardize heart rate variability measurements (Task Force of the European Society of Cardiology & the North American Society of Pacing and Electrophysiology, 1996), there is need for further standardization of these measurements in the light of the importance of respiration control when inferring vagal activity from heart rate variability measures (Brown et al., 1993; Jennings & McKnight, 1994).

- Investigation of the effects of exercise regimens on heart rate variability measures taken in sedentary subjects and athletes during sleep would be useful.

- There is also a need to investigate the potential use of heart rate variability in the detection and prevention of overtraining, and possibly the monitoring of recovery.

- Further investigation is warranted to elucidate the complexity of cardiorespiratory dynamics using analysis techniques other than power spectral analysis. These could include the use of techniques like the smoothed pseudo Wigner-Ville distribution, the complex demodulation, and the wavelet decomposition analysis (Hayano et al., 1994b; Monti, Médigue, & Mangin, 2002; Verlinde, Beckers, Ramaekers, & Aubert, 2001).

- Innovative research, including knowledge discovery and artificial intelligence techniques, is needed to detect subjects at risk of cardiovascular disease so that appropriate intervention may be provided (Azuaje et al., 1997, 1999; Dubitzky et al., 1996).

- HRV measurements could be used not only to assess autonomic function changes in cardiovascular disease but also to monitor the recovery patterns of the patients with CHD.

# five

# Blood Lactate, Respiratory, and Heart Rate Markers on the Capacity for Sustained Exercise

Carl Foster, PhD
*University of Wisconsin at La Crosse*

Holly M. Cotter, MS
*University of Wisconsin at La Crosse*

The observation of abrupt increases in blood lactate concentration during incremental exercise is not new (Bang, 1936; Christansen, Douglas, & Haldand, 1914; Hill, Long, & Lupton, 1924; Owles, 1930), nor is the observation of disproportionately increased ventilation at the same workload that elicits blood lactate accumulation. As early as the mid-1920s, Hill et al. (1924) suggested that the increase in blood lactate was attributable to an inadequacy of oxygen delivery to the working muscles, the beginning of anaerobic metabolism. Although the issue of whether anaerobic metabolism in the muscle is really the causative agent (Brooks, 1985; Davis, 1985; McLellan & Gass, 1989a) remains debatable, there is wide agreement that the exercise intensities associated with the discontinuities in both blood lactate accumulation and ventilation are highly related to the exercise intensity that can be tolerated during sustained exercise. Accordingly, these data are considered widely useful both diagnostically and for activity prescription (Urhausen, Coen, & Kindermann, 2000).

We will not attempt to fully review the topic, but it seems possible to agree that blood (and, for that matter, muscle) lactate accumulation can be attributed to disparities in the rate of lactate production and removal (Donovan & Brooks, 1983; Mazzeo,

Brooks, Schoeller, & Budinger, 1986), whether or not lactate accumulation is related to an $O_2$ deficiency (e.g., anaerobic conditions) in the muscle. The more important of the causes of lactate accumulation is probably a failure in blood lactate removal as blood flow to the renal and splanchnic lactate "sinks" decreases with increasing exercise intensity (Donovan & Brooks, 1983; MacRae, Dennis, Bosch, & Noakes, 1992). Further, although several specific conditions have been shown to uncouple the accumulation of blood lactate and the disproportionate hyperpnea of exercise, the mechanistic link between the buffering of lactate by bicarbonate leading to an increased need to ventilate to maintain $P_aCO_2$ within normal limits is attractive as a general principle.

The accumulation of lactate in the blood and the appearance of excess ventilation both appear to be related to the capacity to perform prolonged exercise. During steady-state exercise, at intensities equal to or greater than those that during incremental exercise would be associated with increased lactate or disproportionate ventilation, there is progressive accumulation of blood lactate, progressively increasing ventilation, and a relatively short time to exhaustion. This represents a z-axis in addition to the usual x-axis (exercise intensity) and y-axis (blood lactate

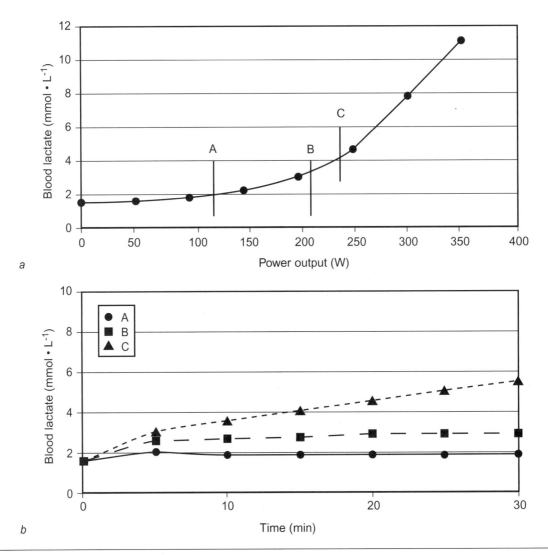

**Figure 5.1**    Representation of the pattern of blood lactate accumulation in relation to power output during *(a)* incremental exercise and *(b)* constant intensity steady-state exercise. During incremental exercise, blood lactate remains unchanged through a wide range of exercise intensities (A) and then above a certain power output (B) increases progressively (C). Decisions about how to evaluate when blood lactate begins to increase are discussed throughout this chapter. The highest intensity associated with a steady state of blood lactate concentration is referred to as the maximal lactate steady state (MLSS), which is B in this example.

concentration or ventilatory pattern) that is usually the focus of incremental exercise studies (figure 5.1). Thus, the "anaerobic threshold" is related in some way to the highest exercise intensity consistent with the steady state of blood lactate concentration, the maximal lactate steady state.

## Relationship of Blood and Muscle Lactate

The blood lactate concentration may be thought of as a shadow of muscle lactate concentration. Figure

5.2 compares simultaneously collected muscle and blood lactate concentrations from literature studies. Except for a brief period late in the recovery period, blood lactate is always less than the associated muscle lactate. The magnitude of difference between muscle and blood depends on the circumstances of measurement. The studies, indicated by filled symbols, represent either peak postexercise blood lactate concentrations compared with end exercise muscle lactate or with simultaneous muscle and blood lactate concentrations obtained during long (>5 min) exercise stages. The best-fit line representing these studies is the dotted line. The data from Jacobs (1981), shown by the solid line, represent simultaneous muscle and

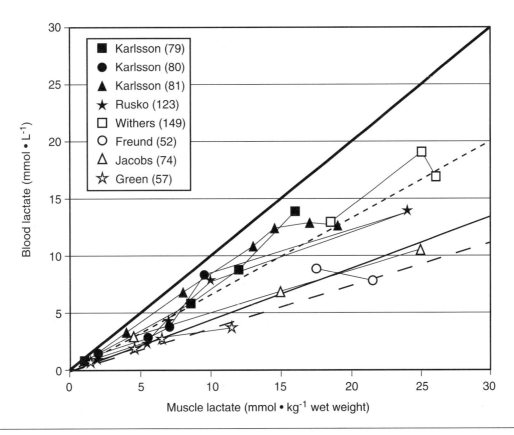

**Figure 5.2**   Comparison of blood and muscle lactate from literature studies.

blood lactate values after 4 min stages. Data from Green, Hughson, Orr, & Ranney (1983) (dashed line) represents simultaneous muscle and blood lactate values during rapidly incremented exercise (1 min stages). Clearly, during longer exercise stages, which have more time for lactate to "percolate" to the body water pool represented by the blood lactate concentration, there is less difference between muscle and blood lactate concentration.

Blood lactate depends on the presence of a net positive gradient for lactate between muscle and blood and is tempered by dilution in the body water, removal in the lactate "sinks," and the temporal lag before lactate produced in the muscle appears in the blood. The rate of release of lactate from muscle to blood is only partially dependent on concentration and is maximal in the range of 4 mmol · kg⁻¹ wet weight of muscle, which is a comparatively low muscle lactate concentration during heavy exercise. Thus, with high muscle lactate concentrations, a substantial time lag may occur before lactate equilibrates in the blood (Jorfeldt, Juhein-Dannfeldt, & Karlsson, 1978). A variety of studies have provided direct comparisons between muscle (vastus lateralis) and blood lactate during cycle ergometer exercise. In the case in which peak

postexercise lactate is measured (Freund et al., 1990; Freund & Zouloumian, 1981a; Karlsson, Diamant, & Saltin, 1971; Karlsson et al., 1983; Karlsson, Nordesjo, Jorfeldt, & Saltin, 1972; Weltman et al., 1987; Withers et al., 1991) or with simultaneous sampling after relatively long (about 5 min) exercise stages (Rodriguez, Banquells, Pons, Dropnic, & Galilea, 1992), there appears to be a reasonably consistent, although not 1:1, relationship between muscle and blood lactate. In the case of shorter exercise stages (<4 min), there appears to be progressive underestimation of the degree of muscular lactic acidosis by simultaneously measured blood lactate values (Jacobs, 1981). With rapidly incremented exercise tests, the difference between muscle and blood lactate concentrations may be rather large (Green et al., 1983). Tesch, Daniels, and Sharp (1982) have pointed out that inconsistencies in the magnitude of gradient between muscle and blood lactate can be resolved by noting the magnitude of increase in blood lactate 1 min following cessation of exercise, which is not unlike the practice of taking serial samples to ascertain the peak blood lactate concentration. The individual anaerobic threshold technique of Stegmann, Kindermann, and Schnabel (1981) addresses this problem conceptually.

The evolution kinetics of blood lactate after exercise have been discussed in terms of a two-compartment model by Freund et al. (1989, 1990) and by Freund and Zouloumian (1981a, b). Their data suggest that, with relatively brief periods of exercise (approximately 3 min), blood lactate may be expected to increase during the first moments of recovery. With longer exercise stages (approximately 6 min), little or no postexercise increase in blood lactate may occur.

## Practical Significance of the Anaerobic Threshold

Given the controversy that has surrounded the term *anaerobic threshold* (and its clones), why should we be interested in measuring it? There are three practical reasons for this interest. First, $\dot{V}O_2$max increases with the onset of training in previously sedentary individuals but is comparatively insensitive to the substantial and prolonged training undertaken by athletes (Daniels, Yarbrough, & Foster, 1978; Foster et al., 1982). The percent of $\dot{V}O_2$max that a person can sustain for prolonged exercise, which is conceptually equivalent to anaerobic threshold, increases with training beyond the point where $\dot{V}O_2$max fails to increase (Costill, Branam, Eddy, & Sparks, 1971; Davis, Frank, Whipp, & Wasserman, 1979; Denis, Fouquet, Poty, Geyssant, & Lacour, 1982; Fohrenbach, Mader, & Hollman, 1987; Freund & Zouloumian, 1981a; Hollman et al., 1981; Hoogeveen & Schep, 1997; Hoogeveen, Schep, & Hoogsteen, 1999; Hurley et al., 1984; Jacobs, 1986; Karlsson & Saltin, 1970; MacRae et al., 1992; Mazzeo et al., 1986; Mujika & Padilla, 2001; Poole & Gaesser, 1985; Poole, Ward, & Whipp, 1990; Reybrouck, Ghesquiere, Cattaert, Fagard, & Amery, 1983; Saltin, Hartley, Kilbom, & Åstrand, 1969; Sjodin, Jacobs, & Svedenhag, 1982; Svedenhag & Sjodin, 1985; Tanaka et al., 1986; Weltman et al., 1992). Thus, information provided by the anaerobic threshold may be much more reflective of training-induced changes in fitness than changes in $\dot{V}O_2$max.

Second, anaerobic threshold is highly related to performance in a variety of endurance activities. In most cases the relationship is better than that between $\dot{V}O_2$max and performance (Allen, Seals, Hurley, Ehsani, & Hagberg, 1985; Bunc, Heller, Leso, Sprynarova, & Zdanowicz, 1987; Conconi, Ferrari, Ziglio, Droghetti, & Codeca, 1982; Costill, Thomason, & Roberts, 1973; Farrell, Wilmore, Coyle, Billing, & Costill, 1979; Fell, Rayfield, Gulbin, & Gaffney, 1998; Iwaoka, Fuchi, Higuchi, & Kobayashi, 1988; Iwaoka,

Fuchi, Atomi, & Miyashita, 1988; Jacobs, 1981, 1986; Karlsson, Diamant, & Saltin, 1971; Kumagai et al., 1982; LaFontaine, Londeree, & Spath, 1981; Lehmann, Berg, Kapp, Wessinghage, & Keul, 1983; Londeree & Ames, 1975; Martin & Whyte, 2000; Nemoto, Iwaoka, Funato, Yoshioka, & Miyashita, 1988; Olbrecht, Madsen, Mader, Liesen, & Hollmann, 1985; Orok, Hughson, Green, & Thompson, 1989; Peronnet, Thibault, Rhodes, & McKenzie, 1987; Rusko, Rahkila, & Karvinen, 1980; Sjodin, Jacobs, & Karlsson, 1981b; Sjodin & Svedenhag, 1985; Smekal et al., 2000; Svedenhag & Sjodin, 1984; Weltman et al., 1987). Thus, anaerobic threshold may represent a superior index of endurance capacity compared with the traditional gold standard, $\dot{V}O_2$max. The continuing problem is that the term *anaerobic threshold* has a dozen or more definitions.

Third, anaerobic threshold may allow a more relevant index of exercise intensity by which to provide guidelines for exercise training (Coen, Schwarz, Urhausen, & Kinnderman, 1991; Foster, Fitzgerald, & Spatz, 1999; Heck et al., 1985; Hollmann, 1985; Hollmann et al., 1981; Jacobs, 1986; Kindermann, Simon, & Keul, 1979; Ljunggren, Ceci, & Karlsson, 1987; Nagel et al., 1970; Olbrecht et al., 1985; Schnable, Kindermann, Schmitt, Biro, & Stegmann, 1982; Simon, Young, Gutin, Blood, & Case, 1983; Sjodin, Jacobs, & Svedenhag, 1982; Stegmann & Kindermann, 1982; Swensen, Harnish, Beitman, & Keller, 1999; Weltman et al., 1987; Yoshida, Suda, & Takeuchi, 1982). Many physiologists and coaches feel that the greatest intensity compatible with steady state for blood lactate might represent some practical upper-limit intensity for exercise training, in that the rate of aerobic metabolism is high but the negative consequences of disturbed acid-base status are absent. At the least, at exercise intensities greater than the anaerobic threshold, the total volume of training is likely to be limited. Accordingly, training at about the intensity associated with the anaerobic threshold may optimize the intensity–duration relationship. Several authors have suggested that performance may stagnate if training is conducted too frequently at intensities associated with elevated blood lactate concentrations, although direct evidence for this is lacking (Heck et al., 1985; Mader & Heck, 1986; Sjodin et al., 1982).

## Laboratory Approaches to Measurement

One of the greatest problems in dealing with the blood lactate and ventilatory markers of limits to the capacity to do prolonged exercise, often broadly referred to as anaerobic threshold, is deciding exactly what we mean by the use of any given term. Analogous to the four blind men describing an elephant—one thinks an elephant is like a fire hose, one like a large leaf on a giant tree, one like a tree trunk, and one like a malodorous swamp—each marker is measuring something that is not part of a smooth transition from rest to maximal exercise, something that has to do with the ability to sustain exercise. What is being measured?

## Definitions

Even texts in English (which do not include many important works in this area) use a plethora of terminology to refer to discontinuities in blood lactate or ventilation during incremental exercise and to the presence or absence of steady-state conditions during constant-load exercise. These terms include:

optimal ventilatory efficiency (Hollmann, 1985; Hollmann et al., 1981),

lactate threshold (Ivy, Costill, van Handel, Essig, & Lower, 1981),

anaerobic threshold (Wasserman, Hansen, Sue, & Whipp, 1987; Wasserman & McIlroy, 1964),

respiratory compensation threshold (Wasserman & McIlroy, 1964),

aerobic-anaerobic threshold (Kindermann et al., 1979),

onset of blood lactate accumulation (Jacobs, 1986),

onset of plasma lactate accumulation (Farrell et al., 1979),

individual anaerobic threshold (Coen, Urhausen, & Kindermann, 2001; Stegmann et al., 1981),

heart rate deflection (Conconi et al., 1982),

and aerobic threshold and anaerobic threshold (Skinner & McLellan, 1980).

Doubtless we have left out some important terms, but the length of this list should suggest that this is a complicated topic. Skinner and McLellan (1980) and Kindermann et al. (1979) have independently suggested that there are at least two apparent points of discontinuity in the blood lactate–ventilatory response to incremental exercise that may serve as general concepts for many of the terms used by other authors. Although their work is now relatively old, we believe that it still provides a valid conceptual underpinning for understanding this issue. The first

of these two discontinuities is associated with the first sustained increase in $\dot{V}_E$ relative to $\dot{V}O_2$. It is also associated with the first evidence of an increase in blood lactate concentration above that seen at rest. But because there is considerable noise in blood lactate concentration data, this point is often chosen either as 1 mmol $\cdot$ L$^{-1}$ above the baseline concentration or at a lactate concentration of about 2.5 mmol $\cdot$ L$^{-1}$. In the nomenclature of Skinner-McLellan-Kindermann, this point is the *aerobic threshold*. The second of these discontinuities is represented by a rapid increase in blood lactate concentration and an increase in ventilation relative to both $\dot{V}O_2$ and $\dot{V}CO_2$. It is also associated with the blood lactate concentration of about 4.0 mM, although there is growing evidence that a single reference at a blood lactate concentration of 4.0 mmol $\cdot$ L$^{-1}$ is too simplistic (Foxdal, Sjodin, Ostman, & Sjodin, 1991; Foxdal, Sjodin, & Sjodin, 1996; Foxdal, Sjodin, Sjodin, & Ostman, 1994; Myburgh, Viljoen, & Tereblanche, 2001) and does not well represent the intensity at the maximal lactate steady state (Beneke, 1995; Beneke, Hutler, & Leithauser, 2000; Beneke & von Duvillard, 1996). In the nomenclature of Skinner-McLellan-Kindermann, this point is the anaerobic threshold. Acknowledging arguments concerning the use of the term *anaerobic* and the problems of whether a true threshold exists, we will use the terms $VT_1$, $VT_2$, $LT_1$, and $LT_2$ to represent the concepts put forth by Skinner and McLellan (1980) and Kindermann et al. (1979).

## Fixed Blood Lactate Concentration

As a strategy for obviating the problem of biological noise associated with detecting the inflection of blood lactate, fixed blood lactate concentrations of about 2.0 to 2.5 mM and 4.0 mM are often used to detect the aerobic and anaerobic thresholds, respectively. Figure 5.3 shows a schematic computation of $LT_1$ and $LT_2$ for fixed blood lactate concentrations during incremental exercise with 4 min stages. $LT_1$ is defined as the exercise workload associated with a blood lactate concentration of 2.5 mM (some authors prefer 2.0 mM). $LT_2$ is defined as the power output associated with a blood lactate concentration of 4.0 mM. In this example, because 4 min stages are used, $LT_2$ is equivalent to the "onset of blood lactate accumulation" (OBLA) described by Jacobs (1981, 1986). Although rapidly incremented exercise stages may be used (Green et al., 1983; Rieu, Miladi, Ferry, & Duvallet, 1989; Tesch, Sharp, & Daniels, 1981), exercise stages of ≥3 min are most often used (Heck et al., 1985; Hollmann, 1985;

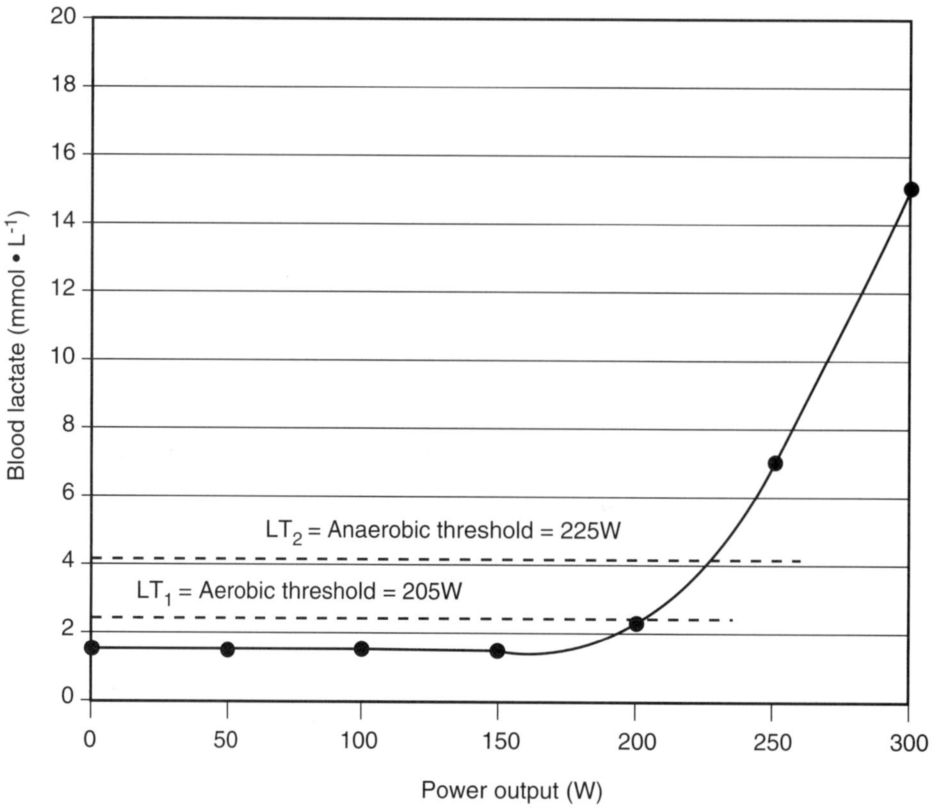

**Figure 5.3**   Computation of aerobic ($L_1$) and anaerobic ($L_2$) threshold from fixed blood lactate concentrations during incremental exercise with 4 min stages. In this example $L_1 = 2.5$ mM and $L_2 = 4.0$ mM. In practice, there is both individual and mode specificity for the actual blood lactate concentrations at $L_1$ and $L_2$.

Hollmann et al., 1981; Jacobs, 1986; Kindermann et al., 1979; Sjodin, Jacobs, & Karlsson, 1981a; Sjodin et al., 1981b, 1982; Sjodin & Svedenhag, 1985; Stegmann & Kindermann, 1982; Stegmann et al., 1981; Weltman et al., 1992, 1987; Yoshida, 1984). The actual workload or oxygen uptake associated with fixed blood lactate concentrations is determined by interpolation from visual plots of power output, velocity or $\dot{V}O_2$ versus blood lactate. Simple curve-fitting programs available for most personal computers can help standardize interpolation.

Heck et al. (1985) has discussed the importance of stage duration and intrastage interval relative to measurement of $LT_2$. With longer stages (6 min versus 3 min) $LT_2$ moves to lower values for velocity/power output (e.g., $\dot{V}O_2$). With longer (90 s versus 30 s) breaks between stages in discontinuous protocols, such as the interruption of running required to allow sampling, $LT_2$ moves to higher values. In view of the data of Freund et al. (1989, 1990), Freund and Zouloumian (1981a, b), Oyono-Euguelle et al. (1989), Rusko et al. (1986), and Yoshida et al. (1982), 4 min represents a practical minimum stage duration for

the determination of both $LT_1$ and $LT_2$, at least unless some information is going to be gleaned from the pattern of recovery blood lactate concentration. The data of Tesch et al. (1982) demonstrating an inconsistent relationship between blood and muscle lactate with 4 min stages as well as the muscle versus blood lactate data presented in figure 5.2 support this concept, as do the data of Foxdal et al. (1991, 1996).

Farrell et al. (1979) and others (Coyle, Coggan, Hopper, & Walters, 1988; Hagberg & Coyle, 1983) have used very long (10 min) stage durations to define the onset of plasma lactate accumulation, which is conceptually equivalent to $LT_1$. Several investigations have used comparatively long stage durations to define $LT_2$, particularly in field settings (Foster et al., 1993; Foxdal et al., 1991; Heck et al., 1985).

The decision regarding stage duration, number of stages, and interval between stages is probably mostly a matter of convenience and logistics within the laboratory. Certainly, longer stages are preferred over shorter stages, and breaks between stages should be brief. We have shown the technique to be robust, such that the differences in stage durations caused by using fixed-distance increments during field testing do not systematically affect the calculated value for either $LT_1$ or $LT_2$ (Foster et al., 1993). At the same time, briefer stages (approximately 3 min) make for more practical exercise examinations, which allow evaluation of several subjects during a day.

## Ventilatory Markers

Wasserman and associates (Wasserman et al., 1987; Wasserman and McIlroy, 1964) and Hollmann and associates (Hollmann, 1985; Hollmann et al., 1981) independently noted that $\dot{V}_E$ increased disproportionately ($VT_1$) during exercise at about the same time as blood lactate began to increase ($LT_1$). Wasserman et al. suggested the mechanistic link of the buffering of lactate by bicarbonate leading to excess $\dot{V}CO_2$ and, accordingly, disproportionate $\dot{V}_E$ in relation to $\dot{V}O_2$. Given that the increase in lactate is matched by a near 1:1 decrease in bicarbonate, maintenance of both $P_aCO_2$ and pH at near resting levels, the period immediately following that $VT_1$ is characterized by a period of isocapnic buffering. At somewhat greater exercise intensity, when bicarbonate is no longer able to defend the pH in the face of increasing blood lactate concentration, $\dot{V}_E$ increases again, this time out of proportion to both $\dot{V}O_2$ and $\dot{V}CO_2$, leading to a decrease in $P_aCO_2$. This has been termed the respiratory compensation threshold (Wasserman et al.) and within the Skinner-McLellan-Kindermann nomenclature is the anaerobic threshold, or $VT_2$.

Ventilatory markers are usually easier to detect with fairly rapidly incremented protocols, the $VT_2$ being particularly difficult to detect during slowly incremented exercise. Various detection strategies have been recommended, varying in their complexity in relation to the technical capabilities of the laboratory producing them. The simplest approach has been to plot $\dot{V}_E$ as a function of either power output or $\dot{V}O_2$. The initial nonlinear increase in $\dot{V}_E$, at about 50% $\dot{V}O_2max$, represents the $VT_1$. If a second discontinuity is evident, usually at about 85% $\dot{V}O_2max$, it represents $VT_2$. Wasserman et al. (1987) and others (Beaver, Wasserman, & Whipp, 1985, 1986; Davis et al., 1983) have recommended using breath-by-breath measurements to increase the precision of detection. But for comparatively crude measurements, like $\dot{V}_E$, samples integrated over convenient time intervals of 10 to 60 s can be just as useful in laboratories not equipped to make breath-by-breath measurements. The ventilatory equivalents for oxygen ($\dot{V}_E/\dot{V}O_2$) and for carbon dioxide ($\dot{V}_E/\dot{V}CO_2$) have also been widely used as markers of $VT_1$, on the premise that at $VT_1$, the $\dot{V}_E/$

$\dot{V}O_2$ will increase while $\dot{V}_E/\dot{V}CO_2$ will remain constant, because the excess $V\dot{V}_E$ is largely attributable to an excess $\dot{V}CO_2$ (Davis, Vodak, Wilmore, Vodak, & Kurtz, 1976; Davis et al., 1982; Davis, Whipp, & Wasserman, 1980; Hollmann, 1985; McLellan, 1985; Wasserman et al., 1987; Yamamoto et al., 1991; Yeh, Gardner, Adams, Yanowitz, & Crapo, 1983). The subsequent nonlinear increase in $\dot{V}_E/\dot{V}CO_2$ can be used as a marker of $VT_2$ (Caiozzo et al., 1982; Gladden, Yates, Stremel, & Stamford, 1985; Hughes, Turner, & Brooks, 1982) (figure 5.4). Beaver, Wasserman, and Whipp (1985) have suggested that simple plots of $\dot{V}CO_2$ versus $\dot{V}O_2$ and $\dot{V}_E$ versus $\dot{V}CO_2$ can also serve as useful markers of $VT_1$ and $VT_2$, respectively. Changes in the concentrations of $O_2$ or $CO_2$ in the mixed-expired—or better, end-tidal—gas may also be convenient markers of $VT_1$ and $VT_2$ (figure 5.4). In our experience, given the impressive graphic capabilities of most computerized gas-exchange systems, it is best to plot several variables to establish candidates for $VT_1$ and $VT_2$ thresholds and then to depend on agreement between two or more candidates to arrive at the best answer.

Several authors, most notably Wasserman's group (Wasserman et al., 1987), have suggested collecting about 2 min of resting data and 2 min of unloaded exercise before beginning the incrementation of the protocol. After incrementation begins, the rate of increase should be rapid, with many small steps rather than fewer large steps. In the ideal world, an essentially infinite number of infinitely small steps could be taken (ramping) (Hughson & Green, 1982). Ideally, the subject should be fatigued within 8 to 12 min after incrementation begins. This protocol seems to enhance the visualization of isocapnic buffering between $VT_1$ and $VT_2$ and make detection of both points easier. With this technique, it is important to recognize that the $VT_1$ and $VT_2$ will be expressed in terms of $\dot{V}O_2$ rather than velocity or power output, as is common with fixed blood lactate concentration protocols. If the intent of measuring $VT_1$ and $VT_2$ is to enhance the exercise prescription, then the $\dot{V}O_2$ will have to be translated into velocity or power output.

## Individual Anaerobic Threshold

Stegmann, Kindermann, and Schnabel (1981) have noted that the absolute values of steady-state blood lactate concentrations vary widely across individuals. On this basis, as well as on the basis of arguments founded on the diffusion of lactate from muscle to blood, they have proposed the concept of individual anaerobic threshold. The individual anaerobic threshold is defined in terms of blood lactate responses

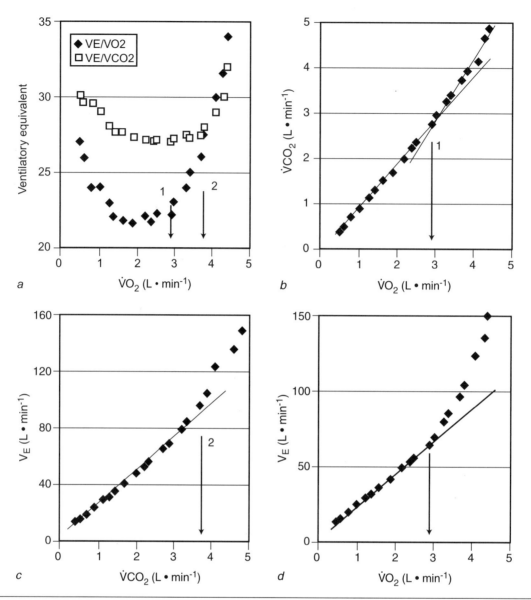

**Figure 5.4** The aerobic (AerT) and anaerobic thresholds (AnT), or first (VT₁) and second (VT₂) ventilatory thresholds, may be computed from respiratory responses during an incremental exercise test on the cycle ergometer with the power output incremented in a ramp fashion. In *(a)*, a candidate for AerT (1) is identified at a $\dot{V}O_2$ of 2.90 L · min⁻¹ from changes in the $\dot{V}_E/\dot{V}O_2$. A candidate for AnT (2) is identified from changes in the $\dot{V}_E/\dot{V}CO_2$ at a $\dot{V}O_2$ of 3.70 L · min⁻¹. In *(b)*, using the V-slope technique of Beaver et al. (1986), plots of $\dot{V}CO_2$ versus $\dot{V}O_2$ are made. By fitting straight lines through the lower and upper segments of the data points (in this case 30 s data collection increments), the point of intersection (indicated by arrow) may be taken as a candidate for VT₁. In both figures the intersection is a $\dot{V}O_2$ of 2.90 L · min⁻¹, in agreement with the candidate from $\dot{V}_E/\dot{V}O_2$ data. Accordingly, VT₁ is defined. By plotting $\dot{V}_E$ versus $\dot{V}CO_2$ *(c)*, we have another method for defining VT₂ (AnT). In this case (indicated by the arrow), the $\dot{V}CO_2$ is 3.70 L · min⁻¹. We note from the $\dot{V}CO_2$ versus $\dot{V}O_2$ plot *(d)* that a $\dot{V}CO_2$ of 3.70 L · min⁻¹ corresponds to a $\dot{V}O_2$ of 3.70 L · min⁻¹, which is in agreement with the $\dot{V}_E/\dot{V}CO_2$ data from the upper left panel. Accordingly, VT₂ is defined. The numerical correspondence of $\dot{V}O_2$ and $\dot{V}CO_2$ at VT₂ in this example is coincidental and has no special relevance.

both during and following exercise (figure 5.5). Kindermann et al. (1979) have suggested that the individual anaerobic threshold may be significantly better than a fixed blood lactate concentration in defining the maximal lactate steady state. The con-

cept of the individual anaerobic threshold has been supported in data from other laboratories (McLellan & Cheung, 1992; McLellan, Cheung, & Jacobs, 1990). Coen et al. (1991, 2001) have demonstrated the value of using the velocity associated with the individual anaerobic threshold in designing training programs for athletes. The topic has also been well reviewed recently by Urhausen, Coen, and Kindermann (2000). Typically, stage durations for individual anaerobic threshold tests are at the short end of the continuum (3-4 min) compared with those used in fixed blood lactate concentration tests (3-10 min).

Blood lactate is sampled during a brief pause at the end of each stage, or without pause if a cycle ergometer is used. Blood lactate is also sampled during the first several minutes of recovery. A plot of blood lactate versus test time is made, and separate curves are plotted for the exercise and recovery portions of the test. The time during the recovery period when the blood lactate concentration equals that at the end of the exercise portion of the test is used as an anchor for fitting a tangent to the lactate curve during exercise. The velocity (power output) and blood lactate associated with the time-versus-blood-lactate and tangent curves represent the individual anaerobic threshold. This value is conceptually equivalent to $VT_2$ or $LT_2$ and should represent a reasonable index of the intensity associated with the MLSS.

Figure 5.5 presents a schematic computation of the individual anaerobic threshold according to Stegmann, Kindermann, and Schnabel (1981) from blood lactate responses during and following an incremental exercise test with stages of 50 W every 3 min. The exercise blood lactate response curve is fitted using either a second- or third-order polynomial. The recovery blood lactate curve is fitted using a third-order polynomial. The correspondence between peak exercise blood lactate and the same blood lactate during recovery is noted. A tangent is fit between the time of the equivalent blood lactate point and the exercise blood lactate curve. The point of contact of the tangent represents the individual anaerobic threshold (e.g., $LT_2$). The blood lactate concentration at the individual anaerobic threshold, 3.8 mM in this example, is noted. The time during the test at which the individual anaerobic threshold occurred is GI. In this example the individual anaerobic threshold occurs at 18 min, which represents a power output of 300 W.

The individual anaerobic threshold is somewhat dependent on the degree to which that test is truly maximal (McLellan & Cheung, 1992), the nature

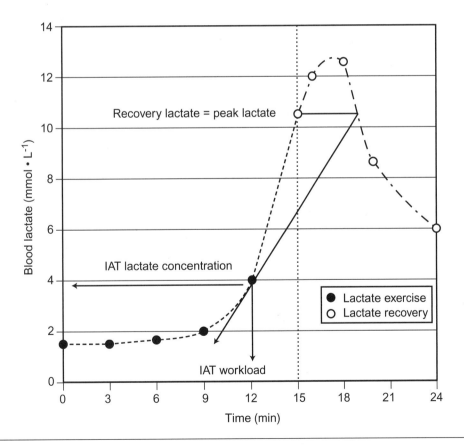

**Figure 5.5**   Schematic computation of the individual anaerobic threshold.
From Stegmann, Kinderman, and Schnabel, 1981.

and intensity of activity during the recovery period (McLellan & Skinner, 1982), and the duration of exercise stages. Coen et al. (2001) have recently reviewed the influence of these factors. In general, however, it is a remarkably robust measurement. The exercise intensity and blood lactate concentration associated with the individual anaerobic threshold are more or less equivalent to $LT_2$ or $VT_2$. The absolute blood lactate concentration at the individual anaerobic threshold is highly individual. Beyond intrinsic individual patterns, highly trained individuals often have lower concentrations (2-3 mM) of blood lactate at the individual anaerobic threshold and less fit individuals have higher concentrations (4-7 mM). This is consistent with data demonstrating that the absolute blood lactate concentration at the MLSS is also highly variable (Beneke, 1995; Beneke et al., 2000; Beneke and von Duvillard, 1996).

In most settings, the individual anaerobic threshold is a good measure of the capacity for sustained exercise, the maximal steady state. Whether it is substantially superior to $LT_2$ or $VT_2$ remains to be determined. The one direct comparison of the two methods, by Stegmann, Kindermann, and Schnabel (1981), suggested that the "anaerobic threshold" (4 mM lactate, e.g., $LT_2$) significantly overestimated the sustained exercise tolerance and that the individual anaerobic threshold provided a valid estimate of the maximal steady state. In fairness, however, anaerobic threshold was estimated using short (2 min) exercise stages, a procedure that tends to inflate the power output at anaerobic threshold.

## Heart Rate Deflection

Conconi et al. (1982) proposed that the linearity of the velocity–heart rate relationship is lost at high velocities in well-trained runners, and that the velocity at the deflection of heart rate (Vd) was associated with the beginning of blood lactate accumulation. The Vd was well correlated with, and more or less equal to, the velocity in races of 1 h duration (e.g., $LT_2$ and $VT_2$). The Vd was also well correlated with performance in 5,000 m and marathon races. This technique is particularly appealing because it can be conducted in the field using the athletes' own sport idiom. Some authors have suggested that Vd can be used to control training in athletes (Janssen, 1987). The Vd technique can also be used with athletes in other sports, such as swimming (Cellini et al., 1986), canoeing, skiing, cycling, skating, rowing, and race walking (figure 5.6) (Droghetti et al., 1985). In activities like cycling and skating, in which air provides the primary resistance, the Vd relationship is better appreciated by using velocity squared in the x-axis. For water sports (swimming, rowing), velocity cubed is the appropriate term on the x-axis. Figure 5.6 presents a schematic representation of the velocity of deflection (Vd) described by Conconi et al. (1982), which probably corresponds to $VT_1$ or $LT_2$ (Ribeiro et al., 1985). The heart rate at the end of successively faster 200 m stages is plotted. Straight-line relationships are noted during the early and late portions of the test. The intersection of the lines represents Vd, in this case 4.45 m · s$^{-1}$.

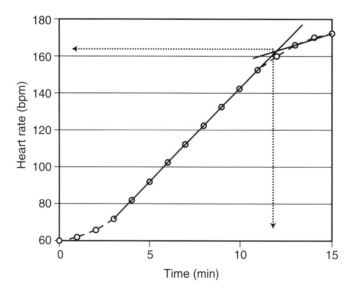

**Figure 5.6** Schematic representation of the velocity of heart rate deflection (Vd).

Others have failed to demonstrate the departure from linearity of the relationship between velocity (power output) and heart rate when fixed-duration stages are performed in the laboratory (Francis, McClatchey, Sumsion, & Hansen, 1989). Others have shown that Vd or power output at the heart rate deflection is better related to $VT_2$ (anaerobic threshold, in the Skinner-McLellan-Kindermann nomenclature) than to $VT_1$ (Ribeiro et al., 1985). The same group has reported difficulty with the reproducibility of the technique, although Conconi et al. (1982), Cellini et al. (1986), and Droghetti et al. (1985) report excellent reproducibility. The technique has been recently reviewed (Bodner & Rhodes, 2000).

The mechanism that allows the Vd technique to work is probably the slowdown in oxygen uptake kinetics, including heart rate, with increasing exercise intensity (Wasserman et al., 1987). This effect is accentuated with the field methods of Conconi et al. (1982) because they use stages of fixed distance and thus progressively shorter duration. The combination of slowed $\dot{V}O_2$ kinetics and progressively shorter stages acts to magnify the effect. In the original protocol, Conconi et al. recommended stages of 200 m. For well-conditioned athletes capable of running at over 20 km/h, the Vd may be observed during a stage that is only 36 s in duration.

For the technique to work well, the subject must complete several stages at velocities well below Vd to allow for a well-developed velocity–heart rate relationship to be established. Devices such as pacing tapes or pacing lights may be useful to help control velocity. The primary limitation on obtaining good data seems to be the ability to make several measurements of heart rate with a large number of evenly incremented stages in between. Strategies that facilitate this are likely to make the technique work better.

The power output and heart rate at the point of deflection of the heart rate performance curve (HRPC) have independently been shown to be associated with the second threshold (Ribeiro et al., 1985), and with left ventricular function during incremental exercise (Foster, Spatz, & Georgakopoulos, 1999; Hofman et al., 1994; Pokan et al., 1993). Ordinarily about 80 to 90% of healthy individuals will demonstrate a negatively accelerated HRPC similar to that described by Conconi (1982). However, there are concerns that the relationship between the deflection of the HRPC and either $LT_2$ or $VT_2$ may not be entirely stable (Francis et al., 1989; Thorland, Podolin, & Mazzeo, 1994).

## Percentage of Maximal Heart Rate

Several authors have examined heart rate responses associated with the aerobic threshold (Bunc et al., 1987; Farrell et al., 1979; McLellan & Skinner, 1981; Rusko et al., 1980), anaerobic threshold (Kindermann et al., 1979; Ljunggren et al., 1987; Ribeiro, et al., 1985; Rusko et al., 1980; Stegmann & Kindermann, 1982), individual anaerobic threshold (Stegmann & Kindermann, 1982), and maximal lactate steady state (Snyder, Woulfe, Welsh, & Foster, 1994). Beyond the statement of Dwyer and Bybee (1983) that heart rate could be used to control subsequent training, there is little agreement about general heart rate guidelines. However, the heart rate is approximately 80% HRmax at $LT_1/VT_1$ and approximately 90% HRmax at $LT_2/VT_2$.

Presuming that an individually unique heart rate associated with the maximal lactate steady state can be identified, one still must account for cardiovascular drift. Heart rate can be expected to increase from 5 to 15 beats $\cdot$ min$^{-1}$ over the course of a 30 min steady-state exercise bout when the work bout is performed in the range of intensities between the $LT_1/VT_1$ and $LT_2/VT_2$ (Heck et al., 1985; Karlsson et al., 1972; Ljunggren et al., 1987; Schnable et al., 1982; Snyder et al., 1994).

Recent studies by Lucia, Hoyos, Carvajal, and Chicharro (1999); Lucia, Hoyos, Perez, and Chicharro (2000); and Fernandez-Garcia, Perez-Landaluce, Rodriguez-Alonso, and Terrados (2000) using elite cyclists have suggested that the % HRmax associated with the $LT_1/VT_1$ and $LT_2/VT_2$ is somewhat variable, although the individual heart rate associated with $LT_1/VT_1$ and $LT_2/VT_2$ is comparatively stable across time (Foster et al., 1993; Lucia et al., 2000), and may be used to interpret physiological loading during training and competition.

## Dmax Method

As a practical strategy to solve the problems of widely varying absolute blood lactate accumulation, Cheng et al. (1992) have proposed the Dmax method, which is a graphic approach to solving power output versus blood lactate curves. The individual data points are plotted and then smoothed with a third-order polynomial curve. Then the maximal distance from a straight line connecting resting and peak exercise values is plotted. The maximum distance to the smoothed curve is identified and represents Dmax. This method supposedly works well with either lactate or ventilatory data. This parameter probably represents an approximation of $LT_1/VT_1$, although other data suggest that it may represent $LT_2/VT_2$ (Bishop, Jenkins, McEniery, & Carey, 2000).

## Lactate Minimum

Lactate accumulation in the blood depends on interaction between lactate production and removal. To try to accentuate this factor and to clarify the point at which lactate begins to accumulate, some authors have tried to use a preceding exercise test to produce high concentrations of lactate and then to use an incremental exercise bout to observe lactate falling and then rising. The lowest value in this curve, the lactate minimum, has been proposed as a marker of MLSS. Unfortunately, the best contemporary data suggest that the technique does not work any better than simpler techniques (Jones & Doust, 1998).

# Laboratory Concerns With Aerobic and Anaerobic Thresholds

Laboratory studies are intended to be both representative of the real world and highly reproducible. In athletes, this is often difficult as it is difficult to standardize their diet and exercise patterns before testing. It is often difficult to make laboratory tests that effectively mimic real-world situations. Several issues, however, are important to standardize as much as possible.

## Sampling Techniques

In the ideal world, arterial blood samples would be obtained to define variables related to blood lactate. Because it is impractical to secure arterial blood, most investigators use arterialized venous blood, capillary blood, or venous blood. Certainly some of the confusion surrounding variables relating to blood lactate is due to the nature of blood sampling superimposed on other variables. The classic paper of Forster, Dempsey, Thomson, Vidruk, and DoPico (1972), demonstrating that near-arterial blood concentrations for pH, lactate, and $pCO_2$, but not $pO_2$, could be obtained in arterialized (by hand warming) venous blood, opened the possibility of systematically easier testing in many laboratories.

In general, venous blood sampling, without arterialization, results in lower blood lactate concentration at any point in time compared with arterial blood

(Foxdal et al., 1994; Robergs et al., 1990; Yoshida, Takeuchi, & Suda, 1982). Capillary blood also usually underestimated arterial values slightly (Foxdal et al.).

Automated enzyme electrode systems are considered reliable and to give only slightly different values than bench chemistry reference methods (Karlsson & Jacobs, 1982; Karlsson et al., 1983), although important differences may exist (Bishop, Smith, Kime, Mayo, & Murphy, 1992; Bishop, Smith, Kime, Mayo, & Tin, 1992; Rodriguez et al., 1992). Automated dry chemistry analyzers have become popular within the last few years, with the Accusport and Lactate Pro being most widely used. Both Accusport (Fell et al., 1998) and Lactate Pro (McNaughton, Thompson, Philips, Backx, & Crickmore, 2002) have been shown to be accurate relative to standard chemistry methods. With the Accusport, some attention must be directed toward standardizing the sample volume, and pipetting of 20 to 50 µL is considered superior to merely allowing a "drop" of blood to fall onto the analyzer. The numerical values of blood lactate vary somewhat among analyzers, probably depending on whether blood or plasma lactate is being analyzed. But because the interpretative value of absolute concentrations of blood lactate is already questionable, this does not present an overwhelming problem relative to their use.

## Reproducibility and Nutritional Status

As might be expected, the groups with the most experience with the measurement of aerobic and anaerobic thresholds report excellent reproducibility of blood (Jacobs, 1981, 1986; Karlsson & Jacobs, 1982; Karlsson et al., 1983), respiratory (Davis, 1985), and heart rate (Cellini et al., 1986; Conconi et al., 1982) methods of detection. Others (Aunola & Rusko, 1984; Heitcamp, Holdt, & Scheib, 1991; Ribeiro et al., 1985) have reported problems with the reproducibility of various methods.

Several authors have demonstrated that blood lactate at any given exercise load is decreased in response to factors that might be expected to be associated with muscle glycogen depletion (Dotan, Rotstein, & Grodjinovsky, 1989; Foster, Snyder, Thompson, & Kuettal, 1988; Hughes et al., 1982; Hurley et al., 1984; Jacobs, 1981; Karlsson, 1971; Maassen & Busse, 1989; Yoshida, 1984). These findings are easy to understand, in that synthesis of lactate without its ultimate precursor, muscle glycogen, is virtually impossible. Some have suggested that the low blood lactate values found in athletes are as much a function of chronic muscle glycogen depletion as of enhanced ability to metabolize lactate. Certainly, the problem of relative glycogen depletion needs to be considered when evaluating exercise responses with blood lactate methods. Given that in athletes the requirements of training and competition may make standardization of preevaluation diet and activity patterns difficult, we have proposed a strategy of using the maximal postexercise blood lactate concentration as a method of indexing submaximal blood lactate responses to the relative momentary nutritional status of the individual. Use of about 30% of the maximal blood lactate concentration in glycogen-depleted individuals predicts about the same power output as does OBLA (e.g., $4.0 \text{ mmol} \cdot \text{L}^{-1}$) in rested athletes (Foster et al., 1988).

Others have noted that the ventilatory and blood lactate markers of aerobic and anaerobic threshold move in opposite directions in response to acute muscle glycogen depletion (Gaesser, Poole, & Gardner, 1984; Hughes et al., 1982; Yoshida, 1984), although this uncoupling of ventilatory and blood lactate responses has not been observed in other laboratories (McLellan & Gass, 1989b). They have interpreted these observations as indicating that the discontinuity of $\dot{V}_E$ and blood lactate at the aerobic threshold is only coincidental and not related by any mechanistic link. Regardless of what these studies may say about the regulation of respiration by blood lactate, they reinforce the case for standardization of nutritional status before laboratory evolutions.

# Summary

Discontinuity in blood lactate accumulation and ventilation during exercise may be used as an index of endurance capacity and as a guideline for exercise training. In general, two main areas of discontinuity, or thresholds, seem to be present despite the substantial amount of methodological variation observed in the literature. These points of discontinuity—aerobic threshold ($LT_1/VT_1$) and anaerobic threshold ($LT_2/VT_2$), in the nomenclature of Skinner-McLellan-Kindermann—are generally related

to (for $LT_1/VT_1$) the first appearance of an elevated blood lactate concentration (up to a concentration of about 2.5 mM) and a disproportionate increase in $\dot{V}_E$ and $\dot{V}CO_2$ relative to $\dot{V}O_2$, and (for $LT_2/VT_2$) the rapid accumulation of blood lactate (a concentration of about 4.0 mM is often used) and a disproportionate increase in $\dot{V}_E$ relative to $\dot{V}CO_2$. This point might also be associated with a negative deflection in the relationship in the heart rate performance curve. A unique exercise intensity, the individual anaerobic threshold described by Stegmann, dependent on blood lactate responses during both exercise and recovery, might be intermediate between the $LT_1$ and $LT_2$ thresholds and potentially might be a better marker of sustainable exercise tolerance.

Markers related to blood lactate are often better appreciated with competitively longer stage durations (>4 min), whereas ventilatory markers are often better appreciated with fairly rapidly incremented stages. During the longer stage durations associated with blood lactate measurement, relevant thresholds may be related to workload (power output or velocity), whereas with the shorter stages used for respiratory measurement, relevant thresholds may be better related to $\dot{V}O_2$.

Various factors can act to modify the aerobic and anaerobic thresholds determined by either blood lactate or respiratory methods. In particular, the short-term nutritional status of the individual and the details of sampling of blood may be important.

Our perspective suggests that much of the confusion relative to the aerobic and anaerobic threshold might be related to the use of mix-and-match technology in various laboratories. For example, respiratory markers do not work particularly well with long exercise stages. If, in trying to combine blood lactate and respiratory markers, a lab uses stages that are too long, the respiratory markers are not going to be revealing. A similar case is found with the use of short exercise stages for methods based on blood lactate. As a practical matter, unless one is doing studies to address unique methods, it makes sense to choose one of the well-established techniques and use it with minimal modification. Otherwise, one is likely to get the worst of each method, rather than the best result, by combining two or more methods for detecting aerobic and anaerobic threshold.

# Testing for Anaerobic Ability

Peter J. Maud, PhD, and Joseph M. Berning, PhD
*New Mexico State University*
Carl Foster, PhD, Holly M. Cotter, MS, and Christopher Dodge, MS
*University of Wisconsin at La Crosse*
Jos J. deKoning, PhD, Floor J. Hettinga, MS, and Joanne Lampen, MS
*Vrije Universiteit, Amsterdam*

Although the focus of much of the literature in exercise physiology is on the aerobic energy contributions to muscular exercise, humans perform much of their exercise under circumstances that cannot be attributed solely, or even primarily, to aerobic metabolism. At the onset of exercise, a lag in oxygen transport occurs. The result of this lag is that during the first several minutes of any rest-to-exercise transition, a considerable amount of the muscular work accomplished cannot be attributed to aerobic metabolism. The work during this period was labeled the oxygen deficit as long ago as 1913 by Krogh and Lindhard. Further, under conditions associated with exercise intensities greater than $\dot{V}O_2max$, the energy source for muscular contraction must also include anaerobic sources. Thus, anaerobic work may be attributable to a delay in the oxygen uptake kinetics at the beginning of work, to exercise intensities greater than those supported by aerobic metabolism, or to a combination of both. The pattern of how muscular work is accomplished during relatively brief periods, with high initial power outputs, often places a particular burden on anaerobic energetic resources. Work during periods of extensive anaerobic metabolism is associated with profound changes in intramuscular phosphagen and lactate concentration (Jacobs, Tesch, Bar-Or, Karlsson, & Dotan, 1983; Jones, McCartney, & McComas, 1986; Karlsson, 1971; Wasserman, Hansen, Sue, & Whipp,

1987), which may or may not contribute to changes in intramuscular pH, which, in turn, may or may not contribute to fatigue (St. Clair Gibson, Lambert, & Noakes, 2001).

Unlike the classical plateau of $\dot{V}O_2$, which may or may not occur with increases in the rate of external muscular work, and which led to the definition of $\dot{V}O_2max$ (Bassett & Howley, 1997; Mitchell, Sproule, & Chapman, 1958; Noakes, 1997), there has not been a single laboratory measurement historically associated with anaerobic work. The concept of the oxygen debt, the continued elevation of $\dot{V}O_2$ after exercise, was outlined in an early attempt to quantitatively account for the magnitude of muscular work performed anaerobically (Hill, 1924; Margaria, Aghemo, & Rovelli, 1966). This model, however, has several problems (Brooks, 1971, 1973; Hill & Smith, 1989), and the concept, particularly relative to describing the proportional contribution of aerobic and anaerobic energy production, is now largely discounted.

## Measurement of Peak and Mean Anaerobic Power

The dominant approach to determine both peak and mean anaerobic power has been to measure the rate and quantity of work performed under circumstances in which aerobic metabolism is assumed to contribute very little. As noted by Inbar, Bar-Or, and Skinner (1996) there has been discussion about the appropriate terminology to use when discussing anaerobic performance-based tests. In this chapter the term *peak anaerobic power* will be used for tests in which the predominant energy source is from the ATP/PC system and the term *mean anaerobic power* for tests in which the anaerobic breakdown of carbohydrate is presumed to be the prime energy contributor.

A complete discussion of the various performance-based tests, both continuous and discontinuous, involving many different modes of activity that have been developed to measure peak and mean anaerobic abilities is beyond the scope of this chapter. We will discuss and describe only six tests, two that measure only peak anaerobic power output and four that measure, or can be used to measure, both parameters.

# Performance-Based Peak Anaerobic Power Tests

Data collection for two popular peak anaerobic power tests, the vertical jump test described by Fox and Matthews (1974) and the stair-climbing test developed by Margaria, Aghemo, and Rovelli (1966), occurs over a brief period of 1 s or less, whereas other tests using cycle ergometers tend to measure peak power output over 3 to 5 s periods. Differences in time over which data is collected, muscle involvement, equipment, and energy source utilization make it difficult to equate these tests because they may be measuring different entities.

## Single Vertical Jump Test

The single vertical jump test, often used as a measure of explosiveness in physical fitness testing, has been used as an index of peak anaerobic power output. The subject stands and reaches upward as high as possible. This height is marked. The subject then performs a maximal jump, and the highest reach is recorded. The vertical jump distance is the difference between standing reach height and highest jumping height. Generally, three jumps are used, usually from a crouched start or sometimes with a preliminary countermovement, with the highest jump used for determination of anaerobic power output. Starting position for the jump has been either with the hands on the hips, where they remain until completion of

the jump, or with use of an arm swing. More detail concerning this test is available in Fox and Matthews (1974). Power is derived from the following formula:

$$\text{power}\left(\text{W}\right) = \\ 21.67 \times \text{body mass}\left(\text{kg}\right) \times \sqrt{\text{vertical displacement}\left(\text{m}\right)}$$

## Stair-Climbing Test

Stair climbing has also been used in an attempt to measure peak anaerobic power production. The basic protocol in the Margaria et al. (1966) test, as described by Fox and Mathews (1974), involves ascending a nine-step staircase of about 1.575 m in height as rapidly as possible, taking three steps at a time. The test has a preliminary run-up to the staircase of about 6 m. A timing mat or photocell activated by the subject's passage is usually used to measure time elapsed during vertical displacement. The subject leaps to the third step, where the timing device is activated, lands his or her second foot on the sixth step, and concludes the test by placing the first foot on the ninth step, which terminates the timing period. Time of the data collection is usually less than 1 s. The test result is computed according to the following formula:

$$\text{power}\left(\text{W}\right) = \\ \frac{\text{body mass}\left(\text{kg}\right) \times \text{vertical displacement}\left(\text{m}\right) \times 9.8}{\text{time(s)}}$$

Bouchard, Taylor, Simoneau, and Dulac (1991) have reviewed published normative data. Power output can range from less than 700 W (12 W/kg) in untrained females or female endurance athletes to as much as 1,500 W (18 W/kg) in some male athletes.

## Continuous Jumping Test

The 60 s continuous, and maximal, vertical jump test developed by Bosco, Luhtanen, and Komi (1983), or the modified 30 s version, has also been used to assess anaerobic power. The highest 5 s power output, usually the first 5 s period, is used to represent peak anaerobic power output. This test takes advantage of the potential for using stored elastic energy in addition to chemical–mechanical energy conversion. A timer connected to a jump mat (switch mat), force plate, or infrared sensor is used to record cumulative flight time during the 30 s or 60 s duration of the test. The subject is instructed to jump continuously, as rapidly, and as high as possible on each jump. Knees are preflexed to 90°. Hands are on hips to minimize horizontal and lateral movement. The average peak power output during any single 5 s segment of the trial can be computed as follows:

$$\text{power}\left(W\right) = \frac{9.8 \times \text{total flight time(s)} \times 5}{4 \times \text{number of jumps} \times \left|5 - \text{total flight time(s)}\right|}$$

Typical 5 s peak anaerobic power output data obtained by Maud and Atwood (unpublished data) for physically active males and females, aged 18 to 30 years, are shown in table 6.1 as absolute values, relative to body weight and relative to fat-free mass.

## Wingate Peak Anaerobic Power Output Cycle Ergometer Test

Cycle ergometer tests have also been widely used as measures of anaerobic power output. In the popular 30 s Wingate anaerobic test (Bar-Or, 1987), the power output during the early part of the test (usually the first 5 s period) may be used as an index of peak anaerobic power output. The test is usually performed on a mechanically braked cycle ergometer (Monark). The resistance or load to be applied is based on individual body weight. The original study used a resistance of 0.075 kp/kg. Optimal resistance for maximum power output, however, has been shown to vary depending on gender, age, and fitness characteristics of the individual or group being

Table 6.1   **60-Second Jump Mat Test—Typical Values for Peak Anaerobic Power Output for Physically Active Males and Females Aged 18 to 30 Years**

| | Absolute | | Relative | | | |
|---|---|---|---|---|---|---|
| | Watts | | $W \cdot kgBW^{-1}$ | | $W \cdot kgFFM^{-1}$ | |
| Percentile rank | Males | Females | Males | Females | Males | Females |
| 95 | 3,033 | 1,313 | 38.00 | 20.28 | 41.25 | 23.94 |
| 90 | 2,314 | 1,190 | 29.60 | 18.88 | 33.40 | 22.68 |
| 85 | 2,091 | 1,078 | 26.40 | 17.70 | 31.00 | 21.76 |
| 80 | 2,059 | 1,016 | 25.10 | 17.06 | 28.10 | 21.06 |
| 75 | 1,963 | 999 | 24.00 | 16.70 | 26.85 | 20.50 |
| 70 | 1,875 | 982 | 22.80 | 16.52 | 25.50 | 20.14 |
| 65 | 1,789 | 975 | 22.45 | 16.14 | 24.75 | 19.68 |
| 60 | 1,748 | 952 | 21.80 | 15.52 | 24.10 | 19.46 |
| 55 | 1,705 | 914 | 21.50 | 15.06 | 24.05 | 19.28 |
| 50 | 1,616 | 905 | 21.30 | 14.90 | 23.90 | 19.10 |
| 45 | 1,569 | 873 | 21.10 | 14.72 | 23.40 | 18.90 |
| 40 | 1,550 | 860 | 20.40 | 14.42 | 22.50 | 18.72 |
| 35 | 1,481 | 857 | 19.70 | 14.04 | 22.00 | 18.40 |
| 30 | 1,458 | 840 | 19.50 | 13.36 | 21.50 | 18.26 |
| 25 | 1,412 | 834 | 19.15 | 13.10 | 21.15 | 17.50 |
| 20 | 1,355 | 806 | 18.40 | 12.84 | 20.80 | 16.60 |
| 15 | 1,329 | 752 | 17.65 | 11.94 | 19.95 | 16.10 |
| 10 | 1,250 | 698 | 15.90 | 11.52 | 18.80 | 14.52 |
| 5 | 1,061 | 618 | 15.80 | 10.70 | 17.85 | 13.38 |
| $\overline{X}$ | 1,719 | 919 | 22.13 | 15.02 | 24.78 | 18.95 |
| N | 29 | 35 | 29 | 35 | 29 | 35 |

studied (Inbar et al., 1996). After warming up, the subject is instructed to pedal as rapidly as possible against zero resistance. The force load is then applied as quickly as possible, with simultaneous activation of the timing process. After the load is applied, the subject continues to pedal, remaining in the seated position, as hard and fast as possible for 30 s. Less time could be used if peak anaerobic power is the only measurement of interest. Flywheel revolutions are counted (usually over a 1 to 5 s interval), ideally by a photocell that can resolve each flywheel revolution into several segments. The power output for the highest 5 s period, invariably the first 5 s period, is used to represent peak power. The typical power output decline over the 30 s period is shown in figure 6.1.

Using the Monark cycle ergometer, in which the distance traveled by the flywheel is 1.615 m per revolution, and converting resistance to newtons, peak power may be determined using the following formula:

$$\text{peak anaerobic power}\left(W\right) =$$
$$\left(\text{flywheel rpm for highest 5 s period} \times 1.615\ \text{m}\right) \times$$
$$\left(\text{resistance in kg} \times 9.8\right)$$

The authoritative text, written by Inbar et al. (1996), presents additional information regarding this test, including a detailed description of the test and discussion of issues such as pedal crank length, optimal resistance, test duration, effects of training, reliability and validity of the test, comparison to histochemical data, factors that may influence performance of the test, and typical values for different age, gender, and fitness groups. Generally, peak power outputs will range from about 6 W/kg in young sedentary females to about 16 W/kg in some male athletes. Absolute peak anaerobic power values, those related to body weight, and those related to fat-free mass obtained by Maud and Shultz (1989) for physically active males and females ranging from 18 to 28 years of age are shown in table 6.2. Table 6.3 shows typical values for untrained Israeli males and females aged 18 to 25 years as reported by Inbar et al. (1996). Both sets of data were derived using a resistance of 0.075 kp/kg as originally used in the development of the test.

As would be expected, the values obtained from untrained Israeli subjects are lower than those reported for physically active young Americans in which the criteria for participation in the study required that they had been exercising at least three times per week for six weeks before testing.

## Quebec 10-Second Test

In recent years it has become recognized that the flywheel inertia related to a high preloading pedaling rate introduces a spurious elevation of calculated power output. The Quebec 10 s test (Simoneau, Lortie, Boulay, & Bouchard, 1983) has been developed to provide a reasonable compromise by having

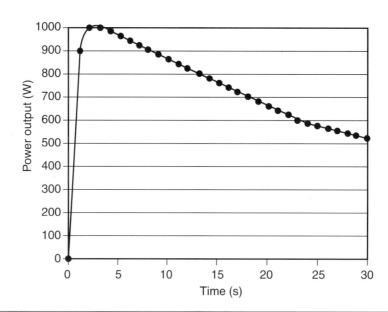

**Figure 6.1**   Pattern of power output during 30 s all-out cycle ergometer test, similar to the widely used Wingate anaerobic test.

Table 6.2    **Percentile Norms for the Wingate Test: Peak Anaerobic Power Output for Physically Active Males and Females Aged 18 to 28 Years**

| | Absolute | | Relative | | | |
| | Watts | | $W \cdot kgBW^{-1}$ | | $W \cdot kgFFM^{-1}$ | |
| Percentile rank | Males | Females | Males | Females | Males | Females |
|---|---|---|---|---|---|---|
| 95 | 866.9 | 602.1 | 11.08 | 9.32 | 12.26 | 11.87 |
| 90 | 821.8 | 560.0 | 10.89 | 9.02 | 11.96 | 11.47 |
| 85 | 807.1 | 529.6 | 10.59 | 8.92 | 11.67 | 11.28 |
| 80 | 776.7 | 526.6 | 10.39 | 8.83 | 11.47 | 10.79 |
| 75 | 767.9 | 517.8 | 10.39 | 8.63 | 11.38 | 10.69 |
| 70 | 757.1 | 505.0 | 10.20 | 8.53 | 11.28 | 10.39 |
| 65 | 744.3 | 493.3 | 10.00 | 8.34 | 11.08 | 10.30 |
| 60 | 720.8 | 479.5 | 9.80 | 8.14 | 10.79 | 10.10 |
| 55 | 706.1 | 463.9 | 9.51 | 7.85 | 10.30 | 9.90 |
| 50 | 689.4 | 449.1 | 9.22 | 7.65 | 10.20 | 9.61 |
| 45 | 677.6 | 447.2 | 9.02 | 7.16 | 10.10 | 9.41 |
| 40 | 680.8 | 432.5 | 8.92 | 6.96 | 10.00 | 8.92 |
| 35 | 661.9 | 417.8 | 8.63 | 6.96 | 9.90 | 8.83 |
| 30 | 656.1 | 399.1 | 8.53 | 6.86 | 9.51 | 8.73 |
| 25 | 646.3 | 396.2 | 8.34 | 6.77 | 9.32 | 8.43 |
| 20 | 617.8 | 375.6 | 8.24 | 6.57 | 9.12 | 8.34 |
| 15 | 594.3 | 361.9 | 7.45 | 6.37 | 8.53 | 8.04 |
| 10 | 569.8 | 353.0 | 7.06 | 5.98 | 8.04 | 7.75 |
| 5 | 530.5 | 329.5 | 6.57 | 5.69 | 7.45 | 6.86 |
| $\overline{X}$ | 699.5 | 454.5 | 9.18 | 7.61 | 10.18 | 9.54 |
| SD | 94.7 | 81.3 | 1.43 | 1.24 | 1.46 | 1.51 |
| N | 62 | 68 | 62 | 68 | 62 | 68 |

the subject pedal at a relatively modest pedal rate (50-60 rpm). The test requires a photocell capable of resolving a minimum of each one-third revolution of the flywheel and a potentiometer capable of sensing the momentary loading in the flywheel. The subject begins pedaling at about 60 rpm, and the investigator loads the flywheel (about 1 N/kg body weight). Remaining in a seated position, the subject then completes as many revolutions as possible in a 10 s period. The load on the flywheel is adjusted during the test to allow the subject to maintain a high pedaling rate. Power output is computed as the highest output observed during any 1 s of the

test, in contrast to the 5 s periods usually recorded for the Wingate test. More information about this test is available (Bouchard et al., 1991; Boulay et al., 1985; Serresse, Lortie, Bouchard, & Boulay, 1988; Simoneau et al., 1983). Although about 1 N/kg body weight is the commonly recommended flywheel resistance for a test of this nature, some authors have suggested a heavier load, particularly for athletes. Valuable discussion of the effect of loading variations is available (Dotan & Bar-Or, 1983; Evans & Quinney, 1981; Gastin, Lawson, Hargreaves, Carey, & Fairweather, 1991; Inbar et al., 1996; Patton, Murphy, & Fredrick, 1985).

Table 6.3   **Typical Wingate Test Peak Anaerobic Power Output Values for Healthy Israeli Untrained Males and Females Aged 18 to 25 Years.**

| | Absolute | | Relative | |
| | Watts | | $W \cdot kgBW^{-1}$ | |
| Category | Males | Females | Males | Females |
| --- | --- | --- | --- | --- |
| Very poor | 339-440 | 319-400 | 5.4-6.8 | 6.28-7.27 |
| Poor | 441-492 | 400-441 | 6.8-7.5 | 7.28-7.77 |
| Below average | 493-544 | 441-482 | 7.5-8.1 | 7.78-8.27 |
| Average | 544-596 | 482-524 | 8.2-8.8 | 8.28-8.78 |
| Good | 596-648 | 524-565 | 8.8-9.5 | 8.79-9.28 |
| Very good | 648-700 | 565-605 | 9.5-10.2 | 9.29-9.78 |
| Excellent | 700-802 | 606-686 | 10.2-11.6 | 9.79-10.78 |
| $\overline{X}$ | 570.41 | 503.0 | 8.50 | 8.5 |
| SD | 103.94 | 93.2 | 1.38 | 1.07 |
| N | 21 | 18 | 21 | 18 |

Reprinted, by permission, from O. Inbar, O. Bar-Or, and J.S. Skinner, 1996, *The Wingate Anaerobic Test* (Champaign, IL: Human Kinetics), 79-90.

## Air-Braked Cycle Ergometer

Withers et al. (1991) and Withers and Telford (1987) have reported data on responses during heavy exercise on an air-braked cycle ergometer. Although they did not report very short-term peak power outputs, it would not be difficult to fit such an ergometer with a sensor for calculating power output per revolution of the flywheel (fan). This ergometer is arranged somewhat differently, yet it is conceptually similar to the Schwinn Air-Dyne ergometer popular in the United States. The simplicity of the air-braked approach has much to offer over the relatively clumsy process of rapidly applying resistance to mechanically braked ergometers. The momentary power output is a simple function of the revolution rate of the flywheel. Resistance will vary, however, in relation to the local barometric pressure and the characteristics of the specific flywheel. A correction for conditions other than 760 mmHg barometric pressure and 20 °C temperature can be computed with the following equation:

$$\text{corrected watts} =$$
$$\text{observed watts} \times P_B / 760 \times 293 / (273 + T)$$

Interpretive standards are available for this instrument (Withers et al. 1991; Withers & Telford, 1987).

## Performance-Based Mean Anaerobic Power Tests

Laboratory approaches to measuring mean anaerobic power output using muscular performance tests have been well reviewed by Bouchard et al. (1991); Noakes (1997); Nummela, Alberts, Rijntjes, Luhtanen, and Rusko (1996); Nummela, Anderson, Hakkinen, and Rusko (1996); Nummela, Mero, and Rusko (1996); Nummela, Mero, Stray-Gundersen, and Rusko (1996); Rusko, Nummela, and Mero (1993); Tossavainen, Nummela, Paavolainen, Mero, and Rusko (1996); and Inbar et al. (1996). Discussion of issues related to the measurement of human muscular power output can also be found in Jones et al. (1986).

Two tests, one a continuous jumping test and the other a cycle ergometer test, as previously mentioned, merit specific description because of their wide use. Discussion and description of two additional cycle ergometer tests are also included because they may be more appropriate than the Wingate test for determination of anaerobic power output when using a cycle ergometer.

### Continuous Jumping Test

In the previously described test of Bosco et al. (1983), both 60 s and modified 30 s continuous jumping tests have also been used to assess mean anaerobic power output. Mean anaerobic power output for the 60 s version of the test may be calculated using the following formula:

$$power\ (W) = \frac{9.8 \times total\ flight\ time(s) \times 60}{4 \times number\ of\ jumps \times \left[60 - total\ flight\ time(s)\right]}$$

Typical mean anaerobic power output data developed by Maud and Atwood (unpublished data) for subjects previously described for peak anaerobic power are shown in table 6.4 as absolute values, relative to body weight and relative to fat-free mass. As with other tests, the pattern of fatigue can be measured, and a fatigue index determined, by noting the power output decline from the first period (usually 1 to 5 s) to the final period.

## Wingate Mean Anaerobic Power Output Test

In the Wingate anaerobic test (Bar-Or, 1987) as previously described, the subject pedals a mechanically braked cycle ergometer at maximal possible rate against a heavy resistance, usually specific to the body weight of the subject, for 30 s. Mean anaerobic power output can be calculated from the following formula:

$$mean\ anaerobic\ power\left(W\right) = \left(flywheel\ rpm\ for\ 30\ s\ period \times 1.615\right) \times \left(resistance\ in\ kg \times 9.8\right)$$

Total work accomplished can be calculated from this formula:

$$work\ in\ joules = average\ watts \times duration(s)$$

Table 6.4   **60-Second Jump Mat Test—Typical Values for Mean Anaerobic Power Output for Physically Active Males and Females Aged 18 to 30 Years**

| Percentile rank | Absolute Watts | | Relative W·kgBW⁻¹ | | Relative W·kgFFM⁻¹ | |
| --- | --- | --- | --- | --- | --- | --- |
| | Males | Females | Males | Females | Males | Females |
| 95 | 2,385 | 961 | 29.85 | 15.32 | 32.45 | 19.81 |
| 90 | 1,556 | 885 | 19.90 | 13.34 | 22.40 | 17.24 |
| 85 | 1,481 | 848 | 18.80 | 13.46 | 21.80 | 16.60 |
| 80 | 1,464 | 810 | 17.80 | 13.26 | 19.70 | 16.02 |
| 75 | 1,395 | 746 | 17.35 | 12.80 | 19.25 | 15.70 |
| 70 | 1,367 | 740 | 17.30 | 12.52 | 19.20 | 15.32 |
| 65 | 1,309 | 730 | 16.35 | 11.94 | 18.90 | 15.24 |
| 60 | 1,267 | 723 | 16.10 | 11.80 | 18.40 | 14.96 |
| 55 | 1,249 | 705 | 16.05 | 11.60 | 18.25 | 14.78 |
| 50 | 1,223 | 703 | 15.90 | 11.60 | 17.90 | 14.40 |
| 45 | 1,203 | 698 | 15.55 | 11.42 | 17.70 | 14.22 |
| 40 | 1,172 | 667 | 15.30 | 11.04 | 17.60 | 13.98 |
| 35 | 1,140 | 639 | 15.30 | 10.76 | 17.25 | 13.54 |
| 30 | 1,120 | 623 | 15.10 | 10.08 | 17.10 | 13.28 |
| 25 | 1,101 | 619 | 14.70 | 9.70 | 16.80 | 13.10 |
| 20 | 1,083 | 594 | 14.70 | 9.52 | 16.30 | 12.90 |
| 15 | 1,060 | 583 | 14.45 | 9.40 | 16.05 | 12.54 |
| 10 | 986 | 547 | 14.10 | 9.20 | 15.80 | 11.42 |
| 5 | 922 | 470 | 12.50 | 8.48 | 14.45 | 10.58 |
| $\bar{X}$ | 1,289 | 700 | 16.69 | 11.47 | 18.81 | 14.47 |
| N | 29 | 35 | 29 | 35 | 29 | 35 |

Normative data for physically active males and females aged 18 to 28 years and for healthy Israeli males and females aged 18 to 28 are shown in tables 6.5 and 6.6. Consult Inbar et al. (1996) or Bouchard et al. (1991) for other normative data.

A fatigue index, which indicates the power decline from peak power to lowest power output over the 30 s duration of the test, can be determined from the following equation:

$$\text{fatigue index} =$$
$$\left[\text{peak power} - \left(\text{lowest power}/\text{peak power}\right)\right] \times 100$$

The mean decline for men (37.7%) and for women (35%) in the study by Maud and Shultz (1989) falls

well within the range of fatigue indices for athletes as shown in the table developed from the literature by Bouchard et al. (1991), in which values range from 25% for male speed skaters to 47% for gymnasts.

## Quebec 90 Second Test

The Quebec 90 s (Simoneau et al., 1983) test is fundamentally similar to the Wingate anaerobic test, but it continues longer and uses a potentiometer to monitor flywheel resistance constantly. Based on data concerning the relationship between work duration and anaerobic capacity estimated from accumulated oxygen deficit, an argument has been made that the duration of the Quebec test may be much more suit-

Table 6.5   **Percentile Norms for Wingate Test: Mean Anaerobic Power Output for Physically Active Males and Females Aged 18 to 28 Years**

| | Absolute | | Relative | | | |
| | Watts | | $W \cdot kgBW^{-1}$ | | $W \cdot kgFFM^{-1}$ | |
| Percentile rank | Males | Females | Males | Females | Males | Females |
|---|---|---|---|---|---|---|
| 95 | 676.6 | 483.0 | 8.63 | 7.52 | 9.30 | 9.43 |
| 90 | 661.8 | 469.9 | 8.24 | 7.31 | 9.03 | 9.01 |
| 85 | 630.5 | 437.0 | 8.09 | 7.08 | 8.88 | 8.88 |
| 80 | 617.9 | 419.4 | 8.01 | 6.95 | 8.80 | 8.76 |
| 75 | 604.3 | 413.5 | 7.96 | 6.93 | 8.70 | 8.68 |
| 70 | 600.0 | 409.7 | 7.91 | 6.77 | 8.63 | 8.52 |
| 65 | 591.7 | 402.2 | 7.70 | 6.65 | 8.50 | 8.32 |
| 60 | 576.8 | 391.4 | 7.59 | 6.59 | 8.44 | 8.18 |
| 55 | 574.5 | 386.0 | 7.46 | 6.51 | 8.24 | 8.13 |
| 50 | 564.6 | 381.1 | 7.44 | 6.39 | 8.21 | 7.93 |
| 45 | 552.8 | 376.9 | 7.26 | 6.20 | 8.14 | 7.86 |
| 40 | 547.6 | 366.9 | 7.14 | 6.15 | 8.04 | 7.70 |
| 35 | 534.6 | 360.5 | 7.08 | 6.13 | 7.95 | 7.57 |
| 30 | 529.7 | 353.2 | 7.00 | 6.03 | 7.80 | 7.46 |
| 25 | 520.6 | 346.8 | 6.79 | 5.94 | 7.64 | 7.32 |
| 20 | 496.1 | 336.5 | 6.59 | 5.71 | 7.46 | 7.11 |
| 15 | 484.6 | 320.3 | 6.39 | 5.56 | 7.28 | 7.03 |
| 10 | 470.9 | 306.1 | 5.98 | 5.25 | 6.83 | 6.83 |
| 5 | 453.2 | 286.5 | 5.56 | 5.07 | 6.49 | 6.70 |
| $\overline{X}$ | 562.7 | 380.8 | 7.28 | 6.35 | 8.11 | 7.96 |
| SD | 66.5 | 56.4 | .88 | .73 | .82 | .88 |
| N | 62 | 68 | 62 | 68 | 62 | 68 |

Table 6.6    **Typical Wingate Test Mean Anaerobic Power Output Values for Healthy Israeli Untrained Males and Females Aged 18 to 25 Years**

| | Absolute | | Relative | |
| | Watts | | $W \cdot kgBW^{-1}$ | |
| Category | Males | Females | Males | Females |
|---|---|---|---|---|
| Very poor | 302-380 | 177-247 | 5.1-6.0 | 4.31-4.90 |
| Poor | 380-418 | 247-282 | 6.0-6.4 | 4.91-5.20 |
| Below average | 418-456 | 282-317 | 6.4-6.9 | 5.121-5.20 |
| Average | 456-494 | 317-352 | 6.9-7.3 | 5.51-5.80 |
| Good | 495-533 | 352-387 | 7.3-7.7 | 5.81-6.10 |
| Very good | 533-571 | 387-422 | 7.7-8.2 | 6.11-6.40 |
| Excellent | 571-649 | 422-492 | 8.2-9.0 | 6.41-7.00 |
| $\overline{X}$ | 475.38 | 334.3 | 7.07 | 5.66 |
| SD | 76.35 | 58.8 | 0.86 | 0.59 |
| N | 21 | 18 | 21 | 18 |

Reprinted, by permission, from O. Inbar, O. Bar-Or, and J.S. Skinner, 1996, *The Wingate Anaerobic Test* (Champaign, IL: Human Kinetics), 79-90.

able than the shorter duration of the Wingate test for estimating mean anaerobic power. The Quebec 90 s test is performed on a Monark or other mechanically braked cycle ergometer. After warming up, the subject begins pedaling at 60 rpm while the workload is adjusted (usually to about 0.5 N/kg body weight). The subject then pedals as rapidly as possible for 90 s while a photocell capable of resolving each one-third revolution counts the revolutions of the flywheel and a potentiometer monitors the resistance on the flywheel. During the test, subjects commonly perform an initial pedaling rate above 130 rpm for the first 20 s. Adjustment of the resistance on the flywheel to maintain a high pedaling rate is usually necessary. Integration of the data can give moment-to-moment power output, give some information regarding the pattern of fatigue, and quantify the total work done in 90 s. Normative data regarding the test are available (Bouchard et al., 1991; Boulay et al., 1985; Simoneau et al., 1983). But given that aerobic metabolism is considerable during a test of 90 s duration, attributing all the work accomplished to anaerobic sources is unrealistic.

Using an air-braked cycle ergometer, Withers et al. (1991) and Withers and Telford (1987) have presented a variation of the Quebec 90 s test. They have demonstrated that accumulated oxygen deficit and muscle lactate concentration reach maximal values after 60 s of exercise if the subject goes all out from the beginning of the test. Their data argue that a 60 s version of either the Wingate anaerobic test or the

Quebec test may be ideal, an option chosen by Szogy and Cherebetiu (1974). Unfortunately, a significant amount of the work done during a test of even 30 s duration may be accounted for aerobically (Fox & Mathews, 1974), so quantification of mean anaerobic power cannot be satisfactorily accomplished from the performance data alone.

## Testing Issues

Laboratory tests are designed to be reproducible and to reflect real-world events. This is often difficult with athletes, particularly standardizing diet and exercise before testing. A few issues that need to be carefully controlled are discussed below.

### Peak Anaerobic Power Output Tests

As has been discussed, several different power output tests have been used to reflect alactacid anaerobic energy utilization. The time over which data are collected for assessment of peak anaerobic power has ranged typically from 1 to 5 s. As the time from the start of the exercise bout to data collection varies, then the specific energy supply utilized may vary. For example, in a single vertical jump test, stored ATP would be the prime source of energy production. But if data were collected from a later period, as for example in the Margaria-Kalamen test following the initial approach to the steps, or in the Quebec 10 s test in which the highest power output for 1 s is extracted from the total 10 s time period making up the test, or over periods in excess of 1 s as in the Wingate or Bosco 5 s tests, then restoration of ATP from CP stores would theoretically be the substrate primarily responsible for energy output. Not surprisingly, then, correlations between these tests are not particularly high. For example, Maud and Shultz (1986) found correlations of r = .74 (males) and r = .72 (females) when comparing the Lewis and the Margaria-Kalamen test but only r = .35 (males) and r = .59 (females) between the Lewis and the Wingate 5 s test and an even lower correlation of r = .24 (males) and r = .26 (females) between the Margaria-Kalamen and the Wingate 5 s tests.

Other considerations include the type of resistance to be used, air-braked or mechanical, and time of collection of data relative to time of commencement of the test.

## Cycle Ergometer Tests

As already discussed, certain issues arise relative to which cycle ergometer protocol is the most appropriate. The Wingate test has been used far more extensively than either of the other two cycle ergometer tests reviewed. The Quebec 10 s and 90 s tests, however, have the ability to measure moment-to-moment variations in pedal resistance, allowing adjustment to resistance to maintain desired pedal revolutions and to try to eliminate the problem of high preloading pedal rate that elevates power output. The use of air-braked cycle ergometers has the advantage of not requiring rapid application of resistance to mechanically braked ergometers.

Another important consideration is the duration of the test when used to assess mean anaerobic power output. The Wingate test duration is 30 s, the Quebec 90 s test is 90 s, and Withers et al. (1991) and Withers and Telford (1987) have suggested that a compromise of 60 s, as used by Szogy and Cherebetiu (1974), may be more appropriate. Certainly, as the test time increases so does the contribution of the aerobic system, but, as already stated, even in a 30 s test significant aerobic contribution occurs.

## Wingate Test

For the Wingate anaerobic cycle test, it has been argued that to achieve the highest possible peak and mean anaerobic values, an optimal resistance load relative to body weight needs to be in excess of the $0.075$ kp $\cdot$ kg$^{-1}$ of body mass as used in the original studies by the Wingate investigators. As an example, Evans and Quinney (1981) reported an optimal resistance of $0.098$ kp $\cdot$ kg$^{-1}$ for active individuals and athletes, Patton et al. (1985), $0.098$ kp $\cdot$ kg$^{-1}$ for soldiers, and Dotan and Bar-Or (1983), $0.087$ kp $\cdot$ kg$^{-1}$ for male and $0.085$ kp $\cdot$ kg$^{-1}$ for female physical education students.

If these higher resistances are to be used, however, they do create some problems such as excessive stress being placed on the equipment. On one occasion during testing of athletes in our laboratory, the handlebar stem of an older model Monark ergometer snapped, and on another occasion the sprocket was stripped. In addition, when heavy subjects are being evaluated it may be impossible to apply the required high resistance setting. A further consideration is that in any test in which heavy accumulation of lactate may occur, the accompanying discomfort of the subject may require higher motivational effort. This problem is more likely to be a factor in tests using the higher resistance setting and a longer period. For these reasons it is suggested that the original load setting of 0.075 might be more realistic, particularly if there was a high rank order correlation between the 0.075 setting and the higher resistance setting. With the lower load, peak and mean anaerobic power abilities could be compared between groups, could be used for general or sport-specific fitness assessments, and certainly could be employed to quantify anaerobic fitness improvements in response to training programs.

## Continuous Jumping Tests

In the Bosco test (Bosco et al., 1983) the recommended duration of the test is 60 s, although some investigators (Fabian et al., 2001; Maud & Atwood, unpublished data) have used a 30 s version. If the purpose of the test is to elicit a high lactate concentration response, then the 60 s version would be preferred even though this test magnifies the aerobic percentage concentration to energy requirements. But if motivational aspects and pain accompanying the test are considered, particularly if multiple tests are required of a subject, then the shorter version may encourage subjects to give maximal effort, particularly if the subjects are nonathletes and less likely to put out maximum effort. Here a high correlation between the two protocols would help justify this recommendation. In obtaining the data shown in table 6.6, relatively high correlation coefficients of .93 for males, .91 for females, and .98 when the test scores were combined were found between the 60 s and 30 s data scores. This probably provides a sound rationale for use of the shorter version of the test.

## Comparisons Between the Wingate Test and the Continuous Jumping Test

Numerous practical considerations could be addressed when trying to determine which anaerobic performance-based test to administer. We have already discussed some of these issues, but some warrant further consideration.

The Wingate anaerobic cycle test, despite the limitations previously addressed, is by far the most popular performance-based anaerobic test. The Bosco test has been less extensively used and studied, but it does provide a method of assessment particularly relevant to groups of individuals who are involved in activities in which jumping plays a major role, such as volleyball and basketball.

In direct comparative studies, the calculated power output from the jumping test has been higher than during cycle ergometer tests of the same duration. For example, in a study by Fabian et al. (2001) relative peak anaerobic power and relative mean anaerobic power outputs were significantly higher (21.29 W/kg versus 11.69 W/kg and 18.30 W/kg versus 6.86 W/kg) for a 30 s modified Bosco jump mat test (Bosco et al., 1983) compared with a 30 s Wingate test. The pattern of decay on power output during jump mat tests is similar, however, to that observed during cycling.

The 30 s time version of the Bosco test (Bosco et al., 1983) is identical to the normal time requirement for the Wingate test, which makes comparisons between these two tests more appropriate. As has been noted, both mean and peak anaerobic power values are significantly higher for the Bosco test with the suggestion made by Fabian et al. (2001) that this is due to "the stretch-shortening cycle and myotatic stretch reflex." The latter factor would tend to suggest that perhaps these two tests are measuring different anaerobic energy contributions and abilities. A strong correlation between these tests may suggest that the energy requirements are similar and that the prime reason for the higher values found for the Bosco test is due primarily to the suggested elastic recoil effect involved in the jumping test. But when a Spearman rank order correlation was used to compare the Wingate and Bosco test results for the 71 subjects studied by Fabian et al. (2001), the correlation coefficients were low and negative for mean power output (rho = −.32) and for peak power output (rho = −.40). This finding hardly supports the notion that these two tests are measuring the same attributes.

Another point of interest when comparing these two tests concerns differences found between genders. In the Wingate anaerobic cycle test (Chicharro et al., 1998) differences in values between genders are reduced when related to body weight and further reduced to insignificant levels when related to fat-free mass. In contrast, data from the Bosco test (Maud & Atwood, unpublished data), although showing a reduction in differences between genders when related to body weight and particularly to fat-free mass, still exhibited significant differences in power output, with males being 45% higher for gross values, 30% higher for values related to body weight, and 23% higher for values relative to fat-free mass. (Similar results were found in peak anaerobic power scores, with a 46% difference between gross scores, a 31% difference between values related to body weight, and a 23% difference between values corrected for fat-free mass.)

## Conclusions on Testing Issues

When consideration is given to which test is the most suitable for comparing individuals or groups, or when evaluating results of training programs, specificity of mode of performance needs to be addressed, just as when testing for maximum oxygen uptake. This factor does not seem to have been of prime consideration in the majority of tests previously used to assess various athletic and nonathletic groups. Obviously, in evaluating cyclists one of the cycle ergometer tests would be the mode of choice. When elastic recoil is involved, however, as in such sports as volleyball and basketball, then the Bosco test would appear to be more appropriate.

Another consideration when conducting these tests is the athletic experiences of the individual or group and the number of trials to which they are to be subjected. As with maximum oxygen uptake testing, pain and discomfort is involved when assessing mean anaerobic power output. The athlete generally has been exposed to these conditions and is trained to accept the pain or discomfort involved. For the nonathlete, pain or discomfort may be a relatively new experience, and the individual may not be prepared to contribute a maximum effort. For these anaerobic tests, unlike tests of maximum oxygen uptake, no currently established benchmarks are available to indicate the degree of effort given.

Based on a review of the studies using peak and mean anaerobic power output to estimate anaerobic contributions to short, maximum exercise bouts, the conclusion must be that because of the differences in values obtained for peak and mean anaerobic tests when using different exercise modalities and time periods for collection of data, no one test can be considered to provide a true reflection of the anaerobic contribution to the activity.

# Accumulated O₂ Deficit

The validity of performance-oriented approaches to the measurement of anaerobic metabolism, particularly mean anaerobic power, can be challenged on at least three fronts.

First, performance-oriented tests usually involve very high rates of muscular power generation and substantial reductions in power output (fatigue) throughout the course of the test. Outside the laboratory, humans almost never work at their absolute maximal rate of momentary energy expenditure for more than about 5 s. They usually use some element of pace, designed to minimize the time required to complete a task. Certainly, few examples in nature produce a power output pattern similar to that imposed by many of the contemporary tests of peak and mean anaerobic power. Even athletic events as short as 20 s (running 200 m) involve some element of relaxing and sustaining speed so that the net duration of the task is minimized. Second, under conditions of very heavy muscular work, the kinetics of the oxygen transport system are far from slow, and aerobic metabolism may account for a substantial portion of the energetic requirement in tests as short as 30 s in duration. For example, aerobic metabolism has been shown to account for as much as 40% of the energetic requirements of the widely used Wingate test (Hill, 1924). This observation suggests that the assumption of a negligible oxidative contribution during tests of anaerobic power and capacity is fundamentally incorrect. Third, the demonstration of considerable muscle lactate accumulation, even during very brief exercise (Jacobs et al., 1983), suggests that partitioning the alactic and lactic components of anaerobic work based on time is intrinsically invalid. Further, the demonstration of significant rates of lactate production even at rest, the largely oxidative fate of produced lactate, and the importance of lactate turnover to the magnitude of lactate accumulation (Åstrand, Hultman, Juhlin-Danfelt, & Reynold, 1986; Brooks, 1991; Stainsby & Brooks, 1990) suggest that the accumulation of muscle and blood lactate during heavy exercise is not a particularly good index of the contribution of anaerobic metabolism during any given exercise

bout, although many authors continue to recommend it as a crude index of mean anaerobic power (Nummela et al., 1996).

From the foregoing it may be justified to suggest that although many of the widely used performance-based tests of peak and mean anaerobic power might be useful indices of the peak rate of achievable muscular power output and of the quantity of muscular work achievable during comparatively brief periods of time, they might not satisfy their stated goal of measuring muscular power output and energetic capacity attributable to anaerobic metabolism alone.

Medbø and Burgers (1990), Medbø et al. (1988), and Medbø and Tabata (1989), following early suggestions by Hermansen (1969) and Karlsson (1971), have suggested that the magnitude of oxygen deficit accumulated during exhausting exercise may serve as a marker of mean anaerobic power. Others have applied this general concept with success (Bangsbo et al., 1990; Buck & McNaughton, 1999; Gastin, 1994; Gastin, Costill, Larson, Krzeminski, & McConell, 1995; Green, Dawson, Goodman, & Carey, 1996; Katz, Snell, & Stray-Gundersen, 1989; Scott & Bogdonfly, 1998; Spencer & Gastin, 2001). A modification of this concept, which includes the magnitude of blood lactate accumulation, has been proposed by Camus and Thys (1991). Serresse et al. (1988) and Serresse, Simoneau, Bouchard, and Boulay (1991) have suggested a conceptually similar approach in which the total quantity of work accomplished is adjusted for the quantity of work attributable to aerobic metabolism. The test may be conducted either on an electrically braked cycle ergometer, during treadmill running, or as a time trial on a wind-braked ergometer (Foster et al., 2003; Spencer & Gastin, 2001). Similarly, the net deficit during a ramp-type test has been calculated and proposed as a measure of anaerobic capacity (Scott & Bogdonfly, 1998). Treadmill studies are often conducted up a rather steep grade (5%-10%) to provide greater safety for getting on the already moving treadmill.

We have found that a 2 mph (3.2 kph) greater velocity than that which the subject perceives as hard (or 5 on the category-ratio scale of Borg) often gives a convenient test duration (unpublished results). If the accumulated O₂ deficit is to provide a satisfactory marker of the mean anaerobic power, at least three criteria must be met.

First, the accumulated O₂ deficit should level off with exercise duration. If there is a practical limit on the depletion of muscle phosphagens and the accumulation of metabolites from anaerobic metabolism, then the amount of work performed anaerobically should increase with exercise duration until the subject achieves some limit (the mean anaerobic power).

To use a practical example, if one were to perform a single vertical jump (or even two or three jumps), the $O_2$ requirement might be very high (supramaximal) relative to the $\dot{V}O_2$max, and the proportion of energy from anaerobic sources might be very high, but the total energy expenditure from nonoxidative sources would be very low. If one were to jump up and down for a minute, as suggested by Bosco, Luhtanen, and Komi (1983), the $O_2$ requirements might be significantly in excess of $\dot{V}O_2$max (supramaximal), and the absolute amount of energy liberated from anaerobic sources would be an important source of energy. If one were to jump up and down for several hours at a submaximal intensity, however, the oxidative contribution exercise bout would be very high and the net anaerobic energy release would probably not be greater than that released during 1 or 2 min of maximal jumping.

The second criterion that must be met is that mean anaerobic power should be fundamentally independent from the $\dot{V}O_2$max. Given the different mechanisms involved in supporting aerobic and anaerobic metabolism, one would not expect much relationship between the two.

The third criterion is that the measure of accumulated $O_2$ deficit should substantially agree with other methods of measuring the mean anaerobic power. Although the performance-based tests can be criticized about their ability to partition aerobic and anaerobic contributions quantitatively, they probably do measure some element of anaerobic capacity. The data of Withers et al. (1991), Withers and Telford (1987), and Szogy and Cherebetiu (1974) provide support for this concept.

Medbø and Burgers (1990), Medbø et al. (1988), and Medbø and Tabata (1989) have described an approach to measuring the accumulated $O_2$ deficit during heavy exercise. The magnitude of the accumulated $O_2$ increases with exercise duration, with maximal values observed at workloads causing exhaustion in 2 to 5 min. They have further demonstrated that the accumulated $O_2$ deficit is larger in sprint-trained athletes than in endurance-trained athletes or in the untrained, is larger in males than in females, and is responsive to training. Bangsbo et al. (1990) have demonstrated a quantitative relationship between accumulated $O_2$ deficit and short-term changes in muscle metabolites during one leg exercise. Others have demonstrated the feasibility of this technique and have begun to describe normally expected values for the magnitude of the accumulated $O_2$ deficit, 30

to 80 ml $O_2$/kg (Foley et al., 1991; Foster, Kuettal, & Thompson, 1989; Karlsson, 1971; Withers et al., 1991). An example test showing the computation of the accumulated $O_2$ deficit is presented in figure 6.2. The $O_2$ requirement is first estimated based on several steady-state submaximal runs according to Medbø and Burgers (1990), Medbø et al. (1988), and Medbø and Tabata (1989). The subject then exercises to exhaustion at a fixed workload, with continuous measurement of $\dot{V}O_2$. Accumulated $O_2$ deficit is computed by subtracting the measured $O_2$ uptake from the $O_2$ required of a specific time interval (30 s in the present example) and summing these measures.

One of the limitations of the accumulated $O_2$ deficit method as described by Medbø and associates (Medbø & Burgers, 1990; Medbø et al., 1988; Medbø & Tabata, 1989) is the requirement that the subject continue an arbitrary workload to exhaustion. The extent to which this requirement is met depends on the subject's level of motivation. We have had subjects perform a series of cycle ergometer tests in which the total work requirement varied from 100 to 700 J/kg (Foster et al., 1989) and with instructions to finish the total work bout as quickly as possible (as in running a race). The pattern of accumulated $O_2$ deficit versus work accomplished followed the predicted relationship by Medbø and associates. The temporal relationship also suggests that exercise duration of 2 to 5 min is associated with maximal values for the accumulated $O_2$ deficit. In a subsequent study, the accumulated $O_2$ deficit during a cycle ergometer ride to exhaustion at a power output requiring 110% of $\dot{V}O_2$max was compared with the accumulated $O_2$ deficit during a 2 km time trial on a bicycle attached to a wind-load

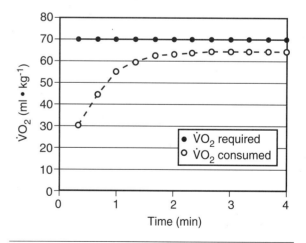

**Figure 6.2** Conceptual approach to the summated $O_2$ deficit based on difference between the estimated $O_2$ requirement for pedaling at a certain power output and the measured $O_2$ uptake.

simulator (Foley et al., 1991). Accumulated $O_2$ deficit was similar during the trials. Reproducibility studies with both techniques suggested significant day-to-day variability in the accumulated $O_2$ deficit. These data suggest that these two approaches to measuring accumulated $O_2$ deficit can be used with more or less equivalent results.

The measurement of the accumulated $O_2$ deficit requires that the measured $\dot{V}O_2$ be subtracted from the required $\dot{V}O_2$ at several periods during the work bout and that the measures of $O_2$ deficit be summated (figure 6.2). In our experience, measurement periods of 20 to 30 s usually give satisfactory results, although in principle either breath-by-breath measurements or a single collection could be made. Remember that accumulated $O_2$ deficit is a quantity, not a rate. The raw data for $\dot{V}O_2$ required and $\dot{V}O_2$ measured are rates. If the collection period is exactly 1 min, rate and quantity will be numerically the same. If, on the other hand, one uses 20 s collections, the deficit during that period is one-third of the difference between the rate of $\dot{V}O_2$ required and $\dot{V}O_2$ measured. The most easily understood presentation is to express the accumulated $O_2$ deficit as an equivalent quantity of $O_2$ that was not consumed, normalized for the size of the individual, in millimeters per kilogram (ml/kg). Dividing the total work accomplished (joules) attributed to aerobic and anaerobic metabolism is also reasonable (Foster et al., in press).

Medbø and associates (Medbø & Burgers, 1990; Medbø et al., 1988; Medbø & Tabata, 1989) have discussed in some detail the importance and difficulty in estimating the $\dot{V}O_2$ requirement. Because the $\dot{V}O_2$ requirement is greater than $\dot{V}O_2$max, $\dot{V}O_2$ measured during submaximal steady-state exercise must be extrapolated. Accordingly, even small measurement errors in submaximal $\dot{V}O_2$ can lead to considerable errors in computing $\dot{V}O_2$ requirement and, accordingly, in the summated $O_2$ deficit. In the original study, Medbø et al. (1988) performed approximately 20 submaximal measurements, each involving a 2 min collection of expired air at the end of a 10 min work bout, a laborious process. Medbø et al. (1988) have discussed an abbreviated approach to the computation of the $O_2$ requirement, which others have used (Katz et al., 1989), including ourselves (Foley et al., 1991; Foster et al., 1989). The validity of the extrapolation method to compute aerobic requirements while using wind-load simulators that have nonlinear response characteristics, as is required for our time-trial method (Foster et al., 2003) or for the air-braked ergometer method of Withers et al. (1991) and Withers and Telford (1987), has not been independently verified. Similarly, suggestions

that running may have nonlinear aerobic requirements at higher velocities presented a problem for the use of running tests. The concept of steady-state $\dot{V}O_2$ is itself not a trivial problem. Wasserman et al. (1987), in discussing the concept of the ventilatory threshold, have suggested that above the exercise intensity associated with the ventilatory threshold, a progressive increase in $\dot{V}O_2$ occurs throughout the duration of a constant-power-output work bout. Accordingly, even if one uses rather long steady-state bouts, as suggested by Medbø and Burgers (1990), Medbø et al. (1988), Medbø and Tabata (1989), and Buck and McNaughton (1999), one either has to restrict the measurements to comparatively low intensities and assume the risk of a greater degree of extrapolation or has to accept the risk that steady-state measurements made above the ventilatory anaerobic threshold systematically underestimate the aerobic requirements and accordingly lead to underestimation in the subsequently calculated $O_2$ deficit.

The validity of using steady-state $\dot{V}O_2$ requirements to estimate the aerobic cost of the beginning moments of exercise is far from clearly established. Medbø and Burgers (1990) and Katz et al. (1989) had subjects drop onto an already running treadmill. It is unclear whether the aerobic requirement of accelerating to high velocity is reflected by the $\dot{V}O_2$ requirement extrapolated from submaximal measurements. During studies on the cycle ergometer, the torque requirements associated with accelerating to the required power output are subjectively very high. Thus, the effective aerobic requirement at the onset of tests of this nature may be poorly represented by extrapolation of submaximal aerobic requirements. Serresse et al. (1988, 1991) have opted for the conceptually similar, although much simpler, approach of measuring the aerobic requirements of a single submaximal workload and then assuming that this efficiency is maintained across the range of power outputs employed in the exhaustive exercise bout. For cycling, this is probably a justified assumption; for running, it may be problematic. With the time-trial studies, we have measured the average velocity over segments of the trial (100-200 m) and have used the steady-state $\dot{V}O_2$ requirement for this velocity in the computation of accumulated $O_2$ deficit. The deficit can then be calculated by summating across the entire ride the pattern of aerobic and anaerobic energetic expenditure (Foster et al., 2003; Spencer & Gastin, 2001) (figure 6.3). Medbø and Tabata (1989) have further discussed the necessity for correcting for desaturation of oxygen stores in the computation of accumulated $O_2$ deficit, although in practice many do not do this because of the necessity for making

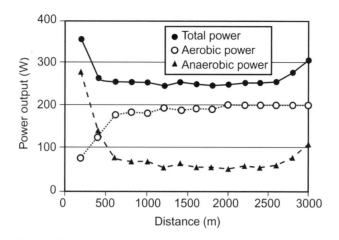

**Figure 6.3**   Power profile during 3,000 m time trial. Calculation of work attributable to anaerobic metabolism by subtracting the work attributable to aerobic metabolism from the total work output accomplished in a 3,000 m cycle time trial. In this approach, the subject is free to vary the power output on a moment-to-moment basis, with work accomplished calculated every 100 m.

assumptions about the magnitude of desaturation of oxygen stores. The $O_2$ stored in muscle and blood is thought to contribute approximately 10% to the maximal accumulated $O_2$ deficit.

## Summary

The measurement of peak and mean anaerobic power remains a difficult area. Several approaches, some as old as the stair-climbing test of Margaria et al. (1966), may provide an estimate of the highest achievable muscular power output during very brief (<5 s) time intervals. If one is willing to accept assumptions regarding the lack of contribution from aerobic metabolism, it may be argued that these tests can provide reasonable estimates of peak anaerobic power. In view of the data of Jacobs et al. (1983) that demonstrate how soon lactate begins to accumulate, partitioning this power output into alactic- and lactic-acid-dependent metabolic pathways may be inappropriate. With the development of nuclear magnetic resonance spectroscopy, perhaps this problem can be profitably readdressed in the near future.

Mean anaerobic power is more difficult to measure. Clearly, most of the performance-oriented tests that have traditionally been used for this purpose have serious deficiencies relative to the basic assumptions that must be made for the test to be used. Attempts to estimate mean anaerobic power using the accumulated $O_2$ deficit or other methods that rely on subtraction of work attributed to aerobic metabolism from total work accomplished are conceptually attractive. Many technical problems remain with the collection of this type of data and with the assumptions necessary to claim that the accumulated $O_2$ deficit represents a meaningful estimate of anaerobic capacity. Combining the ergometric approach adopted in the Quebec 90 s test (Bouchard et al., 1991) or the air-braked ergometer as suggested by Withers et al. (1991) and Withers and Telford (1987) with the accumulated $O_2$ deficit approach suggested by Medbø and associates (Medbø & Burgers, 1990; Medbø et al., 1988; Medbø & Tabata, 1989) might offer the best compromise and allow reasonable simultaneous estimates of both peak and mean anaerobic power, although the technique also appears to be suitable to competitive simulations (Foster et al., 2003; Spencer & Gastin, 2001).

# seven

# The Measurement of Human Mechanical Power

## Everett Harman, PhD

*Military Performance Division, USARIEM*

The ability to generate mechanical power is an important component of physical fitness. Evaluating it requires a special set of assessment tools. The purpose of this chapter is to describe the methodology for generating high-resolution records of human power output during various physical activities. Such measurement must be distinguished from standard anaerobic power testing, such as the Wingate test, the scores of which represent power output averaged over 1 s or longer. Many movements critical to sport, work, or ordinary daily life activities have extremely fast components characterized by high peak power levels over very short time intervals. Thus, an understanding of human physical capability and physical performance during a wide range of activities depends on the ability to generate high-resolution records of human power output. Achieving this goal requires the type of fast measurement systems and advanced data-processing techniques described in this chapter.

The ability to generate force is an important component of performance in most sports and other physically demanding activities (Fisher & Jensen, 1990; Komi & Hakkinen, 1988). Because the acceleration of an object is directly proportional to the force acting on the object divided by the mass of the object (Meriam & Kraige, 2002), an athlete who can generate high ground-reaction forces relative to body mass can change speed or direction quickly. Likewise, the ability to accelerate an external implement (e.g., golf club, baseball bat, javelin) is directly related to the athlete's ability to apply force to the implement. The ability to exert force is critical to "explosive" sports (those in which acceleration is of primary importance).

A logical conclusion is that because acceleration is proportional to applied force and "strength" describes the ability to generate force, "stronger" people should be faster. The fact that this is not necessarily true, particularly in sports activities involving high body segment velocities (e.g., tennis, baseball pitching, football kicking), reflects the limitations of methods commonly used to measure strength. Strength is usually defined as the maximum force produced during an isometric exertion or the maximum weight lifted in a particular movement (Atha, 1981; Enoka, 2002; Fisher & Jensen, 1990). Yet isometric or low-speed strength tests do not quantify the ability to exert force at high speeds, diminishing the relevance of such tests to the many sports and life activities in which the body, body segments, or implements must be accelerated while they are already moving rapidly.

The fact that strength is commonly evaluated by isometric or slow weightlifting tests most likely reflects the limitations in equipment available to physical educators, teachers, coaches, and others commonly engaged in strength testing. Even most sports science laboratories are not equipped to measure force exerted at relatively high speeds. Defining strength in easily testable terms has allowed a wide range of scientists and nonscientists alike to carry out strength testing. Unfortunately, the resulting strength scores have had limited usefulness for prediction of sports performance.

Knuttgen and Kraemer (1987) have suggested a broader definition of the term *strength*, that is, "the maximal force a muscle or muscle group can generate at a specified velocity." According to this definition, testing would have to be conducted over a wide range of velocities, during both concentric and eccentric muscle action, to get a complete picture of a subject's strength. Relatively expensive and sophisticated equipment would be needed either to test strength at a specific range of velocities or to monitor limb velocities during lifting movements. Despite the difficulty and expense involved, such testing would provide a set of strength scores more meaningfully related to sports ability. A major implication of the new definition of strength is that most commonly used strength tests provide only a partial view of the spectrum of an individual's strength.

The limited applicability of strength test scores obtained isometrically or at slow speed has led to a heightened interest in quantifying human power output, thereby accounting for both the force exerted and the speed at which it is applied. Outside the scientific realm, *power* as a physical attribute of living beings is considered synonymous with "vigor, energy, and the capacity for exerting mechanical force or doing work" (Abate, 1996). In the fields of science and engineering, however, *power* is specifically and precisely defined as "the time rate of doing work" (Meriam & Kraige, 2002). In the latter context, *work* is the product of the force exerted on an object and the distance the object moves in the direction in which the force is exerted. In keeping with that definition, work can be calculated as the area under a curve of force versus distance. This method is particularly useful when force or speed varies over time. Average power over the time interval is the calculated work divided by the time interval over which the force is measured.

The discrepancy between the general and scientific definitions of the term *power* has fostered misunderstanding and even conflict between sport researchers and practitioners. For example, the sport of powerlifting is an athletic competition in which heavy weights are lifted without regard to the rate of lifting. During powerlifting (squat, deadlift, and bench press), however, considerably less mechanical power is generated than during Olympic lifting (snatch and clean-and-jerk) or several other sports (Garhammer, 1989). To avoid ambiguity, researchers should use the term *power* strictly in its scientific sense, while acknowledging lay usage of terms such as *power* and *powerful*.

Unfortunately, the concepts of strength and power have become dichotomized, in the sense that *strength* is usually associated with slow speeds and *power* with high speeds of movement. *Strength* means maximal force, and both force and power can be measured at low speed, high speed, or any speed in between. Both relate to an individual's ability to do work at the speed tested. Power is a direct mathematical function of force and velocity. Therefore, if at any instant, any two of the variables force, velocity, and power are known, the third can be calculated. If we say that during a test of maximal knee extension at 200° per second, an individual can generate high force or high power, we are describing precisely the same ability, that is, the ability to exert force at that particular movement speed. If Knuttgen and Kraemer's concept of strength were accepted, power could be calculated from the strength score as the product of the force measured and the velocity at which it occurs. Strength and power testing would thus be concurrent.

This chapter describes how mechanical power and force can be precisely and instantaneously measured during a wide variety of human activities through use of electronic transducers, high-speed video cameras, and computers. Power output tests that do not involve instantaneous measurement, such as the Margaria and Wingate tests (Bar-Or, Dotan, Inbar, Rotstein, & Tesch, 1980; Margaria, Aghemo, and Rovelli, 1966), involve calculation of power averaged over a few seconds. Such tests lack the resolution to measure instantaneous peak power or power that varies rapidly over time (Lakomy, 1987). Neither can they measure rapidly varying forces.

This chapter includes a description of the quantitative foundation of power output measurement, a discussion of the necessary instrumentation and data-processing procedures, and a number of examples of how power output during various physical activities has been measured.

# Quantitative Foundation of Power Testing

The definitions of work and power in equation form follow:

$$\text{work} = \text{force} \times \text{distance} \tag{1}$$

$$\text{power} = \frac{\text{work}}{\text{time}} = \frac{\text{force} \times \text{distance}}{\text{time}} \tag{2}$$

Because the preceding equation can be rewritten as

$$\text{power} = \text{force} \times \frac{\text{distance}}{\text{time}} \tag{3}$$

power can also be defined as force times velocity,

$$\text{power} = \text{force} \times \text{velocity} \tag{4}$$

or more precisely, as the product of the force exerted on an object and the velocity of the object in the direction in which the force is exerted. The same result is obtained when power is calculated as the product of the velocity of an object and the force exerted on it in the direction of its movement.

For the equations to be correct, we must use consistent units. In the International System of Units (SI) (Taylor, 2002), which is the worldwide standard for science and engineering, force is in newtons, distance is in meters, time is in seconds, work is in joules (newton-meters), and power is in watts (joules/second). When necessary, the appropriate input units can be obtained from those of other measurement systems using the factors listed in table 7.1.

Table 7.1   **Factors for Conversion of Common Measures to SI Units of Force and Distance**

| To get | Multiply | By |
| --- | --- | --- |
| Newtons | Pounds | 4.448 |
| Newtons | Kilogram force[1] | 9.807 |
| Meters | Feet | 0.3048 |
| Meters | Inches | 0.02540 |

[1]The kilogram of force is a customary unit in countries using the metric system and not a scientific unit. The newton remains the scientific unit of force. Yet scientific journals have frequently allowed the kilogram to be used as a unit of force. A kilogram reading on a spring or electronic scale is equivalent to 9.807 N of force. However, an actual kilogram of mass registers exactly 1 kilogram of force only if the weighing occurs where the acceleration of gravity happens to be 9.807 m/s$^2$. A balance scale accurately measures mass rather than force because variations in gravity affect the object being assessed as well as the balance masses. For example, a person who registers 70 kg on both a spring and balance scale on the earth if weighed on the moon will register 70 kg on the balance scale but only about 12 kg on the spring scale.

The above work and power equations apply to an object moving through space whose path is traced by the center of mass of the object, a point at which all the mass of the object could be concentrated without changing the path of the object in response to external forces. But the generation of work and power does not require that the center of mass of the object move through space at all, because work can result in rotation without translation. When force is applied to an object, translational work occurs when the object moves through space and rotational work occurs when the object rotates. Both kinds of work can occur at the same time. If the object does not move at all, the applied force results in no work. The equation for rotational work is the following:

$$\text{work} = \text{torque} \times \text{angular displacement} \qquad (5)$$

where torque is the product of a force acting on the object and the perpendicular distance from the line of action of the force to the point about which the object rotates, and angular displacement is the angle through which the object rotates. The equation for rotational power is the following:

$$\text{power} = \frac{\text{work}}{\text{time}} = \frac{\text{torque} \times \text{angular displacement}}{\text{time}} \qquad (6)$$

Because the above equation can be rewritten as

$$\text{power} = \text{torque} \times \frac{\text{angular displacement}}{\text{time}} \qquad (7)$$

power can also be defined as the product of torque and angular velocity.

$$\text{power} = \text{torque} \times \text{angular velocity} \qquad (8)$$

In SI, torque is in newton-meters, and angular displacement is in radians. Just as for translational movement, the work done in rotating an object is in joules, and power is in watts (Taylor, 2002). Angles in degrees can be converted to radians by dividing by 57.30°/radian.

To calculate power output during translational movement, force and velocity must be either monitored directly or calculated. Force can be measured with various types of transducers described in the instrumentation section that follows. Because velocity transducers are not as commonly available as are position transducers or accelerometers, velocity is usually calculated from position or acceleration. In the following sections we will explain indirect calculation of parameter values.

• **Calculating velocity from position or acceleration.** Position, velocity, and acceleration at any juncture can each be calculated from any of the others if position and velocity at the start of data collection are known. Velocity, a basic component of power, can be readily calculated from a time record of position because velocity equals change in position divided by the time interval of measurement. In the data collection system described later in this chapter, the time interval between successive data samples is known. Therefore, velocity throughout the movement can be easily calculated for each sample time interval using position transducer data and equation (9).

$$\text{velocity} = \Delta \, \text{position} / \Delta \, \text{time} \qquad (9)$$

If initial velocity is known, as when an object starts moving from a standstill, velocity can also be calculated from a time record of acceleration. Because acceleration equals change in velocity divided by the time interval of measurement, the change in velocity equals the product of acceleration and the time interval of measurement. Absolute velocity equals velocity at the start of the measurement interval plus the change in velocity. Thus

$$\text{velocity} = \text{initial velocity} + \left(\text{acceleration}\right)\left(\Delta \text{ time}\right) \quad (10)$$

To calculate the velocity throughout an entire movement, the movement must start from a known velocity. The simplest way to start with a known velocity is to begin data sampling before the subject has begun moving, when the velocity equals zero. The velocity at the end of the first sample interval is the velocity at the start of the interval plus the change in velocity over the interval. Similarly, the velocity at the end of each successive sample interval is the velocity at the end of the previous interval plus the change in velocity over the interval.

• **Calculating force from acceleration.** When the mass of an object is known and the movement of the object can be accurately monitored, it is not necessary to directly transduce force exerted on the object. The applied force can be determined from Newton's second law, which states that

$$\text{Force} = \text{mass} \times \text{acceleration} \quad (11)$$

• **Calculating acceleration from force.** If the mass of an object is known and the magnitude and direction of force on the object can be measured, the acceleration of the object can be determined through rearrangement of equation (11) to solve for acceleration:

$$\text{acceleration} = \frac{\text{force}}{\text{mass}} \quad (12)$$

• **Calculating torque from force.** If the magnitude and direction of the force exerted on an object can be measured, the resulting torque about any pivot point of interest can be determined. For example, figure 7.1 shows the force (F) exerted on a pivoted lever. If r is a line from the point of application of F to the pivot point, torque about the pivot can be calculated as r times the component of F perpendicular to r, or $r \cdot (F \cdot \cos\theta)$. When it is more convenient to do so, torque can also be calculated as the product of the force acting on the object and the perpendicular distance from the line of action of the force to the point about which the object rotates.

• **Calculating torque from angular acceleration.** The angular equivalent to equation (11) is

$$\text{torque} = \left(\text{moment of inertia}\right) \times \left(\text{angular acceleration}\right) \quad (13)$$

Therefore, if the moment of inertia is known, and angular acceleration (in radians/sec²) is measured, torque can be calculated. Standard methods are

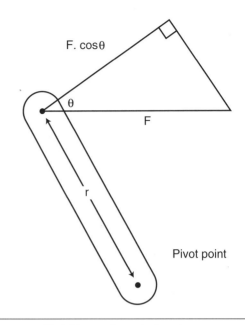

**Figure 7.1**   Variables used to calculate the torque resulting from an applied force (F). Torque about the pivot point is equal to $r \cdot (F \cdot \cos\theta)$.

available for calculating the moment of inertia of a limb (Winter, 1990) or an object (Meriam and Kraige, 2002). Devices are available that can directly measure moment of inertia, such as the Moment of Inertia Device XR-250 (Space Electronics, Berlin, Connecticut).

# Testing Strategy and Test Results

An electronic transducer is a device that produces an electrical signal, most often a voltage, proportional to the magnitude of a parameter of interest. To calculate power output for translational movements (e.g., jumping, running, lifting), transducers are needed to monitor force and velocity. For rotational movements (e.g., cycling) transducers are needed to monitor torque and angular velocity. As described in the previous section, not all these parameters must be measured directly.

The transducer output signal is passed through a signal conditioning device if amplification is necessary or to filter out unwanted electrical noise. If the transducer emits a strong, clean signal this stage is not necessary.

The signal, coming directly from the transducer or from the signal conditioner, is converted by a computer interface device into information compatible with the computer. The interface device may be a freestanding box connected to the computer by cable, but it is often a board that plugs directly into an expansion slot of the computer. The better computer interface devices have some amplification capacity, which can eliminate the need for separate amplifiers, and signal conditioning can be done by software, eliminating the need for a separate hardware signal conditioning stage.

The computer takes the data from the interface device and uses calibration values supplied by the transducer manufacturer or laboratory calibration values to calculate power output with the equations presented earlier. The results may be displayed directly on the screen, printed, or stored in the computer for further processing and statistical analysis.

The system is set up to monitor or "sample" all transduced parameters at known frequencies (e.g., 500 Hz [500 times per second]). The time interval between samples is the mathematical inverse of the sampling frequency (e.g., at a sampling frequency of 500 Hz, the intersample interval is 1/500th of a second). Using calibration coefficients, the computer converts the output of the computer interface device into real units of force, distance, and so on. The directly measured values are usually used to calculate the values of additional variables not directly measured at each point in time. The computer produces a history, rather than a single score, for power output and, if desired, for the following variables as well:

| Linear | Angular |
| --- | --- |
| Force | Torque |
| Linear position | Angular position |
| Linear velocity | Angular velocity |
| Linear acceleration | Angular acceleration |
| Force rate of change | Torque rate of change |

Peaks and their times of occurrence can be determined for power and the other variables. Average magnitude for any of the variables can be calculated over an entire movement or for various movement phases (e.g., push-off phase of the vertical jump). In addition, linear and angular work can be calculated and joint range of motion can be determined. Linear impulse (the area under the force-versus-time curve) and angular impulse (the area under the torque-versus-time curve) may be of interest as well.

# Instrumentation

The following is a more detailed description of components that can be used to assemble the type of power output measurement system described earlier.

This compilation is not intended to be exhaustive. Many electronic and mechanical components not mentioned can be adapted for power output measurement. After becoming familiar with the basic methodology, readers should be able to develop power measurement systems suited to their specific needs.

# Transducers

The calculation of human mechanical power using equations (4) and (8) requires knowledge of force and velocity for translational movement, and torque and angular velocity for rotational movement. Transducers must be able to monitor these variables directly or provide information from which they can be calculated.

Most transducers produce voltages proportional to the magnitude of the parameter being transduced. Several factors must be considered when selecting a transducer:

• **Resolution.** Resolution is the smallest change in the measured parameter that can be detected by the transducer. For human power output testing, a minimum resolution of 1/10 of 1% of full scale is desirable. For example, a force transducer that can measure up to 2,000 newtons should be able to distinguish changes in force as small as 2 newtons.

• **Measurement range.** The transducer must be able to register the minimum and maximum parameter value that might be encountered. Overestimating the range is safer than underestimating it. Transducers can either be damaged or produce meaningless results when subjected to conditions beyond the range for which they are designed. On the other hand, using a transducer whose range is much greater than that expected (e.g., using a 2,000 g accelerometer to measure expected accelerations in the range of 5 g) can result in poor resolution.

• **Accuracy.** Accuracy is the maximum amount the transducer signal can be expected to deviate from the correct output. Standards for accuracy are similar to those for resolution. A transducer can have high resolution but poor accuracy if it can register small changes but produces absolute values that deviate from real-world values.

• **Thermal effects.** Change in the output signal may be caused by variations in temperature rather than by changes in the measured parameter. If, within the range of temperatures expected under laboratory conditions (including the heat produced by the electronic measurement circuitry), the thermal effects are greater than the desired level of accuracy, special temperature compensation circuitry may be necessary. Frequently, plans for such circuitry come with the transducer. Temperature compensation circuitry may not be necessary under typical laboratory conditions in which ambient temperature remains relatively constant, electronic devices are turned on at least an hour before the experiment to allow temperatures to stabilize, and any necessary calibration adjustments are made just before the experiment.

• **Excitation voltage.** Most transducers require a voltage source to excite them. It is convenient if the transducer accepts a voltage level typical of common power supplies, such as 5, 12, or 15 V. The lower the required voltage, the easier it is to use batteries instead of AC power. If batteries are used instead of a commercial power supply, some regulation circuitry is required to make sure that the excitation voltage remains constant.

• **Output voltage.** The full-scale transducer output signal should approach but not exceed 5 V because the most common input voltage range for devices used to interface the transducer to the computer is ±5 V. If the transducer output can exceed the allowable input range of the interface device, then the signal must be reduced by a fixed proportion before being fed into the computer interface device. If the signal is much lower than the allowable input limit, it must be amplified to prevent loss of resolution. The better computer interface devices provide choices of input voltage ranges that can be specified in software or by manipulating switches or jumper wires on the board itself, reducing the need for an extra stage for amplifying or reducing the signal.

• **Frequency response.** Frequency response refers to the highest rate at which the transducer is capable of keeping up with changes in the measured parameter. For power measurement, frequency response should be at least 100 Hz, and even higher for capturing very rapid bursts of power, as in high jumping. "Response time," which is the mathematical inverse of frequency response, is sometimes used to describe the same characteristic (e.g., a frequency response of 100 Hz is comparable to a response time of 0.01 s).

## Specific Transducers

The following are some transducers particularly useful for the measurement of human mechanical power. They either directly transduce the parameters needed for equations (4) and (8) or transduce variables from which torque and velocity or torque and angular velocity can be calculated.

### Load Cell

A load cell (figure 7.2) is a small device that can measure tension or compression force exerted on it. Two sources of these are Omegadyne (Stamford, Connecticut) and Entran Devices (Fairfield, New Jersey). Load cells suitable for power testing can be obtained for about $500.

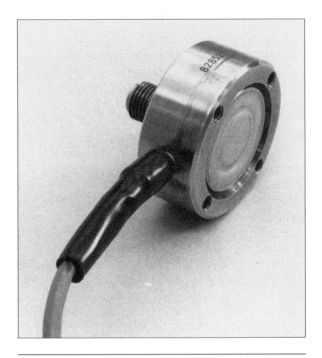

**Figure 7.2**    A load cell which, when provided with electrical excitation, emits a voltage proportional to the tension or compression force to which it is subjected.

### Force Platform

The force platform (figure 7.3) is a standard tool in biomechanics research, most often used for gait analysis. A typical force platform provides voltage signals proportional to forces exerted on the surface of the platform in the up–down, front–back, and left–right directions. Signals allowing the center of pressure to be located are usually provided as well. The devices are well suited to monitoring human power output during jumping. The most widely used force platforms are from Kistler (Amherst, New York) and AMTI (Watertown, Massachusetts). Prices for an AMTI platform with amplifier and other necessary hardware range from about $10,000 for a smaller platform to about $18,000 for a larger one. Software for the analysis of various activities including lifting and jumping adds $4,000 to $5,000. A competent computer programmer, however, can write software

**Figure 7.3** A set of two force platforms, each of which emits six voltage signals proportional to the forces exerted on it by the foot in the vertical, left–right, and fore–aft directions and the torques exerted on it by the foot about axes through the platform origin in the same three directions.

to do human power output analysis using the methods described throughout this chapter.

### Strain Gauges

Some force platforms and most load cells are based on the strain gauge, which is a small, inexpensive, foil-thin device that changes its electrical resistance as it is stretched or compressed. Scientists can have custom-tailored transducers made using strain gauges, or they can order such devices ready-made (Harman, 1989; Sargeant & Davies, 1977). To make a force transducer, the strain gauge is bonded to a piece of metal that bends imperceptibly when force is exerted on it. The bending stretches or compresses both the metal and the strain gauge, changing its electrical resistance. Specially designed electronic circuitry produces a voltage in proportion to the force exerted. Although strain gauge implementation can be somewhat difficult, proficiency can be developed with some effort and patience. Sources for strain gauges include Omegadyne (Stamford, Connecticut) and Entran Devices (Fairfield, New Jersey). The suppliers generally sell strain gauge kits that include the solutions for cleaning the metal to which the gauges are bonded and the bonding cement as well as step-by-step instructions. The gauges generally cost less than $10, so making mistakes while

learning to bond them is not costly. Advantages of making force transducers from strain gauges include low cost and great flexibility of design. The low cost of strain gauges makes them ideal for studies by students and for laboratories with limited budgets. Disadvantages include time and effort involved in making functional transducers that produce accurate, linear response and the difficulty of replicating the precision of commercial products.

### High-Speed Camera

High-speed cameras that can capture images at rates of from 50 to 1,000 frames per second (Hz) are standard equipment in biomechanics laboratories. Obviously, the faster units are considerably more expensive. Cameras that operate at 100 Hz are generally adequate for human power output measurement and other biomechanical analysis. Film-based systems have been largely superseded by video-based systems mainly because the latter include semiautomated data processing that greatly speeds analysis of the data. The process of handpicking joint centers from each frame of film has for the most part fallen into disuse because modern video systems automatically determine the location, in two-dimensional or three-dimensional coordinates, of reflective markers attached to the person or implement. In addition, videotape is much less expensive than film and there are no developing costs or delays. Companies that provide video-based motion analysis systems include Qualisys (Columbiaville, Michigan), Vicon (Lake Forest, California), Peak Performance (Englewood, Colorado), and Motion Analysis (Santa Rosa, California). Some of the devices store only the coordinates in space of reflective markers affixed to the body or implement, rather than entire images. Because these coordinates are stored directly in the computer, no videotape is required. Although researchers performing film analysis usually write their own programs, the video-based systems come with well-developed software packages that make the systems usable for people without extensive training. Disadvantages of video analysis systems in comparison with film-based systems include greater initial expense and lower resolution than that of film.

### Angular Position Transducer (Electrogoniometer)

An angular position transducer provides an effective and economical means of monitoring rotary movement (Chaffin, Chaffin, & Andersson, 1999). When affixed either to a device that rotates (exercise machine) or to a hinge-type body joint (knee, elbow), an angular position transducer can provide the angular position information necessary to calculate angular velocity for power output determination. An angular position transducer is particularly adaptable for custom-designed power output testing, and, as with film or video analysis equipment, it can be used without a force transducer to determine power output during the movement of an object of known mass or moment of inertia.

Electrogoniometers can be simply constructed from rotary potentiometers, which can be purchased in any electronics supply store. Rotary potentiometers are the familiar devices used to raise or lower volume on radios or make adjustments on other electrical implements by turning a knob. As its shaft is turned, the electrical resistance of the device changes. Radios use audio taper potentiometers, which are characterized by nonlinear relationships between electrical resistance and shaft knob position. The nonlinearity is needed because perceived loudness is not a linear function of sound energy. A linear taper potentiometer, however, is preferable for goniometry because it is desirable for electrical resistance to change in direct proportion to the amount of shaft rotation no matter where in the range of motion the movement occurs, allowing a linear equation to convert the digital values corresponding to transducer voltages to actual position values.

Standard computer interface devices accept voltage input signals but are not designed to directly measure the electrical resistance provided by variable resistors. Fortunately, a circuit can be constructed based on Ohm's law, equation (14), to put out a voltage signal directly proportional to the electrical resistance of a variable resistor (Diefenderfer & Holton, 1993).

$$\Delta\text{voltage} = \text{current} \times \text{resistance} \qquad (14)$$

The units for voltage, current, and resistance are respectively volts, amperes, and ohms. Using the analogy of a fluid, voltage can be thought of as pressure, current as flow rate, and resistance as restriction to flow provided by the diameter and length of a pipe.

Figure 7.4 represents a rotary potentiometer. The curved zigzag line represents the resistor itself. The arrow represents a contact arm that rotates when the shaft is turned. The electrical current flowing from point A to point C does not change because the resistance and voltage between the two points are constant. But as the shaft is turned so that the arm contacts the resistor closer to or farther from C, the resistance and thus the voltage between B and C changes. The setup is called a voltage divider because the voltage drop between A and C is divided into one drop between A and B and another drop between B and C. The ratio of each of these voltage drops to the voltage drop across the entire resistor is the same as the ratio of the resistance across which the voltage

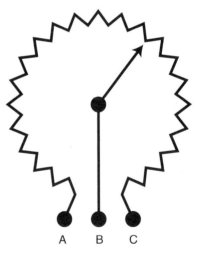

A　　　B　　　C

**Figure 7.4**　A linear taper rotary potentiometer used as an angular position transducer. The electrical resistance between B and C is directly proportional to the position of the potentiometer shaft. If the positive and negative terminals of a voltage source are connected to A and C respectively, the measured voltage between B and C can be used to determine shaft position from equation (15).

drop is measured to the entire resistance. Because the resistance between B and C is directly proportional to the position of the potentiometer shaft, the position of the shaft can be determined from the voltage between B and C. For example, if the voltages at points A and C are constant at 10 and 0 V respectively, a voltage reading at B of 3.8 V can be used to determine the potentiometer shaft angular position ($\mu$) relative to the right horizontal as follows:

$$\theta = 360° \times \frac{(3.8v - 0.0v)}{(10.0v - 0.0v)} = 136.8° \qquad (15)$$

In actuality, any standard potentiometer has a dead zone, which is an electrically nonconductive segment of the angular range of the potentiometer. The dead zone occurs in the shaft angular range where potentiometer resistance changes from highest to lowest, and usually subsumes at least 3°. To find the extent of the dead zone, the potentiometer should be connected to a resistance meter to determine the angular range over which resistance actually changes (e.g., 354°). The actual range can then be substituted for 360° in equation (15). If possible, the dead zone should be avoided. For example, for testing the knee-extension movement, which cannot occur over more than 180°, the potentiometer can be arranged so that the movement never crosses the dead zone.

Angular position transducers are available that directly emit digital signals, without passing through analog-to-digital converters, and may be a good choice in our increasingly digital world. These transducers can be highly precise and have no dead zones but are generally more expensive than rotary potentiometers. Also available are rotary transducers best adapted to measuring change in angle rather than absolute angle. Some of them incorporate an internal disk of alternating light and dark bands. As the shaft is turned, the transducer puts out a pulse each time a band is passed. When connected to a frequency-to-voltage converter (described later), this type of transducer is good for monitoring angular velocity, although it is not easily used for encoding absolute angular position.

Over the past few years commercially made electrogoniometers have become readily available. Although considerably more expensive than homemade units, they can assure precision within 1° of rotation and save the time involved in fabricating units in-house. Commercial sources include Biometrics Ltd. (Ladysmith, Virginia), Signo Motus (Messina, Italy), and MIE Medical Research Ltd. (Leeds, United Kingdom).

## Linear Position Transducer

A useful series of position transducers is made by Celesco (Canoga Park, California; figure 7.5). A model adequate for a variety of human power output testing costs about $500 and consists of a compact box containing a spooled, thin, flexible steel cable, the end of which protrudes from a hole in the box. The device puts out a voltage proportional to the distance that the cable is pulled out. A spring on the spool keeps enough tension on the cable so that it recedes back into the box when an external force is not pulling on it. The device can easily be used to monitor the location of anything moved in a straight line. If cost is a factor, a similar device can be constructed based on an inexpensive linear taper potentiometer described earlier and a cable wrapped around a spring-loaded spool. However, obtaining the same level of precision and durability in a homemade transducer as in a commercially produced unit can be difficult.

**Figure 7.5** A linear position transducer that emits a voltage signal proportional to the distance that the cable is pulled out.

## Accelerometer

When provided with excitation voltage, an accelerometer (figure 7.6) produces a voltage proportional to the acceleration that it experiences. The accelerometer is usually less than 1 cubic in. (16.4 cubic cm) in size, and many current models are much smaller. Some accelerometers contain a small mass supported by a tiny beam. When the container is accelerated, the inertia of the mass bends the beam in proportion to the acceleration. A strain gauge circuit translates the bending into a voltage. Other accelerometers are based on piezoelectric crystals, which generate a charge when force is exerted on them. High-quality accelerometers that measure acceleration in three axes run in the $500 to $2,000 range and can be obtained from sources such as Entran (Fairfield, New Jersey), Omega (Stamford, Connecticut), and Columbia Research Laboratories (Woodlyn, Pennsylvania). Less expensive accelerometers, often used for student research, can be obtained from Vernier (Beaverton, Oregon) and IC Sensors (Milpitas, California). For applications in which very small accelerometers are needed, those made with microelectro mechanical system (MEMS) technology are most appropriate. In such devices, the accelerometers are machined right onto silicon chips. Vendors include Crossbow Technology (San Jose, California) and Silicon Designs (Issaquah, Washington).

**Figure 7.6**   A precision triaxial accelerometer.

### Velocity Transducer

Transducers that produce voltage proportional to velocity are available. For power output determination, the advantages of transducing velocity directly rather than calculating it from position data include (1) reduction in computer programming requirements, (2) faster results, and (3) greater accuracy. Sources of velocity transducers include Trans-Tek (Ellington, Connecticut) and UniMeasure (Corvallis, Oregon). In addition, Lakomy reported that an electrical generator can be used as a velocity transducer. He observed excellent linearity of response, with an $r^2$ of .998 for the output voltage of the

generator versus the angular velocity of its shaft (Lakomy, 1986).

## Signal Conditioning Devices

If a transducer produces a strong, clean signal, no signal conditioning is necessary. A low-amplitude transducer signal, however, must be boosted to the input range of the analog-to-digital converter board, and a signal with significant electronic noise must be filtered. Whenever possible, it is best to avoid the need for conditioning by selecting transducers with desirable specifications and by shielding them and their connecting wires from sources of electronic noise.

Each transducer that requires signal conditioning must be attached to its own amplifier or filter. Multichannel chart recorders with built-in amplifier–filter units are sometimes available in laboratories and can be used to condition transducer signals without using the chart recorder itself. Multichannel instrumentation amplifiers can be purchased from companies such as National Instruments (Austin, Texas) and Analog Devices (Norwood, Massachusetts).

A good way to avoid buying expensive filter–amplifiers is to perform filtering in software rather than hardware and to use small, low-cost single-chip instrumentation amplifiers. Computer software libraries that contain digital-filter, cubic-spline, or other data-smoothing routines appropriate for filtering somewhat noisy data are available from sources such as MathWorks (Natick, Massachusetts) and Visual Numerics (San Ramon, California). In addition, software designed to facilitate the collection of data from analog-to-digital converter boards, such as LabVIEW from National Instruments (Austin, Texas), generally include filtering software. Filtering in software rather than hardware has the advantages of (1) allowing the experimenter to keep the unfiltered raw data, which may be subject to new analysis later, (2) providing great flexibility in degree and type of filtering, and (3) costing less than hardware. Instrumentation amplifiers on a chip, from sources such as Analog Devices (Norwood, Massachusetts), Texas Instruments (Dallas, Texas), and Linear Technologies (New York, New York), are economical and effective. To make these units functional, some electronics experience is helpful for assembling a box containing the chip, power supply, and input-output wiring.

## Computer Interface Devices

Information is transmitted to and received from the computer by interface devices, the most common of which are keyboards and monitors. Because the entry of long streams of data into computers by keyboards is prohibitively slow and labor intensive, other interface devices have been developed to meet the needs of modern scientific laboratories. The interface devices described in the sections that follow greatly enhance the speed and facility of entering data from experimental devices and, given the appropriate experimental setup, allow immediate processing and feedback.

### Analog-to-Digital Converter

The analog-to-digital converter translates a voltage into a numerical value that the computer can read. The device is an essential component of a computerized laboratory because a great majority of transducers put out voltage signals proportional to measured parameters. Although the converter is available in the form of a freestanding box that connects to the computer by cable, a more economical and compact form is that of a board that plugs directly into the expansion slot of a desktop computer (figure 7.7). The plug-in boards are fully functional and as a rule are more compact and less expensive than the standalone units.

All standard computers are digital, meaning that information is transmitted and processed in bits, including all information entering or exiting the central processing unit of the computer. Each bit can assume only two possible values, represented by either a higher or a lower voltage on an electrical wire. The two voltage levels are logically represented as 0 for low and 1 for high. In a typical desktop computer, a character is coded by 8 bits and an integer number by 32 bits.

A set of $n$ wires, each carrying a high or low voltage representing a logical 1 or logical 0, can represent in binary code $2^n$ different numbers. For instance, 16 bits can form $2^{16}$ = 65,536 different combinations of ones and zeros. Most desktop computers today transmit bits in groups of 32, although 64-bit desktop machines are currently on the market. The computer can handle very large numbers and numbers requiring many decimal places of precision by using more than one group of bits to encode a number. The analog-to-digital converter serves the function of converting a voltage, which can vary in infinitely small increments, into a binary number that the computer can read. Table 7.2 shows how the computer represents some numerical values.

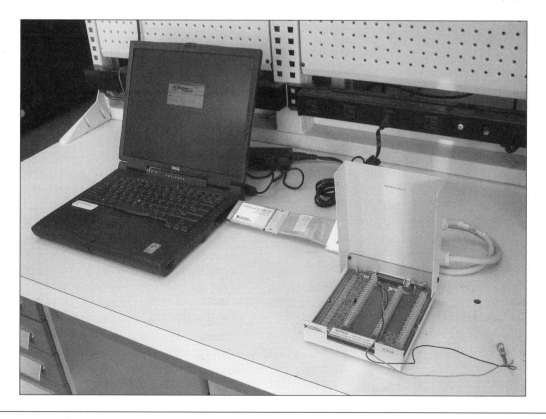

**Figure 7.7**    An analog-to-digital converter card ready to be plugged into a laptop computer. The breakout box on the right links the transducer output wires to the appropriate pin on the ribbon cable connector that leads to the analog-to-digital converter board.

Table 7.2 **Binary (Base 2) Equivalents of Decimal (Base 10) Numbers Using 16 Digits[1]**

| Decimal | Binary |
| --- | --- |
| 1 | 0000000000000001 |
| 2 | 0000000000000010 |
| 3 | 0000000000000011 |
| 4 | 0000000000000100 |
| 5 | 0000000000000101 |
| 6 | 0000000000000110 |
| 7 | 0000000000000111 |
| 8 | 0000000000001000 |
| 9 | 0000000000001001 |
| 10 | 0000000000001011 |
| 65,535 | 1111111111111111 |

[1]In digital electronic devices each 1 represents a bit at a higher voltage and each 0 represents a bit at a lower voltage. A one in the $n^{th}$ bit from the right has the value $2^{(n-1)}$, and n bits can represent $2^n$ different values.

The following are some characteristics that should be considered when purchasing an analog-to-digital converter board:

• **Number of inputs.** The number of transducers that can be connected to one board generally ranges from 1 to 16. For general use, it is best to have 16 channels. If economy is an important consideration for a dedicated system involving only one transducer, a single-channel board may be appropriate.

• **Data collection speed.** The overall speed at which the board can sample and translate voltages is the data collection speed. For example, each of 8 transducers connected to a board that is rated at 10,000 Hz can each be sampled as fast as 10,000 ÷ 8 = 1,250 times per second. For most laboratory testing, each transducer need not be sampled more than 500 times per second so that a board with an overall speed of 5,000 Hz can adequately monitor 10 transducers.

One advantage of a very fast board is that all transducers can be sampled almost simultaneously. For instance, with a 50,000 Hz board, the time between sampling of adjacent channels can be as little as 1/50,000th of a second. A convenient way to use such a fast board is to sweep all the channels at maximum speed but wait 1/100th of a second between sweeps. Because the time between sampling of adjacent channels is only 1/500th as long as the

time between successive sweeps of the channels, the different channels can be considered to be sampled simultaneously. Faster but more expensive boards are available if needed.

When a relatively slow board is used, the time delay between sampling of the different channels becomes significant, and mathematical interpolation should be incorporated into the analysis program to determine what the values of all the transduced variables would have been had they been sampled simultaneously.

• **Input voltage range.** The more flexibility a board has in allowable voltage input, the better. A basic board might have a single input voltage range, say 0 to 5 V, whereas a more flexible board could have multiple ranges including −5 to +5 V and −10 to +10 V.

• **Resolution.** The most important consideration in a board may be resolution. Resolution is defined in bits. For example, 8 bits of resolution means that the full range of input voltage is translated into $2^8 = 256$ different discrete numbers. For example, when a transducer that emits 0 to 5 V in response to 0 to 5,000 newtons of force is connected to an 8-bit analog-to-digital converter board with a 0 to 5 V input range, the smallest change in force that can be registered is 5,000 N ÷ 256 = 19.5 N. If the input range of the board is −10 to +10 V, then only one-quarter of the range would be used, so that resolution would be limited to 5,000 N ÷ (256 ÷ 4) = 78.1 N. For most laboratory purposes, it is best to buy an analog-to-digital converter board with 12 to 16 bits of resolution, which translates the input voltage to between $2^{12}$ (4,096) and $2^{16}$ (65,536) discrete numbers. In addition, to make best use of the resolution of the board, the voltage output range of the transducer should match the voltage input range of the board. More precisely, the highest and lowest voltages expected in the experiment should be fairly close to the highest and lowest voltages that the board could translate. This objective can be accomplished by providing the appropriate excitation voltage to the transducer or amplifying its signal before feeding it into the board.

• **Overvoltage protection.** Even if care is taken, it is not unusual for a board to be accidentally exposed to a higher input voltage than it can translate. Ideally, then, a board should be able to withstand input at least 50% above the highest translatable voltage without being damaged.

• **Gain.** Gain allows the input signal to be amplified to match the input range of the board. A basic board does not have this feature. A full-featured board might be able to multiply the input signal by 1, 10, 100, or 1,000. In combination with multiple input voltage ranges, this gives the experimenter much flexibility and greatly reduces the chance that separate amplifiers will be needed.

- **Digital inputs and outputs.** Several boards have the additional capability of sensing and emitting the digital numbers described previously. This feature is useful for allowing the computer to sense when an event marker switch is tripped or to turn on an external device at a particular time.

- **Input impedance.** A voltage measuring device with a high-input impedance allows only a tiny amount of current to flow into its inputs. This attribute is important because the greater the flow of current into the measuring device, the greater the error in the measured voltage. In general, an input impedance of at least 1 megohm is desirable.

- **Triggering.** A board starts sampling and translating voltages as soon as it is triggered. Some boards can be triggered only by a computer program. Others can also be triggered by an external switch. External triggering is a useful feature because it allows more flexibility in how an experiment or testing device is set up and it facilitates automation. An experimenter may wish to configure a light beam switch or contact pad so that the movement of the test subject triggers data collection.

Some companies that produce analog-to-digital converter boards for PC-compatible computers are National Instruments (Austin, Texas), Data Translation (Marlboro, Massachusetts), Microstar (Bellevue, Washington), and Real Time Devices (State College, Pennsylvania). The boards generally vary in cost between $400 and $1,800. The more expensive models generally have 16-bit resolution, higher overall data acquisition speed, more input channels, and some other features that enhance the power and flexibility of data collection. Most of the less expensive boards, however, have good accuracy and functionality. The boards can usually be purchased with software, which minimizes the time and expense of developing custom data collection programs. The better programs allow mathematical manipulation of the data as they are collected.

### Digitizer Tablet

Digitizer tablets were essential equipment for biomechanics laboratories when 16 mm film was the standard medium for capturing human motion data, but they have been used much less because video-based motion analysis systems have become standard. They can still be used, however, when film of activities of interest is available. To get information from the film record of an activity into the computer, the film is projected frame by frame onto a digitizing tablet or table (figure 7.8). When the researcher places

**Figure 7.8** A rear-projection digitizing table. A slide or film image is projected from the rear onto the translucent glass. When the researcher places a cursor or stylus over each point of interest in the image and presses a button, the x and y digitizer coordinates of the point are fed to a computer.

a cursor or stylus over each point of interest in the film image and pushes a button, the x and y digitizer coordinates of the point are fed to a computer. The computer file containing the x and y coordinates of all points digitized in every frame can be processed by a program to produce histories of various parameters, including power output.

Digitizing tablets can also be used to analyze power output using a multiple-image still camera slide or print taken of an activity of interest using a light strobe. For example, if a subject throws a gel ball as hard as possible while a strobe light flashes several times in succession in front of a camera with its shutter open, a photographic image is created showing multiple images of the ball on one slide or print. The photograph can be placed or projected onto the digitizing tablet or table, and the successive images of the ball can be digitized. By also digitizing the image of an object of known length in the photograph, the position of the ball in each image can be determined. By knowing the mass of the ball, power output during the throw can be determined. The velocity of the ball between successive images based on change in ball position is calculated using equation (9). Acceleration of the ball between successive velocities is calculated as the change in velocity divided by the change in time. Force is determined as the product of the mass of the ball and its acceleration using equation (11).

Sources of digitizing tablets include Numonics (Montgomeryville, Pennsylvania) and GTCO Calcomp (Columbia, Maryland). Prices generally range from $1,500 to $10,000 depending mainly upon the surface area and accuracy of the tablet. Tablet size ranges from about 12 in. by 17 in. (30 cm by 43 cm) to 44 in. by 62 in. (112 cm by 155 cm). Accuracy of at least 1/100th of an inch (0.25 mm) is desirable. A tablet with a translucent surface is desirable because it allows images to be projected from behind the tablet rather than from above it, thus eliminating shadows cast by the operator's hand or body during digitizing.

### Frequency-to-Voltage Converter

A frequency-to-voltage converter puts out a voltage proportional to the rate at which the input signal switches cycles between a higher and lower level. This device is useful for translating the output of pulse-emitting transducers, such as the light and dark band rotary transducer described earlier, into a voltage to be fed into an analog-to-digital converter. The faster the shaft of the rotary transducer is turned, the higher the frequency of the output pulses. The frequency-to-voltage converter can be used to monitor velocity in any experimental setup in which the testing equipment is designed to put out one pulse for each specific increment of movement. Cycles, rowing ergometers, and other devices can be instrumented without difficulty in this way. The frequency-to-voltage converter is available as a flexible, ready-to-use desktop device costing several hundred dollars from companies such as Encore Electronics (Saratoga Springs, New York) or as a very economical (under $10) microchip from companies such as National Semiconductor (Santa Clara, California) or Analog Devices (Norwood, Massachusetts). The latter option requires some additional circuitry to make it operational.

### Mathematical Circuitry

Velocity, a convenient quantity for power output determination, is often calculated indirectly from position or acceleration. Although not difficult, such calculations do require computer time. In cases where very rapid turnaround is needed, the translation of position or acceleration data to velocity can be accomplished in hardware. Velocity can be obtained from position data through mathematical differentiation or from acceleration data through integration. Electronics textbooks describe simple circuits that accomplish both differentiation and integration by using an inexpensive and widely available electronic component called an operational amplifier (Diefenderfer and Holton, 1993), available from companies such as National Semiconductor (Santa Clara, California) or Analog Devices (Norwood, Massachusetts).

## Computer

Many computer hardware and software options are available to the scientist, some of which are appropriate for laboratory testing of human power output.

In terms of hardware, any computer capable of being interfaced to transducers can be used for power output testing. Both Apple and PC-compatible computers are appropriate for general laboratory use because of their power, low cost, and available software and interface devices. Most modern desktop computers are fully capable of data collection through the use of interface cards inserted in expansion slots, and their speed is adequate to process the data without much delay. Functional interface cards are even available for laptop computers.

Limited commercial software is available for human power output calculation, and it is generally sold along with relatively costly video motion analysis or force platform systems. But it is not difficult to write the necessary programs using the steps described in the following section.

## Data Collection and Analysis

When collecting data with an analog-to-digital converter board, the user must specify various parameters such as

- channels to be sampled,
- order of sampling,
- degree of amplification of each channel,
- sampling speed,
- total time of sampling, and
- type of triggering.

The number of setup options varies according to the individual board. The most expensive boards often have the most options as well as the best specifications. Improved software has made it increasingly easy to set up computerized data collection, and some boards come with extensive software. Many systems use pull-down menus and thus do not require the user to write a program to set up and communicate with the analog-to-digital converter board. Stand-alone programs can be purchased separately. Popular programs with wide-ranging capabilities include LabVIEW (National Instruments, Austin, Texas) and MATLAB (MathWorks, Natick, Massachusetts).

When experimental data collection is triggered, the analog-to-digital converter board samples voltages from the specified channels and converts them to binary coded integers to be read by the computer. A file is produced in which the number of values equals the number of channels times the sampling rate per channel (Hz) times the total sampling time. Files can become quite large. For example, if two channels are sampled at 500 Hz each for 3 s, the data file contains 3,000 numbers. The most convenient way to arrange the data file is to have each column represent a different transducer and each row represent a time point.

Analysis of the data in terms of power output is easiest when (1) the transducers directly monitor force and velocity, (2) the transducer signals emit negligible electronic noise, and (3) the analog-to-digital converter board is set up to take readings from the two transducers virtually simultaneously. In such cases the computer program need only implement a loop structure to read each succeeding line, convert the analog-to-digital converter board output units into meaningful values of force and velocity, and calculate power at each time point as the product of force and velocity. Implementing such calculations and graphing the results on the computer screen can be easily implemented using LabVIEW (National Instruments, Austin, Texas) or MATLAB (MathWorks, Natick, Massachusetts).

If position is transduced instead of velocity, the program must determine the velocity between each pair of successive positions as the change in position divided by the change in time. Each resulting velocity can be considered to occur at a time midway between the times corresponding to the two positions from which it was calculated. For example, if force and velocity were monitored at 100 Hz, the data file would contain values for both variables at 0.000 s, 0.010 s, 0.020 s, 0.030 s, and so on. The velocity calculated between 0.020 s and 0.030 s would be considered to occur at 0.025 s, which is a time at which force was not measured. Adjacent velocities would have to be averaged to get velocities corresponding to times at which force was actually measured. For example, the velocities for times 0.015 and 0.025 would be averaged to get the velocity at 0.020 s. As alternatives to averaging, curve fitting and nonlinear interpolation are sometimes used to find the velocities at the desired times, but this is unnecessary for rapid rates of data collection.

Electronic noise should be avoided, if possible, by careful selection of transducers. In addition, some degree of noise can be eliminated in software through mathematical smoothing. Descriptions of the programming steps needed to write a digital filter smoothing routine are available (Press, Teukolsky, Vetterling, and Flannery, 2002). Alternatively, the smoothing and curve-fitting computer subroutines, included in the software from the companies mentioned above, can be used.

If the transducers are not sampled virtually simultaneously, linear or nonlinear interpolation can be used to estimate what the transducer readings would be if all transducers were sampled at the same time. For example, if channels are sampled at the times depicted in table 7.3, then interpolation would result in values for all channels occurring at 0.020, 0.050, 0.080 s, and so on. Channel C is actually measured at those times and does not require interpolation. Using linear interpolation, the reading for channel A at 0.020 s would be two-thirds of the way between its readings at 0.000 and 0.030 s. The reading for channel B at 0.020 s would be one-third of the way between its readings at 0.010 and 0.040 s. The same steps would produce simulated simultaneous readings over the entire sampling period except for very short periods at the start and end of data collection lost in the interpolation process. The small data loss is no problem because data collection always starts somewhat before and ends somewhat after the monitored activity. Alternatively, for nonlinear interpolation, a curvilinear equation can be fit to the data from each channel and used to solve for readings at the desired times.

Table 7.3 **Analog-to-Digital Board Sampling Pattern Requiring Interpolation for Simulation of Simultaneous Sampling**

| Time (s) | Channel |
|----------|---------|
| 0.000 | A |
| 0.010 | B |
| 0.020 | C |
| 0.030 | A |
| 0.040 | B |
| 0.050 | C |
| 0.060 | A |
| 0.070 | B |
| 0.080 | C |

## Computer Languages

The programming languages most widely used on microcomputers are C, C++, and Java. These languages are powerful but require considerable time to learn to use well. Also available are computer programming tools such as MATLAB (MathWorks, Natick, Massachusetts) that have many functions that the user can call up for mathematical computation, analysis, and visualization. Although such systems are not true languages, they do require some learning and practice to be able to write the proper instructions to apply the tools. A set of tools like LabVIEW (National Instruments, Austin, Texas), designed to facilitate the collection of data from analog-to-digital converter boards, gives the user the opportunity to do some data processing without writing programs. In general, computer programming languages such as C++ provide the greatest flexibility of function but also require the greatest amount of time to learn to use and to develop effective programs. Software toolboxes require less learning and development time but have some limitations in what they can do. For most laboratory applications, the toolboxes are more than adequate.

Data collection and analysis can be performed without knowledge of computer programming. Software for data collection can generally be purchased along with an analog-to-digital converter board. After the data are collected, the resulting computer file can be processed to produce meaningful information using commercially available spreadsheet programs such as Excel (Microsoft, Redmond, Washington) and Lotus 1-2-3 (IBM, White Plains, New York).

In the experimental data file, each column usually corresponds to a different transduced variable, and each row corresponds to a point in time. The spreadsheet analysis consists of specifying mathematical formulas to calculate new columns of data from the data in existing columns. Such processing can produce records of force and power from the transduced data.

## System Calibration

Some transducers are precisely calibrated in the factory. Often, each individual transducer comes with its own set of factors for conversion of output to meaningful units. For example, an accelerometer might output 5.8 mV per g of acceleration per volt excitation. Another one with the same model number might output 6.1 mV per g per volt. The different figures result from limitations in the precision of the manufacturing process. Factory calibration of transducers is useful because (1) companies usually have the resources to provide elaborate and accurate calibration devices, (2) the user is spared the time and effort of calibration, and (3) factories have the capability to perform dynamic calibration, whereas most laboratories are equipped only for static calibration. Dynamic calibration is best because, for power measurements, transducers must monitor rapidly changing phenomena.

The following example shows how to develop an equation for translation of transducer output into meaningful units if factory calibration figures are provided. Let us say that the force transducer used in a particular experiment is factory calibrated at 15 $\mu$V per newton. The particular 12-bit analog-to-digital converter board has user-accessible jumper wires that allow the board to be set to accept voltage ranges of 0 to 5, 0 to 10, –5 to +5, or –10 to +10. It can also multiply the signal amplitude by 1, 10, 100, or 1,000. If the maximum force anticipated in the experiment is 5,000 newtons, then the maximum transducer output voltage would be 5,000 × 0.000015 V = 0.075 V. Choosing a board amplification of 100 would bring the maximum signal to 7.5 V. The best approach would be to set the input voltage range of the board to 0 to +10 V, which would most closely match but not be exceeded by the anticipated input. Twelve bits encode $2^{12}$ = 4,096 different binary numbers. But only part of the input voltage range is used. The following equation can be used to produce a factor to convert from the analog-to-digital converter board output values read by the computer, which are in machine units ($\mu$), to actual units of force:

$$\frac{5{,}000 \text{ newtons}}{7.5 \text{ volts}} \times \frac{10.0 \text{ volts}}{4{,}096\,\mu} = 1.63 \text{ newtons per } \mu \quad (16)$$

The resulting number is both the resolution of the system and a conversion factor. To calculate the force in newtons exerted on the transducer, the program must multiply the $\mu$ values read by the computer by 1.63. For example, a computer value of 1,756 $\mu$ corresponds to a force on the transducer of $(1.63\ \text{N}/\mu)(1{,}756\mu) = 2{,}862$ N.

For transducers that must be calibrated by the user, trials can be carried out in which known inputs are provided to the transducer either over its entire range or over the maximum range in which it will be used. If the response of the transducer is fairly linear (i.e., its output is proportional to the transduced parameter), the set of real and machine unit pairs can be processed by a statistical regression program to produce a linear equation with slope and intercept, the purpose of which is to convert machine units into meaningful measurement values. Having a zero intercept is best so that the machine units need merely be multiplied by a constant to produce meaningful results, as is generally the case for factory-calibrated transducers. If the correlation between real and machine units is high (above .95), then a linear equation can provide relatively accurate measurement values. If possible, a transducer should be designed to have a linear response. Although nonlinear output of a transducer is generally considered undesirable, nonlinear curve fitting, a feature of the commercial data collection and analysis software mentioned earlier, may effectively compensate for transducer nonlinearity.

Sometimes a transducer can be highly linear but have a slope and intercept that vary slightly. In such case, calibration should be performed before each testing session, with only two calibration measurements needed to determine the equation for converting machine units into meaningful values. The calibration procedure involves subjecting the transducer to two known stimuli, one at the low end and one at the high end of the anticipated measurement range, and recording the machine unit readings. The equation to convert from machine units to real values is then

$$Q_t = Q_L + \frac{\mu_t - \mu_L}{\mu_H - \mu_L} \times \left(Q_H - Q_L\right) \quad (17)$$

where $Q_t$ = real quantity measured at time $t$

$Q_L$ = real quantity to which transducer is subjected during low calibration measurement

$Q_H$ = real quantity to which transducer is subjected during high calibration measurement

$\mu_t$ = machine unit reading at time $t$

$\mu_L$ = machine unit reading during low calibration measurement

$\mu_H$ = machine unit reading during high calibration measurement

The equation must be used to convert each $\mu$ value into meaningful units. Each conversion involves an addition, three subtractions, and a multiplication. Because the computer may have to convert thousands of values, processing can be slow. To speed up the calculations, it is best to rearrange the equation into the slope and intercept form below so that only one multiplication and one addition need be done per conversion. The program must be written so that the slope and intercept are not recalculated each time a $\mu$ value is converted, but calculated only once, before the conversions of hundreds, thousands, or even millions of $\mu$ scores to meaningful values. In the following equation, the calculations within the parentheses are made first and the calculated values are written directly into the computer program so that the calculations involve only one multiplication and one addition.

$$Q_t = \left(\frac{Q_H - Q_L}{\mu_H - \mu_L}\right)\mu_t + \left(Q_L - \frac{\mu_L\left(Q_H - Q_L\right)}{\mu_H - \mu_L}\right) \quad (18)$$

## Specific Applications

The components and procedures described in the previous section can be used to measure power output in a variety of human activities. Sample applications such as video analysis, weightlifting, vertical jumping, isokinetic dynamometry, nonmotorized treadmill running, and cycling are described in the following sections.

## Video Analysis of Human Movement

The power output used in moving a sports implement such as a baseball or javelin is easily determined from video analysis. Force on the implement need not be measured directly because it can be determined from acceleration of the known mass of the implement.

The first step is to locate the center of mass of the implement. For a typical ball, the center of mass is at dead center of the ball. For a javelin, tennis racket, or baseball bat, the lengthwise location of the center of mass is the point at which the object balances along a knife edge. For an asymmetrical object such as a hockey stick, the center of mass can actually fall outside of the object. In that case, a somewhat more involved procedure is needed to locate the center of mass (Winter, 1990). Besides locating the center of mass of the implement, the mass of the object must be determined, preferably using a kilogram balance scale, which accurately measures mass no matter what the location.

The use of a spring or electronic scale, which actually measures force rather than mass, results in an error in measurement of up to a half percent of mass on the surface of the earth because of local variations in the gravitational force of the earth, an error level generally considered acceptable for this type of measurement. But precision mass measurement can be carried out anywhere on the earth using an accurate spring or electronic scale by first calculating the weight of the object in newtons from the pound or kilogram scale readings using the factors in table 7.1. The mass of the object in kilograms is then determined as the weight in newtons divided by the acceleration of gravity in m/s² at the location of testing (table 7.4). For experimentation in space, scales cannot be used at all, and special devices must be employed to measure mass.

A record of the power output generated in throwing an object can be obtained from the velocity and acceleration of the center of mass of the object while the implement is in contact with the hand. If the object has a reflective coating, a video analysis system can automatically track the object. Otherwise, the image can be projected frame by frame onto the monitor, and the mouse can be used to identify the center of mass of the object. For the time corresponding to each film or video image, net force applied to the implement is calculated from the mass of the implement and acceleration using equation (11). The horizontal force exerted by the hand equals the net horizontal force calculated from the mass of the implement and horizontal acceleration. The vertical force exerted by the hand is equal to that calculated from the vertical acceleration plus the weight of the implement. Power transmitted to the implement at each instant is then calculated as the product of the component of force applied in the direction of the travel and velocity of the implement.

A test was developed in our laboratory to determine the external force and power generated during very rapid elbow flexion and extension. While lying supine, subjects were videotaped at 60 Hz as each horizontally threw balls of 0.91, 1.81, 3.63, and 5.44 kg at maximal speed. Physical restraint ensured that either elbow flexion or elbow extension was the only movement used.

Using the video images of the ball and reflective markers on the shoulder, elbow, and wrist, video analysis software calculated, at intervals of a 60th of a second, the x and y position, velocity, and acceleration of the ball, and the elbow joint angle. A commercial spreadsheet program was used to calculate power output from the data file. First, the direction and absolute velocity of the ball during each video frame were determined from its x and y velocities. Then, the acceleration in the direction of ball travel was determined from the x and y accelerations. Force in the direction of ball travel was calculated as the product of acceleration and ball mass. Power on the ball was the product of the velocity of the ball and the force in the direction of ball travel. Figure 7.9 shows ball velocity, joint angular velocity, and force and power exerted on the ball during an elbow flexion throw of the 0.91 kg ball.

The total power output of the human throwing the implement can be considerably greater than the power transmitted to the implement itself because of the work required to move the athlete's body. One can calculate the total power output generated during body movement by digitizing the video image locations of all major body joint centers and processing the resulting file of x and y joint coordinates using standard biomechanical methodology (Winter, 1990). To summarize the

Table 7.4 **Acceleration Because of Gravity at Sea Level by Latitude**

| Latitude (°) | Acceleration due to gravity (m/s²) | Sample locations |
|---|---|---|
| 0 | 9.780 | Ecuador, Kenya |
| 15 | 9.784 | Philippines, Guatemala |
| 30 | 9.793 | Texas, Israel |
| 45 | 9.806 | Oregon, France |
| 60 | 9.819 | Alaska, Sweden |
| 75 | 9.829 | Greenland, Antarctica |
| 90 | 9.832 | North Pole, South Pole |

Gravitational variations are largely accounted for by the equatorial bulge of the earth rather than altitude above sea level.

A more complete table can be found in the *CRC Handbook of Chemistry and Physics* (Lide, 2003).

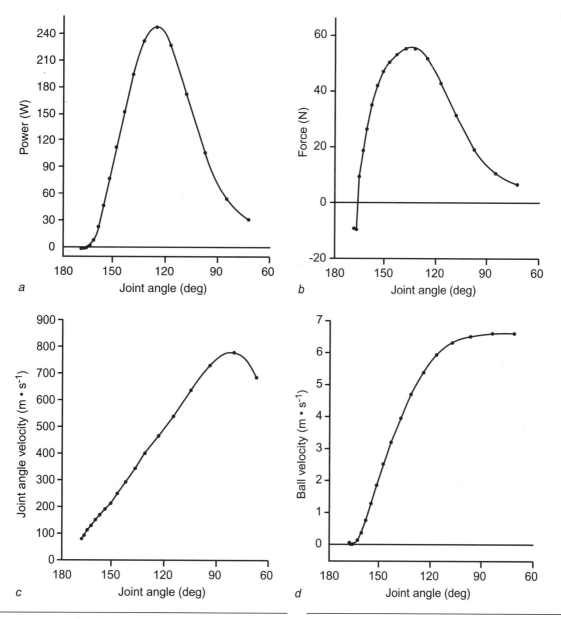

**Figure 7.9**  Power, force, and velocity generated by a supine subject throwing a 0.91 kg baseball-sized medicine ball horizontally using elbow flexion only. Physical restraints prevented extraneous body movements.

process, the video analysis system determines the coordinates of reflective markers attached over the center of each major body joint. Then, a computer program calculates for each video image the location of the center of mass of each body segment using standard body proportions. The location of the center of mass of the total body is calculated as the average of the locations of the centers of mass of the individual segments weighted according to the mass of each segment. From the resulting record of total-body center of mass position, curves of velocity and acceleration are derived. The net force propelling the body during each interframe time interval

equals the mass of the body times the acceleration of its center of mass. When the force of gravity is taken into consideration, force and power exerted on the center of mass of the body can be calculated, as in the analysis of the implement described earlier. More complex analyses are sometimes used to take into account the energy transferred within and between the body segments (Winter, 1978; Winter, 1979). A major advantage of video analysis for the determination of power output is that it can be used to monitor a wide range of physical activities and need not interfere with performance. The cost, however, can be considerable.

## Free-Weight Lifting

A system was developed in our laboratory to test power output during various body movements. The subject, wearing an electrogoniometer on the knee or elbow, held a dumbbell or wore a weighted iron shoe (Rosenstein, Harman, Frykman, and Johnson, 1989). The body was stabilized so that movement could only occur about the joint in question and the location of that joint would not change during the movement. On cue, the subject raised the weight as quickly as possible. The electrogoniometer output was fed into an analog-to-digital converter board of a computer, where the voltage was converted to digital values every hundredth of a second.

Based on the calibration values of the goniometer, a record of joint angle at every hundredth of a second was determined, and from that, records of angular velocity and angular acceleration were established. Power output was calculated as the product of joint angular velocity and torque. Torque at all time points was calculated as the sum of torque attributable to gravity and the torque attributable to the linear and angular acceleration of the weight and limb. Figure 7.10 depicts the power output test results. Figure 7.11 and the following equations explain these calculations for the dumbbell lifts.

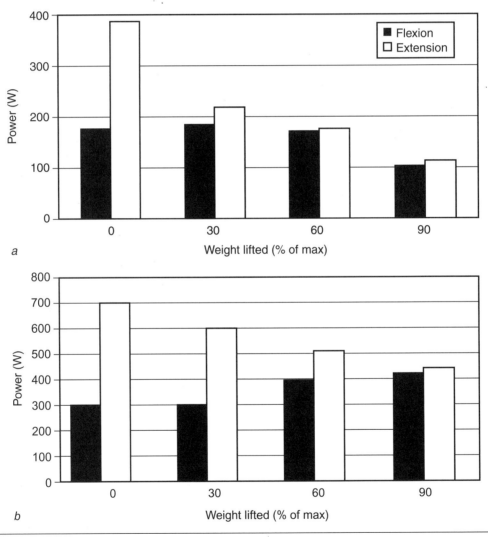

**Figure 7.10**   Peak power generated by 21 male subjects flexing and extending *(a)* one elbow and *(b)* one knee at maximal speed against the resistance of a dumbbell and iron shoe, respectively. Each of the four movements was tested using a different body position so that gravity provided resistance during a major portion of each range of motion. For the elbow, peak power during extension was the highest with no weight, whereas peak power for flexion was the highest with the 30 lb (13.6 kg) weight. For the knee, although peak power during extension was again the highest with no weight, peak power for flexion was the highest with the 90 lb (40.9 kg) weight. The differences between the four movements are attributable to both biomechanical and physiological factors.

$$T_g = \left(W_f\, r_f + W_d\, r_d\right)\cos\theta \qquad (19)$$

where $T_g$ = torque attributable to gravity (N·m)

$W_f$ = weight of the forearm-and-hand (N)

$r_f$ = distance from the elbow joint center to the center of mass of the forearm and hand (m)

$W_d$ = weight of the dumbbell (N)

$r_d$ = distance from the elbow joint center to the center of mass of the dumbbell (m)

$\theta$ = angle of the forearm relative to the horizontal (radians)

$$T_a = \left(I_f + m_f\, r_f^2 + I_d + m_d\, r_d^2\right)\alpha \qquad (20)$$

where

$T_a$ = torque attributable to acceleration (N·m)

$I_f$ = moment of inertia of the forearm and hand about its own center of mass (kg·m²)

$m_f$ = mass of the forearm and hand (kg)

$I_d$ = moment of inertia of the dumbbell (kg·m²) about its own center of mass (kg·m²)

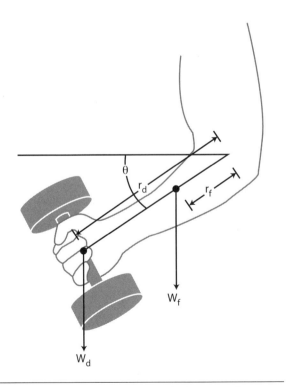

**Figure 7.11** Variables used to calculate muscle torque during a dumbbell lift. The total torque is the sum of torque attributable to gravity and the torque attributable to acceleration of the weight and limb.

$m_d$ = mass of the dumbbell (kg)

$\alpha$ = angular acceleration of the limb and dumbbell

$r_f$ and $r_d$ are defined earlier.

Because it is more accurate, the reaction board method (Hay, 1978) was used to estimate limb weight rather than calculate limb weight as a proportion of total body weight using standard tables. The reaction board method involves laying the subject on a horizontal board supported at one end by the scale and taking scale readings when the limb is held both horizontally and vertically. Based on the change in the scale reading and measurement of body position relative to the board, weight of the limb can be calculated. The limb moment of inertia needed for the previous equations was calculated according to a method described by Winter (1990). Moments of inertia of the weights were estimated from their shapes and masses using standard equations (Meriam and Kraige, 2002).

## Weight-Stack Machine Lifting

A linear position transducer was used as part of a system (figure 7.12) used to monitor power output produced by a subject while lifting on a leg-extension weight-stack machine (Frykman, Harman, Rosenstein, and Rosenstein, 1989). A model PT-101 Celesco linear position transducer was placed on the floor. Its cable was drawn up and connected to a custom-made arm protruding from the top of the weight stack of a knee-extension exercise machine. The transducer produced a voltage linearly related to the height of the stack above the floor. Custom-machined adapters allowed a BLH load cell to be affixed between the weight stack and its supporting cable to detect any forces exerted on the stack. Another custom-made device was used to take up the slack in the cable caused by the addition of the load cell. Power output throughout a lift was calculated as the velocity of the stack times the concurrent cable force.

The addition of an electrogoniometer at the lifter's knee allowed calculation of net instantaneous muscle torque. Assuming negligible friction in the moving parts of the well-lubricated exercise machine, power at the knee had to equal power at the weight stack. Thus

$$\text{power}_{knee} = \text{power}_{weights} \qquad (21)$$

$$\begin{aligned}\text{torque}_{knee} \times \text{angular velocity}_{knee} = \\ \text{force}_{weights} \times \text{velocity}_{weights}\end{aligned} \qquad (22)$$

$$\text{torque}_{knee} = \frac{\text{force}_{weights} \times \text{velocity}_{weights}}{\text{angular velocity}_{knee}} \qquad (23)$$

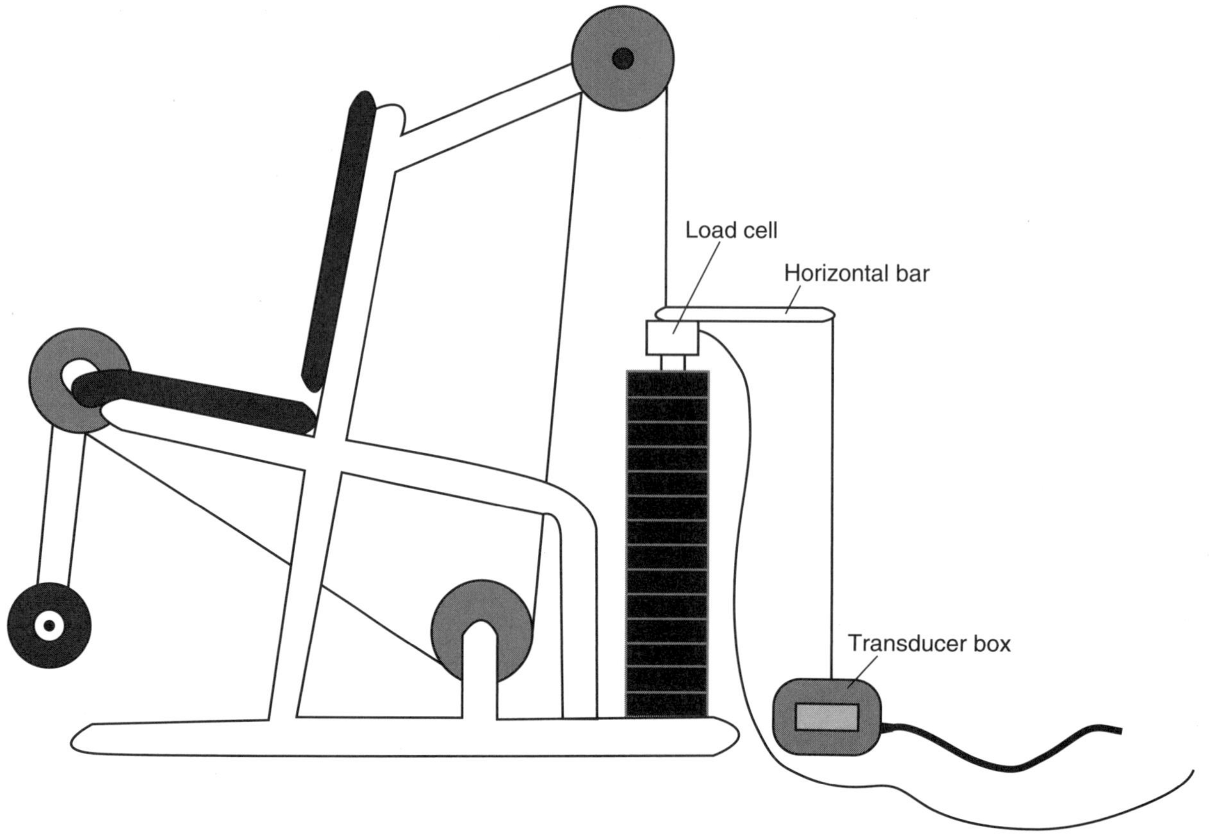

**Figure 7.12** A knee-extension weight-stack machine instrumented for power output measurement. The load cell is attached between the weight stack and its supporting cable. A thin wire extends from the linear position transducer box on the floor to the end of the horizontal bar positioned just above the load cell, allowing the vertical position of the stack to be monitored.

For the equation to be correct, it is essential to use the appropriate units (torque in newton-meters, force in newtons, velocity in meters per second, and angular velocity in radians per second).

With knowledge of the mass of the stack, the force transducer was not really necessary because force could have been derived using equation (11) from the mass and acceleration of the stack (determined from position transducer data). The masses of plates in weight-stack machines, however, have been found both to be variable and to differ from labeled values. The force transducer makes it unnecessary to disassemble exercise machines upon which the system is mounted to weigh each plate in the stack.

## Vertical Jumping

The force platform has often been used to measure human power output during the vertical jump (Davies and Rennie, 1968; Davies, 1971; Davies and Young, 1983; Harman, Rosenstein, Frykman, and Rosenstein, 1990; Harman, Rosenstein, Frykman, Rosenstein, and Kraemer, 1991; Sayers, Harackiewicz, Harman, Frykman, and Rosenstein, 1999; Driss, Vandewalle, Quievre, Miller, and Monod, 2001; Newton, Kraemer, and Hakkinen, 1999). The procedure involves calculation of power as the product of force and velocity, the latter two of which are obtained from the force platform data. Contractile force exerted by the jumper's muscles results in force exerted by the feet on the platform, which is mirrored by vertical ground reaction force (VGRF) equal in magnitude but opposite in direction to the force exerted by the feet on the platform. The vertical force channel of the force platform puts out a voltage proportional to the VGRF. For calculation purposes, the VGRF can be considered to act at the total-body center of mass (TBCM) to accelerate the body upward. Instantaneous jumping power is then calculated as the product of VGRF and TBCM vertical velocity (TBCMVV).

Although VGRF can be obtained continuously from force platform output, TBCMVV must be calculated, using the principle that impulse (the product of force and time) equals change in momentum (the product of mass and velocity). In jumping, body mass does not change while force is applied so that

$$\text{force} \times \text{time} = m \times \Delta V \qquad (24)$$

and

$$\Delta V = \text{force} \times \text{time}/\text{mass} \qquad (25)$$

Thus, change in TBCMVV during each sampling interval equals the net vertical force acting on the body multiplied by the intersample time period (t) divided by body mass (BM). The force used for the vertical velocity calculation is the VGRF reading from the force platform minus body weight (BW), because it is net vertical force that results in changes in vertical velocity of the jumper's body and VGRF acts in the opposite direction to the force of gravity:

$$\Delta \text{TBCMVV} m/s = \left( \text{VGRF}_N - \text{BW}_N \right) \times t_s / \text{BM}_{kg} \quad (26)$$

Absolute TBCMVV is updated at the end of each time interval by adding $\Delta$TBCMVV to the velocity

at the start of the interval, starting at zero velocity at the beginning of the jump. Instantaneous power is calculated throughout the jumping movement as the product of VGRF and the calculated TBCMVV. Equation (26) is equivalent to calculating TBCM vertical acceleration as VGRF minus body weight divided by body mass according to equation (12), and calculating $\Delta$TBCMVV as the product of acceleration and time (Davies and Rennie, 1968; Davies, 1971; Davies and Young, 1983).

Figure 7.13 shows VGRF, TBCM vertical position and velocity, and muscle power output during a jump. During the landing phase, as in any eccentric muscle activity, muscle power output is negative because the muscles absorb power generated by gravity. Table 7.5 shows the power generated by physically active male subjects jumping four different ways.

The jumping power experimentation in our laboratory (Harman, Rosenstein, Frykman, and Rosenstein, 1990) showed that the Lewis formula, often used for estimation of power output during the vertical jump (Fox and Mathews, 1981; Fox and Mathews, 1974, Kirkendall, Gruber, and Johnson, 1987), does not

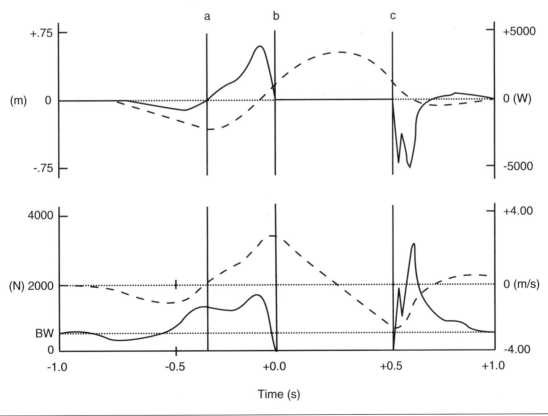

**Figure 7.13** Position, velocity, force, and power during a vertical jump with countermovement. The lower graph shows vertical ground reaction force (solid line) and vertical velocity of the center of mass of the body (dashed line). The upper graph shows vertical position of the center of mass (dashed line) and power output (solid line). Vertical line a = jump low point, b = takeoff, and c = landing.

Table 7.5     **Peak Power Output and Peak Ground Reaction Force[1]**

| | No countermovement | | Countermovement | |
|---|---|---|---|---|
| | **No arms** | **Arms** | **No arms** | **Arms** |
| Power (W) | 3,262 ± 626 | 3,804 ± 684 | 3,216 ± 607 | 3,896 ± 681 |
| Force (N) | 1,562 ± 219 | 1,687 ± 205 | 1,697 ± 308 | 1,725 ± 218 |

[1]Generated by 18 male subjects performing maximal vertical jumps from a force platform with and without arms and countermovement. The no-countermovement jumps began with the knees bent, and no initial downward body movement was allowed. During the no-arms jumps, the arms were kept down at the sides. The peak forces were more than twice body weight.

accurately reflect either peak or average power produced by a jumper. Rather, regression analysis of the jumps of 17 young male subjects showed that vertical jump height and body weight provide good prediction of peak power and fair prediction of average power. A subsequent regression analysis of the jumps of 108 male and female athletes and nonathletes showed that peak power during the squat jump can be predicted with good accuracy ($R^2$ = .88, SEE = 373 W) using the following equation (Sayers, Harackiewicz, Harman, Frykman, and Rosenstein, 1999):

$$\text{Peak power}\left(\text{W}\right) = 60.7\left[\text{jump height}\left(\text{cm}\right)\right] + \atop 45.3\left[\text{body mass}\left(\text{kg}\right)\right] - 2,055 \quad (27)$$

Note that both the Lewis formula and equation (27) were based on squat jump height. The squat jump is performed by squatting into the preparatory jump position, staying in that position for at least a full second, and then jumping as high as possible in a single movement. Countermovement jumps, more characteristic of sports movements, are performed with an almost instantaneous transition between the squat and the jump, thereby utilizing the stretch-shortening cycle to jump higher. Because of the greater variation with which a countermovement jump than a squat jump can be performed, the prediction of peak power during a countermovement jump is not quite as precise ($R_2$ =.78, SEE = 562) as it is for a squat jump (Sayers, Harackiewicz, Harman, Frykman, and Rosenstein, 1999).

$$\text{Peak power}\left(\text{W}\right) = 51.9\left[\text{jump height}\left(\text{cm}\right)\right] + \atop 48.9\left[\text{body mass}\left(\text{kg}\right)\right] - 2,007 \quad (27)$$

The force platform can be used to calculate power output during many activities other than vertical jumping. The activities must be those in which (1) the mass of the athlete (and any implement) is known, and (2) the athlete or implement is not in contact with any surface other than the force platform during the power measurement. Garham-mer (Garhammer, 1989) did extensive research on power output generated during weightlifting using a force platform.

## Isokinetic Dynamometry

Isokinetic dynamometers produce records of torque generated during body movements that are, for the most part, constant velocity. A record of power output can be determined by multiplying the torque value by the constant angular velocity selected by the operator. These machines are designed more for clinical use than for research. They generally come with integral computers and can be expensive. The output information is limited, and it is difficult to modify the internal programs to provide additional processing or to make the raw data available for transfer to another computer for further calculations.

Older Cybex II (Ronkonkoma, New York) isokinetic dynamometers are amenable to custom computerization, and several are still available in laboratories and physical therapy facilities. These systems were not sold with integral computers, and output signals were fed to chart recorders. The heads of the devices frequently came with connectors for monitoring voltages corresponding to angular position and torque. We passed these output signals through the amplifier–conditioner stage of the standard Cybex chart recorder before sending them to an analog-to-digital converter and computer for analysis. Signals were sampled at 500 Hz. A program determined angular velocity from the angular position data and calculated power as the product of torque and angular velocity. Peak torque and power were determined as well as the torque and power produced at selected joint angles. Figure 7.14 depicts peak power produced during isokinetic knee extension at various speeds of movement. Perrine and Edgerton found a similar positive association of power output and isokinetic test speed, but with more of a leveling off of power at the higher test speeds (Perrine and Edgerton, 1978). Based on additional experimentation, they reported

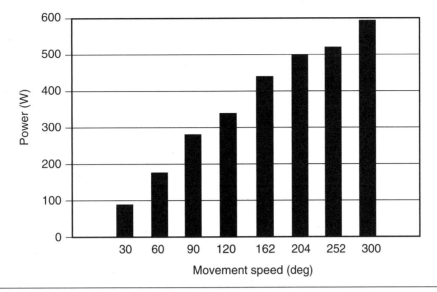

**Figure 7.14**   Peak power (W) generated by 13 male subjects during isokinetic single-leg knee extension at various movement speeds.

an r of .87 between vertical jump height and peak isokinetic leg power normalized for body mass among collegiate volleyball players, and similar correlation between leg power and sprint speed among women track athletes (Perrine, 1986).

## Nonmotorized Treadmill Running

Lakomy (Lakomy, 1984; Lakomy, 1987) employed a small electrical generator in conjunction with a load cell to measure horizontal propulsive power during sprint running on a nonmotorized treadmill. The test subject, rather than a motor, propelled the running surface. The system was based on the fact that, to keep the runner from accelerating horizontally, the strap connecting the runner to the load cell had to exert a force on the runner equal in magnitude to, and opposite in direction from, the horizontal force of the treadmill belt on the runner's feet. Horizontal power output was calculated as the product of the instantaneous force registered on the load cell and the treadmill belt velocity. The method required the assumption that errors because of the following were negligibly small: (1) Force and velocity were not measured at the same point, (2) the tether strap deviated from the horizontal during running, and (3) the strap had some elasticity that could have had a damping effect.

Lakomy used an electrical generator in a similar way to measure power output during ergometer cycling. His study showed that the Wingate test, by averaging power over 5 s and not taking flywheel acceleration into account, underestimated peak cycling power by a mean of 51% and overestimated

the time needed to reach peak power by 3.8 s (Lakomy, 1986).

## Cycling

Cycle ergometers have been instrumented in various ways to continuously monitor the mechanics of pedaling (Daly and Cavanagh, 1976; Frykman, Harman, and Kraemer, 1987; Harman, Frykman, and Kraemer, 1986; Harman, Knuttgen, and Frykman, 1987; Hull and Davis, 1981; McCartney, Spriet, Heigenhauser, Kowalchuk, Sutton, and Jones, 1986; Patterson and Moreno, 1990). Most cycle force transducers have incorporated strain gauge technology and have been custom made for each laboratory. When strain gauges are mounted on the crank arm (Daly and Cavanagh, 1976; McCartney, Spriet, Heigenhauser, Kowalchuk, Sutton, and Jones, 1986), they are capable of measuring only forces perpendicular to the crank arm. When strain gauges are mounted on the pedals, they have been able to monitor forces perpendicular and parallel to the pedal surface (Frykman, Harman, and Kraemer, 1987; Harman, Frykman, and Kraemer, 1986; Harman, Knuttgen, and Frykman, 1987; Hull and Davis, 1981; Patterson and Moreno, 1990).

In our system, seven cycle parameters are continuously monitored:

- The crank angle relative to the cycle frame
- Forces exerted by the feet in the following directions:
  - Perpendicular to the left pedal surface
  - Parallel to the left pedal surface

- Perpendicular to the right pedal surface
- Parallel to the right pedal surface
- The angle of the left pedal relative to the left crank arm, and
- The angle of the right pedal relative to the right crank arm.

Details of the methodology have been presented elsewhere (Harman, Knuttgen, and Frykman, 1987). To summarize, the component of pedal force perpendicular to each crank arm, the only component that can result in power generation, is calculated from the transduced forces and pedal angle. Pedaling torque is determined as the product of the force exerted perpendicular to the crank arm and the length of the crank arm. Angular position data from the crank arm is processed to produce a record of angular velocity. Power application throughout the entire pedal cycle is determined as the product of torque and angular velocity.

At 40 rpm, maximal power averaged over one full crank revolution was 3.07 ± 0.40 W per square cm of midthigh muscle cross-sectional area for a group of 17 physically active males, whereas at 100 rpm, maximal power was 4.68 ± 0.81 W per square cm (Harman, Frykman, and Kraemer, 1986). McCartney et al. (McCartney, Spriet, Heigenhauser, Kowalchuk, Sutton, and Jones, 1986) reported peak instantaneous 100 rpm cycling power of 1,626 ± 102 W for eight male university students.

## Summary

A wide variety of electronic equipment is available today that can be applied to the measurement of human mechanical power. The financial resources and expertise needed to set up such systems are available to most laboratories. The procedures described herein can be applied to the measurement of power output during diverse human activities. Readers should go beyond the examples described in the chapter to develop systems suited to their specific needs.

# eight

# Strength Training: Development and Evaluation of Methodology

**William J. Kraemer, PhD, FACSM, CSCS**
*University of Connecticut*
**Nicholas A. Ratamess, PhD, CSCS**
*The College of New Jersey*
**Andrew C. Fry, PhD, CSCS**
*University of Memphis*
**Duncan N. French, PhD, CSCS**
*University of Connecticut*

The expression of muscular strength is a fundamental property of human performance. The magnitude of the empirical and scientific literature devoted to strength development and evaluation makes it clear that a key aspect to the design of proper resistance training is the systematic inclusion of various testing modalities to evaluate the quality of the program. The evaluation of physiological responses and adaptations (e.g., cardiovascular, endocrine, neuromuscular, metabolic) associated with force production has allowed us to gain greater understanding of the various systems under conditions of high-threshold motor unit recruitment. In evaluating the effects of training, muscular fatigue, injury rehabilitation, muscular balance, or the functional abilities of different individuals, strength testing provides important information regarding human performance. This chapter will give an overview of the various factors involved in understanding, developing, implementing, and evaluating strength-testing protocols.

## What Is Muscular Strength?

*Force* (mass × acceleration) is a push or pull that has the ability to accelerate, decelerate, stop, or change the direction of an object typically measured in newtons (N) or pounds (lb). Of particular interest for our purposes is the peak force or peak torque (e.g., torque = force × moment arm length in newton-meters [Nm]) exerted during muscular activity. A number of biomechanical and physiological factors contribute to the development of maximal strength (see figure 8.1). For the purposes of this chapter, strength will be operationally defined as the maximal force a muscle or muscle group can generate at a specified or determined velocity (Knuttgen & Kraemer, 1987).

Atha (1981) presented and Enoka (1988) endorsed a definition of strength as "the ability to develop force against an unyielding resistance in a single contraction of unlimited duration." This definition pertains only to isometric (static) strength and

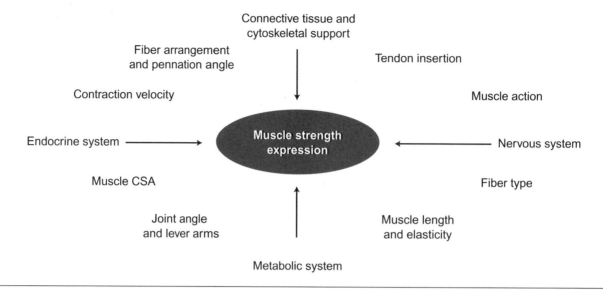

**Figure 8.1**  Physiological and biomechanical factors influencing the acute expression of muscular strength.

avoids consideration of the complex interaction of force development and the velocity of eccentric or concentric muscle actions. Although this definition appears to make the task of strength evaluation simpler, it avoids dynamic force production, which is prevalent in human performance.

Force will vary depending on the velocity of the movement. The expression and ultimate quantification of strength depend on the conditions of the test. Because of the number of variables or conditions involved, the strength of a muscle or muscle group must be defined as the maximal force generated at a specified velocity (Knuttgen & Kraemer, 1987). The velocity may be zero (i.e., isometric action) or may involve a range of velocities of shortening and lengthening for concentric and eccentric actions, respectively. For comparative purposes, the force and torque measurements must be performed when the muscle or muscle groups are at similar lengths. The force measurements can be obtained directly from the muscle or its tendons, from a particular point on one of the body parts, or as torque developed on a testing device. The muscles may perform in isometric, dynamic concentric, or dynamic eccentric muscle actions. In addition, the expression of peak dynamic eccentric and concentric torque at a specified velocity may be determined by an isokinetic dynamometer. Muscles are capable of producing the greatest torque during eccentric muscle actions, followed by isometric and concentric.

Different power outputs are associated with the force-velocity characteristics of the movement. Maximal muscular power of a joint or muscle group should not be confused with maximal aerobic power,

which relates to the cardiorespiratory ability of the body to deliver and utilize oxygen.

The term *isotonic* is frequently and improperly used to indicate dynamic muscle activity when the external resistance is constant. The term actually denotes a dynamic event in which the muscle generates the same amount of force throughout the entire movement. Such a condition occurs infrequently, if at all, in human performance, because of a combination of differences in force generation by muscle at various lengths and the changes in mechanical advantage at different joint angles. Therefore, the term *dynamic constant external resistance* (Fleck & Kraemer, 1997) will be used here to describe the testing of muscle activity using specific external resistances, such as free weights. This type of exercise includes both concentric and eccentric muscle activity.

## Why Is Measurement of Strength Important?

Probably the most important reason for monitoring strength performance is to assist in the evaluation and progression of resistance-training programs. Presently, most segments of the population perform resistance training, from children to the elderly, and the American College of Sports Medicine (1998, 2002) recommends resistance training for inclusion in general health and fitness exercise programs in adults. The programs, as well as the goals for training, are diverse. The amount of strength development depends on the initial level of muscular fitness, exercise prescription, time available, and objectives of the program. Regular assessment of muscular strength enables proper evaluation of the exercise prescription and modifications when appropriate.

The rate of strength increase differs considerably between untrained and trained individuals, with trained individuals showing much slower rates of improvement. A general review of approximately 150 studies revealed that increases in muscular strength, on average, are approximately 40% in untrained individuals, 20% in moderately trained individuals, 16% in trained individuals, 10% in advanced individuals, and 2% in elite individuals over periods ranging from 4 weeks to 2 years (American College of Sports Medicine, 2002). Although the training programs, durations, and testing procedures of these studies differed considerably, these data clearly show a specific trend toward slower rates of progression of strength development with training experience. This has recently been shown by meta-analysis of 140 studies (Rhea, Alvar, Burkett, & Ball, 2003). In this study, statistically significant effect size (ES) differences were observed between resistance-trained individuals and untrained individuals for training intensity (ES range of 0.65 to 1.80 for trained versus 1.60 to 2.80 for untrained), frequency (ES range of 0.70 to 1.40 for trained versus 1.20 to 1.90 for untrained), and volume (ES range of 0.47 to 1.17 for trained versus 1.16 to 2.28 for untrained).

Decisions must be made regarding the exercise prescription as to the cost–benefit ratio of putting additional attention, time, and effort into strength improvement. In certain circumstances small changes in strength require large amounts of training time because the "window of adaptation" is close to the theoretical genetic ceiling. Furthermore, the increase observed might be less than 5% compared with the higher percentage gains observed earlier in a person's training history. Such small gains may be the difference between winning and losing in certain types of elite athletic competitions but may be less important for other situations. The increased time needed to obtain small gains in strength might not be the best use of available training time, unless the small gains are directly related to needed performance abilities. For example, a 5% gain by a competitive lifter in a competitive lift could mean the difference between winning and placing fifth in a competition. Conversely, an individual with adequate strength in the biceps might not need to put in the extra time to make a 5% gain in the arm curl 1RM and might better use the additional training time to focus on another exercise movement. That is, he or she may need a "strength cap" for that particular exercise (i.e., maintaining the present level

of biceps training to focus on other areas). Thus, so-called clinical judgments often have to be made regarding the exercise prescription. Making such decisions requires solid knowledge of the individual's strength profile for a variety of muscles. Furthermore, one must understand the basic physiological adaptations associated with strength-training programs.

# Physiological Adaptations Associated With Strength Training

The physiological adaptations to resistance training have been previously reviewed in great detail (Kraemer, 1990; Kraemer, Deschenes, & Fleck, 1988; Sale, 1988; Kraemer, Fry, Frykman, Conroy, & Hoffman, 1989; Kraemer, Fleck, & Evans, 1996; Kraemer & Koziris, 1992; Kraemer & Ratamess, 2000; Kraemer, Ratamess, & French, 2002). Although this chapter will not address this concept in detail, table 8.1 presents some of the basic muscular adaptations resulting from strength training. The remainder of this section will discuss rehabilitation and safety, physiological benefits, and program evaluation.

## Rehabilitation and Safety of Strength Training

For over 65 years, resistance training has been used to intervene effectively in postinjury rehabilitation programs to enhance patient recovery (Fees, Decker, Snyder-Mackler, & Axe, 1998; Morrissey & Brewster, 1986; Smidt, Albright, & Densingerm, 1984). Evidence now indicates that strength-training programs reduce injury risk profiles (Kraemer et al., 1988; Wathen et al., 1983). Although injuries do occur during weightlifting, injuries occur less commonly during supervised strength training than they do during other sporting activities. In a 4-year study, resistance training has been shown to be safe (0.13 injuries per 1,000 h of athlete exposure, or 0.35 per 100 players per season), regardless of the training methods used (Zemper, 1990). The most common site of injury during weightlifting is the lower back, and the type of injury mostly encountered is a muscle strain (Mazur, Yetman, & Risser, 1993). Injuries are often attributable to improper exercise technique, improper loading (i.e., attempting to lift more weight than capable of lifting), overtraining, weight room accidents, and unsupervised training (Brown & Kimball, 1983; Mazur et al., 1993; Shankman, 1984).

Table 8.1    **Muscle Fiber Adaptations With Resistance Training**

| Variables | Adaptational responses |
|---|---|
| **Morphological** | |
| Pennation angle | Increases |
| Muscle volume | Increases |
| **Contractile** | |
| Peak force | Increases |
| Time to peak tension | No change, decreases |
| Half relaxation time | No change, decreases |
| **Muscle fiber type (%)** | |
| Type I | No change |
| Type IIA | Increases |
| Type IIB | Decreases |
| **Fiber cross-sectional area** | |
| Type I | Increases |
| Type IIA | Increases |
| Type IIB | Increases |
| **Structural** | |
| Capillary density | No change, increases, decreases |
| Mitochondrial density | Decreases |
| Sarcoplasmic reticulum network | Increases |
| T-tubule network | Increases |
| **Protein constituents** | |
| Actin filament number | Increases |
| Myosin filament number | Increases |
| DNA | Increases |
| RNA | Increases |
| **Enzymatic** | |
| RNA polymerase | Increases |
| Succinate dehydrogenase | No change, decreases |
| Malate dehydrogenase | No change, decreases |
| Citrate synthase | No change, increases |
| 3-hydroxyacyl-CoA dehydrogenase | No change, decreases |
| Creatine phosphokinase | No change, increases |
| Myokinase | No change, increases |
| Phosphofructokinase | No change, decreases |
| Lactate dehydrogenase | No change, increases |
| $Ca^{2+}$ ATPase | No change, increases |
| $Na^+$-$K^+$ ATPase | Increases |
| **Substrates** | |
| Stored ATP | Increases, no change |
| Stored PC | Increases |
| Stored glycogen | Increases, no change |
| Stored triglycerides | Increases |
| Myoglobin | Decreases |
| **Receptor regulation** | |
| Androgen receptors | No change, increases |
| Sarcolemmal $Ca^{2+}$ receptors | Increases |
| **Connective tissue** | |
| Collagen fibril diameter, packing density | Increases |
| Fibroblast activity | Increases |
| Tendon cross-sectional area | Increases |
| Fascia cross-sectional area | Increases |

Note: Specific changes depend on the nature of the exercise program used and are consequent to the manipulation of acute training variables including load and volume. Furthermore, the characteristics of individual muscle fibers will also affect the physiological responses to resistance training. In addition, changes related to protein content, enzymes, and mitochondrial and capillary density depend on the magnitude of hypertrophy (i.e., an increase in hypertrophy may lead to a decrease of the selected variable in relative terms).

However, competitive lifting (weightlifting, powerlifting, bodybuilding, and strength competitions), which involves the lifting of maximal weights for complex exercises for maximal performance or muscle hypertrophy, may pose a greater risk for injury compared with noncompetitive lifting. In examination of 358 bodybuilders and 60 powerlifters, Goertzen, Schoppe, Lange, and Schulitz (1989) reported that muscle injuries were common among most participants (84%) and that major shoulder or elbow injuries had occurred at some point in the athletes' training history in 40% of the participants, with the incidence twice as high in the powerlifters. Likewise, Raske and Norlin (2002) reported higher injury rates in lifters. In comparing elite weightlifters and powerlifters, weightlifters exhibited higher frequency of low back and knee injuries whereas shoulder injuries were more prominent in power lifters (Raske & Norlin, 2002).

Of interest regarding injuries encountered during strength training is the use of free weights and machines. Free-weight exercises, particularly those involving multiple joints and several major muscle groups, may pose a greater risk of injury because they require the individual to control the exercise completely (i.e., maintain body stability and displacement of the resistance), whereas machines may reduce the risk of acute injury because they assist in stabilizing the body and movement of the resistance. A recent study that examined the injuries related to resistance training in 12 athletes reported that 9 of the injuries occurred using free weights and 3 injuries occurred during machine exercises (Van der Wall et al., 1999). Although both types of lifting may cause injury, it appears that proper program design and supervision are especially warranted with use of free weights.

## Psychological Benefits

Resistance training can improve several psychological factors such as self-concept, self-efficacy, self-esteem, and reduce social physique anxiety (Brown & Kimball, 1983; Brown & Harrison, 1986; Brown et al., 1998; Williams & Cash, 2001). In the athletic community, strength testing is often used as a motivator (Allerheiligen et al., 1983). Visually observing an increase in strength performance motivates individuals to continue training vigorously and increases adherence to the resistance-training program. Furthermore, strength testing allows for quantification of specific training goals. In general, a properly designed resistance-training program has numerous benefits that have practical applications for many populations (Fleck & Kraemer, 1988).

## Evaluation of Strength-Training Programs

A strength-training program can have many roles. An optimal training program requires periodic assessments of the variables of interest. Assessment, in turn, requires accurate and objective methods of evaluating strength, which also allows one to predict athletic performance (Cisar, Johnson, Fry, & Ryan, 1987), evaluate physical performance in an industrial setting (Legg & Pateman, 1984), assess strength fitness (Fry & Kraemer, 1991;Hoffman, Fry, Howard, Maresh, & Kraemer, 1991), identify the physical capabilities associated with varying levels of performance (Chmelar, Shultz, Ruhling, Fitt, & Johnson, 1988), and characterize physical demands and characteristics of various occupations or athletic positions in a particular sport (Housh, Johnson, Hughes, Cisar, & Thorland, 1988; Housh et al., 1988). Furthermore, changes in muscular strength during an injury rehabilitation program can be closely monitored (Morrissey & Brewster, 1986; Smidt et al., 1984).

## Testing Modalities

We will now examine some of the modalities used in strength testing. The modality of testing needs to be specific to the training mode. In addition, factors such as familiarization and practice, safety/spotting, standardization of procedures, and proper positioning of the individual are key elements to proper testing regardless of the modality.

### Free Weights

Perhaps the most common tools for assessing strength are free weights—barbells, dumbbells, Olympic-style weights, and related types of equipment. Interpreted broadly, free weights can also include body weight when used as resistance in callisthenic-type activity (e.g., push-ups, chin-ups, dips, body-weight squats). In strength tests using free weights, the velocity of movement is typically not controlled but could be determined through laboratory methods (e.g., video analysis) (Atha, 1981; Enoka, 1988; Hoffman, Maresh, & Armstrong, 1992). Intentionally attempting to decrease concentric lifting velocity or control velocity during testing may limit motor unit recruitment and result in less force production (Keogh, Wilson, & Weatherby, 1999). When a person performs more than one repetition (e.g., 5RM) in a strength test, each subsequent repetition may be slower (Mookerjee & Ratamess, 1999) until the individual is unable to complete another repetition (Knuttgen & Kraemer, 1987). Demanding a specific power-output performance during multiple-repetition strength tests by implementing a specific time interval for the performance of the movement creates another type of strength test for a given number of repetitions.

During free-weight strength testing, the movements involved can be somewhat similar to those found on the athletic field or in other areas of motor performance because constant external resistance that requires balance is involved. Free weights present a number of different testing conditions compared with weight machines (Fleck & Kraemer, 1988; Nosse & Hunter, 1985). Free weights require greater motor coordination than do machines (Nosse & Hunter, 1985; Rutherford & Jones, 1986), primarily because the individual must control free weights through all spatial dimensions, whereas machines generally involve control through only one plane of movement (Fleck & Kraemer, 1988). This attribute can be an advantage or a disadvantage, depending on the motor skill level of the individuals being tested. Individuals with poor balance and impaired motor function (e.g., frail elderly, those with neuromuscular disease, people with arthritis, and so on) may require machine-based testing initially until sufficient improvement in physical function occurs. Another more practical reason for using free weights is their low cost and availability. The test specificity of free weights may be more appropriate for certain tasks being evaluated (Fleck & Kraemer, 1997; Rutherford & Jones, 1986). Test and training-mode specificity are vital for optimal expression of true strength gains. That is, using a free-weight testing modality is the most effective method for determining strength gains associated with free-weight training.

Total-body strength or power is easier to assess with free weights, because these total-body multiple-joint movements (e.g., power clean, snatch) require large muscle mass recruitment, which cannot be readily assessed within the constructs of machines. Strength assessment for multiple-joint basic strength exercises such as the squat, bench press, and bent-over row may be performed with free weights or machines. The choice of machines or free weights for these exercises depends on the training status and motor skill development of the individual, goals of the resistance training program, and testing specificity (i.e., the test should correspond to the individual's modality during training). Single-joint exercises (e.g., leg curls, arm curls) may be performed with free weights or machines. However, these exercises may be performed more strictly with the use of a machine when isolated muscle function is of interest. Free weights also involve both concentric and eccentric muscle activity. Eccentric muscle activity is not possible on some weight-training machines (e.g., typical hydraulic machines) or with some isokinetic (e.g., controlled constant velocity) dynamometers (Nosse & Hunter, 1985). Free weights almost always allow the desired range of motion, whereas other strength-testing modalities can be somewhat limited if it is not possible to get a proper fit between the individual and the machine configuration (Fleck & Kraemer, 1988).

## Familiarization

Obtaining baseline muscular strength data at the beginning of a program or scientific investigation is important. This task entails proper familiarization with each type of strength-testing protocol, testing apparatus, and proper lifting technique (Sale, Fleck, & Kraemer, 1988; Stone & O'Bryant, 1987). Without proper familiarization, values for strength gains can be inflated and not representative of the true physiological adaptations that have occurred from training. This concern is also evident in highly trained individuals if their lifting form in training deviates from what will be used for testing. Certain free-weight exercises may require a relatively long familiarization period before 1RM testing because of their complexity. Furthermore, some training is necessary for certain total-body exercises such as the Olympic-style lifts and variations before testing is performed. Two weeks of resistance training, including three or four training sessions for each exercise and practice of the tests to be performed, will achieve appropriate familiarization and reliability for maximal strength testing (Fry et al., 1991; Ploutz-Snyder & Giamis, 2001). A training effect from testing alone rarely occurs if the number of maximal efforts and volume of exercise are kept to a minimum (Fry et al., 1991).

The familiarization period is more important if the individual has no prior experience with resistance exercise or if several exercises will be tested. Individuals must learn how to produce maximal effort during the test, and doing so requires practice. In many instances, individuals naturally tend to misinterpret submaximal effort for maximal effort because of lack of experience. Proper familiarization minimizes submaximal efforts and avoids increases in performance that are due solely to improved motor coordination, a phenomenon commonly observed when using free-weight exercises. Familiarization is crucial in some populations, such as children or older adults. Elderly women require more familiarization than younger women do (e.g., approximately eight or nine sessions versus three or four) before performing the 1RM knee-extension exercise (Ploutz-Snyder & Giamis, 2001). Considering the simplicity of the machine knee-extension exercise, further familiarization may be warranted for multiple-joint free-weight exercises because they require a longer neural adaptational period (Chilibeck, Calder, Sale, & Webber, 1998). Special patient populations often require specialized instructions for strength testing, particularly with proper breathing, because hemodynamic or orthopedic limitations pose additional concerns. For example, cardiac patients are capable of performing 1RM tests if properly instructed (Barnard, Adams, Swank, Mann, & Denny, 1999; Faigenbaum, Skrinar, Cesare, Kraemer, & Thomas, 1990). But 1RM testing with free weights or machines in these special populations is questionable because other methods of evaluation (e.g., submaximal tests, analysis of training logs) may be more appropriate.

The conditions of the test are always unique to the methods used and developed for the population being evaluated, and the familiarization period must have specific goals related to achieving the type of strength test needed for the individual. This period is important for the development of professional rapport between the testing staff and the individual. The level of motivation and encouragement can be established during this time. Furthermore, before establishing any test profile, the test–retest reliabilities for each exercise should be determined within the sample population studied. Familiarization includes the following:

1. Supervised practice of the exercises using correct technique

2. A "dry run" through the test protocol

3. Repeated practice of the test protocol under actual test conditions and maximal efforts until a stable, reproducible baseline for strength performance is achieved

### Safety Considerations

Safety is an important consideration when performing strength evaluations with free weights. All equipment used must be in proper working order (Fry, 1985; Huegli, Richardson, Graffis, Kroll, & Epley, 1989). Olympic barbells must rotate properly, barbell collars must function properly, and weights must be solidly secured on both barbells and dumbbells. Any related equipment used (benches, weight racks, and so on) should be properly functioning and capable of withstanding the use of maximal resistance. Thorough inspection of all equipment to be used and calibration of all weights to be used should precede

each test session. Posted weights are not always correct and may be off by 10% or more. This discrepancy usually occurs with weights made from cast metals; weights that have been machined are generally more accurate. Any equipment malfunctions can lead to accidental injury.

The presence of properly trained spotters enhances the safety of the lifting environment (Earle & Baechle, 2000; Fry, 1985; Stone & O'Bryant, 1987; Thomson et al., 1989). Spotters are individuals who assist in the execution of an exercise to reduce the risk of injury and who motivate the individual to exert maximal effort (Earle & Baechle, 2000). Even if safety equipment such as power racks and other specially designed devices are available, an adequate number of spotters must always be present. Injuries and accidents are improbable during strength testing, but spotters must always be aware of the potential dangers of the test protocols. Emergency procedures should be formulated before testing is performed (Kraemer & Fleck, 1993). Although it may seem that an excessive number of precautionary measures have been suggested, all possible steps must be taken to ensure the safety of your participants, athletes, or patients. Prudent methods and procedures will help avoid unwanted litigation due to negligence and enhance approval of strength-testing protocols by human use review boards. Maximal strength tests have been shown to be quite safe if proper technique is used. Note that these safety considerations are by no means limited to the use of free weights but concern all strength-testing modalities (weight machines, isokinetic dynamometers, and so on).

### Proper Spotting

As discussed previously, the use of spotters is mandatory for free-weight strength testing. Spotters must be experienced and knowledgeable about correct exercise execution. The primary responsibility of a spotter during strength testing is injury prevention. Thus, the spotter must be able to recognize poor exercise technique and identify hazardous situations (e.g., excessive loading) in which intervening is necessary. Spotters must have strength sufficient to assist in lifting a heavy weight when necessary. Thus, several spotters (two or three) may be necessary when spotting a strong individual or when spotting an exercise in which high loading is common (e.g., back squat).

The procedures for proper spotting are specific to the exercise used during testing. Although not all exercises require a spotter, a spotter should be present during strength testing. Exercises that involve lifting the weight over head (e.g., shoulder press), lying supine on a bench and lifting the weight over the face or trunk (e.g., bench press, incline press), or placing the bar on the rear or anterior aspects of the shoulders (e.g., back squat, front squat) place the individual at greater risk for injury when failure occurs. Thus, spotters are mandatory for these exercises. Spotters are beneficial to other exercises such as the arm curl and upright row, although failure in these exercises does not place the lifter in immediate danger. Spotters are not necessary for total-body Olympic lifts or for exercises involving lifting the weight off the ground (e.g., deadlift, bent-over row) because it is much safer for the individual (and spotter) to drop the weight on failure or move out of the way (for Olympic lifts) rather than risk greater injury lifting it until its full completion or attempting to catch the weight.

The procedures used for spotting during strength testing have both similarities and differences with the procedures used in training. In both, spotters should be located close to the lifter (but not close enough to impede bar or bodily movement), follow the bar trajectory or the individual's body at each segment of the range of motion to prepare for momentary failure, and have his or her hands and body in the correct position to apply force to the bar at any point where failure could occur. When assisting a failed attempt by the lifter, the spotter should use an alternate hand grip (one hand pronated, one hand supinated) for added support, maintain a flat back position, and maintain a stable lower-body base support. A spotter who loses control of the bar or loses balance during the assistance phase of the lift places both the lifter and him- or herself at greater risk of injury. Communication between the spotter or spotters and the lifter is important. Information concerning the individual's lifting capacity (e.g., perceived 1RM), exercise technique, and circumstances for identifying failure should be discussed before testing. In addition, the use of a liftoff (assistance by a spotter in lifting the bar from the racked position to the starting position of the exercise) for certain exercises such as the bench press and shoulder press requires communication between the spotter and lifter to ensure coordinated effort and timing of force application.

The major difference between spotting for strength training compared with spotting for testing is the amount of assistance given to the lifter. During resistance training, a spotter normally applies minimal force to keep the bar moving when the lifter encounters failure. This technique, producing what is known as a forced repetition, ensures that the lifter continues to exert maximal force when he or she is not able to complete the repetition without assistance. This approach, however, is not ideal for maximal strength

testing. Fatigue must be kept to a minimum during strength testing. Having the lifter perform a forced repetition may cause undue fatigue that could affect subsequent trials if the 1RM has yet to be determined. Thus, the spotter may provide substantial assistance to spare the lifter unwanted fatigue. Note that the spotter should not touch the bar until failure has occurred during strength testing. Even the slightest touch gives assistance to the lifter and can overinflate the 1RM value.

## Proper Positioning With Free Weights

Proper positioning of the individual is an often overlooked variable that can affect test–retest reliability. Factors such as grip position (e.g., wide versus narrow), grip style (e.g., thumbless, thumbs around, hook grip, use of wrist straps), foot stance (e.g., wide versus narrow, heels elevated versus heels flat, toes straight ahead versus toes flared out), bar position (e.g., high trap versus low trap for barbell positions in the squat), and posture (Wilson, Murphy, & Walshe, 1996) can all affect the mechanics and results of the lift. In these respects, the biomechanical aspects of the lift can influence the expression of strength.

The starting position of the limbs is important. In the bench press, the lifter can perform an eccentric movement of the musculature before the concentric phase of the exercise. Conversely, the individual can start with the concentric phase, first lifting the weight from the chest. Whether one has muscle lengthening (the eccentric phase) before muscle shortening (the concentric phase) or vice versa is an important characteristic of the test and needs serious consideration in the development of a strength-test protocol. Performance of only the concentric phase, both eccentric and concentric phases, or including a pause between the eccentric and concentric phases significantly affects the amount of force produced and subsequent 1RM (Wilson, Elliott, & Wood, 1991). Thus, the muscle actions and cadence used should be standardized and controlled over all testing situations (Gregor, Edgerton, Perrine, Capion, & DeBus, 1979; Kraemer et al., 1991; Kraemer et al., 1990). This also applies to the testing range of motion. Strength testing should be performed in full exercise range of motion (unless specific testing is required for a limited range of motion exercise). Standardization of the range of motion is crucial to obtaining accurate data and for making legitimate comparisons with previous scores. Muscular strength has been shown to vary throughout the joint range of motion (Kulig, Andrews, & Hay, 1984); thus, exercises performed in a limited range of motion may overestimate the true full range of motion 1RM. Mookerjee and Ratamess (1999) compared the full and partial range of motion (descent to an elbow angle of 90°) 1RM bench press in highly trained lifters and reported an approximately 10.7% higher 1RM value for the partial bench press. Similar data have been obtained during comparisons of the isometric bench press at elbow angles of 90 and 120° (Murphy & Wilson, 1996). The 1RM squat is one exercise in which a lack of control over the testing range of motion can make a significant difference in the results obtained. Performing the exercise above the parallel position can result in a 1RM that is much greater than that obtained during a parallel squat (Verdera, Champavier, Schmidt, Bermon, & Marconnet, 1999; Weiss, Fry, Wood, & Melton, 2000). This technique can lead to an overinflation of the strength increase associated with a resistance-training program. Steps should be taken to ensure a standard range of motion. Some methods of standardization used are (1) to have a spotter or research assistant located laterally to the lifter to give an "up" or "start" signal after the lifter descends to the appropriate position, (2) to use a bench, a box, or power rack pins to designate the bottom position of the squat, or (3) to use an electronic device that sounds when the lifter reaches the appropriate position. Consistency of test procedures is vital for establishing good test reliability. The use of free weights can make reliability difficult to attain if confounding variables are not controlled. Test-related variations for 1RM evaluations have been reported as ranging from 1.5 to 20% (Stone & O'Bryant, 1987). Emphasizing the importance of precise testing procedures is vital to accurate assessments.

## Resistance Exercise Machines

Much of the information already presented in this chapter for the development of strength-testing protocols using free weights can also be used in the development of strength tests using weight machines. Although many different types of resistance-training machines are currently available, this section will focus on machines that use a movable external resistance such as a weight stack or plate-loaded apparatus. Later sections will address isometric, isokinetic, and isokinetic-like machines. In general, these machines are somewhat like free weights in that the external resistance is constant. Under certain circumstances, specific quantification of the forces involved with the movement may require more advanced evaluation of the machine. If needed, evaluation of the actual forces produced beyond a plate number or stack weight will involve more sophisticated analysis and breakdown of the machine (e.g., weighing the stack plates, determin-

ing friction coefficients, determining the movement patterns for gravity corrections) (Dudley, Tesch, Miller, & Buchanan, 1991; Kraemer et al., 1991).

Many machines alter the resistance encountered by the muscles with a system of cams, levers, or pulleys, resulting in a variable-resistance system. Even more interesting is that the weight lifted in some machines is not the same as the amount of weight lowered in the exercise movement because of mechanical variations in the direction of movement and the method of loading for the machine (Dudley et al., 1991). Some manufacturers attempt to increase the resistance during a range of motion in an attempt to mimic the human strength curves of various joints or physical movements. Strength curves are variable, however, and in some instances the strength curves of the machines are not identical with those of the human body (Fleming, 1985). Machines of this nature can be useful training tools but may be inappropriate for testing unless their unique characteristics are considered.

Depending on the type of evaluation, strength testing with machines may or may not be appropriate. Machines may not be appropriate as an evaluation tool if training was performed with a different modality or if the strength curve of the machine is significantly different from the normal human strength curve (Fleming, 1985). Thus, strength testing with machines is most appropriate when training was performed with machines (a concept known as training–testing specificity). On most variable-resistance machines, the actual resistance to the muscles is modified by levers, pulleys, or cams. This configuration can create highly variable movement velocities over the range of motion, which is magnified during maximal testing. The lifter must perform a smooth movement for a particular exercise, especially at slow velocities that are typical of 1RM testing. A true 1RM is difficult to assess if the movement pattern is interrupted at different phases of the range of motion. Such individual equipment incompatibility should be screened for in the familiarization process before testing. Lastly, some weight machines may not have enough resistance to determine 1RM on certain exercises for stronger individuals. This problem may be alleviated with plate-loaded machines to a certain extent. Thus, depending on the maximal strength and training history of the individual, machines may not be appropriate in certain situations.

The use of weight machines involves other considerations. Most machines operate in only one plane of motion (Fleck & Kraemer, 1988), resulting in different requirements for motor coordination (Nosse & Hunter, 1985; Rutherford & Jones, 1986) and lifting technique (Fleck & Kraemer, 1988; Kraemer & Fleck, 1993). By ruling out technique factors, strength testing with machines may more accurately assess pure strength changes if the individual strictly maintains the proper movement pattern and exercise form. This consideration is important when evaluating individuals with little resistance-training experience or clinical populations with limited motor function. Machine-based strength testing makes up a large component of the assessment of resistance-training programs in the elderly, previously untrained, and special populations (e.g., those in cardiac rehabilitation, physical therapy, and so on). But there is no guarantee that the individual has applied the optimal direction of force to the device. For example, in performing a machine bench press, an individual only needs to apply force in a general upward direction. But if a lifter uses an improper plane of movement with free weights, a loss of control will occur, resulting in a failed trial. The magnitude of synergistic control of movement desired in a specific test is a major consideration when choosing the testing modalities.

Despite their high cost, machines are readily available at many facilities including health clubs, gyms, weight rooms, and exercise physiology laboratories. Machine strength testing may also provide certain advantages such as the ability to assess exercises difficult to perform with free weights (e.g., lat pulldown, knee extension). As with free weights, close inspection of the equipment is necessary to prevent breakdowns that can interfere with testing or cause injury to the individual (Fry, 1985; Huegli et al., 1989). This examination includes checking for loose screws; worn parts; frayed cables, chains, or pulleys; and torn upholstery and grip supports. Machines must also be adjusted properly to the individual and be operating smoothly; improper lubrication or a misalignment of the device may add unknown resistance to the machine. The inspection should include making sure that the machine plates do not stick together. Devices that are always in optimal operating condition enhance test reliability.

## Safety and Machine 1RM Testing

A popular aspect of machines for many people is the perceived increase in safety. Although injuries do occur with use of both free weights and machines (Van der Wall et al., 1999), it appears that the risk of injury is less with machine-based strength testing. Generally, spotters are not needed during 1RM trials, although someone knowledgeable in the operation of the equipment and members of training or research team should always be present for assistance and motivation (Fleck & Kraemer, 1988).

The safety of maximal strength testing in special populations (e.g., the elderly, children) has been questioned. Studies examining 1RM testing in children and older adults have shown very few to zero injuries when proper supervision and instruction are used (Faigenbaum, Westcott, Loud, & Long, 1999; Schlicht, Camaione, & Owens, 2001). Shaw, McCully, and Posner (1995) examined the safety of machine 1RM strength testing in 83 elderly men and women for five exercises (chest press, leg extension, abdominal curl, arm curl, and calf raise). They reported that only two individuals, neither of whom had previous lifting experience, were injured during testing (back injury and rib fracture). No one with any lifting experience was injured. Similar results were obtained by Barnard, Adams, Swank, Mann, and Denny (1999), who reported no injuries during 1RM testing in 74 men and women enrolled in a cardiac rehabilitation program, and Adams, Swank, Barnard, Berning, and Sevene-Adams (2000), who reported no injuries during 1RM bench press and Smith machine squat testing in 26 women (ages 44-68). These data demonstrate that 1RM strength testing is safe in the clinical setting but also stress the importance of familiarization and proper instruction before testing. The training staff makes the ultimate decision to include 1RM testing in these populations. Useful assessment alternatives include determining a multiple RM, estimation of 1RM by regression equations, or close examination of training logs.

## Proper Positioning in Machines

Proper positioning in machines depends on the exercises chosen for testing. In general, seats should be adjusted so that the resistance bar (or grips) is properly located in relation to the individual (e.g., over the lower-to-middle segment of the chest for the chest press and aligned near shoulder level for the shoulder press). For many limb exercises, such as the leg extension and leg curl, the machine must be adjusted so that the knee (or the joint in question) is aligned with the axis of rotation and freely moveable (i.e., not fixated against the pad of the machine, a position that places the individual at risk for injury). Machines should also be adjusted to minimize gaps between the individual and the machine. Once proper alignment is obtained, the numbers located on the adjustable parts of the machine should be recorded for future use. Body parts must remain securely fixed to the machine to allow for optimal force expression (e.g., properly securing the lower body pad support during the lat pulldown, using a seat belt for the back extension, and firmly holding grip supports for exercises such as the leg extension, leg press, and leg curl). Weight machines often can isolate muscle groups or joints while minimizing extraneous body movements. Achieving this benefit requires that each individual be properly fitted, which may be a problem with certain populations. Children and small adults might not be able to adjust to the dimensions of a machine (National Strength and Conditioning Association, 1985). Currently, many manufacturers make adjustable machines or even design their equipment specifically for use by children or adults of various sizes (from small to large). Simply adjusting seat positions, however, cannot always account for differences in limb lengths in relation to various machine lever lengths. Even with properly adjusted equipment, the individual must be positioned each time according to a predetermined protocol developed by the investigator to meet the needs of the specific test. Small position deviations on a machine can affect resultant force production (Lewis & Spitler, 1989). All such positional data must be recorded and quantified for each test.

Starting positions of the machine during strength testing can affect test results. For example, during a 1RM machine bench press test, the exercise is usually initiated with the machine handles in the lowest position, thus eliminating the eccentric phase of the lift. If the test is initiated with the arms fully extended, eccentric muscle activity precedes the concentric muscle activity of the actual lift. This procedure enhances performance by making greater use of the stretch-shortening cycle. Although either procedure is acceptable, all tests should be consistent. Furthermore, when comparing different test protocols, these factors must be taken into consideration.

The weight increments possible with some machines may also be a limiting factor. When evaluating individuals who are relatively weak or activities in which low levels of force production are possible (e.g., injury rehabilitation), increases of 9 to 12.5 lb (4 to 5.6 kg), as are commonly found with many weight stacks, may be too great. Adaptations to the equipment must be made to allow accurate assessment of the actual strength capabilities for the activity. If a weight stack is used, attaching smaller weights to the stack may be necessary to allow smaller incremental changes.

## Strength-Test Protocols for Repetition Maximums

Tests using free weights and machines for evaluation of muscular strength are commonly cited in scientific literature (Altug, Altug, & Altug, 1987). Various methods have been used, although few studies have reported detailed procedures of 1RM testing. Usually a 1RM effort is considered the gold standard for evaluating strength, although RM testing can involve any number of repetitions (Kraemer, 1983). The 1RM is the maximal amount of weight that can be lifted once for a specific exercise. High intraclass correlation coefficients (e.g., test–retest reliabilities) have been shown for 1RM testing with ranges from $R = .79$ to .99 (see table 8.2). The 1RM is exercise specific; thus values will be different for each exercise tested (Hoeger, Barette, Hale, & Hopkins, 1990). A modification of the 1RM methods of Stone and O'Bryant (1987) and Kraemer and Fleck (1993) is listed here. Although several protocols for strength testing are effective, the following is one that we have commonly used in our laboratories:

1. A light warm-up of 5 to 10 repetitions at 40% to 60% of perceived maximum is performed.

2. After a 1 min rest with light stretching, three to five repetitions at 60% to 80% of perceived maximum is performed.

3. Step 2 will take the individual close to the perceived 1RM. A conservative increase in weight is made, and the individual attempts a 1RM lift. If the lift is successful, a rest period of approximately 3 min is allowed. One of the most costly errors is not allowing enough rest before the next maximal attempt. A 1RM should be obtained within three to five sets to avoid excessive fatigue. The process of increasing the weight up to a true 1RM can be enhanced by prior familiarization and expertise of the investigator in evaluating performance. This process continues until a failed attempt occurs. Weight is then adjusted accordingly.

4. The 1RM value is reported as the weight of the last successfully completed lift.

A great deal of communication must take place between the individual being tested and the testing staff. In some instances, a trained testing staff can recognize an individual's capacity and determine a progression pattern of loading with sufficient accuracy. Feedback from the individual, however, is critical to determining the loading progressions, especially if the lifter has experience and a general idea of the perceived 1RM. This is evident during resistance-training studies with inexperienced individuals in which 1RM testing initially may be more trial and error, whereas subsequent testing sessions with the same group of participants is more efficient after baseline data are obtained. Questions such as "How do you feel?" "Are

## Table 8.2    Test—Retest Reliability for Various 1 RM Testing Protocols

| Exercise | Coefficient | Reference |
|---|---|---|
| Squat | .94 | Sewell & Lander (1991) |
| | .99 | Giorgi et al. (1998) |
| | .95 | Hickson, Hidaka, Foster, Falduto, & Chatterton (1994) |
| | .92 | Sanborn et al. (2000) |
| | .99 | McBride, Triplett-McBride, Davie, & Newton (2002) |
| Bench press | .98 | Sewell & Lander (1991) |
| | .99 | Giorgi et al. (1998) |
| | .94 | Hoeger et al. (1990) |
| | .98 | Kraemer et al. (2000) |
| | .99 | Rhea, Ball, Phillips, & Burkett (2002) |
| | .99 | Hickson et al. (1994) |
| Leg press | .89 | Hoeger et al. (1990) |
| | .99 | Kraemer et al. (2000) |
| | .99 | Rhea et al. (2002) |
| | .69-.91 | Patterson, Sherman, Hitzelberger, & Nichols (1996) |
| Lat pulldown | .79-.98 | Hoeger et al. (1990) |
| | .92-.98 | Patterson et al. (1996) |
| Shoulder press | .98 | Kraemer et al. (2000) |
| | .97-.98 | Patterson et al. (1996) |
| Leg extension | .92-.98 | Hoeger et al. (1990) |
| | .74-.97 | Patterson et al. (1996) |
| Leg curl | .93-.97 | Hoeger et al. (1990) |
| Sit-up | .98 | Hoeger et al. (1990) |
| Arm curl | .86-.97 | Hoeger et al. (1990) |
| | .98 | Sale et al. (1988) |
| Triceps extension | .98 | Sale et al. (1988) |

you ready to go?" "How close to your 1RM do you think you are?" and "Can you lift 5 more pounds?" are vital to the interaction with people as they attempt to exert maximal force. Showing concern for the individual as the test is under way is also vital to getting a maximal effort.

Another consideration is the length of the rest period. The number of preliminary repetition attempts (warm-up) is also important. For all tests of this type, the number of repetitions must be adequate for proper warm-up but few enough to avoid fatiguing the individual before he or she attempts a

maximal effort (Berger, 1967). To some extent these factors need to be individualized because a person lifting very heavy weights (e.g., a 700 lb [320 kg] squat) will require more warm-up sets and longer rest intervals than someone lifting a lighter weight (e.g., a 200 lb [90 kg] squat). Sufficient recovery for enhanced maximal performance needs to be given during the test. Some individuals may require at least 5 min rest between attempts, whereas others need only 1 or 2 min of rest, depending on the loading. In particular, Behm, Reardon, Fitzgerald, and Drinkwater (2002) have shown that complete recovery does not occur within 3 min of rest following a set of a 5RM, 10RM, or 20RM. Before testing, the testing staff must stress to each individual to rest 48 h before testing (if the individual has control over day-to-day activities) and observe proper nutrition practice (e.g., adequate kilocalorie intake, proper hydration, and so on) (Schoffstall, Branch, Leutholtz, & Swain, 2001). If rest time appears restrictive and elongates testing protocols, testing can often be performed with more than one individual. This method allows one individual to lift while another rests. If a greater number of repetitions are to be attempted, such as a 5 or 10RM, fewer warm-up sets should be used. Research has recently shown that higher RMs (e.g., 10 and 20RM) produce greater muscle disruption, thereby indicating that one trial may be necessary or that additional trials will require a much longer recovery period (Behm et al., 2002). The following protocol is suggested for testing a 6RM and could be modified to test other RM numbers (e.g., 5RM, 10RM).

1. The individual warms up with 5 to 10 repetitions with 50% of the estimated 6RM.

2. After 1 min of rest and some stretching, the individual performs 6 repetitions at 70% of the estimated 6RM.

3. Step 2 is repeated at 90% of the estimated 6RM for 3 to 6 repetitions.

4. After at least 2 min of rest, depending on the effort required for the previous set, 6 repetitions are performed with 100% to 105% of the estimated 6RM.

5. After at least 3 min of rest, if step 4 is successful, increase the resistance by 2.5 to 5% for another 6RM attempt. If 6 repetitions were not completed in step 4, subtract 2.5 to 5% of the resistance used in step 4 and attempt another 6RM.

6. If all sets performed through step 5 were successful, retesting should occur after 24 h of rest because fatigue would greatly affect performance

on additional sets. If weight was removed for step 5 and 6 repetitions were performed, this is the individual's 6RM. If the individual was not successful with this reduced resistance, retesting should occur after 24 h of rest.

Testing procedures for RMs on machines are essentially the same as the procedures for free weights. In many situations, the test may require less time because a simple adjustment of a weight stack is all that is necessary. Altug et al. (1987) have compiled a sample of machine-test protocols. Special care must be taken with machine testing to make sure that proper technique is used at all times. When a reduction in the velocity of the movement with heavy resistances occurs, lifters tend to alter lifting form in an attempt to complete the lift. The testing staff must require lifters to adhere to proper exercise technique and discard any repetitions not accurately performed, because "cheating" can increase the risk of injury and overinflate the actual RM.

## Isometric Testing

Another method of assessing muscle strength is isometric testing, in which the length of the active muscle remains constant. Isometric testing removes variation in the velocity of movement, because all tests are performed at $0° \cdot s^{-1}$. Factors such as the angle

to be evaluated, anatomical point force recordings, standardization between individuals, feedback, and motivation make isometric testing as demanding as any other strength-testing procedure. Force varies significantly throughout the joint ROM; therefore, careful consideration is needed for standardizing joint angles (see figure 8.2).

Devices for isometric strength assessment have been in existence for many years. Early work in this area commonly used devices such as the hip and back dynamometer to test multiple-joint isometric capabilities (Clark, 1967); these devices are still in use (Amusa & Obajuluwa, 1986). Strength of muscles of isolated joints can be tested using cable tensiometers. Still popular is the handgrip dynamometer, which is simply a variation of a dynamometer designed specifically for grip force evaluation. Modern technology has developed force transducers (load cells) and strain gauges that allow interfacing with computerized data storage and analysis systems. Furthermore, isokinetic devices can be used for isometric testing by simply selecting a velocity of zero.

An innovative method of isometric testing involves using a force plate and an immovable resistance. This technology has been used successfully in several laboratories including our own (Blazevich, Gill, & Newton, 2002; Giorgi, Wilson, Weatherby, & Murphy, 1998; Murphy & Wilson, 1996; Rahmani, Viale, Dalleau, & Lacour, 2001; Van der Wall et al., 1999; Wilson, Murphy, & Pryor, 1994; Wilson & Murphy,

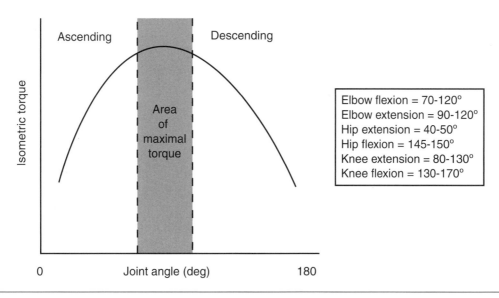

**Figure 8.2** Human strength curve. A typical strength curve that reflects a single-joint movement. The curve is an ascending–descending curve that is characteristic of most joints. Torque increases through the ascending region, reaches a peak at the area of maximal torque, and decreases through the descending region. The shape of the curve will vary depending on the joint selected. Several joint actions and their corresponding area of maximal torque production are listed. Data reported are ranges obtained from numerous investigations.

1996). A force plate measures ground reaction force while the individual exerts maximal isometric force at a predetermined angle. The force plate is positioned within a Smith machine that has adjustable mechanical stops. The precise joint angle of the individual is measured with a goniometer, and the Smith machine bar is locked at that exact location. Mechanical stops are subsequently placed superior to the bar to eliminate any movement, and this location is recorded. The individual then exerts maximal force to the bar, based on the desired testing protocol, and the force plate measures the resultant ground reaction force. The major utility of this setup is that isometric force can be measured for multiple-joint exercises, which may have greater relevance to athletic performance. Similar setups have been used for other exercises on machines such as the back squat (Blazevich et al., 2002). In our laboratory, we have effectively used this system to measure isometric force during the squat, bench press, incline bench press, shoulder press, deadlift, upright row, bench row, and arm curl. This system has been shown to be very reliable ($R$ = .96 to .98) for the isometric squat (Blazevich et al., 2002; Rahmani et al., 2001; Van der Wall et al., 1999; Wilson & Murphy, 1996) and the isometric bench press ($R$ = .85 to .87) (Wilson et al., 1994). Note with this system that lifters must be carefully supervised to eliminate any "cheating" or unwanted contributions from other muscles not involved in the exercise. Therefore, proper familiarization and instruction are critical to collecting accurate and reliable data.

Many test protocols have been developed for isometric testing (Altug et al., 1987). Although peak force assessment is often the primary purpose for isometric testing, other data have been generated. The rate of isometric force development is important, because it provides information about the functional abilities of the neuromuscular units being tested and reflects physiological characteristics of the muscle fibers (e.g., fast-twitch versus slow-twitch fibers). The rate of isometric force development can be quantified in several ways (Sale et al., 1988). If the testing equipment has a timing mechanism, the time to peak force can be measured in milliseconds. This measurement is sometimes difficult because of the oscillations found in the resulting strength curve. If an isokinetic dynamometer is used for the test, the damping control can be set to filter out these oscillations. An alternative method is to select an arbitrary percentage of the peak force of an individual (e.g., 50% maximal force production) as the cutoff for timing.

For various reasons, some test protocols call for a gradual increase of force, in which case the time to peak force measures would be inappropriate.

Note that the rate of force production may affect the resulting peak force, with faster muscle actions eliciting slightly greater peak values (Bemben, Clasey, & Massey, 1990). Furthermore, it is suggested that the isometric muscle action must be held for at least 5 s to allow a maximal effort. Shorter periods might not permit a maximal effort. Whatever protocol is selected, be sure that all individuals are clearly instructed in, and are allowed to practice, the appropriate procedures to assure high test–retest reliability.

## Fatigue Tests

Fatigue tests are commonly used with isometric devices to assess local muscular endurance. These tests generally involve maintaining a certain level of muscular force over a specified period (e.g., 60 s). If a fatigue curve is to be developed, individuals must exert maximal force and not pace themselves because doing so makes interpretation of the fatigue curve difficult and less accurate. Evaluation of the mean force over a specific period (e.g., 30 or 60 s) does provide a measure of isometric endurance and the total amount of fatigue. Mean force over the selected period can give valuable information that can more accurately be compared among participants. Other isometric performance tests can be quantified in several ways, with fatigue being defined as the time interval for which the individual can maintain a certain percentage of the maximum voluntary isometric muscle action. The percent decline can also be determined over a set period. The decline is calculated simply as the percentage of initial force that is being maintained at the end of the selected period.

## Pretest Considerations

Calibration of the isometric testing device should be performed before testing and at periodic intervals. Calibration should be performed through the entire range of measurement of the device, because a nonsignificant error at light resistances might be magnified at greater resistances (Sale et al., 1988). Before testing, all individuals should warm up with activities using the musculature to be tested (i.e., light stationary cycling before testing peak isometric torque of the knee extensors or flexors). The individual must not become fatigued; therefore, an activity with little or no resistance should be used. Following the general warm-up, some practice trials may be performed with submaximal effort (50-60% maximal voluntary contraction [MVC]). This activity serves as both a specific warm-up and a preparatory tool before exerting maximal effort. The actual test requires only two or three maximal voluntary efforts (Sale et al., 1988). Additional trials may be necessary if, in fact, it appears that the individual is still improving. Generally, the peak force for the highest trial is recorded because averaging the scores does not improve reliability and penalizes the individual for a top voluntary effort (Alderman & Banfield, 1969).

## Body Position, Test Specificity, and Visual Feedback

Every test protocol requires proper isolation of the joint and position of the individual. For isometric testing, positioning must be carefully quantified with regard to the specific joint angles used in each individual setup. Correct positioning enhances the reliabilities of isometric strength tests. First, each individual must be comfortable because discomfort may lead to submaximal performance (Sale et al., 1988). The surface of the dynamometer must be comfortably padded for optimal force application. Care must be taken, however, not to use too much padding, which can result in less than maximal forces because of absorption of the forces by the padding material. The individual must also be securely stabilized in the desired position. Otherwise, the data might be contaminated because different body positions can result in different forces (Lewis & Spitler, 1989). During maximal isometric efforts, restraining straps may possibly give way or shift, resulting in altered body positions. The angles of nontested joints are often overlooked. For example, although a test may be concerned only with knee extensor capabilities, the associated hip angle can affect the resulting knee-extension forces (Nosse, 1982) because of the effects of different levels of muscle stretch on force production (Thomson & Chapman, 1988). Therefore, careful attention must be given to isolating the joints tested as well as adjacent areas that could contribute to observed peak force measurement.

All isometric testing is specific to a velocity of zero, so it is not always appropriate to compare these results with those of dynamic testing (Fry & Powell, 1987a; Fry, Powell, & Kraemer, 1992; Murphy & Wilson, 1996). Although the relationship between dynamic and isometric strength is positive when body positions between the two are similar ($r = .57$ to $.77$) (Baker, Wilson, & Carlyon, 1994; Blazevich et al., 2002), the relation between dynamic strength increases and isometric strength increases during resistance training is not high. Following 12 weeks of dynamic resistance training, significant increases in 1RM squat and bench press have been reported with smaller, unrelated

increases in isometric strength (Baker et al., 1994). The concept of specificity of testing must be considered. That is, test protocols should be identical to the training modality when possible. Research that utilizes dynamic training but assesses strength changes only with isometric methods is not evaluating (and may underestimate) specific dynamic training improvements (Amusa & Obajuluwa, 1986). Therefore, the testing tool must be identical to the training tool to serve as a valid assessment of functional changes resulting from the mode-specific training.

Isometric testing can provide valuable joint-specific data to evaluate various sport-specific movement angles. Furthermore, isometric testing can help evaluate joints in rehabilitation when dynamic movements are not possible. Isometric tests can also provide information on carryover effects from dynamic training. Thus, the magnitude and pattern of strength changes at the various joint angle forces can be compared with 1RM data. Additionally, unique isometric tests can be developed to help evaluate injury and sport-specific gains in muscular strength. In one study, sport-specific volleyball tests for the two-arm overhead block, two-arm underhand bump, and single straight-arm dig were developed and shown to be sensitive to dynamic training and detraining (Fry et al., 1991). Isometric testing must be properly integrated and used in a strength-testing battery. Isometric test data can certainly provide valuable information on force production characteristics of muscles in various joints and should not be overlooked in a testing profile.

Individuals performing an isometric test should be provided concurrent feedback concerning their performance. Feedback is especially important for unfamiliar tasks (Graves & James, 1990). Local muscular endurance may actually decrease in the absence of feedback (Graves & James, 1990). Typically, visual feedback is the most important type of feedback, but oral cues of the magnitude of force can also be given along with encouragement. Oral cues are important in many maximal efforts during which individuals may tend to close their eyes during exertion. Familiarization and specifics of the test protocol must be thoroughly explained such that visual cues, oral cues, and breathing patterns are accounted for. These factors can contribute to optimal reproducibility of isometric tests. Reproducibility is quite good with many types of isometric performance characteristics such as maximum force, plateau force, inflection force, and maximal rate of force increase. Variables such as time to percent maximum force and force curve shapes, however, have greater variability, resulting in poor reproducibility (Going, Massey, Hoshizaki, & Lohman, 1987).

## Physiological Responses to Isometric Testing

Physiological responses to isometric strength measures pose further considerations. Research has shown that heart rate and blood pressure increase during tests of this type (Nagle, Seals, & Hanson, 1988). Interestingly, these increases were independent of time of muscle activity but were dependent on the magnitude of force produced and the muscle mass involved. Valsalva maneuvers can also affect these variables. All individuals should be instructed not to hold their breath during tests of long duration such as isometric fatigue tests. The health profiles of the individuals must be carefully inspected to avoid any testing that might be harmful to them. For example, caution is required when testing cardiac patients who are not clinically stable (Faigenbaum et al., 1990).

## Isokinetic Testing

Isokinetic assessment of muscle strength has been used in research, athletic, and clinical settings since its rise in popularity in the late 1960s. Assessing isokinetic muscle strength gives the researcher or practitioner valuable information regarding mostly open kinetic chain torque production during different muscle actions and at various angular velocities, which may be used in the rehabilitation of injuries and identification of strengths and weaknesses between limbs or between agonist and antagonist muscle groups. These data can be compared with norms (i.e., gender, age, activity level, and so on), which serve as the basis of strength-training programs designed to increase muscle strength, balance, endurance, hypertrophy, and power to increase motor performance and reduce the risk of injury.

Isokinetic testing is performed with a dynamometer that maintains the lever arm at a constant angular velocity (Sale et al., 1988). This type of strength evaluation accounts for velocity of movement that is uncontrolled with free weights and machines, and that is lacking in isometric evaluations. Most isokinetic dynamometers resist generated forces proportionally to that which is applied. That is, it acts passively as an accommodating resistance device. Other dynamometers also use a motor for resistance (i.e., active), enabling isokinetic eccentric loading (Osternig, 2000). The measurement of isokinetic eccentric torque is unique because it has been used only over the past decade. Isokinetic assessments have become quite popular in recent years as technology has advanced, and a wide variety of protocols are being used (Altug et al., 1987; Kellis & Baltzopoulos, 1995). The cost of an isokinetic dynamometer can be prohibitive, but many laboratories, training rooms, and clinical facilities use them extensively. Similar to machine testing and different from assessing strength with free weights, isokinetic testing does not monitor changes in motor coordination because the individual is restricted to the direction of movement dictated by the dynamometer. Thorough reviews of isokinetic testing have been previously reported (Gaines & Talbot, 1999; Osternig, 1986; Sale et al., 1988).

A number of isokinetic-like devices (e.g., hydraulics) are available on the market. Although these machines are able to control velocity of movement to a certain extent, they are not identical to an isokinetic dynamometer. The reliability of hydraulic devices has been found acceptable (LaChance, Katch, Mistry, & Hortobagyi, 1988), but their results are unique and specific to the instrument and should not be confused with those of true isokinetic modalities (Hortobagyi, LaChance, & Katch, 1987). Other devices on the market are isokinetic in nature but utilize multiple-joint movements. As a result, the velocity of the machine lever arm rather than the individual's joint angular limb velocity quantifies exercise velocities. The use of multiple-joint movements would make isokinetic testing more applicable to many athletic movements. Fry, Webber, Weiss, Fry, and Li (2000) reported high reliability of an isokinetic squat dynamometer (coefficient of variation = 7.0%). In another study, isokinetic squat tests were performed at 0.4 $m \cdot s^{-1}$, whereas knee extensions were performed at 1.05, 2.09, and 3.14 $rad \cdot s^{-1}$. Intraclass correlations ranged from .89 to .96 (with coefficients of variation <9%) (Wilson, Walshe, & Fisher, 1997). Interestingly, the squat tests had a higher correlation to cycling performance ($r = .57$ to .65) than knee extensions did ($r = .45$ to .51). The squat tests were also superior for discriminating between different levels of cycling ability. These results demonstrate the importance of testing specificity as it relates to the isokinetic assessment of multiple-joint movements and their relationship to athletic performance.

### Preliminary Considerations

Because of the complex nature of isokinetic assessments, a number of considerations are important:

selection of the angular velocity and order;

selection of the number of repetitions to be performed;

selection of rest intervals;

selection of muscle action or actions;

range of motion;

test position;

equipment calibration and test–retest reliability;

feedback, instruction, and familiarization; and

gravity compensation.

Other considerations include determining maximum and minimum torque limits of the dynamometer before testing, testing the uninvolved side first, and ensuring that the testing staff is experienced with isokinetic testing and the protocol used for the investigation or clinical assessment (Davies, Heiderscheit, & Brinks, 2000).

## Angular Velocity

Of primary importance is the specificity of angular limb velocities. The accepted SI unit for angular limb velocity is radians per second ($rad \cdot s^{-1}$), although it is common to report the velocities in degrees per second with $1 \, rad \cdot s^{-1} = 57.29578° \cdot s^{-1}$ (Sale et al., 1988). The velocity used for concentric testing should be carefully considered because the resulting torques can increase with decreasing velocities, that is, force/torque-velocity relationship (Cisar et al., 1987; Fry & Powell, 1987a; Fry et al., 1992; Walmsley & Szybbo, 1987). The testing angular velocity should also reflect the training velocities. Studies examining isokinetic resistance exercise have shown strength increases specific to the training velocity with some spillover above (as great as $210° \cdot s^{-1}$) and below (as great as $180° \cdot s^{-1}$) the training velocity mostly at moderate and fast velocities, but not at very slow velocity (e.g., $30° \cdot s^{-1}$) (Fleck & Kraemer, 1997). Several researchers have trained individuals at $30° \cdot s^{-1}$ (Knapik, Mawdsley, & Ramos, 1983), $36° \cdot s^{-1}$ (Moffroid & Whipple, 1970), $60° \cdot s^{-1}$ (Coyle et al., 1981; Kanehisa & Miyashita, 1983), $100° \cdot s^{-1}$ (Tomberline et al., 1991), $108° \cdot s^{-1}$ (Moffroid & Whipple, 1970), $120° \cdot s^{-1}$ (Housh, Housh, Johnson, & Chu, 1992), $179° \cdot s^{-1}$ (Kanehisa & Miyashita, 1983), $240° \cdot s^{-1}$ (Going et al., 1987), and $300° \cdot s^{-1}$ (Coyle et al., 1981; Kanehisa & Miyashita, 1983) and reported significant increases in muscular strength. Training at moderate velocity ($180$-$240° \cdot s^{-1}$), however, appears to have produced the greatest strength increases across all testing velocities (Kanehisa & Miyashita, 1983). Therefore, a spectrum of velocities (e.g., $30$-$360° \cdot s^{-1}$) may be selected, typically performed in order from slowest to fastest.

## Number of Repetitions

At least 3 to 5 repetitions, and at least 5 for elderly individuals, should be performed for isokinetic strength measurements (Capranica, Battent, Demarie, & Figura, 1998; Davies et al., 2000). Tests of local muscle endurance will require more repetitions, up to 50 (Brown & Whitehurst, 2000). Several submaximal warm-up efforts should precede the test itself. Although these warm-ups do not necessarily enhance the resulting performance (O'Conner, Simmons, & O'Shea, 1989), they serve as a safety precaution (Osternig, 1986; Sale et al., 1988). For the actual test, individuals should be instructed to perform their repetitions as forcefully and as quickly as possible (Sale et al.). The greatest torques often come on the second or third repetition (Sale et al., 1988).

## Rest Intervals

Rest intervals between efforts are often approximately 1 min in length. Rest intervals as short as 10 s between sets (Caiozzo, Perrine, & Edgerton, 1981; Dudley & Djamil, 1985; Edgerton & Perrine, 1978) or as long as 3 min between single repetitions (Conroy, Stanley, Fry, & Kraemer, 1984) have been used. A rest interval of 15 s appears to be too short to allow optimal recovery between sets (Parcell, Sawyer, Tricoli, & Chinevere, 2002), but no differences have been observed between 60, 180, and 300 s rest intervals (Parcell et al., 2002). Full recovery should occur between efforts, but 3 min rests between repetitions did not enhance recovery when compared with three consecutive repetitions (Conroy et al., 1984). Thus, at least 40 to 90 s between sets has been recommended (Davies et al., 2000; Parcell et al., 2002; Wrigley & Strauss, 2000).

## Muscle Actions

Isokinetic equipment enables use of numerous protocols including isometric, isokinetic concentric, isokinetic eccentric, isokinetic concentric and eccentric, isokinetic eccentric and concentric, and endurance testing. Isokinetic torque is greatest during eccentric muscle actions, followed by isometric and concentric (Fleck & Kraemer, 1997). Torque of the concentric phase is increased by a preceding eccentric phase or isometric preload (Aydin, Yildiz, Yildiz, & Kalyon, 2001). Most dynamometers allow reciprocal test performance (e.g., knee extension followed by flexion), which is useful for determining agonist-to-antagonist ratios. Another option is to perform single movements in one direction, that is, three to five repetitions of knee extensions. Limited data are available to compare these approaches, but it appears that similar results may be obtained with each for torque (Wrigley & Strauss, 2000), although a greater acceleration rate is observed with reciprocal repetitions because of antagonist muscle activation of agonists (Brown et al., 1998).

## Range of Motion

Range of motion during isokinetic testing is important to record because it relates to torque production. For example, if a trial is performed at 90° for a joint with a physiological ROM of 180°, the torque curve will be different when examining 0 to 90° versus 90 to 180°. As previously discussed, torque varies considerably throughout the full joint ROM, with many joints yielding ascending–descending curves (Kulig et al., 1984). Thus, identifying the corresponding ROM with torque produced is important for accurate assessment. Standardizing ROM may be difficult at times, especially when comparing different dynamometers. In addition, the actual ROM where maximal force is exerted is affected by the use of a preload. Preloading the limb before measurement increases torque production and allows exertion through a greater ROM because of reduced acceleration range (Brown & Whitehurst, 2000).

## Test Position

To collect meaningful data, the position of the individual during isokinetic testing needs to be standardized. As with other types of strength testing, various limb positions can result in altered torque production (Lewis & Spitler, 1989). Joints that have extremely large ranges of motion in many directions, such as the shoulder, must be carefully secured in the proper position. For example, assessments of internal and external rotator cuff torques depend on the position assumed at the shoulder (Walmsley & Szybbo, 1987). In addition, contributions from other muscles such as the trunk rotators must be eliminated. Generally, only one joint is tested at one time using an open kinetic chain, so it is necessary to stabilize all body parts that may affect the torque measurement. A test of the knee extensors and flexors requires stabilization at the ankle, thigh, hips or waist, and upper torso, but it is not necessary to stabilize the nontested limb because movement there does not affect the results (Patteson, Nelson, & Duncan, 1984). All straps used for securing must be snug but comfortable, without interfering with the proper range of motion of the exercise. All securing devices should be regularly inspected as a safety precaution. Devices such as Velcro straps will wear out with usage. Some test protocols require participants to grasp handles by their sides (Caiozzo et al., 1981; Dudley & Djamil, 1985). This method has been shown to enhance isokinetic knee-extension torque in men (Stumbo et al., 2001). If this procedure is chosen it should be consistent for all individuals. Another problem with positioning occurs when utilizing chairs and devices, such as an upper-body exercise table, that are not directly attached to the dynamometer. Very large or strong individuals can cause the whole device to move, resulting in altered joint angles. Individuals must be completely stabilized in these instances.

Secure stabilization must occur when proper positioning is determined. Some of the more common testing protocols include muscle strength assessment of knee extension and flexion, ankle plantar and dorsiflexion, internal and external shoulder rotation, hip extension and flexion, elbow flexion and extension, and trunk flexion and extension. For example, the recommended test position for knee extension is a flexed hip position at 80° with secure stabilization of the thigh, pelvis, and upper torso; for knee flexion in a prone position with the hip in the neutral position; for prone ankle plantar and dorsiflexion with knee and hip at 0°; and for seated shoulder internal and external rotation with 45° humeral abduction and 90° elbow flexion (Wrigley & Strauss, 2000). Thus, secure positioning within the constructs of the dynamometer is mandatory to prevent slipping, extraneous movements, and contribution from other muscles not involved in the movement.

## Alignment of the Axis of Motion and Equipment Calibration

Proper torque measurements can occur only if similarity exists between the limb of the individual and the lever arm of the dynamometer. Considering that torque is the product of force and lever arm length, misalignment of the lever axis and the joint axis will lead to erroneous data collection. Thus, the rotational axis of the dynamometer and the axis of rotation of the joint tested must be aligned for proper data collection. Attaining the correct position is not difficult for many joints, but achieving it for joints such as the shoulder and trunk may be difficult.

Maintenance of the equipment is critical to proper isokinetic strength testing. Isokinetic testing equipment must be calibrated regularly according to the manufacturer's guidelines for proper data collection. Calibration is typically performed with verified calibration load.

## Test–Retest Reliability

The reliability of the testing dynamometer and protocol used must be high for acceptable data to be collected. In fact, intraclass coefficients ($R$) should be determined and reported within scientific publications. Isokinetic data should be obtained over several testing sessions to establish reliability of the equipment and procedures. When determining dynamometer reliability that includes eccentric muscle actions, postexercise muscle soreness should be considered before scheduling subsequent sessions (Emery, Maitland, & Meeuwisse, 1999). Test–retest reliability of isokinetic dynamometers and protocols has been shown to be high for numerous exercise protocols (Aydin et al., 2001; Dvir & David, 1996; Dvir & Keating, 2001; Emery et al., 1999; Gaines & Talbot, 1999). Reliability significantly increases when standardized protocols are used, equipment is properly calibrated, visual feedback is given to the individuals, healthy individuals are tested, and familiarization has occurred (Kellis & Baltzopoulos, 1995; Kim & Kramer, 1997). Some studies have shown higher reliability in women than in men (Dvir & Keating, 2001). Some isokinetic dynamometers have exhibited high day-to-day reliability of $R = .99$ when assessed with inert weights (Osternig, 1986). All isokinetic dynamometers should be monitored for test–retest reliability to assure accurate data (Seger, Westing, Hanson, Karlson, & Ekblom, 1988). Data from two different brands or models of dynamometers cannot always be compared because their results may be significantly different (Francis & Hoobler, 1987). Furthermore, the actual testing apparatus, such as the mechanism to anchor the limb to the machine, can also affect the results (Epler, Nawoczenski, & Englehardt, 1988).

A key to producing high equipment and protocol test–retest reliability is obtaining maximal effort by the participants. When determining test–retest reliability over several testing occasions, it is critical that individuals provide maximal effort. Submaximal efforts on one day, for example, increase the coefficient of variation and decrease test–retest reliability. In fact, protocols have been developed to differentiate submaximal from maximal effort in healthy individuals during testing of knee extensor (Dvir & David, 1996) and trunk extensor strength (Dvir, 1997). Proper familiarization and education in the testing protocol increase the likelihood of obtaining maximal effort.

## Verbal Instructions, Familiarization, and Feedback

Instructions given to participants should be consistent from one testing session to the next. Familiarization sessions assist in teaching participants the protocol and procedures for isokinetic strength testing. During the actual test, verbal instructions should be given to the individual to ensure maximal performance (because unfamiliarity may limit effort produced). Verbal instructions regarding proper breathing, test performance, number of repetitions to perform, and encouragement to exert maximal force during the test should be given.

Optimal isokinetic testing protocols allow test participants to observe their performance results as they perform the test. This type of feedback is beneficial (Graves & James, 1990), but all individuals should be made aware of the feedback in advance of each test. This practice helps assure similar efforts for each test. Augmented sensory feedback is critical to maximal motor unit activation and subsequent strength performance, as well as to obtaining higher reliability coefficients, particularly at slow angular velocities and, to a lesser extent, at high velocities (Kim & Kramer, 1997). Torque has been reported to be 3 to 19% higher when visual feedback is used during testing (Kim & Kramer, 1997).

## Correction for Gravity

The torque produced during isokinetic testing is the algebraic sum of the net torque applied by the participant, inertial torque because of acceleration or deceleration of the limb–lever arm system, and gravitational torque (Wrigley & Strauss, 2000). Gravity, that is, the weight of the individual's limb, can have a large effect on some isokinetic tests (Kellis & Baltzopoulos, 1995; Thorstensson, Grimby, & Karlsson, 1976). Some studies (Poulmedis, 1985) correct for gravity because gravitational effects can significantly alter the results. Current machines have gravity correction capabilities. Errors of over several hundred percent have been documented in flexion at the knee in certain situations (Thorstensson et al., 1976). Because this error enhances results for only one of the muscle groups tested, muscle balance characteristics can also be greatly altered (Appen & Duncan, 1986). Some tests, such as fatigue tests, might not require this correction factor because absolute torque development is not of primary concern (Sale et al., 1988).

## Miscellaneous Items Related to Isokinetic Testing

Several important aspects of isokinetic assessments are often overlooked. Three phases have been associated with a maximal isokinetic trial: an acceleration phase in which the machine makes adjustments for the preset limb velocity, a constant velocity phase in which torque measurement occurs, and a deceleration phase (Osternig, 1986). These phases are affected by limb velocity and range of motion, with faster velocities requiring longer acceleration (Osternig, Sawhill, Bates, & Hamill, 1983) and deceleration phases. Some experts have used a protocol in which maximal effort was delayed until approximately one-half the range of motion had passed, which resulted in consistently greater torques (Edgerton & Perrine, 1978). Although attributed to a fatigue effect (Osternig, 1986), this phenomenon may be influenced by the muscle fiber contraction velocities or dynamometer characteristics as well. Note also that a decreased range of motion for maximal effort or an increased limb velocity would also decrease the constant-velocity portion of the curve, thus making it more difficult to measure torque during this phase. Increased limb velocity has also been associated with a decreased angle of peak torque for extension at the knee (Thorstensson et al., 1976). It was theorized that this was due to the time necessary for muscle contractile force to reach its maximum, which at greater limb speeds lagged slightly behind the time to reach the anatomically optimal joint angle (Osternig, 1986). Care must also be taken to avoid measuring torque during the deceleration phase, late in the range of motion. In general, angle-specific torques can provide valuable information for isokinetic testing.

Early phases of an isokinetic effort are characterized by large oscillations in the strength curve (Edgerton & Perrine, 1978; Osternig, 1986). This result is primarily due to the mechanics of the dynamometer as it makes constant adjustments to assure that the preset velocity is attained. Velocities of up to 200% of the selected speed have been recorded during the initial acceleration (Sapega, Nicholas, Sokolow, & Saraniti, 1982). A braking mechanism within the dynamometer slows the limb to the desired velocity. Velocity overshoot occurs as the limb accelerates past the desired velocity. The subsequent braking results in a torque overshoot that is mild at slow velocities but increases in magnitude with greater velocities (Brown & Whitehurst, 2000). To avoid this problem area of the curve, some investigators have measured torque only at a later point in the strength curve, such

as 20 to 50° of knee flexion (Caiozzo et al., 1981; Dudley & Djamil, 1985). The use of the damping control will smooth out this portion of the curve, but caution must be used not to filter out important characteristics of the curve as it shifts the torque curve to the right. Furthermore, the damping must be set consistently from one test to the next.

## Data Analysis and Interpretation

Each isokinetic system has its own software program and format for data collection and reporting. Along with graphic printouts, numerous performance-related variables are measured and can subsequently be used for data analyses. The variables available will depend on the brand and model of dynamometer used as well as the software package accompanying it. Figure 8.3 depicts several of these variables. Although a number of measures can be made, care must be taken to avoid redundancy. Note also that different training programs can have differing effects on the shape of the torque curve (Caiozzo et al., 1981). This effect can be due to the training velocity, the range of motion, or the motor patterns used in training. This specificity of testing to training can also explain why isokinetic performances can lag behind other strength performances in a training program (Fry et al., 1991).

Data obtained from isokinetic testing can be interpreted in many ways. Bilateral comparisons involve evaluating one limb in relation to the other and are common in data analysis. These properties depend on the test conditions and will vary considerably between different joints and individuals. In general, a difference of 10 to 15% is considered asymmetrical, and higher deficits may reflect an imbalance (Davies et al., 2000). This deficit appears mostly at slow testing velocities and decreases as the velocity increases (Wrigley & Strauss, 2000). In some instances these data may suggest that the individual is more predisposed to an injury, but this is probably not the case. For example, an athlete may have a deficit (i.e., limb dominance) because of the unilateral nature of his or her sport and reliance on one limb more than the other. Bilateral comparisons can give the researcher some useful information, but the interpretation needs to consider the individual's activity level and history, training status, and health and injury status.

Reciprocal ratios (agonist-to-antagonist ratios) are also commonly used. These ratios examine the relationship between opposing muscle groups, which may be useful for identifying weaknesses. The weaker muscle group is expressed as a percentage

**Peak torque:** Highest point of the torque curve. The most commonly used parameter during isokinetic testing. Corresponds to the area of maximal torque on a strength curve. Measured in newton meters (N·m). May also be expressed relative to body weight.

**Average peak torque (N·m):** The average of all peak torques for each repetition performed.

**Average torque (N·m):** The average of torque produced for the entire torque curve.

**Angle-specific torque (N·m):** Represents torque produced at a specific joint angle of the range of motion.

**Total work (J):** The total amount of work performed based on the number of repetitions. Total work is the product of average torque (N·m) and angular displacement (rad). May be expressed relative to body weight and as the average work performed per repetition.

**Peak power (W):** The maximum product of torque and velocity (highest power attained during the trial).

**Average power (W):** Total work divided by the time taken to perform the work. Average power may be expressed relative to body weight.

**Torque acceleration energy:** The amount of work performed in the first 1/8 of a second.

**Acceleration time:** The time to the variable of interest (i.e., peak torque or joint angle) has been determined.

**Average points variance:** A guideline determining consistency set forth to ensure that all repetitions performed are within 10% above and below the average.

**Rate of peak torque development:** Time it takes to reach peak torque during a trial.

**Endurance (fatigue) ratio and index:** Determined by endurance protocols in which either maximal number of repetitions is performed until a specific reduction (i.e., 50%) is observed, percent loss of torque is calculated over a test, or total work is measured for a specified number of repetitions and is compared with another specified group.

**Force decay rate:** Down slope of a torque curve.

**Torque-velocity relationship:** Useful tool for analysis of torque produced at a specified velocity.

**Reciprocal innervation time:** Time between agonist and antagonist muscle actions.

**Isomap:** Multiple dimension analyses.

**Figure 8.3**   Common parameters obtained from isokinetic testing.

of the stronger group. As the velocity of assessment changes, the characteristics of the two muscle groups can result in variable ratios. For example, hamstring-to-quadriceps ratios increase with greater velocities (Fry & Powell, 1987a; Nosse, 1982), whereas internal-to-external rotator ratios are not as affected because torque for internal rotator muscles is not affected by limb velocity (Walmsley & Szybbo, 1987). The determination of muscle balance ratios is an area full of potential pitfalls. For example, the value of 0.60 for hamstring–quadriceps torques was widely accepted yet is dependent on limb velocity and joint angle (Fry & Powell, 1987a; MacDougall, Tuxen, Sale, Moroz, & Sutton, 1985).

Eccentric–concentric torque ratios have also been used regularly. Eccentric torque production is greater than concentric torque production, especially at higher velocities of movement. This ratio should be greater than 100% for each muscle group tested (Wrigley & Strauss, 2000).

# Overview of Testing Considerations

Strength testing is ideal for determination of one's maximal force-producing capacity. However, certain circumstances make it difficult to test strength. Therefore, other methods of assessment/estimation are necessary to assess the efficacy of a training program. The following section discusses these alternative testing approaches.

## RM Testing

Testing of maximal effort with resistance exercise usually involves assessment of an individual's RM for a desired number of repetitions. This procedure has been used for a number of years (Clark, 1967) and is usually associated with 1RM testing (Berger, 1962), but it can involve any number of repetitions.

As previously described, 1RM testing involves several one-repetition trials with increasing resistance until failure. Multiple RM testing, such as 5 or 10RM, is more difficult to assess because the individual will often be able to perform greater or fewer repetitions. Furthermore, the greater number of repetitions involved can contribute to the onset of fatigue. Making more than one effort at 5 or 10RM can be exhausting and can introduce greater chance for error. Highly trained individuals, however, will generally be able to self-report their 5 or 10RM capabilities, making this type of testing easier with this population (Fry, Schmidt, Johnson, Tharp, & Kraemer, 1993). Two nonconsecutive days of 1RM testing have been used in some cases, with the best performance of the two recorded (Hill, Collins, Cureton, & DeMello, 1989). This method helps eliminate random poor performances and helps establish greater reliability. Note that as one deviates from 1RM to higher RM values (e.g., 10-12RM or higher), the measurement of local muscular endurance becomes the primary focus. Therefore, use of a 1RM or near 1RM (e.g., 3RM) will be most accurate at determining maximal dynamic strength.

## Estimating RM

Many test situations simply estimate 1RM capabilities from a multiple RM performance trial or from muscle cross-sectional area (CSA) assessments (Mayhew, Piper, & Ware, 1993; Mayhew et al., 1995; Morales & Sobonya, 1996). Table 8.3 depicts some of the popular prediction equations that have been investigated among athletes (Mayhew et al., 1995; Mayhew et al., 1999), men and women with little experience (Cummings & Finn, 1998; Mayhew et al., 1992), and older adults (Knutzen, Brilla, & Caine, 1999). Determination of a specific RM has been discussed previously in this chapter. The selection of the load in these studies was arbitrary but has ranged from 55 to 95% of 1RM (i.e., 2 to 20 repetitions) with the mean in most studies approximately 68 to 80% of 1RM (Mayhew et al., 1992; Morales & Sobonya, 1996; Ware, Clemens, Mayhew, & Johnston, 1995). Once a specific RM and actual 1RM are determined for a large number of participants, the load and repetition number are entered into a regression analysis to develop a regression equation for 1RM prediction. This procedure is attractive from an administrative standpoint, but its validity can be seriously questioned because these equations have been shown to underestimate and overestimate 1RMs in certain populations in cross-validation studies (Cummings & Finn, 1998; Knutzen et al., 1999; Walmsley & Szybbo, 1987; Mayhew et al., 1995). In fact, each equation could potentially generate a different 1RM value when examining a single selected sample of the population (Chapman, Whitehead, & Binkert, 1998). Conversion factors have been developed that allow, for example, a 10RM resistance to be equated with a 1RM resistance (Abdo, 1985; Lander, 1985; Rose & Ball, 1992), but these can lead to erroneous results because different individuals and different exercises exhibit variable relationships when comparing 1RM capabilities with multiple-repetition tests (Hoeger, Barette, Hale, & Hopkins, 1987). Inaccuracy may result from differ-

Table 8.3    **Formulas Used to Estimate 1RM Lifting Performance**

| Reference | Equation |
|-----------|----------|
| Brzycki (1993) | 1RM = Wt./1.0278 − 0.0278(# reps)<br>%1RM = 102.78 − 2.78(# reps) |
| Epley (1985) | 1RM = 0.033(Wt.)(# reps) + Wt. |
| Lander (1985) | 1RM = Wt./1.013 − 0.02671(# reps)<br>%1RM = 101.3 − 2.67123(# reps) |
| Mayhew et al. (1999) | 1RM (lb) = 226.7 + 7.1(# reps w/225) (used in college football players) |
| Cummings & Finn (1998) | 1RM = Wt.(1.149) + 0.7119<br>1RM = Wt.(1.175) + # reps(0.839) − 4.2978 (used in untrained women) |
| Mayhew et al. (1992) | $1RM = Wt./\{[52.2 + 41.9e^{-0.055(\# reps)}]/100\}$<br>$\%1RM = 52.2 + 41.9e^{-0.055(\# reps)}$ |
| O'Connor et al. (1989) | 1RM = Wt. (1 + 0.025 × # reps) |
| Wathen (1994) | $1RM = 100 × Wt./[48.8 + 53.8e^{-0.075(\# reps)}]$ |
| Abadie, Altorfer, & Schuler (1999) | 1RM = 8.8147 + 1.1828(7-10RM) |

ences in the neuromuscular and metabolic demands associated with low (i.e., 1-6RM) and moderate to high RMs (i.e., 10-12RM and higher). When fewer than 10 repetitions are performed, accuracy of these estimations does appear greater when using nonlinear equations (Chapman et al., 1998; Mayhew et al., 1995; Mayhew et al., 1999; Morales & Sobonya, 1996), but not all studies produce that conclusion (Mayhew et al., 1992). For example, Mayhew et al. (1999) reported a standard error of estimate (SEE) of 17.1 lb (7.8 kg) when more than 10 repetitions were performed versus a SEE of 11.4 lbs (5.2 kg) for fewer than 10 repetitions during the 225 lb (102 kg) bench press test. In comparisons between different athletic groups and lesser-trained men, Mayhew et al. (1995) reported a range of SEEs of 4.0 to 4.5 kg when fewer than 10 repetitions were performed versus a range of 6.5 to 16.7 kg when more than 10 repetitions were performed. For field settings involving large numbers of athletes and limited time, clinical populations, or untrained individuals with little resistance-training experience, a case may be made for this type of estimation of a 1RM, but definitely never for research purposes. Such scales can help target the actual 1RM test and may be used with training logs as an indirect form of evaluation without specific 1RM testing.

Many of the aforementioned studies generating prediction equations of 1RM have used the bench press exercise; thus the accuracy appears greater during this exercise (LeSuer, McCormick, Mayhew, Wasserstein, & Arnold, 1997). The error rate increases significantly when these equations are applied to other exercises. When examining the 1RM squat in athletes, prediction equations were found to overestimate 1RM squat by 11 to 34% (Ware et al., 1995). All these equations, however, underestimated 1RM squat in college students by 5 to 17 lb (2.3 to 7.7 kg) (LeSuer et al., 1997). Larger differences (i.e., underestimations) have been reported for the deadlift (9 to 14%) in this same sample (LeSuer et al., 1997). Morales and Sobonya (1996) performed an interesting study of Division I-A track and field throwers and football players. The participants performed the bench press, squat, and power clean at 70, 75, 80, 85, 90, and 95% of their 1RMs. Morales and Sobonya reported that the best predictor for 1RM bench press was 95% of 1RM (about two repetitions), 80% of 1RM (about eight repetitions) for the 1RM squat, and 90% of 1 RM (four to five repetitions) for the 1RM power clean. These data demonstrate the variability between exercises and the need for exercise-specific equations for improved accuracy.

Prediction tables based on regression equations have been developed (Abdo, 1985; Lander, 1985;

Wathen, 1994). As with equations themselves, these tables should be used only as estimates. The utility of these tables is that they not only enable 1RM prediction but also are useful for estimating other RMs from either a 1RM or multiple RM. These approximations may be useful in estimating training loads during resistance training. Accuracy, however, is questionable. The number of repetitions performed relative to the 1RM for different exercises is highly variable (see table 8.4). For example, Hoeger et al. (1990) reported that at 80% of 1RM, double the number of repetitions could be performed for the leg press compared with the leg curl. Hoeger et al. (1990) also reported that training status and gender can affect the number of repetitions performed. The amount of muscle mass used (i.e., large muscle mass may lead to a higher number of repetitions per intensity) and the selection of either a free weight or machine for the same exercise may also affect the accuracy of these tables (Wathen, 1994). These factors need to be considered when examining a prediction table that is "one size fits all" in relation to exercise selection, training status, gender, and health status.

## Multiple RMs and Muscle Endurance

Local muscle endurance tests are performed specifically to evaluate repetitive muscular capabilities of individuals. Although there is some relationship between 1RM capabilities and muscle endurance (Berger, 1967), it is generally poor. Some settings, however, such as industrial and military (Legg & Pateman, 1984; Sharp, Harman, Vogel, Knapik, & Legg, 1988), are primarily interested in local muscular endurance. Lifting endurance tests have even been used to estimate $\dot{V}O_2$ with satisfactory results (Sharp et al., 1988). An important consideration is what resistance to use. Relative loading is based on a percentage of the individual's RM capability (e.g., 60% 1RM), whereas absolute loading has the same resistance for all individuals and all tests (e.g., 50 kg or 100% of body weight) (Anderson & Kearney, 1982; Biddle, 1986). In a research investigation using a relative testing procedure, the amount of weight lifted posttraining will be greater, so the number of repetitions performed to failure may not change drastically (Mazzetti et al., 2000). But a significant increase in the number of repetitions performed will be observed using absolute loading (i.e., 50 kg pre and post). This entails specific examination of either high-intensity or submaximal endurance. For most individuals, the relationship

Table 8.4    **Number of Repetitions Performed Relative to 1 RM Capacity**

| Exercise | 40% | 60% | 70% | 75% | 80% | 85% | 90% | 95% |
|---|---|---|---|---|---|---|---|---|
| **Leg press** | | | | | | | | |
| UT | 80.1 | 33.9 | *** | *** | 15.2-20.3 | *** | *** | *** |
| TR | 77.6 | 45.5 | | | 19.4-21.0 | | | |
| **Lat pulldown** | | | | | | | | |
| UT | 41.5 | 19.7 | *** | *** | 9.8 | *** | *** | *** |
| TR | 42.9 | 23.5 | | | 12.2 | | | |
| **Bench press** | | | | | | | | |
| UT | 34.9 | 19.7 | 13.4-14.0 | 10.9-10.6 | 9.6-9.8 | 6.0 | 4.0 | 2.2 |
| TR | 38.8 | 22.6 | | 14.1 | 9.2-12.2 | | | |
| **Leg extension** | | | | | | | | |
| UT | 23.4 | 15.4 | *** | *** | 9.3 | *** | *** | *** |
| TR | 32.9 | 18.3 | | | 11.6 | | | |
| **Sit-up** | | | | | | | | |
| UT | 21.1 | 15.0 | *** | *** | 8.3 | *** | *** | *** |
| TR | 27.1 | 18.9 | | | 12.2 | | | |
| **Arm curl** | | | | | | | | |
| UT | 24.3 | 15.3 | *** | *** | 7.6 | *** | *** | *** |
| TR | 35.3 | 21.3 | | | 11.4 | | | |
| **Leg curl** | | | | | | | | |
| UT | 18.6 | 11.2 | *** | *** | 6.3 | *** | *** | *** |
| TR | 24.3 | 15.4 | | | 7.2 | | | |
| **Squat** | | | | | | | | |
| UT | *** | *** | 13.5 | 10.6 | 8.4 | 6.5 | 4.6 | 2.4 |
| TR | | | | | | | | |
| **Power clean** | | | | | | | | |
| UT | *** | *** | 13.6-17.0 | 11.5 | 9.3 | 7.0 | 4.6 | 2.6 |
| TR | | | | | | | | |

Data from Hoeger et al. 1990; Morales and Sobonya 1996; Mayhew et al. 1992; Ware et al. 1995; Kraemer et al. 1999; Mayhew et al. 1999.

between strength and absolute muscular endurance is positive, whereas the relationship between strength and relative endurance is negative (Biddle, 1986). Women have been shown to fatigue less with relative loads (Clarke, 1986; Sale & Delman, 1983), whereas men fatigue less with absolute loads (Clarke, 1986). An example of absolute load testing is the 225 lb (102 kg) bench press test currently performed by professional American football scouts as they evaluate prospective players. The validity of this test has been questioned, but it has been shown to correlate highly to the 1 RM bench press ($r = .96$), especially when fewer than 10 reps are performed (Chapman et al., 1998; Mayhew et al., 1999). The 1 RM bench press has the best predictive value for playing ability (Fry & Kraemer, 1991). Thus, careful consideration must be given to the purpose, validity, and appropriateness of this type of test for football players.

## Individual Training Status

For a strength-testing protocol, the training status of the participants is of primary importance (Häkkinen, 1989). Gains in muscular strength become more difficult to attain as one gains experience in training (American College of Sports Medicine, 2002). Because this window of adaptation decreases as one progresses, the resultant 1 RM scores will reflect the smaller gains associated with long-term training. Previous training experience affects the function of the nervous system (Sale & MacDougall, 1979), motor coordination (Nosse & Hunter, 1985; Rutherford & Jones, 1986), muscle CSA (Fleck & Kraemer, 1997), and the hormonal profile (Kraemer, Noble, Clark, & Culver, 1987). These effects all depend on the type (Jenkins, Thackaberry, & Killian, 1984; Thomee, Renstrom, Grimby, & Peterson, 1987), intensity, and duration of previous training (Anderson & Kearney, 1982).

Different intensities and durations of exercise provide different physiological stresses (Häkkinen, Kauhanen, & Komi, 1987). Furthermore, fiber-type composition can contribute to physical performance (Gregor et al., 1979). The effect of prior training differences can be seen in the different muscle performance characteristics of various athletes (Appen & Duncan, 1986; Cook, Gray, Savinar-Nogue, & Medeiros, 1987). Different strength profiles are also evident in similarly trained individuals of various calibers (Chmelar, 1988; Fry & Kraemer, 1991). Even motivational factors may contribute, with highly trained individuals affected the most (Ikai & Steinhaus, 1961). Depending on the variables being measured, short-term prior training may (Sale & MacDougall, 1979) or may not (Fry & Powell, 1987b) affect the monitored variables. Thus, the amount of strength improvement exhibited by an individual depends greatly on pretraining status.

## Age

The role of age in muscle strength has received increasing attention over the past decade. Youth and resistance training has received some interest (Faigenbaum et al., 1999; Kraemer & Fleck, 1993; Kraemer et al., 1989; National Strength and Conditioning Association, 1985), but much greater attention has been given to resistance training in the elderly in light of the many positive health benefits, such as increased muscle strength and power, bone mass, and muscle mass (DiBrezzo & Fort, 1987; Häkkinen, Alen, Kallinen, Newton, & Kraemer, 2000; Kraemer et al., 2002; Viitasalo, Era, Leskinen, & Häkkinen, 1985). Considering the vast benefits of resistance training in the elderly, the importance of strength testing has grown in this population. Using 1RM strength testing in this population has been shown to be safe if proper familiarization, supervision, and instruction are used (Adams et al., 2000; Hunter et al., 2001; Shaw et al., 1995). The decision to use a 1RM in this population is up to the discretion of the training or research team because multiple RMs (i.e., 3RM) have also been used effectively (Hurley et al., 1995). Of primary importance in this population is the need for greater familiarization (Ploutz-Snyder & Giamis, 2001). Individuals seem to be reluctant initially because most are not accustomed to exerting maximal effort for a movement that they may have never performed previously. Motivation factors may be different with older individuals; they may require more encouragement to achieve maximal efforts. For that reason, more familiarization sessions are needed before strength testing of the elderly to obtain accurate data. Care must also be taken to limit joint compression with any test and

to allow adequate recovery between testing sessions so as not to compromise results because of confounding muscle soreness or joint soreness.

In the past, it was believed that preadolescents could not increase strength apart from maturational increases, because of their immature hormonal profiles (Smith, 1984). But it has been shown that strength increases can be achieved through resistance training with this age group (Kraemer & Fleck, 1993; Sailors & Berg, 1987). Factors such as increased fat-free mass (Housh, Johnson, Hughes, et al., 1988; Kraemer et al., 1989) and neural development (Housh et al., 1989; Kraemer et al., 1989) can contribute to such strength gains in youth. When compared with adolescents, preadolescent males exhibit different muscle forces and contraction times, but they are able to achieve full activation of their motor units as early as 6 years of age (Belanger & McComas, 1989). A further observation is that muscle balance characteristics, including velocity specificity and joint specificity, are similar for youth and adults (Weltman et al., 1988). Reliability of strength testing of young age groups has been high, that is, $R = .93$ to $.98$ (Faigenbaum et al., 1999; Going et al., 1987), but testing must be carefully controlled. Injury to young individuals is an important concern during strength testing, but recorded injuries are few and occur most often during unsupervised training, when improper lifting techniques are used, when teaching has been inadequate, or when excessive loads are used (Brown & Kimball, 1983; Guy & Micheli, 2001; National Strength and Conditioning Association, 1985). Research has shown that strength testing is safe in this population if proper precautions are taken (Faigenbaum et al., 1999). The following protocol from Faigenbaum et al. (1999) has been used effectively for 1RM testing of children between the ages of 5 and 12:

1. Six repetitions are performed with a light load.
2. Three repetitions are performed with a heavier load.
3. One repetition is performed with approximately 95% of predicted 1RM.
4. Subsequent single-repetition trials are performed with weight increases of 1.0 to 2.5 kg until 1RM is obtained.

## Gender

Strength performance differences because of gender must be carefully considered when comparing sexes (Atha, 1981). Although men typically exhibit greater absolute strength, upper-body strength differences between men and women are greater than lower-body strength differences (Bishop, Cureton, & Collins, 1987; Heyward, Johannes-Ellis, & Romer, 1986). This circumstance is most likely due to fat-free mass levels, which have been reported to account for 97% of the gender strength differences (Bishop et al., 1987). The difference in fat-free mass levels may also explain why upper-body strength is sometimes more difficult to develop and maintain in women (Fry et al., 1991). In addition, women have different muscle fatigue patterns, with greater muscle endurance with relative loads and poorer muscle endurance with absolute loads (Clarke, 1986; Sale & Delman, 1983). Other variables, such as muscle balance characteristics, are similar for both genders (DiBrezzo & Fort, 1987). Some neural or structural differences may exist because the integrated electromyographic activity and peak torque are associated with different knee joint angles for men and women (Brownstein, Lamb, & Mangine, 1985). Isokinetic strength testing has shown this to occur because women have a higher coefficient of variation in their torque curves at higher velocities than men do (Akebi, Saeki, Hieda, & Goto, 1998). Some evidence also exists for anatomical gender differences contributing to altered lifting technique with untrained individuals (Fry, Bibi, Eyford, & Kraemer, 1990). Although anatomical position setups for all tests need to be individualized and recorded, care must be taken when testing women (e.g., in the squat, proper alignment may require a potentially wider spread of the feet to maintain foot hip alignment for proper exercise performance).

## Mode of Muscle Activity

An often overlooked aspect of strength testing is what type of muscle activity is tested—concentric, eccentric, or isometric. The issue is specificity of testing. Some types of training, such as isokinetic, might not have an eccentric component (Nosse & Hunter, 1985), thus affecting tests that do include eccentric activity (Hunter & Culpepper, 1988). Note also that eccentric activity generates greater peak torque than concentric activity does (Rizzardo, Bay, & Wessel, 1988). If different types of muscle activity are used for testing as compared with training, the test may not be sensitive to the physiological alterations that have occurred (Fry et al., 1991, 1992). Therefore, extensive familiarization and practice are needed if non-training-specific tests are used. Training using one mode of muscle activity should typically be assessed with an identical type of muscle activity within a strength-testing profile.

## Joint Angle

The angle of each joint will determine the amount of force produced. Outside factors, such as the type of equipment used for testing (e.g., isokinetic dynamometer), can affect the joint angles to be monitored (Caiozzo et al. 1981; Dudley & Djamil, 1985; Edgerton & Perrine, 1978). Force variation throughout the range of motion is depicted as strength curves (see figure 8.2). Most isolated joint movements produce ascending–descending curves, although some studies have reported either ascending or descending strength curves (Kulig et al., 1984). Thus, most single joints will produce maximal force midway through the range of motion and have areas where force development is limited (typically known as sticking points when referring to resistance exercise). The shape of the strength curve differs, however, when multiple-joint movements are assessed. For example, exercises such as the squat and bench press (assessed isometrically at various areas of the range of motion) have their area of maximal force at the end of the range of motion near full joint extension, that is, at 120° elbow angle for the bench press (Weiss et al., 2000; Wilson et al., 1991). Thus, all joint angles need to be monitored and kept constant for all testing. Changing joint angles can alter the contribution of various muscles or muscle groups, thereby affecting the results (Walmsley & Szybbo, 1987). Although the exact relationship is not clear, gender differences in joint-angle characteristics may exist (Brownstein et al., 1985). Testing that does not involve eccentric muscle actions of antagonistic muscles (e.g., some isokinetic or hydraulic instruments) can also affect force production because the stretch-shortening cycle is not emphasized (LaChance, Gabriel, Hortobagyi, & Katch, 1987).

## Velocity of Movement

Some test modalities, such as isokinetic or isometric, control the velocity of movement, whereas dynamic constant external resistance exercise does not. To select the appropriate velocity at which to test, several important factors should be considered. Velocity characteristics of prior training can affect velocity-specific test results (Caiozzo et al., 1981; Dudley & Djamil, 1985), with the greatest improvements occurring at the training-specific velocities (Thomee et al., 1987). Peak torque will vary at different limb velocities (Cisar et al., 1987) and will decrease as velocity increases (Fry & Powell, 1987a; Fry et al., 1992), but power increases as velocity increases (Rizzardo et al., 1988). Concentric muscle activity is affected by this more than is eccentric activity (Hageman, Gillaspie, & Hill, 1988). Greater limb velocities for training have been associated with preferential development of fast-twitch fibers (Thomas, 1984), which may explain why the velocity at which torque decreases is related to fiber composition (Gregor et al., 1979). Velocity considerations are also important when determining muscle balance characteristics (Fry & Powell, 1987a; Klopfer & Greij, 1988). Still, not all muscle groups exhibit altered torque production with different limb velocities (Cook et al., 1987; Walmsley & Szybbo, 1987).

## Nutritional Concerns and Recovery

Adequate kilocalorie intake, including proper macronutrient and micronutrient intake and correct timing of meals, is important for the optimal expression of muscular strength. Although it is beyond the scope of this chapter to discuss optimal nutritional strategies for resistance training, deviation in diet or malnutrition may negatively affect performance during strength testing. The optimal nutritional strategy includes proper hydration and carbohydrate intake before strength and endurance (multiple RM) testing. Schofstall et al. (2001) have recently shown that dehydration resulting in a 1.5% loss of body mass results in a significant reduction in 1RM bench press performance (approximately 6%) in resistance-trained men. After a 2 h rehydration and rest period, 1RM bench press returned to baseline. Leveritt and Abernethy (1999) reported that 2 d of carbohydrate restriction significantly reduced multiple RM performance of the squat over three sets but did not affect isokinetic strength. Thus, normal dietary patterns before testing appear warranted because large restrictions may limit strength performance.

Recovery from prior workouts or testing sessions is critical to optimal strength testing. That is, the time between workouts or testing sessions for rest and nutrient intake is crucial. Elite weightlifters typically train using workouts of short to moderate duration with high frequency to optimize recovery and performance during the workout (Häkkinen, Pakarinen, Alen, Kauhanen, & Komi, 1988). A similar scenario may be applied to strength testing. Research has shown that after a 1RM testing session of the squat and bench press, recovery was full within 2 h of the testing session, as well as 6 and 24 h posttesting (Sewall & Lander, 1991). These results show that 1RM testing reliability could be high when testing in a short period. But 1RM testing following a training ses-

sion poses a different stress. This issue was examined in resistance-trained men during the initiation of a 4-week overreaching program. Ratamess et al. (2003) had participants adhere to a 4-week base resistance-training program followed by a 4-week overreaching (significant increase in volume and frequency) period. Participants lifted 4 d consecutively, training the entire body, and were tested for 1RM strength on the following day (24 h postworkout). After the initial week of overreaching, 1RM bench press and squat were significantly reduced. Interestingly, a second group that consumed an amino acid supplement was able to maintain strength performance during this stressful alarm phase reaction to the program. This finding demonstrates a benefit of amino acid supplementation during the alarm phase of overreaching when a drastic increase in volume occurs. During the subsequent 3 weeks, participants were able to strength test successfully (i.e., maintain and improve 1RMs) within 24 h of the previous workout. These data demonstrate that resistance-trained men can effectively be strength tested within 24 h of a strenuous workout when adapted to the program demands.

## Test Specificity

An essential consideration in strength testing is test specificity. Various types of exercise impose different metabolic demands (Häkkinen et al., 1987). Different types of muscle actions will also have specific effects. For example, dynamic constant external resistance exercise will increase dynamic strength more than it will isometric strength (Fry et al., 1992; Rutherford, Greig, Sargeant, & Jones, 1986). Conversely, in some individuals dynamic constant external resistance exercise training will increase isometric strength more than it will dynamic strength (Fry et al., 1991). Such variability in strength relationships to training underscores the need for the inclusion of tests specific to the mode used in training. Therefore, proper assessment of a training program should include a similar type of muscle action, not a different type, as is common in many investigations (Amusa & Obajuluwa, 1986). Adaptations are also specific to limb velocity in training (Jenkins et al., 1984). Training velocities will preferentially affect specific portions of the force-velocity curve as well as specific populations of muscle fibers (Thomee et al., 1988). Other training modalities, such as hydraulic machines, will also present different physiological stresses and cannot always be equated with dynamic constant external resistance or isokinetic exercise (Hortobagyi et al., 1987; Hunter & Culpepper, 1988).

## Technique

Some types of strength evaluations require use of a specific movement pattern. This concern is not typical for machine testing, because the machine usually dictates the movement pattern, although proper positioning is required (Kraemer & Fleck, 1993; Lewis & Spitler, 1989). Regardless of whether free weights or machines are used, proper exercise technique is required for safety of the individual. Lifting injuries are often associated with unsupervised exercise sessions (Brown & Kimball, 1983; National Strength and Conditioning Association, 1985) and are often avoidable with knowledgeable supervision. Proper exercise technique also ensures that an appropriate range of motion is used (Fleck & Kraemer, 1988). Changes in the range of motion used can affect test results. If proper technique is not used, the test should be terminated. Criteria for termination of the test should be established before the testing. This includes standardization of grip widths, foot stances and position, movement patterns, range of motion, and body position. For example, a bench press test may be terminated if the individual raises the buttocks off the bench or if excessive twisting motion of the body is evident. Good technique is essential for any strength-testing exercise and has been previously described in detail (Kraemer & Fleck, 1993).

## Pretest Exercise and Ambient Temperature

The scientific efficacy of exercise before performance is not completely clear, but individuals typically perform some type of warm-up exercise before actual strength testing occurs (Safran, Garrett, Seaber, Glisson, & Ribbeck, 1988). Most protocols use the specific mode of testing at submaximal efforts to accomplish this before the test. This practice also familiarizes the individual with the test. Research has also shown that there is an optimal muscle temperature for force and power development (Armstrong, 1988). Whether all pretest exercise accomplishes this objective remains unknown. Temperature of the surrounding environment can also affect performance (Meese, Schiefer, Kustner, Kok, & Lewis, 1986) and should be held constant during test sessions. Additional measures should be taken when testing in an uncontrolled environment such as the outdoors, including extra warm-up clothing or cancellation of testing because of variable conditions.

## Breathing and Blood Pressure

Although a seemingly simple matter, the role of breathing during resistance exercise is a major concern (Austin, Roll, Kreis, Palmieri, & Lander, 1987). Beginning exercisers are often instructed to breathe continuously during resistance training (e.g., inhale on the eccentric phase, exhale on the concentric phase), but advanced lifters have found that breath holding during certain phases of some lifts may enhance their performance on maximum efforts (Austin et al., 1987). Breath holding results in a Valsalva maneuver, with extremely high blood pressures resulting in some instances (MacDougall et al., 1985). The amount of blood pressure increase depends in part on the resistances used, with greater weights and forces eliciting greater pressure increases (Harman, Frykman, Clagett, & Kraemer, 1988; Nagle et al., 1988). The increased blood pressure is related to increased abdominal and thoracic cavity pressures, thus contributing to body stability (Harman et al., 1988). The increase in intra-abdominal pressure (IAP) is thought to reduce disc compressive force by up to 40% (Harman et al., 1988). Harman et al. (1988) demonstrated that IAP increases as training load increases and is high for exercises requiring trunk stability or bending (e.g., deadlift). IAP increases with abdominal training (Cresswell, Blake, & Thorstensson, 1994) and with use of a weight belt (Hunter et al., 1984). Thus, it is important that IAP increase during strength testing of exercises requiring trunk stability (e.g., squat, deadlift, bent-over row, and so on) to reduce the risk of spinal injury. The amount of muscle mass used in the exercise (Nagle et al., 1988) as well as the lifting cadence (Fry et al., 1993) can contribute to the blood pressure response. For the monitoring of cardiovascular variables during resistance exercise, breathing patterns must be constant for all individuals and compatible with health considerations.

## Arousal and Encouragement

Thorough instructions should be given to all individuals before the actual strength test (Sale et al., 1988). Performance on strength tasks can be enhanced if the individuals are optimally aroused (Biddle, 1986). Verbal encouragement is suggested for all strength testing to help ensure optimal results. Recent research has shown that having an audience present can enhance 1RM performance during lifting competition (Rhea, Landers, Alvar, & Arent, 2003). Care must be taken that the form of this encouragement is consistent for all individuals. For some tests, such as isometric, isokinetic, or fatigue tests, simultaneous visual feedback concerning performance should be provided (Graves & James, 1990). Factors such as yelling during a strength task or prior training experience can affect arousal and the resulting strength scores (Ikai & Steinhaus, 1961).

## Rest Intervals

Adequate recovery time must be allowed for all individuals to obtain a true RM. The time required will depend on the amount of musculature involved and the forces (e.g., weight lifted) developed during the exercise. Thus, high-velocity, low-force activity (e.g., isokinetic testing) may require shorter recovery times than low-velocity, high-force activity (e.g., 1RM barbell squat) does. Rest intervals as short as 10 s have been successfully used between isokinetic efforts (Dudley & Djamil, 1985; Edgerton & Perrine, 1978; Gregor et al., 1979). Rests of up to 3 min between single repetitions for isokinetic leg extension and flexion testing have resulted in no significant performance differences compared with 1 min rests between sets of three repetitions (Conroy et al., 1984). Rest intervals during free-weight exercise may be more critical. In one study, 1 min of rest was adequate for 1RM testing of the bench press (Weir, Wagner, & Housh, 1994). Rests of 5 min between 90% and 100% of 10RM squat have appeared adequate and necessary (Fry et al., 1993). Thus, it appears that 1 min may be adequate for single attempts for sets leading up to the 1RM, and for some individuals 1 min may be long enough throughout the testing protocol (Weir et al., 1994). But 1RM assessment of strong or highly trained individuals for multiple-joint exercises (e.g., exercises that enable high loads to be lifted) will likely require additional time to recover during 1RM or multiple RM sets (i.e., at least 2-3 min). Behm et al. (2002) compared one set of 5, 10, or 20RM and reported that maximal voluntary contraction was depressed 21.4% after 30 s of recovery and was still reduced by 12.3% after 3 min of recovery. In addition, muscle inactivation was lower by 3.2% at 30 s and 1.4% after 3 min. Integrated electromyogram (IEMG) was depressed by 30% after 2 min but recovered by 3 min of recovery. In addition, muscle antagonist activity was approximately 12% greater up to 1 min of recovery and returned to baseline by 2 min of recovery. Note that recovery is necessary to obtain a true RM, so lifters should not be hurried to complete the protocol in a short period. Consistency of the rest interval may be important for some types of strength testing, especially if related physiological parameters are also being monitored (e.g., blood pressure or heart rate).

## Reliability

Test reliability is the ability of a strength test (e.g., equipment, protocol, and effort) to give the same results repeatedly for identical performances. Test reliability should not be confused with test validity—that is, whether the test actually assesses the desired variable. Reliability is determined from several test sessions with similar test performances using the same individuals, conditions, and calibration procedures. Test–retest reliability is quantified by the standard deviation or method error of repeated trials (coefficient of variation). The coefficient of variation (SD/mean) is also used as an indicator of reliability. Further information is provided by the intraclass correlation coefficient $(R)$ of repeated trials (Sale et al., 1988). The closer the $R$ value is to 1, the greater the reliability. For research purposes, high reliability is $R > .90$, moderate is $.80 < R < .90$, and poor is $R < .80$. Error in testing typically results from biological variation (i.e., exertion, effort, overtraining, and so on) and experimental error. Experimental error can result from lack of experience among the testing staff, insufficient familiarization, or improper calibration or function of the testing equipment. Before testing begins, all equipment should be evaluated for reliability (Kraemer et al., 1991; LaChance et al., 1988; Seger et al., 1988; Smidt et al., 1984). Furthermore, reliability cannot be assured among different types of equipment and their various attachments (Epler et al., 1988; Francis & Hoobler, 1987). The reliability of test technicians should also be periodically monitored to ensure that their methods and techniques are similar. The testing staff should be aware of norms when data is collected so that they have a greater chance of identifying a problem by recognizing abnormal values. Although reliability may have been established for certain measures of a strength test, the reliability may not extend to other variables of the same test (Going et al., 1987).

## Muscular Balance

A variable often monitored is muscular balance, in which strength levels of antagonistic muscles are assessed relative to each other or bilateral strength measures are compared. Nosse (1982) summarizes the substantial amount of research that has been performed in this area. Attempts have also been made to identify the muscle balance requirements of a variety of populations (Hemba, 1985). Muscle balance characteristics are not different when comparing prepubertal males with adults (Weltman et al., 1988), but age may be a factor for other populations

(Thomas, 1984). Other factors to consider include gender (Christensen, 1975) and body size (Rankin & Thompson, 1983).

The strength ratios (agonist to antagonist) are affected by limb velocity (Fry & Powell, 1987a; Hageman et al., 1988; Nunn & Mayhew, 1988), because different muscle groups have different force-velocity characteristics. Thus, limb velocities of muscle balance properties must be identified (Klopfer & Greij, 1988). The ratios are also joint-angle specific (Fry & Powell, 1987a; Walmsley & Szybbo, 1987), which is an important consideration when using angle-specific isokinetic torques or isometric testing. Furthermore, different joints have unique muscle balance ratios (Cook et al., 1987; Hageman et al., 1988; Poulmedis, 1985), and corrections for gravity are necessary before examining a ratio (Poulmedis, 1985).

## Time of Day of Strength Testing

Strength testing should be performed at a similar time of day during multiple testing sessions. Diurnal variations in force production have been shown (Cappaert, 1999). Wyse, Mercer, and Gleeson (1994) reported that peak isokinetic torque was produced between 1800 and 1930, and Lundeen et al. (1990) reported that peak grip strength was produced between 1400 and 1516 and peak quadriceps strength was produced between 1144 and 1536. Although much further research is warranted in this area, it appears that differences may exist at different times of day. Thus, strength testing of individuals should occur at a consistent time of day.

## Ergogenics

Methods and materials used to improve performance can be classified as ergogenics. This broad category includes items such as lifting belts, knee and elbow wraps, wrist straps for enhancing grip, bench press shirts, squat suits and briefs, erector shirts, mental preparation techniques, other accessories (e.g., chalk, shoes), dietary manipulations, performance-enhancing drugs and supplements, and special equipment such as those that enable heavier loads to be lifted during the eccentric phase than during the concentric phase (Doan et al., 2002). The use of any of these can affect performance, so careful consideration must be given to allowing these items to be used when testing. Some of these are easily controlled, such as the use of various lifting apparel. Others, such as performance-enhancing drugs, are more difficult to screen. Acceptable drug-testing procedures can help control this factor, but such measures are not foolproof. In addition, stimulants (e.g., caffeine, thermogenic substances) should be advised against before the strength-testing session because such substances may enhance performance. If acceptable, equipment such as lifting belts should be made available. Standardizing the use of these agents for all testing sessions is important. The same type of accessory should be used, and in the case of apparel, the same level of fit should be used because adjustments in the tightness of accessory may affect performance (Lander, Simonton, & Giacobbe, 1990).

## Units of Measure

The International System of Units (SI) includes the accepted units of measure for strength testing. The SI unit for force is the newton (Sale et al., 1988). In the everyday world, pounds (1 lb = 4.44822 N) and kilograms (1 kg = 9.80665 N) are often reported as units of force. Testing for research purposes should always use SI units, but for field testing—for example, with athletic teams or with large populations for fitness assessments—it may be beneficial to use pounds and kilograms. Several discussions on appropriate units of measure are found in the literature (Knuttgen & Kraemer, 1987; Sale et al., 1988). See table 8.5 for a list of some commonly used SI units.

Table 8.5   **Common Units of Measurement Used in Strength Testing**

| Measure | SI unit | Common conversions |
|---------|---------|--------------------|
| Force | Newton (N) | 1 lb = 4.44822 N |
| Mass | Kilogram (kg) | 1 lb = 0.45359237 kg<br>1 kg = 2.2046 lb |
| Power | Watt (W) | 1 hp = 746 W |
| Work | Joule (J) | 1 ft-lb = 1.355818 J |
| Torque | Newton meter (N·m) | 1 ft-lb = 1.355818 N·m |
| Velocity | Meters per second (m·s$^{-1}$) | 1 mph = 0.44704 m·s$^{-1}$ |
| Acceleration | Meters per second$^2$ (m·s$^{-1}$)$^2$ | |
| Angle | Radian (rad) | 1 rad = 57.29578° |
| Angular velocity | Radians per second (rad·s$^{-1}$) | 1 rad·s$^{-1}$ = 57.29578°·s$^{-1}$ |
| Distance | Meter (m) | 1 ft = 0.3048m |

## Correction for Body Mass

The individual's body mass can influence strength performances. Therefore, correction factors for body mass have been developed. The relevance of correcting for body mass is to make comparisons between lifting performances in individuals of different size. For example, powerlifting and weightlifting competition is based on weight classes. In some competitions, however, comparisons are made across an array of weight classes; thus, an accurate system of comparison is needed. Some methods involve coefficient tables based on body weights (Stone & O'Bryant, 1987), which are quite popular with the competitive lifting sports. An easier method is simply to correct for body weight by dividing the strength score by body weight or fat-free mass, also known as isometric scaling (Challis, 1999). This method allows for comparisons between genders (Bishop et al., 1987), as well as across chronological ages, both intrasubject (Housh, Johnson, Hughes, et al., 1988; Housh et al., 1989) and intersubject (Viitasalo et al., 1985). However, when comparisons are made between lifters, scores tend to favor individuals with less body mass, that is, less than 60 kg body mass (Challis, 1999). Consequently, some authorities have used isometric scaling with body mass expressed $kg^{2/3}$ because of the relationship that muscle CSA is proportional to body mass$^{2/3}$ (Dooman & Vanderburgh, 2000; Jaric, Ugarkovic, & Kukolj, 2002), which tends to favor a higher weight class (e.g., more than 82.5 kg) (Challis, 1999).

Another type of scaling procedure used is nonisometric scaling, or allometric scaling (Challis, 1999; Dooman & Vanderburgh, 2000), which uses the following expression:

$$Y = a \cdot X^b$$

where $Y$ equates to the weight lifted for a specific exercise, $a$ refers to a constant, $X$ is body mass, and $b$ is the scaling exponent, which is typically the variable of investigation (Dooman & Vanderburgh, 2000).

Data from elite Olympic weightlifters and powerlifters have been analyzed using both isometric and allometric scaling procedures, as well as a second-order polynomial model (Batterham & George, 1997), to determine which model has the most accurate fit. Challis (1999) reported that isometric scaling to the 2/3 exponent (0.67) was effective for scaling because the $b$ exponents were determined to be 0.65 for powerlifting and 0.64 for weightlifting. In contrast, Dooman and Vanderburgh (2000) examined allometric scaling in elite powerlifters and reported a body mass bias for the deadlift ($b = 0.46$)

but not for the squat ($b = 0.60$) or bench press ($b = 0.57$). Therefore, it appears that various models may be used when drawing comparisons between lifting performances in athletes of different sizes.

## Total Work

An important consideration in comparing training programs is total work as measured in joules (see table 8.2). Although sometimes defined as the area under a torque curve (Morrissey, 1987), work is actually "force expressed through a distance but with no limitation on time" (Knuttgen & Kraemer, 1987). Therefore, if one is trying to equate training programs, the forces of each repetition and the distances moved must be carefully monitored to control for total work (Kraemer et al., 1991, 1990).

## Normative Data

The development of normative strength data is difficult because of many of the previously mentioned factors. Strength performances are population specific and depend on the testing equipment and methodology used. With a large enough sample size, you can develop norms specific to your particular needs. A small sampling of available strength data includes testing of large population groups (Clark, 1967; Fry & Kraemer, 1991; Harman, Sharp, Manikowski, Frykman, & Rosenstein, 1987), more specific age and gender studies (DiBrezzo & Fort, 1987; Housh, Johnson, Hughes, et al., 1988; Housh et al., 1989; Kindig, Soares, Wisenbaker, & Mrvos, 1984; Poulmedis, 1985; Viitasalo et al., 1985), as well as a compilation of muscle balance characteristics (Nosse, 1982).

## Summary

As evident from the preceding material, accurate strength testing must take many considerations into account. Care must be taken to design and develop a test protocol that controls variables important to the particular testing situation. The tests selected should be specific to the training modality. Simple field tests for general fitness appraisals might not appear difficult, but they require careful consideration of all the variables mentioned in this chapter for accuracy and clinical viability. When strength testing is used as a highly defined research tool, the procedures can be elaborate. The Strength-Testing Checklist that follows lists variables that must be addressed in many test situations. Careful consideration of these factors assists in developing a protocol appropriate for an individual's needs.

## Strength-Testing Checklist

❏ Has the individual been medically cleared to weight train, and can this individual safely perform strength or endurance testing?

❏ Does this individual require any special accommodations?

❏ Is the time of day similar between multiple testing sessions?

❏ Is muscle strength or endurance to be tested?

❏ Is a 1RM or multiple RM to be used?

❏ What is the age and gender of the individual?

❏ Was the individual thoroughly familiarized with the testing protocol? How many familiarization sessions were performed?

❏ Was proper technique explained and demonstrated (including range of motion, grips, stances, body position, and so on)?

❏ Was test—retest reliability of the equipment and protocol performed?

❏ What is the individual's training history?

❏ Was adequate nutrition intake (and hydration) consumed before testing?

❏ Was ambient temperature controlled for?

❏ If the individual is experienced, what is his or her perceived or expected RM?

❏ What type of muscle action is to be tested?

    ❏ Concentric?

    ❏ Eccentric?

    ❏ Isometric?

❏ What type of resistance is to be used?

    ❏ Dynamic constant external resistance?

    ❏ Variable resistance?

    ❏ Isokinetic?

    ❏ Isometric?

❏ For machine-based exercise, what are the appropriate machine settings and starting position of the resistance?

❏ Was the individual properly positioned, and does the equipment accurately accommodate the individual?

❏ For isometric training, what joint angles will be examined?

❏ What is the velocity of movement?

❏ Are knowledgeable spotters present?

❏ Was the equipment calibrated according to manufacturer's guidelines?

❏ Test specificity:

    ❏ Were the movement patterns tested similar to those performed during training?

    ❏ Is there metabolic (energy system) specificity?

❏ Were adequate instructions given?

❏ Was a proper warm-up performed? Did it include submaximal practice repetitions for the testing exercises?

❏ Did the individual use proper technique, and did spotters assist in lifting the weight?

❏ Was adequate rest given between repetitions and sets?

❏ Were proper breathing patterns used?

❏ Was the lifter verbally encouraged throughout the testing protocol, and was a proper lifting environment set for testing?

❏ Was visual feedback given for isokinetic testing?

❏ Were the proper units of measurement used?

❏ What was the individual's 1RM or multiple RM score?

❏ Were ergogenics controlled for? What types (if any) were used?

    ❏ Lifting accessories and apparel?

    ❏ Nutrition supplements?

    ❏ Drugs?

❏ Did the individual give 100% effort?

❏ Were there any other factors that may have affected the test (e.g., illness, injury)?

# nine

# Skeletal Muscle Structure and Function

Michael McGuigan, PhD
*Edith Cowan University*
Matthew Sharman, PhD
*Edith Cowan University*

Several methods have been developed to investigate the morphological, biochemical, and physiological properties of human skeletal muscle. The adaptability of skeletal muscle in response to training has long been the focal point of a large amount of research directed at a better understanding of muscle morphology. This chapter outlines several methods that can be used in the laboratory to look at these responses. The needle biopsy technique for obtaining muscle samples will be described. Methods that can be used to analyze the structural and functional properties of skeletal muscle include mATPase histochemistry, gel electrophoresis, and immunoblotting techniques to investigate muscle fiber type and proteins. Other stains to show glycogen content and capillaries will be outlined. In addition, methods for determining cross-sectional areas of muscle fibers will be discussed.

## Skeletal Muscle Structure and Function

Skeletal muscle is an extremely heterogeneous tissue, which is reflected by the composition of a variety of diverse fiber types (Pette, 1998). These differences in the heterogeneity of muscle tissue reflect its high degree of functional specialization and are the basis of its functional plasticity (Pette & Staron, 1990). Furthermore, skeletal muscle is characterized by a highly adaptive potential; it is a highly plastic tissue capable of altering its contractile proteins and its contractile properties with increased use or with disuse (Pette & Vrbova, 1992).

Skeletal muscle is a highly organized tissue that is designed foremost for force production and movement. The contractile unit responsible for this force production and movement has many varying levels of organization that are tailored in a sophisticated way to perform this task. The functional units of skeletal muscles are the muscle fibers, which are long cylindrical multinucleated cells (Berchtold, Brinkmeier, & Muntener, 2000). They vary considerably in their morphological, biochemical, and physiological properties. The fiber-type composition, varying from muscle to muscle, is the basis of the structural and functional muscular diversity. Fibers are able to change their characteristics in response to a large variety of stimuli leading to muscular plasticity (Berchtold et al., 2000).

Skeletal muscle fiber diversity was identified as early as in the 17th century when Lorenzini noted color differences in animal muscles. In 1873, Ranvier distinguished these differences as white and red muscles (Pette & Staron, 1990). It has also been reported that the red muscles contracted more slowly than the white muscles (Thompson, 1994).

Skeletal muscle responds to altered functional demands by specific qualitative and quantitative alterations in gene expression if the stimuli are of sufficient magnitude and duration (Pette, 1998). Continual heightened neuromuscular activity will induce a series of related changes in gene expression.

During these graded transitions, some of the changes in myofibrillar protein isoforms occur in parallel, suggesting fiber-type-specific programs of gene expression, for example, myosin heavy chain (MyHC) IIb→MyHC IIa. Myofibrillar protein isoforms generally show specific tissue distribution and can be used as markers for fiber types in skeletal muscle (Schiaffino & Reggiani, 1996). The composition of MyHC isoforms has been used to determine the expression of the fast fibers in mammalian muscle. Several studies have also analyzed the relative contribution of MyHC and myosin light chain (MyLC) isoforms to the enzymatic properties of myosin (Barany, 1967; Wagner, 1981; Wagner & Giniger, 1981), with the MyHCs appearing to be the major determinant of mATPase activity. Consequently, a more complete fiber-type classification may be achieved by examining the myosin molecule more directly, such as using sodium dodecyl sulphate-polyacrylamide gel electrophoresis (SDS-PAGE). The continuum of fiber types (IIB ↔ IIAB ↔ IIA ↔ IIAC ↔ IIC ↔ IC ↔ I) is difficult to quantify using a qualitative technique such as mATPase histochemistry (Fry, Allemeir, & Staron, 1994). Histochemical methods for determining fiber-type distribution and fiber-type area yield important information, because the MyHC content alone does not indicate fiber-type composition. The transformational shifts in the MyHC isoforms, however, do allow changes to be detected in the contractile apparatus much earlier than histochemical methods allow. Gel electrophoresis provides a quantitative analysis of MyHC content and hybrid MyHC isoforms, compared with the semiqualitative mATPase histochemical method (Fry et al., 1994). Significant changes in the cross-sectional area (CSA) during resistance training appear to take at least 6 to 8 weeks, whereas changes in the expression of MyHC isoforms occur after 2 weeks (Staron, Karapondo, Kraemer, et al., 1994). Therefore, it is now clear that the MyHC isoforms represent the best possible marker of fiber-type transitions (Pette, 1998). Analysis of myosin expression in skeletal muscle has attracted a great deal of attention because changes in the phenotypic expression appear to be related to muscle function and adaptation to a variety of physiological stimuli such as training (Pette & Staron, 1990).

# Needle Muscle Biopsy

To investigate the various fiber and enzymatic features of skeletal muscle, it is first necessary to obtain a muscle sample. The percutaneous muscle biopsy technique described by Bergstrom (1992) has been commonly used for research in exercise physiology. This technique has been used since the 1860s (Duchenne) and was repopularized by Bergstrom (1992; Bergstom, Hermansen, Hultman, & Saltin, 1967). The technique involves the insertion of a hollow-bored needle under local anesthetic and sterile conditions to obtain specimens (Dubowitz & Brooke, 1973). This method is useful for biochemical, histochemical, and immunohistochemical analyses and is the common and preferred method for obtaining biopsy samples in exercise science research. The muscle groups that are most commonly sampled in exercise and sport science research are the vastus lateralis, triceps brachii, deltoid, and gastrocnemius muscles.

The following section describes proper muscle biopsy procedure.

## MUSCLE BIOPSY PROCEDURE

### MATERIALS

- Biopsy needles
- Betadine
- Local anesthetic (xylocaine)
- 25 g needles
- 18 g needles
- Scalpel blade no. 11
- Large syringe (≥20 cc)
- Rubber tube (IV extension set 48 in.) and 200 μL pipette tip
- 4 in. × 4 in. sterile gauze
- Adhesive bandages
- Razor

### PROCEDURE

1. Locate the site for incision by having the subject contract the muscle to be biopsied. Remove all hair from the site with a razor and clean the skin with surgical preparation. Take the biopsy from a position where the muscle is under the least stretch so that the muscle is relaxed.

2. Anesthetize the skin and subcutaneous tissue with a local anesthetic (xylocaine), avoiding contact with muscle.

3. Incise skin and deep fascia 1 cm with a scalpel blade. Instruct the subject that he or she will feel pressure on the muscle with the biopsy and to keep the muscle as relaxed as possible.

4. Insert the needle with the needle notch closed into the incision.

5.  Insert the needle approximately 0.5 cm into the muscle.

6.  Cut off the sample by ramming the inner (sharpened) cylinder along the needle. Often two to three cuts are made to maximize the biopsy sample. Better samples can be obtained if suction is applied to the needle as the muscle is cut (Evans, Phinney, & Young, 1982). Do this by fitting a large syringe ($\geq$20 cc) with a rubber tube and tip that will insert into the end of the biopsy needle. The increased tissue yield allows for a wider array of analytic analyses to be made from each sample. Note that the use of suction enhanced nipples (SEN) can increase tissue yield two- to fivefold (Hennessey, Chromiak, Della Ventura, Guertin, & MacLean, 1997). These nipples enhance the suction by blocking the passage of air between the cutting and outer trochar, therefore increasing the amount and duration of the negative pressure that can be developed.

7.  Remove the needle and use the central rod to withdraw the specimen. When withdrawing the needle, apply pressure over the wound with sterile gauze. Maintain this pressure for at least 20 min to reduce the amount of bleeding. The amount of muscle soreness experienced by the subject is directly related to the amount of bleeding into the biopsy site.

8.  Close the wound and seal the skin with adhesive bandages. Instruct the subject not to remove the bandages for 3 d. A pressure bandage should be kept over the adhesive bandages for at least 24 h. Ice packs can be applied to help reduce the amount of bleeding into the biopsy site.

The advantages of the needle biopsy procedure are that repeat biopsies are convenient, little scarring occurs, and the method is relatively cost effective. Multiple biopsies can be obtained the same day, before and after exercise, for example, by inserting the biopsy needle into the same incision but angling it away to sample a different part of the muscle. If the biopsy is taken hours or days following the initial procedure, then the incision should be made at least 2 cm from the original site. The disadvantages of the needle biopsy procedure include the relatively small specimens that can be obtained and the fact that it is a blind procedure.

Single biopsies from unspecified muscle depths have been routinely performed to assess capillary density and fiber type. Capillary density can vary in some cases at different muscle depths (Dwyer, Browning, & Weinstein, 1999). When taking measurements from biopsies of skeletal muscles in humans, it is desirable to obtain the thickness of the subcutaneous fat and the underlying muscle to standardize the sampling depth within the muscle (Dwyer et al., 1999).

# Processing Muscle Tissue

The freezing and mounting of the muscle tissue will depend on the type of analysis that is to be performed. In the cases of biochemical analysis and immunoblotting, the sample should be immediately snap frozen in liquid nitrogen.

# Preparation for Histology, Histochemistry, and Morphology

The basic aims of the histochemical preparation are optimal preservation of tissue architecture, minimal artifact formation (disruption of the tissue caused by the formation of ice crystals during freezing), and retention of enzyme activity for metabolic stains.

Standard histological procedures are therefore not recommended for muscle tissue because fiber typing becomes impossible. The preferred method is freezing tissue for cryostat sectioning. Therefore, care is required on the initial orientation of the tissue for freezing. The maximal amount of information is obtained when the sections are taken transversely to the long axis of the fibers.

## HISTOCHEMICAL PREPARATION

### EQUIPMENT AND REAGENTS

- Cryostat or microtome
- Glass cover slips (22 × 22 mm)
- Cryotubes
- Forceps or tweezers
- Scalpel
- Tissue Tek (O.C.T.) or tragacanth gum
- Cork
- Liquid nitrogen
- Isopentane
- Styrofoam container to hold a small amount of liquid nitrogen
- Small beaker to hold isopentane that will fit into the Styrofoam container

## PROCEDURE

1. Orientation: Cut the muscle and orient it so that the direction of the muscle fibers is visible. This procedure is easier on large specimens, but those from needle biopsy may require use of a medium-power dissecting microscope.

2. Mounting and labeling: Pour liquid nitrogen into a Styrofoam container and place a small beaker with approximately 2 cm of isopentane into the container. The isopentane will begin to freeze and will be ready for use when a small amount of unfrozen isopentane is in the center.

   Place a small amount of mounting medium in the end of a piece of cork. Using forceps, place the piece of muscle vertically so that it sits in the mounting medium on the cork. Then place the muscle in the isopentane until it is frozen. Next, put the sample (now mounted on the cork) into a labeled airtight vial (cryotube) and place it in liquid nitrogen or store it at –80 °C.

3. Storage: Store samples at –80 °C until sectioning. For longer-term storage, specimens are best heat sealed into labeled plastic tubing to prevent them from drying out. Avoid even minor thawing during transport of specimens. Do not allow the muscle mount to thaw at any time.

4. Sectioning: Carry out the sectioning of the samples in a cryostat or microtome, optimally at –20 °C. Attach the cork with the sample to the specimen mount with mounting medium and wait for approximately 20 min to adjust to the temperature of the cryostat. Align the blade of the cryostat and cut serial 8 to 10 μm thick transverse sections until sufficient numbers from each biopsy sample are available to carry out recommended staining techniques. Collect the sections on glass cover slips by touching them to the muscle section as it lies on the blade. Place the glass cover slips in a Columbia staining jar for staining. These will generally hold four cover slips and 10 ml of solution.

## Histochemical Analysis

The common approach in the study of the effect of exercise on skeletal muscle fiber type has been based on mATPase activity. Myofibrillar adenosine triphosphatase (mATPase) activity has been shown to be proportional to the speed of contraction of a given muscle (Barany, 1967). Based on these differences in alkali stability and acid lability of mATPase activity in dif-ferent fibers, a relatively simple assay was developed to differentiate between fast-twitch and slow-twitch muscle fibers (Brooke & Kaiser, 1970). Since this earlier work, it has become apparent that a vast continuum of muscle fibers, within a given muscle, exists based on pH sensitivity of mATPase activity (Staron & Hikida, 1992a; Staron, Hikida, & Hagerman, 1983) (figure 9.1). Seven divisions of muscle fibers have been reported, based on this type of analysis (I, IC, IIC, IIAC, IIA, IIAB, IIB) (Staron & Hikida, 1992a). The most commonly reported divisions, however, are I, IIC, IIA, and IIB (Staron & Hikida, 1992a). Cor-

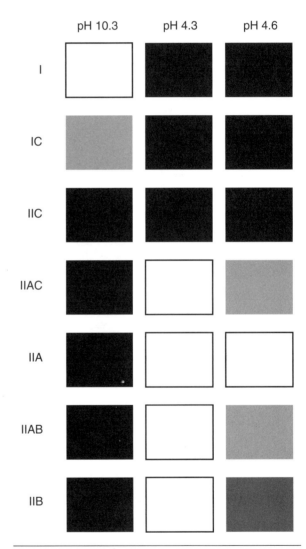

**Figure 9.1**   Seven divisions of muscle fiber type based on myofibrillar adenosine triphosphatase activity (mATPase) after preincubation at pH 10.3, 4.3, and 4.6. A dark square indicates high levels of mATPase activity, whereas a clear square indicates no mATPase activity.

Adapted, by permission, from R.S. Staron and R.S. Hikida, "Histochemical, biochemical, and ultrastructural analyses of single human muscle fibers, with special reference to the C-fiber population," *Journal of Histochemistry & Cytochemistry* 40(4): 563-5638, 1992.

relations between mATPase activity in muscle fibers after various pH preincubations and myosin heavy chain (MHC) content have been reported (Fry et al., 1994). This finding indicates that mATPase fiber type is representative of the relative amounts of the type of protein isoforms within the fiber and therefore is a useful tool in evaluating changes in muscle tissue with exercise (Fry et al., 1994).

# Histochemical Manipulation for Fiber Typing: Altering pH

The calcium method for ATPase demonstration, employing solutions of different pH values, has been used primarily to distinguish muscle fiber types (Brooke & Kaiser, 1970). The preincubation pH inactivates the myosin-ATPase enzyme of specific fiber types. The remaining active enzyme is attached to a calcium atom that is replaced by a cobalt atom and finally precipitated as a black insoluble compound by the ammonium sulfide. One must exercise care in several areas to achieve good fiber-type differentiation with this complicated stain. The most important factors to consider are

pH of the solutions,

temperature, and

timing of the incubations.

These factors will determine the appearance of the different fiber types. The literature discusses several variations of this stain that manipulate preincubation time, temperature, or pH to identify a variety of fiber subtypes. Generally, six different fiber types (I, IC, IIC, IIA, IIAB, and IIB) can be distinguished using routine myofibrillar mATPase histochemistry after preincubation pH values of 4.3, 4.6, and 10.3 (Padykula & Herman, 1955; Staron & Hikida, 1992a). Type I fibers are stable in the acid ranges but labile in the alkaline. Type IIA fibers display the reverse pattern. All fibers stable at pH 4.6 and 10.3 but labile at 4.3 can be classified as either IIAB or IIB depending on the intensity of the stain following preincubation at pH 4.6 (the type IIAB fibers will stain intermediate the IIA and IIB fibers). Fibers classified as type IC or IIC remain stable to varying degrees throughout the pH ranges. Often, researchers will condense the six types into the three major fiber types, I, IIA, and IIB (e.g., Hikida et al., 2000).

Occasionally, depending on the muscle, all three types can be delineated in one stain with a preincubation at a pH of around 4.54. At this pH, type I fibers stain dark, type IIA fibers remain unstained, and the type IIB fibers stain intermediately. The pH needs to be exact to differentiate the different fiber types clearly, and it may be necessary to use a range of pHs from 4.3 to 4.6 to confirm that the stain has worked correctly. If only the two major fiber types, I and II, are required for typing, then a single stain at a pH of either 10.3 or 4.3 would be sufficient.

## FIBER-TYPING METHOD

### SOLUTIONS

The following reagents should be prepared:

1. 1% w/v calcium chloride (5 g $CaCl_2 2H_2O$ distilled water 500 ml), stored at room temperature

2. 2% w/v cobalt chloride (10 g $CaCl_2 6H_2O$ distilled water 500 ml), stored at room temperature

3. 2% v/v solution of ammonium sulfide (0.2 ml stock $NH_4SO_2$ + 9.8 ml D.I. $H_2O$)

4. 10.3 ATP preincubating solution

   0.451 g glycine

   0.351 g sodium chloride

   0.480 g calcium chloride

   54 ml sodium hydroxide (0.1 M) (Store at room temperature. Note that another source for inadequate differentiation is the pH solutions, particularly the sodium hydroxide, which should be not more than 2 weeks old.)

   60 ml deionized water

Adjust the pH to 9.4 just before use.

5. 4.3 and 4.6 ATP preincubating solutions

   0.780 g sodium acetate

   0.740 g potassium chloride

   100 ml deionized water

Adjust pH between 4.30 and 4.60 with 1 M HCl as necessary just before use.

6. ATP incubating solution (volume here is sufficient for three staining jars)

   102 mg ATP powder

   60 ml APB

Prepare just before use and adjust the pH to 9.4 with a few drops of 1 M NaOH. Make enough solution for all staining jars.

### STAINING PROCEDURE

1. Place three or four cover slips for each biopsy sample in a separate, labeled Colombia staining dish for each preincubating solution.

2. Incubate in the 4.6 and 4.3 solutions for exactly 5 min at room temperature. Add the alkaline solution for 10 to 15 min at room temperature.

3. After the appropriate preincubation periods, pour out the solution and rinse well with distilled water.

4. Pour the ATP solution into the staining jars. Incubate for 30 to 45 min at 37 °C.

5. Wash each staining jar with 1% calcium chloride and incubate for 3 min at room temperature.

6. Wash well with distilled water.

7. Add 2% cobalt chloride and incubate for 3 min at room temperature.

8. Wash well with distilled water.

9. Incubate in the 2% v/v solution of ammonium sulfide for 20 to 30 s. Perform this step in a fume hood. Rinse in the fume hood with approximately five changes of distilled water. The sections should now appear dark.

10. Allow the sections to air dry or dehydrate in ascending alcohols and then xylenes three times for 10 s each. Wipe off excess xylene.

11. Mount cover slips onto labeled glass slides with Aquamount or some similar mounting medium. The sections will tend to fade in time, especially if exposed to light.

# NADH Tetrazolium Reductase

The NADH tetrazolium reductase stain is a measure of the ability of the muscle to process NADH, therefore providing a measure of the respiratory capacity (Halkjaer-Kristensen & Ingemann-Hansen, 1979). Diaphorase is the term given to flavoprotein enzymes that have the property of transferring hydrogen from reduced nicotinamide adenine dinucleotide (NADH) to various dyes. The hydrogen transfer reduces the dye. Usually tetrazolium compounds function as the hydrogen acceptor when diaphorases are being demonstrated histochemically, and the product of the reduction is the water-insoluble formazan pigment. Commonly used tetrazoliums include nitro blue tetrazolium (NBT). Enzymatic activity releases hydrogen from the substrate, and the released hydrogen is transferred to the tetrazolium. With the addition of hydrogen, the tetrazolium is converted to purple-blue formazan pigment marking the site of enzyme activity.

This stain involves a 30 min incubation in 1 mM NADH and 0.45 mM P-nitroblue tetrazolium chloride in 20 mM MOPS (3-[N-morpholino] propane sulphonic acid) buffer. The stain is often used together with ATPase staining to classify fibers. In human skeletal muscle, type I fibers will stain dark blue and type II fibers will generally stain light blue. The NADH-tetrazolium reductase stain can also be used to avoid any possible shrinkage and therefore is useful for the analysis of cross-sectional area of the muscle fibers. This stain can be used to determine cross-sectional area of the different fiber types after cross-referencing with the pH 4.6 and pH 10.3 samples. Purple formazan precipitate is deposited at sites of mitochondria in sarcoplasmic network. Therefore, the stain is useful for demonstrating disruptions of the fiber architecture.

## STAINING METHOD

### REAGENTS

- Nicotinamide adenine dinucleotide, reduced (NADH)
- Nitro blue tetrazolium (NBT) (stored at 0 °C to 5 °C)
- MOPS (3-[N-morpholino] propane sulphonic acid) buffer

For every 10 ml of reagent, mix together 2 mg of NBT, 8 mg of NADH, and 10 ml of buffer (pH 7.4).

### STAINING PROCEDURE

1. Incubate cover slips in a Columbia staining jar for 30 min at 37 °C.

2. Rinse with distilled water.

3. Dehydrate in ascending alcohols and then xylenes three times for 10 s each. Wipe off excess xylene or allow to air dry.

4. Mount the cover slips with a mounting medium onto a labeled glass slide.

# Periodic Acid Schiff Stain (PAS)

The PAS stain is used for the visualization of glycogen in cellular material (Andersen, 1975). Tissue sections are first oxidized by periodic acid. The oxidative process results in the formation of aldehyde groupings through carbon-to-carbon bond cleavage. Free hydroxyl groups should be present for oxidation to take place. Oxidation is completed when it reaches the aldehyde stage. The Schiff reagent detects the aldehyde groups. A colorless, unstable dialdehyde compound is formed and then transformed to the colored final product by restoration of the quinoid chromophoric grouping.

The PAS (periodic acid Schiff reagent) stain with diastase or amylase digestion has histochemical specificity for glycogen. Glycogen, basement membranes, collagen fibers, glycolipids, and phospholipids are demonstrated as pink to red to purple color. If diastase or α-amylase is used for a negative control, the glycogen deposits are removed, leaving only the plasma membrane staining pink. The glycogen content can be assessed qualitatively by grouping the fibers as light, medium, or dark. Greater accuracy is obtained by quantitatively measuring the absorbance of each fiber. The average absorbance of each section can then be compared with the glycogen content of that muscle measured biochemically. If absorbance readings will be taken on sections that are to be correlated with total glycogen content, these measurements should be made from sections analyzed together in the same staining jar. Different intensity of staining usually distinguishes the two major types of fibers. After glycogen values are correlated with the absorbance values, glycogen content can be determined for each fiber type.

## STAINING METHOD

### REAGENTS

- Amylase, stored at room temperature
- Chloroform, stored at room temperature in a flammable cabinet
- Periodic acid, stored at room temperature
- Schiff reagent, stored at room temperature

### SOLUTIONS

1. Carnoy's fixative (stored at room temperature)

   Alcohol, 100% 60 ml

   Chloroform, 30 ml

   Glacial acetic acid, 10 ml

2. Periodic acid solution, 0.5 % (w/v). This should be prepared fresh for each stain.

3. Periodic acid 50 mg dissolved in distilled water 10 ml

4. Schiff's reagent (stored in the refrigerator)

This reagent is messy and unstable. Dissolve 1 g of basic fuschin in 200 ml of boiling distilled water. Cool to 50 °C and filter. Add 20 ml of 1-N HCl. Cool to room temperature and add 1 g of sodium metabisulfite and leave in the dark overnight. Then add 2 g of charcoal, shake for 2 min, filter, and store in a dark bottle.

The PAS stain is also available as a standard histochemical kit (Sigma).

## STAINING PROCEDURE

1. Place the cover slip with the section attached into a Columbia staining jar.

2. Add Carnoy's fixative to dish for 10 min.

3. Rinse carefully several times with distilled water. Be careful because the sections may wash off.

4. Add periodic acid solution to the staining jars for 10 min.

5. Rinse carefully several times in distilled water. Add Schiff reagent for 5 min.

6. Carefully wash with distilled water.

7. Air dry or dehydrate in ascending alcohol solutions in the Columbia staining jar.

8. Mount cover slip onto a labeled glass slide with Aquamount, Permount, or other suitable mounting medium.

# Capillary Staining

The stain for capillary density is the same as for glycogen staining, but to visualize the capillaries, the glycogen contained in the fibers can be removed using amylase digestion. The amylase-PAS method for capillary staining was originally described by Andersen (1975).

## STAINING METHOD

### PROCEDURE

1. Prepare a 1 to 3% amylase solution.

2. After the fixation in the Carnoy's solution and rinsing, incubate the sections for 60 min at 37 °C.

3. Following further rinsing with distilled water, incubate sections in 1% periodic acid as previously described.

The walls of capillaries contain a small, relatively stable amount of glycogen. After dissolving other polysaccharides, particularly glycogen, with amylase, the PAS stains all basal laminae. Capillaries will then stand out as small red spots on a white background. Capillaries can be counted as described by Andersen and Henriksson (1977). Sections can then be projected using a light microscope and photographed using a camera or image-capturing system to determine the different measures such as capillary density. An area with at least 50 fibers should be selected and used for the capillary counting as recommended by McCall, Byrnes, Dickinson, & Fleck (1998).

Several immunohistochemical methods have also been used to stain skeletal muscle capillaries including von Willebrand factor, Ulex europaeus agglutinin I lectin, and anticollagen type IV staining. But these methods will not be discussed here. Qu, Andersen, & Zhou (1997) have suggested that these techniques reveal the muscle fiber borders more clearly and are therefore more suitable for computerized image analyses.

## Measures of Tissue Capillarity

Different measurements have been proposed for quantifying tissue capillarity in skeletal muscle. Andersen and Henriksson (1977) proposed that the fiber area supplied per capillary contact allows quantification of the capillary supply relative to the cross-sectional area of the muscle fibers. Capillary-to-fiber ratios, the number of capillary contacts per fiber and per $\mu m^2$ fiber cross-sectional area, have been the common measures used (e.g., McCall, Byrnes, Dickinson, Pattany, & Fleck, 1996; Qu, Andersen, & Zhou, 1997). These measures are scale dependent in that they change with fiber size. Researchers have used several other measures that take into account changes in fiber size. For example, Hepple (1997) has proposed measures of the capillary-to-fiber perimeter exchange index to provide a means of quantifying potential alterations in oxygen flux and delivery of substrates between capillaries and muscle fibers. The CFPE is derived as the quotient of the individual capillary-to-fiber ratio and the fiber perimeter, and quantifies the degree of capillary–muscle fiber contact in transverse sections of needle biopsy sections. Readers are encouraged to investigate articles such as those of Hepple if they wish to use these measurements in their research.

## Fiber Cross-Sectional Area

Several simple methods have been used to measure muscle fiber area including tracing, planimetry, and least fiber diameter. These methods use image magnification and projection to gain an approximation of fiber area. With advanced technology, the images of muscle samples in serial sections can be analyzed with a computerized image analyzer. Several computer programs are now available that can accurately measure the fiber area. Fiber areas of the different fiber types can be measured using Scion Image software beta version 3b (Scion Corporation). The NIH Image for Macintosh or Scion Image program for Windows is one of the simplest that can be used to analyze fiber cross-sectional area. This program is available for download at www.scioncorp.com. This software and digitizing process have been shown to be a valid and reliable method for determining fiber cross-sectional area, in addition to distinguishing between protein bands as separated by gel electrophoresis techniques (Humphries, Newton, Abernethy, & Blake, 1997).

Fiber-type percentages and cross-sectional areas have been calculated from varying numbers of fibers in the muscle sections, but there is no consensus on the number of fibers that must be measured to obtain reliable and valid results. Fiber area data are often used to measure morphological changes in skeletal muscle in response to different training stimuli. Obtaining mean values based on measurements of different populations of fibers is therefore important in providing an accurate reflection of the fiber size. Alway, Grumbt, Gonyea, and Stray-Gundersen (1989) showed that at least 200 type I and II fibers need to be measured in elite bodybuilders to determine fiber area accurately in this population. Lexell and Taylor (1989) showed that to obtain a good estimate of the mean fiber cross-sectional area for a whole muscle, the number of biopsies has a much greater influence on the sampling error than the number of fibers measured in each biopsy. Most studies have based their fiber area measurements on lower numbers of fibers, with one study using measurements from as few as 10 fibers (Tesch, Thorsson, & Kaiser, 1984). For the purposes of training studies, getting an accurate reflection of changes in fiber area of the different subtypes is important. Therefore, it is essential to know the number of fibers for the different subtypes that need to be measured to determine fiber area accurately. A study by McCall et al. (1998) indicated that mean values calculated from at least 50 fibers for each type (I and II) were representative of an individual's muscle biopsy. Because the authors did not differentiate between the major fast fiber subtypes (e.g., IIA and IIB), the appropriate sample size for characterizing specific fiber subtypes could not be determined from the results. We recently showed that different numbers of fibers are needed depending on the fiber type to characterize the mean fiber population accurately (McGuigan, Kraemer, Deschenes, et al., 2002).

Analysis of fiber data from a previous investigation (Kraemer, Patton, Gordon, et al., 1995) showed that the number of fibers that need to be measured for an accurate reflection of fiber size were 150 fibers for type I, 200 fibers for type IIA, and 50 fibers for IIB. Because type IIB fibers represent a smaller percentage of the three major fiber types, a smaller sample size would represent a greater percentage of the total number of type IIB fibers, particularly in moderately active individuals.

# Gel Electrophoresis

Polyacrylamide gel electrophoresis (PAGE) is an extensively used method for the analysis of proteins, making it a frequently used technique in muscle research laboratories. The popularity of gel electrophoresis results from several factors: (1) its high resolving power, (2) its ability to analyze many samples simultaneously, (3) the microgram quantities of protein that are required, (4) the possibility to recover the proteins after electrophoresis, and (5) the modest cost of the method (Barany, Barany, & Giometti, 1998). Electrophoresis refers to the migration of charged particles in a liquid medium under the influence of an electrical field. Tiselius, in 1937, devised the first electrophoresis method used in the study of proteins (Epstein, 1986).

## Background

In PAGE, separation of charged molecules results from differences in both charge and size. More highly charged species move more rapidly in an electric field; larger molecules and those less spherical in shape are retarded by the gel to a greater extent. These two effects may work together or in opposition. The proper choice of a gel system will maximize the differences in electrophoretic mobility or gel retardation to give the best possible resolution of particular macromolecules.

Separation because of size depends on the pore size of the gel matrix. The pore size of polyacrylamide gels, which are polymerized as the gel is formed, decreases as the concentration of gel monomer (expressed as %T) increases; however, the extent of cross-linkage also affects it. Increasing the concentration of cross-linker relative to the total monomer (expressed as %C) up to 5% by weight decreases the pore size. Above 5%C, the pore size increases again because the cross-linker dimerizes with itself to form more expanded gels.

Most proteins are fractionated on polyacrylamide ranging from 5%T to 20%T, containing 2 to 5%C. Very large proteins, such as titin, or various protein aggregates, are resolved on agarose gels or by polyacrylamide gels reinforced with agarose. For optimal resolution, all species of interest should be retarded by the gel but not completely excluded. When the size range of the sample components is too wide to be sieved by a gel of a single pore size, a gradient gel may be used.

## Denaturing Versus Nondenaturing Gels

Electrophoresis of proteins may be carried out under either denaturing or nondenaturing conditions. Nondenaturing buffers are required whenever biological activity must be retained, although denaturing buffers can be used if there is a way to renature biological activity after electrophoresis. A denaturing gel that allows detection of enzymatic activity following removal of the denaturant is called an activity gel. Nondenaturing buffers are also used when charge differences are known to exist that may give greater resolution than separation based on size.

In contrast, if only the molecular size is relevant, the proteins are denatured with sodium dodecyl sulfate SDS (Weber & Osborn, 1969). SDS is an anionic detergent that denatures proteins by wrapping around the polypeptide backbone. In so doing, SDS confers a net negative charge to the polypeptide in proportion to its length. When treated with SDS and a reducing agent, the polypeptides become rods of negative charges with equal "charge densities," or charge per unit length. Because the charge per unit length of the SDS-polypeptide complex is constant, the electrical force exerted on the complex per unit length is also constant. Thus, the sieving effect of the acrylamide matrix totally determines the migration velocity so that proteins are separated according to their molecular weight.

## Continuous and Discontinuous Buffer Systems

A continuous buffer system uses the same buffer throughout the gel and at both electrodes. Continuous buffer systems require only a single layer of gel. They are simple to set up and are often adequate for such purposes as monitoring enzyme purification or for preparative electrophoresis of a partially purified protein. Continuous buffer systems are also used for nucleic acid electrophoresis. The main disadvantage is that the sample must be in concentrated form, because each resolved band will be as wide as the depth of the original sample in the well.

In contrast, a discontinuous buffer system concentrates each sample component into a narrow band known as the stack. The original sample may therefore be much more dilute, and components that band closely together can be resolved more easily.

A discontinuous system uses buffers of different composition and pH to create a discontinuous voltage and pH gradient. At least two different gel layers stabilize the different buffer zones. The upper layer, through which the sample passes first, is known as the stacking gel, a large-pore gel that is nonrestrictive to the protein sample. The buffer in which the stacking gel is made contains an ion (usually an anion) whose electrophoretic mobility is greater than that of the protein, whereas the tank buffer, or electrode buffer, must contain an ion whose mobility is less than that of the protein. As electrophoresis begins, the "leading ion" in the stacking gel moves faster than the protein and leaves behind it a zone of lower conductivity. The higher voltage gradient of this zone causes the protein to move faster and to "stack" at the boundary between the leading and trailing ions.

Below the stacking gel is a deeper layer of gel with a smaller pore size, known as the resolving or separating gel. This gel is prepared in a buffer of higher concentration and pH (in an anionic system). In this environment, the mobility of the trailing ion increases so that its boundary moves ahead of the protein. The protein is resolved into individual bands according to size and, in the case of a nondenaturing gel, according to shape.

### Sodium Dodecyl Sulfate-Polyacrylamide Gel Electrophoresis (SDS-PAGE)

Sodium dodecyl sulfate-polyacrylamide gel electrophoresis (SDS-PAGE), as devised by Laemmli (1970), is perhaps the most popular type of protein gel electrophoresis technique for separating proteins, and noteworthy progress has been made in recent years. The commercial availability of various precast gels, buffers, and staining solutions has also simplified the procedure and increased the quality of the gels. In the Laemmli buffer system, the leading ion is chloride and the trailing ion is glycine. Accordingly, the resolving gel and the stacking gel are made up in Tris-HCl buffers (of different concentration and pH), whereas the tank buffer is Tris-glycine. All buffers contain 0.1% SDS.

The electrophoresis cell has an upper and lower chamber. The positive anode is connected to the lower chamber, and the negative cathode is connected to the upper chamber. When the electrophoresis cell is attached to a power supply, a negative charge flows from the cathode into the upper buffer chamber and moves through the gel into the lower buffer chamber toward the anode, allowing the circuit to be completed.

Therefore, negatively charged molecules, such as SDS-coated proteins, will migrate to the positive electrode. This is termed an anionic system (Ausubel, 1994).

Thus, the purpose of SDS-PAGE is to separate proteins according to their size. Sodium dodecyl sulfate (SDS) is a detergent that can dissolve hydrophobic molecules but also has a negatively charged sulfate attached to it. Therefore, if a cell is incubated with SDS, the membranes will be dissolved, causing the proteins to be unfolded by the detergent and covered with many negative charges. The denaturation of the protein is achieved by heating the protein in excess SDS and a thiol reagent such as 2-mercaptoethanol, which is needed to cleave the disulphide bonds. The extent of the molecular sieving during PAGE depends on how closely the gel pore size approximates the size of the migrating proteins. The effectual pore size of a polyacrylamide gel varies both with the total concentration of acrylamide in the gel mixture and with the proportion of the bisacrylamide cross-linker used. The lower percentage gels have the largest pore sizes and are used to separate larger proteins.

Visualization of proteins in electrophoresis gels may use any of the various staining procedures that have been developed. Coomassie blue is a commonly used stain in SDS-PAGE (Bradford, 1976). Many other stains may also be used in the detection of proteins in SDS-PAGE, such as silver, copper, and zinc stains. The choice of stain depends on the particular proteins in the gel, their concentrations, and the subsequent uses intended for them. Computer-assisted integration of the stained protein bands has also allowed for rapid quantification of tissue composition.

The skeletal muscle contractile protein, myosin heavy chain (MyHC), is routinely separated in vertical slab electrophoresis gels. These gels may be performed in a variety of formats with the typical gel size in the range of 16 × 16 cm, and smaller minigels of 8 × 10 cm are also popular. The identification of human MyHC isoforms has largely been identified using SDS-PAGE and immunoblotting analyses (Bamman, Clarke, Talmadge, & Feeback, 1999; Staron et al., 1994; Talmadge & Roy, 1993; Staron & Hikida, 1992b). These isoforms that have been separated in human muscle using SDS-PAGE have been called MyHC I, MyHC IIa, and MyHC IIb according to the order of decreasing electrophoretic mobility (Smerdu, Karsch-Mizrachi, Campione, Leinwand, & Schiaffino, 1994). Some researchers prefer to rename MyHC IIb as IIx because type IIx myosin heavy chain transcripts are expressed in type IIb fibers of human skeletal muscle (Smerdu et al., 1994).

## AN EXAMPLE OF SDS-PAGE FOR WHOLE MUSCLE HOMOGENATE MyHC USING THE BIO-RAD MINI-PROTEAN II CHAMBER

Note: Always wear gloves when handling reagents or electrophoresis gels.

### REAGENTS AND SOLUTIONS

- Acrylamide–bis acrylamide mix: 40% of 37.5: 1 ratio
- Resolving gel buffer (1.5 M Tris-HCl pH 8.8)
- 10% ammonium persulphate (APS)
- TEMED
- 100 mM EDTA
- 1 M glycine
- Tertiary amyl alcohol, water saturated (1:9 dilution)
- Stacking gel buffer (0.5 M Tris-HCl pH 6.8)
- 100% glycerol
- ddH$_2$O or ROH$_2$O
- 10X running buffer
- Stacking gel buffer: 0.5 M Tris-HCL (pH 6.8) (Tris MW 121.1)

60.55 g of Tris is dissolved in 400 ml of water. Titrate with 2M hydrochloric acid (approximately 240 ml) to pH 6.8. Make up to 1,000 ml with water.

- Resolving gel buffer: 1.5 M Tris-HCl (pH 8.8) (Tris MW 121.1)

181.65 g of Tris is dissolved in 400 ml of water. Titrate with 2M hydrochloric acid (approximately 240 ml) to pH 8.8. Make up to 1,000 ml with water.

- 10X running buffer (store at 4 °C; storage life about 1 month)

  72 g glycine

  15 g Tris base

  5 g SDS

  To 500 ml with distilled water

- Ammonium persulphate solution: 10%

  Ammonium persulphate, 100 mg

  To 1.0 ml with distilled water

  Make up fresh before use.

- Sodium dodecyl sulphate (SDS) solution: 10% (stable at room temperature, precipitates at 4 °C)

  Sodium dodecyl sulphate, 1g

  Add to 10 ml to distilled water

- 1 M glycine MW = 75.07

  Glycine, 7.5 g

  To 100 ml with distilled water

Store at 4 °C (storage life about 1 month)

| Homogenizing solution (pH to 6.8, store at 4 °C) | | |
|---|---|---|
| | 100 ml | MW |
| 20 mM KCl (potassium chloride) | 0.149 g | 74.55 |
| 2 mM K$_2$HPO$_4$ (potassium phosphate dibasic) | 0.035 g | 174.18 |
| 1 mM EGTA (ethyleneglycoltetraacetic acid) | 0.038 g | 380.4 |

| Precipitation solution (pH to 9.5, store at 4 °C) | | |
|---|---|---|
| | 100 ml | MW |
| 40 mM Na$_4$P$_2$O$_7$ (tetrasodium pyrophospate anhydrous) | 0.106 g | 265.9 |
| 1 mM MgCl$_2$ 6H$_2$O (magnesium chloride) | 0.002 g | 203.3 |
| 1 mM EGTA (ethyleneglycoltetraacetic acid) | 0.003 g | 380.4 |

| Glycerol storage buffer (store at 4 °C) | | |
|---|---|---|
| | 100 ml | MW |
| 50% glycerol | 50 ml | 265.9 |
| 100 mM Na$_4$PO$_7$ (tetrasodium pyrophospate anhydrous) | 2.695 g | 265.9 |
| 5 mM EDTA (ethylenediaminetetraacetic acid) | 0.146 g | 292.3 |

| SDS loading buffer | 8 ml | 40 ml | MW |
|---|---|---|---|
| 62.5 mM Tris (pH 6.8) | 1,000 μL (0.5 M) | 4,000 μL (0.5 M) | 121.14 |
| 25% glycerol | 2.0 ml | 10 ml | |
| 6.25% β-mercaptoethanol | 500 μL | 2,500 μL | 78.13 |
| 2.0% SDS | 1.6 ml (10%) | 8 ml (10%) | 288.4 |
| H$_2$O | 2.9 ml | 14.5 ml | |

Use enough bromophenol blue to give a deep blue color (~0.005 mg). Store at −20 °C and freeze and thaw only twice.

- 100 mM EDTA MW = 292.24

  EDTA, 0.3 g

  To 100 ml with distilled water (vacuum filter before use)

Store at 4 °C (storage life about 1 month)

- 2-methyl,2-butanol (tert-amyl alcohol) (1:9 ratio)

1.1 ml tert-amyl alcohol in 8.9 ml of distilled $H_2O$

## TISSUE PREPARATION

The method chosen in the tissue preparation will depend greatly on the amount of sample that you have available. Muscle samples are a valuable commodity, especially samples obtained from biopsies in human subjects. Therefore, a sample may be needed for several different assays.

### A. Serial cross-sections

1. Muscle samples are serially cross-sectioned at –20 °C to a thickness of 20 μm (5-10) and thawed in lysis buffer (30 μL/10 μg of tissue of ice-cold RIPA buffer + inhibitors). Ensure that samples remain frozen and add buffer immediately before use to prevent protein denaturation by proteases.

   Add inhibitors at time of use:

   - 10 mg/ml PMSF (phenylmethylsulfonyl fluoride) in isopropanol (add at 30 μL/ml RIPA).

   - Aprotinin (add at 30 μL/ml RIPA).

   - 100 mM sodium orthovanadate (add at 10μL/ml RIPA).

2. Homogenize tissue using a tissue homogenizer (glass/glass) maintaining temperature at 4 °C and incubate on ice for 30 min. Centrifuge samples at 10,000xg for 10 min at 4 °C. The supernatant fluid can then be mixed with an equal volume of electrophoresis sample buffer.

3. Measure the protein concentration using the Bradford method. Adjust final concentration to 10-50 μg with sample buffer. Vortex samples for 5 min and heat samples at 80 °C for 5 min.

### B. Whole tissue homogenization

1. Remove muscle specimens from the –70 °C freezer and place on ice. Weigh the specimens while still frozen by tarring the balance with an empty microcentrifuge tube.

2. Chop the tissue using a cooled scalpel blade and petri dish and transfer to the microcentrifuge tube.

3. Homogenize the tissue using a tissue homogenizer (glass/glass) on ice in approximately 10 volumes of "homogenizing buffer" (20 mM KCl, 2mM $K_2HPO_4$, and 1mM EGTA ethyleneglycolteraacetic acid) pH 6.8. Centrifuge at 13,000 rpm for 1 min.

4. Decant the supernatant and replace the tube into the ice. Repeat steps with homogenizing buffer until the pellet is white.

5. Add 50 to 100μL of precipitation solution [(40mM $Na_2O_2)]_7$, 1mM $MgCl_2$, and 1mM EGTA] pH 9.5 depending on the size of the pellet. For example, add 100μL if the precipitate covers the bottom of the tube.

6. Resuspend the pellet in the precipitation solution and stand on ice for 15 min.

7. Pipette the supernatant (contains myosin) into a new microcentrifuge tube and dilute 1:1 with glycerol storage buffer.

8. Measure the protein concentration using the Bradford method.

### Gel preparation

| Resolving gel | 10 ml | 20 ml | 30 ml |
| --- | --- | --- | --- |
| 8% acrylamide | | | |
| 100% glycerol | 3 ml | 6 ml | 9 ml |
| 40% acrylamide-bis (37.5:1) | 2 ml | 4 ml | 6 ml |
| Tris 1.5 M (pH 8.8) | 2 ml | 4 ml | 6 ml |
| 1 M glycine | 1 ml | 2 ml | 3 ml |
| 10% SDS | 0.4 ml | 0.8 ml | 1.2 ml |
| Distilled $H_2O$ | 1.5 ml | 3 ml | 4.5 ml |
| 6% acrylamide | | | |
| 100% glycerol | 3 ml | 6 ml | 9 ml |
| 40% acrylamide-bis (37.5:1) | 1.5 ml | 3 ml | 4.5 ml |
| Tris 1.5 M (pH 8.8) | 2.5 ml | 5 ml | 7.5 ml |
| 1 M glycine | 1 ml | 2 ml | 3 ml |
| 10% SDS | 0.4 ml | 0.8 ml | 1.2 ml |
| Distilled $H_2O$ | 1.5 ml | 3 ml | 4.5 ml |

Mix the preceding, vacuum de-gas for 20 min, and then add the following:

| Resolving gel | 10 ml | 20 ml | 30 ml |
| --- | --- | --- | --- |
| APS 10% | 100 μL | 200 μL | 300 μL |
| TEMED | 10 μL | 20 μL | 30 μL |

1. Swirl to mix the solution and pour immediately. Ensure that no bubbles form within the acrylamide solution because bubbles may distort the gel and not allow polymerization to occur. Do not completely fill the space between the glass plates. Allow about 3 cm at the top for the stacking gel.

2. Overlay with water-saturated tertiary amyl alcohol (1:9) and allow to set for 30 min.

3. Discard tertiary amyl alcohol and wash thoroughly with distilled $H_2O$.

4. Remove any excess water using blotting paper.

5. Pour the stacking gel.

| Stacking gel | 10 ml | 20 ml | 30 ml |
|---|---|---|---|
| 5% acrylamide | | | |
| 100% glycerol | 3 ml | 6 ml | 9 ml |
| 40% acrylamide-bis (37.5:1) | 1.25 ml | 2.5 ml | 5 ml |
| Tris 0.5 M (pH 6.8) | 2.5 ml | 5 ml | 7.5 ml |
| 100 mM EDTA (pH 7.0) | 0.4 ml | 0.8 ml | 1.2 ml |
| 10% SDS | 0.4 ml | 0.8 ml | 1.2 ml |
| Distilled $H_2O$ | 2.45 ml | 4.9 ml | 6.1 ml |
| 4% acrylamide | | | |
| 100% glycerol | 3 ml | 6 ml | 9 ml |
| 40% acrylamide-bis (37.5:1) | 1.0 ml | 2 ml | 3 ml |
| Tris 0.5 M (pH 6.8) | 2.5 ml | 5 ml | 7.5 ml |
| 100 mM EDTA (pH 7.0) | 0.4 ml | 0.8 ml | 1.2 ml |
| 10% SDS | 0.4 ml | 0.8 ml | 1.2 ml |
| Distilled $H_2O$ | 2.6 ml | 5.2 ml | 7.8 ml |

Mix the preceding, vacuum de-gas for 20 min, and then add the following:

| Resolving gel | 10 ml | 20 ml | 30 ml |
|---|---|---|---|
| APS 10% | 100 μL | 200 μL | 300 μL |
| TEMED | 10μL | 20 μL | 30 μL |

1. Swirl to mix the solution and pour immediately. Fill to the top of the glass plates. Insert the comb at an angle to avoid collecting air bubbles and ensure that each well is filled with acrylamide solution.

2. Smear approximately 20 μL of TEMED over the comb before inserting between the glass plates to aid in the polymerization of the wells.

3. Allow setting before loading samples. Pour $ddH_2O$ over the gel and allow curing overnight, removing comb before running samples and rinsing thoroughly with distilled $H_2O$ to remove any unpolymerized acrylamide from the wells.

### How to use acrylamide–bis acrylamide solution 40% of 37.5:1 ratio

ml of acrylamide solution = [(final desired gel concentration)(final ml of gel to be prepared)] / (% of acrylamide–bis acrylamide solution)

Polyacrylamide gels are described with reference to two parameters that determine pore size. The total monomer concentration, or %T, is defined as

$$\%T = [(\text{grams acrylamide} + \text{grams cross-linker})/ \text{total volume (ml)}] \times 100$$

The weight percentage of cross-linker, or %C, is defined as

$$\%C = [\text{grams cross-linker}/(\text{grams acrylamide} + \text{grams cross-linker})] \times 100$$

By varying these two parameters, you can optimize the pore size of the gel to give the best separation and resolution for the molecule of interest.

### Determining amounts of cross-linker catalysts

For every ml of acrylamide solution add

- 1 μL of TEMED, and
- 10 μL of APS

## RUNNING CONDITIONS

1. Load 5 μL of sample for each well. Load 3 to 5 μL of a prestained standard (Bio-Rad Kaleidoscope prestained standards) with an empty lane between the standard and next sample.

2. Run at 70 V (constant voltage) for approximately 30 min and then at 200 V for 6 h for 6% gels, or at 70 V (constant voltage) for approximately 30 min and then at 150 V for 18 h for 8% gels. Run all gels at 4 °C or in an ice bath during the electrophoresis run.

3. Remove gel and stain according to instructions (i.e., Coomassie blue or silver stain) or transfer to a membrane for Western blotting.

# Protein Quantification Through Bradford or Lowry Methods

In most current methods used to quantify total protein in solution, a chemical reagent is added to the protein that produces a color change in the protein sample (Stoscheck, 1990). This color change is monitored spectrophotometrically and compared with a standard curve in which samples of known concentration are reacted in parallel with the reagent. The amount of protein in the sample is determined from the standard curve of absorbance versus known protein concentration.

Four spectroscopic methods are routinely used to determine the concentration of protein in a solution. These include measurement of the intrinsic UV absorbance of the protein and three methods that generate a protein-dependent color change: the Lowry assay, the Smith copper-bicinchoninic acid (BCA) assay, and the Bradford assay.

The problem is that amino acids do not absorb light in the same way they do when they are found in complex biomolecular environments. Several approaches have been developed that use chemical reactions with the peptide chain. These chemistries create a color in the visible range that can be measured; hence, they are referred to as colorimetric methods. The amount of colored complex formed is proportional to the amount of protein in the corresponding sample.

These chemistries use redox reactions of metal ions and complexion formation. Copper is the primary metal involved in the redox-complexation chemistry of both the classic Lowry assay and the newer bicinchoninic acid (BCA) assay. Reduced copper interacts with molybdenum and tungsten phosphometal complexes in the Lowry assay to form the deep purple color. Reduced copper complexes directly with the BCA reagent. In dye-binding assays, no redox chemistry takes place but the absorbance of the dye is altered upon binding to protein. This is the basis of the Bradford assay and newer fluorescent dye-binding assays (Bradford, 1976).

Each of these approaches has strengths and weaknesses. You must think carefully about the other chemicals and biomolecules in the sample and account for their effect on any colorimetric assay. The Lowry and copper-bicinchoninic assays are based on reduction of $Cu^{2+}$ to $Cu^{1+}$ by amides (Lowry, 1951). Although this makes them potentially quite accurate, they require the preparation of several reagent solutions, which must be carefully measured and mixed during the assay. This is followed by prolonged, accurately timed incubations at closely controlled temperatures, followed by immediate absorbance measurements of the unstable solutions. These assays may also be affected by other substances frequently present in biochemi-

cal solutions, including detergents, lipids, buffers, and reducing agents.

## BRADFORD PROTEIN ASSAY PROCEDURE

The Bradford method uses a dye called Coomassie brilliant blue G250, which undergoes a color change upon noncovalent binding to proteins greater than 3,000 to 5,000 Daltons dependent on the amino acid composition. The dye binds to basic and aromatic amino acid residues, especially arginine. The Bradford dye assay is based on the equilibrium between three forms of Coomassie blue G dye. Under strongly acid conditions, the dye is most stable as a doubly protonated red form. Upon binding to protein, however, it is most stable as an unprotonated blue form.

The Bradford assay is fast, involves few steps, does not require heating, and gives a more stable colorimetric response than other protein assays do. Like other protein assays, its response is prone to influence from nonprotein sources, particularly detergents, and becomes progressively more nonlinear at the high end of its useful protein concentration range. The response is also protein dependent and varies with the composition of the protein.

### BRADFORD REAGENT

Dissolve 100 mg Coomassie blue G-250 in 50 ml 95% ethanol, adding 100 ml 85% (w/v) phosphoric acid to this solution. Dilute to 1 L when the dye has completely dissolved and filter through Whatman #1 paper just before use.

### BOVINE SERUM ALBUMIN (BSA) (1 MG/ML)

Dissolve BSA in saline and store it frozen in 1 ml aliquots for quick use. The standard should be dissolved in a buffer similar to that which the unknowns will be dissolved in.

### STANDARD ASSAY PROCEDURE, 200 TO 1,500 μG/ML

1. Prepare a series of protein standards using BSA diluted with 0.15 M NaCl to final concentrations of 0 (blank = NaCl only), 250, 500, 750, and 1,500 μg BSA/ml. Also, prepare serial dilutions of the unknown sample to be measured.

2. Add 100 μL of each of the previous to a separate test tube.

3. Add 5.0 ml of Bradford reagent to each tube and mix by vortex or inversion.

4. Wait 5 min and read each of the standards and each of the samples at 595 nm.

5. Plot the absorbance of the standards versus their concentration. Compute the extinction coefficient and calculate the concentrations of the unknown samples.

## MICRO ASSAY PROCEDURE, 10 TO 100 μG/ML PROTEIN

1. Prepare standard concentrations of BSA of 1, 5, 7.5, and 10 μg/ml. Prepare a blank of NaCl only. Prepare a series of sample dilutions.

2. Add 100 μL of each of the previous to separate tubes (use microcentrifuge tubes) and add 1.0 ml of Bradford reagent to each tube.

3. Wait 2 min and read the absorbance of standards and unknowns at 595 nm.

4. Plot the absorbance of the standards versus their concentration. Compute the extinction coefficient and calculate the concentrations of the unknown samples.

## LOWRY PROTEIN ASSAY

The Lowry assay is an often-cited general-use protein assay (Hartree, 1972; Lowry, 1951). For some time it was the method of choice for accurate protein determination for cell fractions, chromatography fractions, enzyme preparations, and so on. Under alkaline conditions, copper complexes with protein. When folin phenol reagent (phospho-molybdic-phosphotungstic reagent) is added, the folin phenol reagent binds to the protein. Bound reagent is slowly reduced and changes color from yellow to blue. Although widely used, the Lowry procedure is a less preferable assay than some other protein assays because it is more subject to interference by a variety of chemicals. Among the chemicals reported to interfere with the Lowry procedure are barbital, CAPS, cesium chloride, citrate, cysteine, diethanolamine, dithiothreitol, EDTA, EGTA, HEPES, mercaptoethanol, Nonidet P-40, phenol, polyvinyl pyrrolidone, sodium deoxycholate, sodium salicylate, thimerosol, Tricine, Tris, and Triton X-100.

## STOCK SOLUTIONS

- Lowry A: 2% $Na_2CO_3$ in 0.1 M NaOH
- Lowry B: 1% $CuSO_4$ in $diH_2O$
- Lowry C: 2% sodium potassium tartrate ($NaKC_4H_4O_6 \cdot 4H_2O$)

## REAGENTS

- Lowry stock reagent
  49 ml Lowry A
  0.5 ml Lowry B
  0.5 ml Lowry C
- Folin reagent: phenol reagent—2N (folin-Ciocalteau reagent). Dilute 1:1 in distilled $H_2O$ before use.

## BOVINE SERUM ALBUMIN (BSA) (1 MG/ML)

Dissolve BSA in saline and store it frozen in 1 ml aliquots for quick use. The standard should be dissolved in a buffer similar to that which the unknowns will be dissolved in.

## PROCEDURE

1. Add 100 μL of sample (sample + buffer = 100 μL) per tube.

2. Add 1.0 ml of Lowry stock reagent to each tube.

3. Incubate 30 min at room temperature.

4. Add 100 μL of folin reagent to each tube.

5. Incubate 30 min at room temperature.

6. Read the absorbance of standards and unknowns at 595 nm.

Note: The Lowry method depends on the presence of tyrosine within the protein to be measured. Unless the standard protein contains approximately the same number of tyrosine residues as the sample, the procedure will be inaccurate. If no tyrosine residues are in the sample to be measured, the Lowry method of protein determination is ineffective and the Bradford assay should be used. In general, the Bradford assay is the method of choice for protein determinations.

# Immunohistochemistry for Steroid Receptor Analysis

Immunohistochemistry is the procedure for detecting antigens with a microscope using antibodies directed specifically against those antigens of interest. Immunohistochemistry may be considered the most powerful and potentially specific of all the localization techniques (Ausubel, 1994).

The basic principle of immunohistochemistry is the use of enzyme-linked antibodies (Ab)(mono- or polyclonal) to detect tissue antigens (Ag). The colorless substrate is converted by the enzyme into a colored product that precipitates on the slide at the site of the reaction. Thus, immunohistochemistry localizes antigens in a tissue section. The procedure may be summarized as follows.

## IMMUNOHISTOCHEMISTRY PROCEDURE

1. Preparation of the antigen—Through a conventional histological section of the tissue of interest.

2. Antigen–antibody reaction—The combination of the Ag and Ab allows them to bind through the addition of an enzyme-linked Ab to the slide. This Ab is incubated and then washed to remove unbound Ab.

3. Visualization—Detect Ag:Ab binding by adding a colorless substrate such as fluorescent materials; coloring by DAB (3-3′ diaminobenzidine) and enzyme (e.g., peroxidase-antiperoxidase or avidine-biotin complex).

This substrate is then incubated and washed to remove remaining substrate. Counterstaining may be used to make nonstaining cells and structures visible. If Ags were present in the tissue, a color reaction indicates its location (in cells or in the extracellular matrix).

The two main methods utilized in immunohistochemistry are direct and indirect.

1. Direct immunohistochemical methods

    antigen + (antibody + marker substance)

2. Indirect immunohistochemical methods

    antigen + primary antibody +
    (secondary antibody + marker substance)

The indirect method is an amplification process whereby large amounts of marker substance can be attached. It is the most widely used method in immunohistochemistry; examples include the peroxidase-antiperoxidase (PAP) and avidine-biotin complex (ABC) methods. By using different enzymes (peroxidase, alkaline phosphatase, and so on) or different cromogens (DAB, AEC, Fast RED, and so on), several possibilities of visualizing such interactions exist. Constantly, new reagents appear capable to amplify and, with that, to improve the expression of these interactions.

## AN EXAMPLE OF IMMUNOHISTOCHEMISTRY OF ANDROGEN RECEPTORS IN SKELETAL MUSCLE

We have found that the commercially available kit from Santa Cruz Biotechnology (Kit #sc-816k, Santa Cruz Biotechnology, Inc., Santa Cruz, California) works well for the immunoperoxidase staining of androgen receptors in skeletal muscle. Santa Cruz Biotechnology, Inc. offers either their ABC or ImmunoCruz staining systems. The ABC staining system uses a preformed avidin-biotinylated horseradish peroxidase (HRP) complex as the detection reagent, whereas the ImmunoCruz staining systems use a streptavidin-HRP complex. In addition, the ImmunoCruz systems offer the advantage of having all reagents prediluted to the optimal concentrations needed for tissue staining.

The following procedure is modified from the Santa Cruz Biotechnology Research Applications manual, which is available for download through the Santa Cruz Web site (www.scbt.com).

### PREPARATION OF FROZEN TISSUE SECTIONS

1. Cut 4 to 10 μm thick sections. Adhere sections to slides at room temperature. Slides may be stored at −70 °C. Thaw slides at room temperature before fixing and staining.

2. Fix slides in cold acetone for 10 min and keep refrigerated. Wash in three changes of PBS.

3. Optional: Incubate for 5 to 10 min in 0.1 to 1% hydrogen peroxide in PBS to quench endogenous peroxidase activity. Wash in PBS twice for 5 min each.

### SANTA CRUZ IMMUNOCRUZ STAINING PROCEDURE

The entire procedure should be carried out at room temperature, and all staining reagents must be allowed to reach room temperature before use. Tissue sections should never be allowed to dry during the procedure. Use suction to remove reagents after each step but avoid drying of specimens between steps. Use sufficient reagents to cover the specimens (approximately 100 μL per slide is usually adequate).

1. Incubate specimens for 20 min in one to three drops of serum block. Aspirate serum from slides.

2. Immediately add one to three drops of prediluted primary antibody. Incubate for 2 h. Rinse

with PBS and then wash in PBS twice for 2 min each on a stir plate. Aspirate excess liquid from slides.

3. Incubate for 30 min in one to three drops of biotinylated secondary antibody. Wash as before.

4. Incubate for 30 min in one to three drops of HRP-streptavidin complex. Wash as before.

5. Add one to three drops HRP substrate mixture. Develop for 30 s to 10 min, or until desired stain intensity develops. Rinse with distilled $H_2O$ and transfer to a distilled $H_2O$ wash for 2 min on a stir plate.

6. If desired, counterstain in Gill's formulation #2 hematoxylin for 5 to 10 s. Immediately wash with several changes of distilled $H_2O$.

7. Dehydrate through alcohols and xylenes as follows: Soak in 95% ethanol twice for 10 s each, then in 100% ethanol twice for 10 s each, and then in xylenes three times for 10 s each. Wipe off excess xylene. Immediately add one to two drops of permanent mounting medium, cover with a glass cover slip, and observe by light microscopy.

# Western Blotting for Steroid Receptor Analysis

Western blotting is a method for identifying a specific protein in a complex mixture and simultaneously determining its molecular weight. DNA blotting was first described by E.M. Southern and hence became known as Southern blotting. RNA blotting was then given the name Northern blotting, and protein blotting became known as Western blotting.

In a Western blot, antigens are first separated according to molecular weight by gel electrophoresis and then blotted onto a membrane (either nitrocellulose or PVDF). The procedure is summarized as follows.

## WESTERN BLOT PROCEDURE

1. Preparation of the antigen—Electrophoresis through PAGE and then blotting (transferring) the Ags from the gel matrix to a membrane (nylon or nitrocellulose) where they can be fixed

2. Combination of the Ag and Ab—Allowing them to bind by covering the membrane with the unknown serum sample separating bound and unbound Abs and Ags

3. Detecting Ag:Ab binding—Adding an antiglobulin that has been covalently conjugated (labeled) with an enzyme, allowing it to bind, removing the unbound conjugated antiglobulin, and finally adding a colorless substrate for the enzyme

If the enzyme-conjugated Ab is present, the colorless substrate is converted to a product with color, indicating the presence of an Ag:Ab reaction in step 1.

The unique feature of Western blotting is that it identifies one or more individual Ags to which the Abs are bound. A positive sample will show binding of the enzyme-labeled antiglobulin at the particular sites where Ags of the infectious agent are known to migrate in electrophoresis. If binding occurs at other sites, the sample is considered negative because the binding is to a contaminating Ag in the preparation.

## AN EXAMPLE OF WESTERN BLOTTING FOR ANDROGEN RECEPTORS IN SKELETAL MUSCLE UTILIZING THE BIO-RAD MINI-PROTEAN II MINI TRANS-BLOT TRANSFER CELL

### REAGENTS AND SOLUTIONS

- 1 X PBS buffer (adjust to pH 7.4), for 2 L $dH_2O$
  9.1 mM dibasic sodium phosphate, 2.5836 g
  1.7 mM monobasic sodium phosphate, 0.4692 g
  150 mM NaCl, 17.529 g
- Transfer buffer, pH 8.3, for 2 L $ddH_2O$
  25 mM Tris, 6.06 g
  192 mM glycine, 28.8
  20% v/v methanol, 400 ml
- TBS, for 2 L $ddH_2O$
  10 mM Tris, 2.422 g
  150 mM NaCl, 17.529 g
  pH 8.0
- TTBS, pH 8.0, for 2 L $ddH_2O$
  10 mM Tris, 2.422 g
  150 mM NaCl, 17.529 g
  0.05% Tween-20, 10 ml of 10% Tween-20
- Blotto A (Santa Cruz), to 500 ml 1 X TBS
  5% milk
  0.05% Tween-20

- RIPA buffer, for 200 ml

  1% Igepal, 2 ml

  0.5% sodium deoxycholate, 1 g

  0.1% SDS, 0.2 g

  1x PBS, 198 ml

- Add inhibitors at time of use:

  - 10 mg/ml PMSF (phenylmethylsulfonyl fluoride) in isopropanol (added at 30 µL/ml RIPA)

  - Aprotinin (added at 30 µL/ml RIPA)

  - 100 mM sodium orthovanadate (add at 10 µL/ml)

- Laemmli buffer (stored at 4 °C), for 20 ml

  Distilled water, 7.25 ml

  0.5 M Tris-HCl buffer (pH 6.8), 2.5 ml

  25% glycerol, 5.0 ml (stock)

  2.0% SDS, 4.0 ml (10% SDS)

  2-mercaptoethanol, 1.25 ml (stock)

  1.0% bromophenol blue, enough to give a deep blue color

## SAMPLE PREPARATION

Muscle samples are separated by loading 10 to 50 µg of homogenate protein onto an 8% SDS-PAGE gel and run at 150 V for 1 h at 25 °C.

## PREPARATION FOR BLOTTING

1. Fill the cooling unit with water and store it at –20 °C until ready to use. After use, return the cooling unit to the freezer for storage. Prepare the transfer buffer (using buffer chilled at 4 °C will improve heat dissipation). Approximately 500 ml of buffer is required for the mini trans-blot cell. Do not add acid or base to adjust pH of the transfer of TTBS buffers. Methanol should be analytical reagent grade, because metallic contaminants in low-grade methanol will plate on the electrodes.

2. Cut the membrane and the filter paper to the dimensions of the gel. Always wear gloves when handling membranes to prevent contamination. Equilibrate the gel and soak the membrane, filter paper, and fiber pads in transfer buffer (15 min to 1 h).

3. All electrophoresis gels should be preequilibrated in transfer buffer before electrophoretic transfer. Preequilibration will facilitate the removal of contaminating electrophoresis buffer salts and neutralization salts. If the salts are not removed, they will increase the conduc-

tivity of the transfer buffer and the amount of heat generated during the transfer. Also, low-percentage gels (<12%) will shrink in methanol buffers. Equilibration allows the gel to adjust to its final size before electrophoretic transfer.

4. Prepare the gel sandwich. Removing any air bubbles that may have formed is important for good results. Use a glass tube to roll out air bubbles gently. Add the frozen Bio-Ice cooling unit. Place in tank and completely fill the tank with buffer. Add a standard stir bar to help maintain even buffer temperature and ion distribution in the tank. Set the speed as fast as possible to keep ion distribution even.

## TRANSFER CONDITIONS

1. Transfer proteins to a nitrocellulose membrane at 25 V in a buffer containing 192 mM glycine, 25 mM Tris, and 20% (vol/vol) methanol at room temperature for 12 h. Handle nitrocellulose membrane with tweezers or gloved hands only.

2. Block nonspecific binding by incubating the membrane in Blotto (1x TBS, 5% milk, 0.05% Tween-20) for 30 to 60 min at room temperature. Alternatively, the membrane may be blocked at 4 °C overnight in a covered container, using Blotto without Tween-20.

3. Incubate the membrane in the primary antibody diluted in Blotto (0.5 µg/ml – 2.0 µg/ml) for 1 h at room temperature (Santa Cruz Biotechnology, Rabbit polyclonal IgG #sc-816).

4. Wash membrane three times for 5 min each with TBS, 0.05% Tween-20, and then incubate for 45 min at room temperature with a HRP conjugated secondary antibody, diluted to 1:500 to 1:2,000 in Blotto (Santa Cruz Biotechnology, Goat anti-rabbit IgG-HRP #sc-2004).

5. Wash membrane three times for 5 min each with TBS, 0.05% Tween-20, and once for 5 min with TBS.

## DETECTION CONDITIONS

- Santa Cruz Biotechnology Western Blotting Luminol Reagent

  1. Mix equal quantities of Luminol Reagent Solution A and Solution B by inversion in a screw cap vial. Use 0.125 ml/cm² total volume. Pour off TBS wash from the nitrocellulose membrane and add the mixed Luminol Reagent Solutions to the membrane, protein side facing

up. Incubate for 1 min at room temperature. (Work quickly and carefully because maximal detection and light emission due to enzymatic reaction occur during the first hour).

2. Pour off excess Luminol Reagent or lift membrane with tweezers to drain excess reagent and lightly dab the edge of the membrane with a Chemwipe to remove excess solution. Use caution not to wipe or smear the membrane surface.

3. Quantify for protein content using an appropriate imaging system (i.e., a phosphor imager) and analysis.

- Bio-Rad Opti-4CN Colorimetric HRP Substrate Kit

1. Add 0.2 ml of Opti-4CN substrate per 10 ml of diluent (e.g., combine 1 ml Opti-4CN diluent concentrate with 9 ml distilled $H_2O$ and 0.2 ml Opti-4CN substrate). Mix well and pour onto membrane.

2. Incubate membrane with gentle agitation in the substrate for up to 30 min or until the desired level of sensitivity is attained. Wash the membrane in distilled $H_2O$ for 15 min.

3. Quantify for protein content by densitometric scanning and analysis.

## Summary

Skeletal muscle is an extremely heterogeneous tissue. A variety of methods are available for researching its structure and function. This chapter does not provide coverage of all techniques that can be used in the laboratory to analyze skeletal muscle structure and function. We have not covered several techniques, such as muscle enzyme analysis (e.g., citrate synthase) and single-fiber analysis. Rather, the chapter is designed to provide readers with several techniques that can be used and adapted to the specific needs of their own laboratories and research. Readers will find variations of the assays that we have covered in different scientific publications. The chapter should provide a convenient starting point for anyone interested in obtaining information about exercise-induced responses of skeletal muscle.

The percutaneous muscle biopsy technique has been commonly used to obtain muscle samples for research in exercise physiology. The freezing and mounting of the muscle tissue will depend on the type of analysis to be performed. A variety of methods are then available to measure muscle fiber type. The common approach has been based on mATPase activity. This method also allows muscle fiber cross-sectional area to be determined. Polyacrylamide gel electrophoresis is also an extensively used method for the analysis of proteins in muscle research laboratories. Immunohistochemistry is the procedure for detecting antigens with a microscope using antibodies directed specifically against those antigens of interest. This method can be used to measure androgen receptors in muscle cross-sections.

We have attempted to provide readers with enough information to undertake a sophisticated analysis of human skeletal muscle that would provide information about the histochemical and biochemical properties. We have also included several key references that provide further information regarding these techniques. As with any biochemical analyses, one must be proficient in several basic chemical techniques before undertaking an analysis. Basic laboratory methods for mixing and diluting chemicals should be reviewed before undertaking the various assays. A wide variety of techniques are available for investigating the structural and functional properties of human skeletal muscle. A combination of different methods will provide important information regarding the responses of this heterogeneous tissue to exercise.

# ten

# The Utility of Near-Infrared Spectrophotometry in Athletic Assessment

Kenneth W. Rundell, PhD
*Marywood University*
**Joohee Im, MS**
*University of Pennsylvania at Philadelphia*

Whole-body oxygen consumption during exercise is considered the most significant measurement for assessment of training status and athletic potential in sports requiring aerobic energy production. Monitoring blood lactate threshold and economy of movement, however, has also received much attention and is used routinely in the design of training programs for endurance sport participants. Lactate threshold is accepted as one of the most important tools for evaluating the effects of training and has been deemed critical to endurance performance. Moreover, improvements in endurance performance have correlated better with peripheral changes than with $\dot{V}O_2$max status (Costes et al., 2001; Coyle, 1995; Foster, Rundell, Snyder, et al., 1999; Kouzaki, Shinohara, Ikeda, Watarai, & Fukunaga, 2001; Neary, McKenzie, & Bhambhani, 2002; Neary, Hall, & Bhambhani, 2001; Rundell, Nioka, & Chance, 1997; Rundell, 1996).

Until recently, evaluation of local oxidative metabolism has required invasive techniques such as artery and vein catheterization or muscle biopsy. Muscle oxygen consumption during exercise has been estimated using invasive measurements of arteriovenous oxygen difference (a-v $O_2$ difference) and has provided valuable information concerning muscle adaptation to exercise and blood flow during exercise. This application is limited, however, because of the invasive nature and the general assumptions that must be made when evaluating large muscle mass (such as the quadriceps) using oxygen measurements in the large femoral vessels of the leg.

Beginning in the late 1980s, in vivo measurement of muscle oxygen utilization by noninvasive spectrophotometry in the near-infrared region of the light spectrum (NIRS) has been used to evaluate training status and muscle oxygen kinetics in athletic performance. NIRS offers the opportunity to evaluate the contribution of specific muscles to whole-body exercise, evaluate oxygen kinetics during exercise and during recovery, gain insight to central and peripheral adaptations to training, evaluate positional blood volume dynamics, and monitor lactate threshold noninvasively. The principles of NIRS are based on the absorbance characteristics of oxygenated and deoxygenated hemoglobin and myoglobin in the near-infrared light region. When light penetrates tissue, it is either absorbed or scattered. Photo detectors measure change in absorption between the oxy- and deoxy-heme molecules, thus providing a relative change in muscle tissue oxygen utilization.

Although quantification of the NIRS signal is difficult because of tissue homogeneity, subcutaneous fat, variable light scatter, and muscle fiber type, new systems and algorithms improve the capability to quantify specific muscle tissue $\dot{V}O_2$. Recent studies using physiological manipulations have demonstrated techniques for NIRS signal quantification. This will be discussed later in this chapter.

Indeed, NIRS technology is improving rapidly and demonstrates exciting noninvasive means to measure tissue oxygen kinetics and blood flow dynamics. The combination of NIRS and [31]P-MRS has validated the use of NIRS (Blei, Conley, Odderson, Esselman, & Kushmerick, 1993; Cerretelli & Binzoni, 1997; Chance, Borer, Evans, Holtom, et al., 1989; Chance et al., 1985; Dulieu, Casillas, Maillefert, et al., 1997; Kemp, Roberts, Bimson, et al., 2001; Kemp, Hands, Ramaswami, et al., 1995; McCully, Iotti, Kendrick, et al., 1994; McCully, 1993; Tran, Sailasuta, Kreutzer, et al., 1999; Zatina, Berkowitz, Gross, Maris, & Chance, 1986). Likewise, multichannel NIRS units and EMG instrumentation have expanded our knowledge on muscle heterogeneity by concomitant measurements during exercise (Alfonsi, Pavesi, Merlo, Gelmetti, et al., 1999; Boushel, Langberg, Green, et al., 2000; Boushel, Langberg, Olesen, et al., 2000; Boushel & Piantadosi, 2000; De Blasi, Almenräder, & Ferrari, 1997; De Blasi, Fantini, Franceschini, Ferrari, & Gratton, 1995; De Blasi et al., 1994; De Blasi, Cope, Elwell, Safoue, & Ferrari, 1993; Hicks, McGill, & Hughson, 1999; Miura, McCully, Hong, Nioka, & Chance, 2001; Miura, Araki, Matoba, & Kitagawa, 2000; Miura, Takeuchi, Sato, et al., 1998; Quaresima, Colier, van der Sluijs, & Ferrari, 2001; Takaishi, Sugiura, Katayama, et al., 2002; Yoxall & Weindling, 1997). Recent papers describe physiological manipulations that allow blood flow measurements using NIRS.

Much documentation concerning medical use of NIRS is available, but this chapter will focus on the role and application of NIRS in the physiological evaluation of athletes. This chapter will begin with NIRS instrumentation and principles of measurement followed by a section examining validation and quantification of NIRS measurements of exercising muscle. Oxygen consumption and blood flow modeling will then be explored. Finally, practical applications to sports medicine and sports science will be presented and discussed.

## NIRS Instrumentation

The relatively new noninvasive optical instrumentation to measure real time muscle oxygenation and blood volume change has received much recent attention. Portable lightweight systems with memory cards for on-board real time measurements during exercise are commercially available (figure 10.1) (Ferrari, Wei, Carraresi, De Blasi, & Zaccanti, 1992; Foster et al.,

1999; Im, Nioka, Chance, Kime, & Rundell, in review; Im, Nioka, Chance, & Rundell, 2000; Nioka et al., 1999; Nioka, Moser, Lech, et al., 1998; Quaresima, Ferrari, Ciabattoni, Cantò, & Colonna, 1999; Rundell et al., 1997; Shiga, Tanabe, Nakase, Shida, & Chance, 1995; Szmedra, Im, Nioka, Chance, & Rundell, 2001). These systems are one-channel photometers that use continuous wave light. Acquisition time for

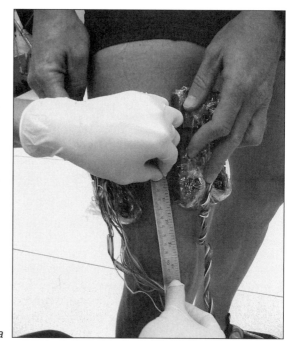

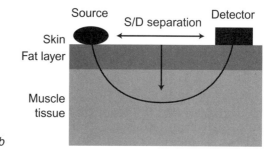

**Figure 10.1** Placement of two eight-channel probes on the medial and lateral aspects of the quadriceps muscle. Probe placement must be consistent when doing repeat measurements on an individual. Measurement of probe placement and outlining the probe with a nonwashable marker will assure consistent placement. The light follows a banana-shaped path from source to detector with a penetration of approximately one-third of the distance between source and detector. Note that the strength of the signal is proportional to the distance of light penetration; therefore, subcutaneous fat will affect the absolute readings of the system. Normalizing to a maximal deoxygenation from a cuff ischemic response is a method of correcting for the thickness of the subcutaneous layer.

this system is 0.2 s. Although the original concept of using near-infrared spectroscopy (NIRS) to measure muscle oxygenation and deoxygenation is credited to Millikan (1942), the current NIRS adapted to monitor hemoglobin and myoglobin deoxygenation in skeletal muscle was developed by Chance and coworkers (1992).

The NIRS includes an LED light source probe in the near-infrared range (700-900 nm) coupled with photodiode detectors linked to a spectrophotometer. The source-to-detector separation of 3.0 cm results in a detectible light penetration of approximately 2 to 3 cm. The pattern of light path from source to detector follows a banana-shaped figure in which the depth of the light penetration is equal to approximately one-half to one-third of the distance between source and detector. A 650 to 1,000 nm wavelength range allows the detection of light through 8 cm of tissue. In the 700 to 900 nm near-infrared range of the light spectrum, the heme group of hemoglobin, myoglobin, and cytochromes are the prominent absorbing compounds in muscle tissue. Because Mb and Hb have similar absorption spectrums, distinguishing the two by optical properties alone is difficult. Results from several spectral studies, however, lead to the conclusion that more than 90% of the NIRS signal comes from Hb (Seiyama, Hazeki, & Tamura, 1988; Wang, Noyszewski, & Leigh, 1990; Wittenberg, 1970).

## Principle of NIRS Measurement

Currently, only qualitative values of tissue oxygenation during dynamic exercise can be obtained with NIRS. Portable continuous wavelength instruments have been shown to produce reliable measurements that correlate with venous blood samples (Costes et al., 1996; Franceschini, Boas, Zourabian, et al., 2002; MacDonald, Tarnopolsky, Green, & Hughson, 1999; Mancini et al., 1994; Sahlin, 1992). The instrumentation is similar in principle to the pulse oximeter. It is a noninvasive method based on differential absorption characteristics of near-infrared light by the oxygenated and deoxygenated forms of heme groups (hemoglobin, myoglobin, cytochrome C) (Chance et al., 1989; Conley, Ordway, & Richardson, 2000; Hampson & Piantadosi, 1988; Piantadosi, Hemstreet, & Jobsis-Vandervliet, 1986).

Near-infrared light has much better penetration into biological tissue than visible light does. The light emitted by the NIRS probe permeates through the skin and subcutaneous fat layer with minimal scatter

and absorption; it then enters the muscle tissue where it is either absorbed or scattered within tissue. Part of the scattered light then returns to the detectors. This light reflectance is then calculated to estimate relative Hb/Mb-deoxygenation in the muscle tissue.

Duel wavelength spectrophotometry in the near-infrared region provides an opportunity to measure hemoglobin/myoglobin (Hb/Mb) deoxygenation in the capillary bed of an exercising muscle (Chance et al., 1992; Hampson & Piantadosi, 1988); the observed signal is derived from muscle capillaries and small vessels because the large blood vessels nearly completely absorb light and reduce the optical signal. Muscle oxygen saturation measured by NIRS reflects the balance between oxygen supply and oxygen consumption in the small vessels of the muscle (arterioles, capillaries, and small veins). This can be accomplished because the absorption of light varies with oxygen binding to the heme group so that relative levels of muscle oxygenation are measured. Deoxy-Hb/Mb can be measured at 760 nm, whereas $HbO_2/MbO_2$ demonstrate peak absorbance at 850 nm (figure 10.2). The oxy-deoxy isobestic point, the wavelength where oxy- and deoxy-heme absorb equally, is approximately 800 nm (Chance, Nioka, Kent, McCully, et al., 1988). Some spectroscopies use this point of 800 nm to measure the changes of blood volume because it is almost free from cross-talk between blood volume and deoxygenation changes. Therefore, 805 nm light may provide more accurate measures of relative blood volume changes. The difference between the

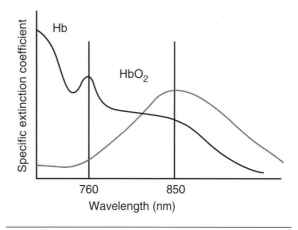

**Figure 10.2** Absorption spectra in the near-infrared region. Deoxygenated hemoglobin and myoglobin demonstrate peak absorption at approximately 760 nm, whereas oxygenated hemoglobin and myoglobin have peak absorption at 850 nm. The oxygenated-deoxygenated isobestic point is at 800 nm; changes in blood volume are best measured at this point.

signals at 760 and 850 is used to determine oxygen saturation (Chance et al., 1992, 1988), and the sum of the signals at 760 and 850 provide a reference to blood volume changes. To calculate the changes in oxy-Hb/Mb, deoxy-Hb/Mb, and total Hb/Mb from two or more wavelength methods, the Beer-Lambert law can be applied:

$$OD = -\log \varepsilon(I/I_0) = \varepsilon L[C]$$

where OD is the optical density, I is the detected light intensity, $I_0$ is the incident light intensity, $\varepsilon$ is the extinction coefficient, L is the photomigration path length, and [C] is the concentration of solution (mM).

The Hb and Mb spectra cannot be resolved by NIRS because of the overlap of Hb and Mb spectra. Using proton $^1$H-NMR, Wang et al. (1990) suggested that about 75% of the NIRS signal is a result of $HbO_2 \rightarrow Hb$ and the remaining 25% of the signal is due to $MbO_2 \rightarrow Mb$. The myoglobin signal is probably insignificant, however, especially at sublactate threshold intensities because the affinity for oxygen by myoglobin is much greater than that by hemoglobin. In fact, Binzoni, Quaresima, Barattelli, et al. (1998) found that the change in light absorption for deoxygenation of one molecule of $HbO_2$ is equivalent to four molecules of $MbO_2$. Moreover, the ratio of hemoglobin to myoglobin in skeletal muscle is 10:1 (Seiyama et al., 1988).

## Validation of NIRS Measurement of Exercising Muscle

In vitro studies have been conducted to evaluate the validity of NIRS measurements. This is accomplished with an intralipid solution as a scatter medium using yeast and blood models for oxygen saturation and hemodynamic measurements. A light source with a detector is placed on the intralipid solution to measure blood at specific concentrations of 10 to 200 μM of hemoglobin, similar to human blood, added serially to simulate the in vivo blood volume change. Yeast and sodium dithionite are used to provide altered oxygen saturation. Oxygen bubbling is used to cause reoxygenation of hemoglobin, and the cessation of bubbling causes deoxygenation of the hemoglobin because of yeast respiration. Chance et al. (1992) found that the subtracted signals at 760 and 850 nm showed a linear relationship to hemoglobin deoxygenation. Wilson et al. (1989) demonstrated a linear relationship between near-infrared spectroscopy measurements and venous hemoglobin saturation in an animal model.

### In Vivo Study

The most common method used to calibrate the concentration changes of Hb/Mb in vivo is the cuff ischemia response. This requires placing a blood pressure cuff proximal to the probe (e.g., upper-arm cuff with probe placement in the forearm) and inflating to suprasystolic pressure for 3 to 5 min or until a plateau in deoxygenation is reached. This procedure enables the researcher to establish a quasi-quantitative scale through the range of maximum deoxygenation and hyperemic reoxygenation of normal muscle (100% desaturation to 100% saturation). To use this method for quantitative measurements, determination of the optical path length (L) is required in the highly scattering muscle tissue. This can be accomplished by using time-resolved or phase-resolved spectroscopy measurements of absorbance changes (optical density) obtained during cuff ischemia. The percent deoxygenation during exercise can then be calculated from the maximal deoxygenation of the ischemic response.

The ischemic response is necessary when comparisons between individuals are being made. Because the NIRS source light path length is inversely proportional to the strength of the signal, a longer path length to the target tissue because of subcutaneous adipose tissue thickness results in an artificially weak signal (Matsushita, Homma, & Okada, 1998; Niwayama, Hamaoka, Lin, et al. 2000; Niwayama, Lin, Shao, Kudo, & Yamamoto, 2000; Rundell, Szmedra, Im, Nioka, & Ploetz, 1998; van Beekvelt, Borghuis, van Engelen, Wevers, & Colier, 2001). The cuff ischemic response provides a means of normalizing NIRS signals across subjects (Rundell et al., 1998; van Beekvelt, et al., 2001).

Studies have examined the use of NIRS methods to examine oxygen saturation during exercise (Hamaoka, McCully, Katsumura, Teruiki, & Chance, 2000; Wilson et al., 1989). In those studies, NIRS-measured oxygen saturation has been shown to decline during incremental exercise. And this decline has been correlated to decreases in directly measured venous oxygen saturation.

### NMR Studies

$^{31}$Phosphorus (P) magnetic resonance spectroscopy ($^{31}$P-MRS) has become a popular tool in the fields of both physiology and medicine since 1980. For the past decades, $^{31}$P-MRS has been used as a gold standard to examine muscle oxidative metabolism noninvasively. Many studies have adapted $^{31}$P-MRS to examine the role of phosphate metabolites in muscle metabolism in human subjects, especially in oxidative phosphorylation during exercise and recovery (Blei et al., 1993; Chance et al., 1985; Dulieu et al., 1997; Kemp et al., 2001; McCully, 1993; Zatina et al., 1986). Studies have reported that PCr recovery rate significantly correlates with mitochondrial oxidative enzymes measured with biopsy specimen (McCully, 1993; Meyer, 1988). Postexercise PCr resynthesis rate measured by [31]P-MRS has been recognized as one of the most reliable parameters for quantifying the rate of oxidative ATP production.

Efforts have been made to correlate the oxidative metabolism responses obtained by [31]P-MRS and NIRS. McCully et al. (1994) simultaneously examined the PCr time constants and $HbO_2$ resaturation rate after exercise and found similar recovery time constants in those two variables as long as the pH did not drop below 7.0. In clinical studies, PCr changes and deoxygenation as well as PCr recovery and reoxygenation responses, obtained by [31]P-MRS and NIRS respectively, are also in agreement where patients with PVD have shown larger PCr changes during exercise and slower PCr recovery, a decrease in functional capacity for oxidative ATP synthesis, as well as faster deoxygenation during exercise and slower postexercise reoxygenation (Kemp et al., 2001, 1995; McCully et al., 1994).

## Quantification of NIRS

Because of the unquantifiable optical path length, NIRS can measure only relative changes in Hb/Mb-deoxygenation. Recent studies, however, have tried to validate and quantify muscle oxidative metabolism using NIRS (Hamaoka, Iwane, Shimomitsu, et al., 1996; Sahlin, 1992; Sako, Hamaoka, Higuchi, Kurosawa, & Katsumura, 2001). Because the changes seen by the NIRS measurements reflect a balance between $O_2$ supply and $O_2$ consumption, it is necessary to know the amount of the $O_2$ supply to the muscle to quantify muscle $O_2$ consumption. This can be accomplished by cuff ischemia in which $O_2$ supply is interrupted. The rate of $Hb/MbO_2$ decline under such a condition can then reflect muscle $O_2$ consumption. Hamaoka et al. (1996) examined muscle oxidative metabolism postexercise relative to the resting values by occluding arterial blood flow and succeeded in quantitatively calculating muscle oxidative metabolic rate postexercise.

To validate the quantitative measure of oxidative metabolic rate using NIRS, Sako et al. (2001) compared the measurements made by the NIRS with the standard absolute values obtained by [31]P-MRS. Postexercise $\dot{V}O_2$, estimated by the NIRS measurement, was compared with the PCr resynthesis rate

measured by [31]P-MRS. Significant correlation was shown between the values obtained by the NIRS and [31]P-MRS (r = .965, p < .001).

## Blood Flow and $O_2$ Consumption Modeling

Changes in hemodynamics can be evaluated using NIRS. The determination of blood flow using NIRS has been previously validated (De Blasi et al., 1994; van Beekvelt, Colier, Wevers, & Van Engelen, 2001). Comparison of the NIRS method to quantify blood flow with more established methods, including venous occlusion plethysmography and plethysmographic flow measurement, has been reported to show high correlations. Using the differential pathlength factor in the Lambert-Beer law,

$$I = GI_0 e^{-(aHbcHb + aHbO_2 cHbO_2) \times L}$$

and applying venous occlusion, blood flow can be quantified using NIRS, at rest and during exercise (De Blasi et al., 1994; Homma, Eda, Ogasawara, & Kagaya, 1996; van Beekvelt, Colier, et al., 2001). The sum of oxy-hemoglobin/myoglobin ($O_2Hb/MB$) and deoxy-hemoglobin/myoglobin (HHb) concentrations reflects the total amount of hemoglobin (tHb) in the muscle region of interest. The linear increasing rate in tHb during venous occlusion within the first seconds can be used to calculate blood flow. The changes in tHb are expressed in micromoles per second ($\mu M/s$) and can be converted to ml blood/100 ml tissue/min by using individual Hb concentration obtained from the blood samples (figure 10.3) (van Beekvelt, Colier, et al, 2001).

Blood flow index = [d(tHb)/dt] $\times$ SaHbO$_2$ $\times$ 100$^{-1}$

Because of nonquantifiable biophysical properties such as the optical path length, absolute concentration changes of hemoglobin or myoglobin (Hb/Mb) cannot be determined. Therefore, NIRS can measure only the balance between supply and demand. By using the venous occlusion method, however, one can also estimate mitochondrial oxidative metabolism (Homma et al., 1996). Mitochondrial oxidative rate is estimated by subtracting the product of the peak rate of increase in tHb and arterial deoxygenation from the peak rate of increase in deoxyHb during venous occlusion.

$O_2$ consumption index =[d(deoxyHb/Mb)/dt] – [d(totalHb/Mb)/dt] $\times$ (100 – SaO$_2$) $\times$ 100$^{-1}$

To observe changes in the Hb/Mb muscle tissue oxygenation/deoxygenation (oxy/deoxy-Hb/Mb)

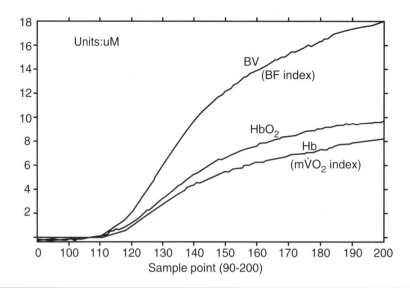

**Figure 10.3** Muscle oxidative rate ($\dot{V}O_2$) and blood flow can be estimated using venous occlusion. During venous occlusion, the total oxygenated and deoxygenated hemoglobin or myoglobin reflects the total amount of hemoglobin in the measured muscle section. The increase in total hemoglobin in the first seconds of venous occlusion can be used to calculate blood flow. Muscle oxidative activity can be calculated by subtracting the product of the peak rate of increase in total hemoglobin and arterial deoxygenation from the peak rate of increase in deoxygenated hemoglobin during venous occlusion.

during and after exercise, a multiple-channel three continuous wavelength LED imager is used. Following exercise test, venous occlusion is performed to estimate blood flow and oxygen consumption. The cwNIRS unit includes a probe consisting of multiple LED light sources and photodiode detectors coupled with fiber optics for magnet use. An 11 cm × 7 cm optical probe with a 3 cm source to detect separation is positioned on the medial head of the target muscle. A double-tuned surface coil is placed simultaneously for MRS measurements. The location of the probe and the coil will be recorded by noting distances from the anatomical landmarks to ensure identical probe placement during all trials.

## Physiological Measures That Can Be Monitored Using NIRS

Whole-body $\dot{V}O_2$ and muscle $\dot{V}O_2$ can be monitored using NIRS. NIRS has been used successfully to monitor muscle oxidative metabolism in athletes (figure 10.4) (Angus, Welford, Sellens, Thompson, & Cooper, 1999; Azuma, Homma, & Kagaya, 2000; Bae et al., 2000; Chance et al., 1992; Foster et al., 1999; Haga, Bae, Hamaoka, et al., 1998; Im et al., in review, 2000; Neary et al., 2002, 2001; Nioka et al., 1999; Rundell et al., 1997; Szmedra et al., 2001; Tamaki, Uchiyama, Tamura, & Nakano, 1994; Thompson, Kemp, Sanderson, et al.,1996; Webster et al., 1998) and in vascular and heart-diseased patients (Cheatle, Potter, Cope, et

al., 1991; Costes et al., 1999; Kemp et al., 2001, 1995; McCully & Hamaoka, 2000; Simonson & Piantadosi, 1996; Wilson et al., 1989). The rate of muscle deoxygenation at the onset of exercise has been used as an indicator of oxidative metabolism in normal subjects (Asanoi et al., 1992; Belardinelli, Barstow, Porszasz, & Wasserman, 1995a, 1995b; Bhambhani, Maikala, & Esmail, 2001; Bhambhani, Buckley, Susaki, 1999; Bhambhani, Maikala, & Buckley, 1998; Colier, Meeu-

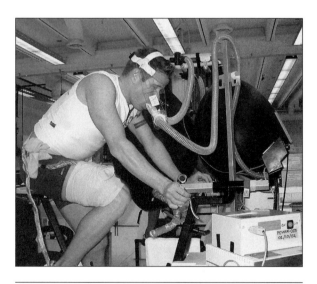

**Figure 10.4** NIRS and whole-body oxygen consumption measurements are being recorded from an incremental cycle ergometer test by an elite cyclist.

wsen, Degens, & Oeseburg, 1995; Delpy & Cope, 1997; Demarie et al., 2001; Ding, Wang, Lei, et al., 2001; Ferrari, Binzoni, & Quaresima, 1997; Jensen-Urstad, Hallback, & Sahlin, 1995; Oda, Yamashita, Nakano, et al., 2000; Quaresima, Komiyama, & Ferrari, 2002; Quaresima et al., 2001, 1999; Quaresima, Homma, Azuma, et al., 2001; Quaresima, Pizzi, De Blasi, Ferrari, & Ferrari, 1996; Simonson & Piantadosi, 1996), and it has been shown that kinetics of mitochondrial oxygen consumption measured by NIRS is different in trained versus normal subjects (Neary et al., 2001, 2002). Most important to the athlete, it has been shown that NIRS can monitor training effects on muscle oxygen kinetics.

Muscle oxygenation recovery reflects a balance between oxygen supply and utilization in the muscles. NIRS measurements of muscle oxygenation have been correlated to the maximal oxygen uptake or blood lactate concentration, depending on exercise mode (Angus et al., 1999; Belardinelli et al., 1995a; Bhambhani et al., 2001, 1998; Chance et al. 1992; Costes et al., 1999; Demarie et al., 2001; Foster et al., 1999; Grassi, Quaresima, Marconi, Ferrari, & Cerretelli, 1999; Im et al., in review, 2000; Neary et al., 2002, 2001; Rundell et al., 1997). Studies have identified a relationship between reoxygenation rate and muscle oxidative capacity (Chance et al., 1992; Hamaoka et al., 1996; Mancini et al., 1994). Some studies reported significant relationships between maximal oxygen uptake or bLA concentration to the reoxygenation rate (Chance et al., 1992; Foster et al., 1999; Mancini et al., 1994; Rundell et al., 1997), whereas

others found no relationship (Im et al., 2000). Im et al. (2000) found no relationship between maximal oxygen uptake or bLA concentration to the reoxygenation rate after a whole-body, dynamic exercise. These investigators found a strong relationship, however, between whole-body $\dot{V}O_2$ and the NIRS signal during incremental cross-country skiing.

Studies (Belardinelli et al., 1995a, 1995b; Bhambhani et al., 2001, 1999, 1998; Foster et al., 1999; Grassi et al., 1999; Im et al., in review, 2000; Kouzaki et al., 2001; Miura et al., 1998; Rundell et al., 1997) have evaluated changes in tissue oxygenation using NIRS while measuring whole-body $\dot{V}O_2$ determined by indirect calorimetry during exercise. Im et al. (2000) examined the percent oxygen desaturation (%OD) response in the vastus lateralis to metabolic changes during submaximal and maximal cross-country ski skating. Whole-body $\dot{V}O_2$ and muscle OD was continuously monitored throughout incremental cross-country ski exercise. At the completion of exercise, a thigh cuff placed proximal to the NIRS probe was inflated to approximately 300 mmHg to occlude the blood flow and determine maximal OD. Relative OD (%OD) for each submaximal stage and maximal exercise was calculated by taking the difference between baseline at rest and greatest OD that occurred during each stage of exercise and dividing by the maximal OD obtained during cuff ischemia (figure 10.5). OD values from the baseline, exercise, and cuff ischemia were obtained by taking 10 s averages. By normalizing to a cuff ischemic response after each trial, potential variability between trial subject

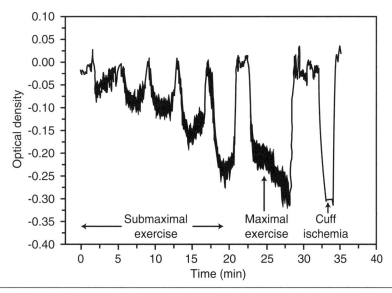

**Figure 10.5** Real time NIRS values during a treadmill roller ski test of five submaximal stages, a maximal stage, and arterial occlusion. Note the maximal muscle deoxygenation (%OD) at the conclusion of maximal exercise.

Reprinted, by permission, from J. Im et al., 2000, "Muscle oxygen desaturation is related to whole body $\dot{V}O_2$ during cross-country ski skating." *Interatinal Journal of Sports Medicine* 22-356-360.

NIR signal was eliminated (Chance et al., 1992, 1988; Foster et al.; Im et al., 2000; Rundell et al., 1998, 1997; Szmedra et al., 2001).

This study demonstrated that skeletal muscle %OD progressively increased as the intensity of exercise increased. Percent oxygen desaturation reached a peak and plateaued as $\dot{V}O_2$max was reached. The strong correlation observed between %OD and $\dot{V}O_2$ (figure 10.6, r = .83) suggests that %OD during submaximal treadmill roller skiing is due to the metabolic component, and not a static load component as seen in speed skaters (Foster et al., 1999; Rundell et al., 1997) or alpine skiers (Szmedra et al., 2001).

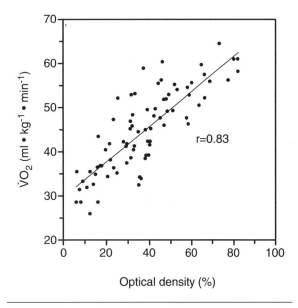

r=0.83

**Figure 10.6** Muscle deoxygenation and whole-body $\dot{V}O_2$ demonstrate a strong correlation during treadmill roller skiing by elite Nordic cross-country skiers.

Data from Im et al. 2000.

This assumption is supported by the low correlation between blood volume change ($\Delta$BV) and $\dot{V}O_2$ (r = .28) observed in this study. As exercise intensity increased, both $\dot{V}O_2$ and %OD increased accordingly while $\Delta$BV was minimal. This was not the case with NIRS studies of speed skating, in which a tight couple between %OD and bLA was identified (Foster et al., 1999; Rundell et al., 1997). In those studies and in a study of alpine skiing (Szmedra et al., 2001), the high $\Delta$BV during skating implicated restricted blood flow to the quadriceps.

Belardinelli et al. (1995a) and Grassi et al. (1999) used NIRS to assess tissue $O_2$ saturation during incremental exercise. In those studies, the researchers observed an initial increase in tissue $O_2$ saturation during light exercise, a progressive fall during the middle work intensities, and a plateau or minimum saturation at maximal exercise. The initial increase in tissue $O_2$ saturation during light exercise may be due to the redistribution of blood flow to the limbs. These studies demonstrate that oxidative rate in the muscle can be successfully monitored by NIRS. The strong relationship observed between %OD and whole-body $\dot{V}O_2$ may be attributed to $O_2$ dissociation in the capillary bed of the muscle to meet aerobic energy demand and is independent of blood flow dynamics.

## Mechanical Constraints, Blood Flow, and Muscle Tissue Deoxygenation

Mechanical restriction in blood flow during specific exercise modalities such as speed skating and alpine skiing have been shown to limit $O_2$ delivery and distribution within the muscle. Blood flow through the muscle during dynamic exercise usually occurs between contractions and should therefore depend on time available between contractions. When muscle contractions approach 50% of maximal voluntary contractions, intramuscular pressure exceeds perfusion pressure and impedes blood flow to the working muscle. Consequently, as the intensity of exercise increases, or the elevation of a treadmill increases and one has to exercise faster and climb harder to keep up with the workload change, the ratio of active tension time to no active tension time, or duty cycle, increases and affects muscle blood flow. Several studies in which different modes of exercise demonstrate disproportionate muscle deoxygenation with lower postures have reported this finding (Foster et al., 1999; Rundell et al., 1997; Szmedra et al., 2001). The resultant static load during these activities compromises blood flow to the working muscles, limits oxygen delivery, and creates an imbalance that leads to muscle hypoxia and increases in deoxy-Hb/Mb.

Research has shown that blood flow and subsequent oxygen desaturation in exercising muscle is coupled to the static component during speed skating (Foster et al., 1999; Rundell et al., 1997). Increased $O_2$ desaturation can be dissociated from whole-body $\dot{V}O_2$ and heart rate when speed skaters increase the static component by "sitting low." When increased intramuscular pressure exceeds perfusion pressure during the static load and local blood flow to the working muscle is impeded, the result is an imbalance of oxygen supply and demand.

Rundell et al. (1997) found a reduced blood flow to working muscle during low-positioned skating. These authors suggested that compromised blood flow to working muscle resulted in increased %OD without the corresponding increase in oxygen uptake.

In contrast to this, Im et al. (2000) observed a strong relation of %OD and $\dot{V}O_2$ but not $\Delta$BV in cross-country skiers. This may indicate that the increase in %OD during incremental exercise observed in cross-country skiing may be due to $O_2$ dissociation in the capillary bed of the muscle to meet aerobic energy demands and is independent of any change in blood flow. Grassi et al. (1999) found similar results to Im et al. in which a slight increase in BV was observed during cycling, while %OD continued to increase as the exercise intensity increased.

These seemingly paradoxical findings demonstrate the importance of sport-specific metabolic demands and blood flow dynamics when interpreting NIRS data. The difference in %OD increase during different exercise modalities suggests that intramuscular pressure can impede blood flow and create a hypoxic condition in the working muscle. As a result, accelerated blood lactate accumulation occurs. This response is similar to that observed by Richardson, Knight, Poole, et al. (1995), who found higher lactate concentration as well as steeper slope of intracellular $PO_2$ during hypoxia when compared with normoxia.

Studies by Foster et al. (1999), Rundell et al. (1997), and Szmedra et al. (2001) showed greater %OD associated with lower postures during speed skating (Foster et al., 1999; Rundell et al.) and the giant slalom event of alpine skiing (Szmedra et al.). These studies suggested that static load compromised blood flow to the working muscles, limited oxygen delivery, and increased the %OD (figure 10.7). Although Im et al. (2000) hypothesized that greater %OD would occur during cross-country skiing on an incline because of a greater static load of the working muscle, no significant differences between incremental incline skiing at a constant speed and skiing at a 5% incline with incremental speed increases were found. The lack of significantly greater %OD at higher inclines suggested that the expected static load may have been attenuated by the increased contribution of poling, redistributing the energy demand so that the upper body made a greater contribution during the later stages of incline skiing as the grade increase occurred (figure 10.8). Alternatively, a longer cycle length (duty cycle) during the speed–constant incline based protocol in that study may have increased the static load in the quadriceps at higher speeds, resulting in a greater than expected %OD, thus masking any effect of static load during the incline skiing protocol. But because of the shorter duty cycles during cross-country skiing, when compared with low sitting speed skating position, blood flow may not be limiting during cross-country ski skating.

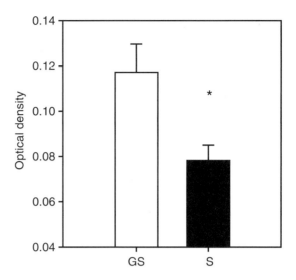

**Figure 10.7**  Positional changes during speed skating and alpine skiing have been shown to affect muscle deoxygenation. This has been assumed to reflect restrictions in blood flow to the working muscle because of the static component of the skating and skiing positions. This figure demonstrates a significantly greater oxygen desaturation during giant slalom (GS) than during slalom (S) alpine skiing. This was thought to result from the greater static load during the GS event.

Data from Szmedra et al. 2001.

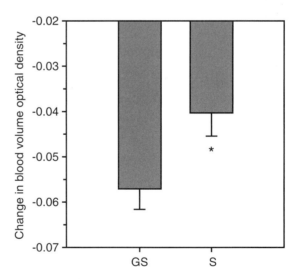

**Figure 10.8**  This figure shows the greater affect on total muscle blood volume change in the quadriceps during the giant slalom (GS) than the slalom (S) event. Szmedra et al. (2001) suggested that this negative change was due to intramuscular impedance of arterial blood flow to the statically contracted muscle during the prolonged turn in GS (compared with S).

## Anaerobic (Lactate) Threshold

NIRS shows great potential for investigating the relationship between blood lactate accumulation and muscle deoxygenation measured by NIRS and has been shown to be a useful noninvasive tool for determining the anaerobic threshold (AT). AT, or lactate threshold, has been defined as "the highest $\dot{V}O_2$ beyond which lactate begins accumulating in the blood, causing metabolic acidosis" during incremental work (Aunola & Rusko, 1986; Coyle, 1995; Foster et al., 1999; Gaesser & Poole, 1986; Grassi et al., 1999; Green, Hughson, Orr, & Ranney, 1983; Rundell, 1996). AT is used as frequently as $\dot{V}O_2$max for evaluation of aerobic performance, and it is often used to monitor athlete training status. In fact, with highly trained athletes, AT or %$\dot{V}O_2$ at AT may be a better indicator of training status than $\dot{V}O_2$max (figure 10.9). Improvements in performance after the first 2 or 3 years of intense training are more associated with improvements in AT $\dot{V}O_2$ than in $\dot{V}O_2$max, which generally increases very little thereafter (Coyle, 1995).

One of the first recent studies that demonstrated a change in muscle oxygenation during incremental exercise NIRS was done by Wilson et al. (1989). Belardinelli et. al. (1995a) further expanded this work by finding a strong relationship between AT (or lactate acidosis) and muscle deoxygenation rate. Their study demonstrated the potential utility of NIRS in the evaluation of athletes. They found a slow decrease in muscle deoxygenation at low levels of exercise followed by rapid muscle desaturation in association with AT and a slowing of desaturation at near maximal exercise above AT as $\dot{V}O_2$max was approached. More recently, Grassi et al. (1999) confirmed this observation by showing that the onset of muscle deoxygenation measured with NIRS was significantly correlated with the onset of blood lactate accumulation ($R^2 = 0.95$).

A slow increase in %OD has been observed during the first stages of incremental exercise (Im et al., 2000). Percent OD in the quadriceps is minimal, similar to blood lactate response, until about 35% of maximal OD (figure 10.10). Foster et al. (1999) reported a progressive lactate accumulation only after desaturation value had exceeded about 30%. Once the AT was reached, the rise in %OD was more rapid as exercise intensity increased. The %OD and $\dot{V}O_2$ relationship to AT during submaximal exercise is remarkably similar whereby accelerated %OD occurs at >75% of $\dot{V}O_2$max, the precise point where onset of blood lactate accumulation occurs. This value is slightly higher than what Belardinelli et al. (1995a) and Grassi et al. (1999) found (~50% $\dot{V}O_2$max and

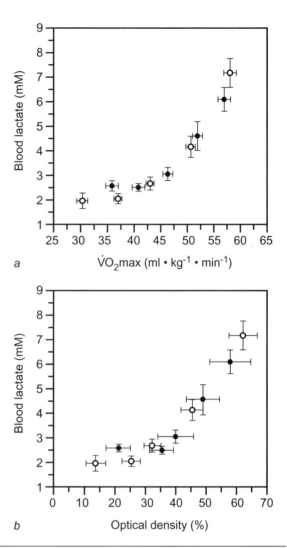

**Figure 10.9** This figure depicts the noninvasive utility of NIRS in evaluating blood lactate threshold in elite athletes. Filled symbols represent treadmill roller skiing at a constant grade with increasing speed stages, and open symbols represent treadmill roller skiing at a constant speed with grade-increasing stages. Subjects were elite cross-country skiers. Note the similarity in curves produced by NIRS deoxygenation values (%OD) (b) and whole-body oxygen consumption ($\dot{V}O_2$) (a).

~65% $\dot{V}O_2$max, respectively) and may be due to the fitness levels of respective subjects. The subjects in the Foster et al. (1999) and Im et al. (2000) studies were highly trained speed skaters and cross-country skiers capable of sustaining a high percentage of maximal work output before reaching AT.

As exercise intensity increases during submaximal stages, the time to resaturate between stages is prolonged. As expected, BV returns to resting level faster than resaturation of hemoglobin, which represents a repayment of oxygen deficit (Chance et al., 1992).

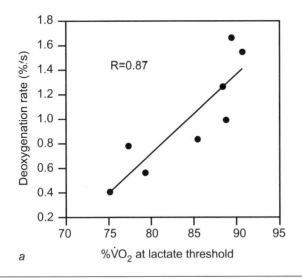 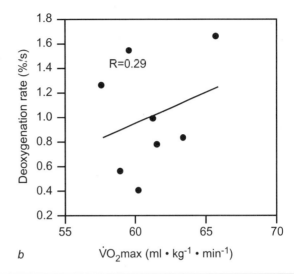

**Figure 10.10**   Im et al. (unpublished data) examined the rate of quadriceps muscle deoxygenation during incremental treadmill cross-country roller skiing. The results of this study demonstrated an accelerated rate of deoxygenation at about 75% $\dot{V}O_2$max, the point at which blood lactate accumulation occurred *(a)*. The authors suggested this response resulted from the enhanced unloading of oxygen from hemoglobin (the Bohr effect). At lower intensities and at $\dot{V}O_2$max, this relationship was not evident *(b)*.

Immediately on completion of a $\dot{V}O_2$max test, a hyperemic response occurs where there is a rapid increase in BV and tissue $O_2$ saturation to a level higher than at rest or during exercise. Thereafter, the $O_2$ saturation level slowly goes back to the rest and preexercise level. A wide range of half-time recovery is observed with no significant relationships to any exercise variables that have been examined to date. As Chance et al. reported, $\dot{V}O_2$max, recovery time, and blood accumulation do not appear to be closely correlated on an individual basis. Half-time recovery may be associated with the level of aerobic fitness, muscle fiber type, or capillary density. The precise nature of this phenomenon is not known.

### IP of ΔBV and AT

In agreement with findings of Belardinelli et al. (1995a) and Bhambhani et al. (2001, 1998), Im et al. (unpublished data) demonstrated that AT can be determined using NIRS. This study is unique in that it evaluated the relationship between AT and ΔBV during whole-body, dynamic exercise rather than Hb/Mb-deoxygenation using cw-NIRS.

During incremental exercise to $\dot{V}O_2$max, there appears to be a linear relationship between changes in BV (ΔBV) and exercise intensity to a certain point, at which the rate of increase in ΔBV is minimal. This point, at which there is no more increase in ΔBV (IP of ΔBV), may be due to greater intramuscular pressure impairing blood flow, at the higher intensity exercise, that limits $O_2$ supply to the working muscle and shifts

the energy pathway to anaerobic glycolysis (Foster et al., 1999; Rundell, 1996; Rundell et al., 1997). Alternatively, at higher exercise intensities (AT), $O_2$ delivery may simply not be able to keep pace with $O_2$ demand. But it is unclear whether ΔBV and muscle deoxygenation rate in trained subjects, during whole-body exercise, indicates oxidative capacity.

Im et al. (unpublished data) found a strong relationship between the time to reach a plateau of increasing ΔBV to the time at which AT occurs (figure 10.11). These results are consistent with several other findings (Asanoi et al., 1992; Kagaya, 1990; Moraine et al., 1993; Moritani, Berry, Bacharach, & Nakamura, 1987) that found high correlations of maximal blood flow and AT $\dot{V}O_2$.

**Time Lag**   Several investigators have reported the disparity of plasma lactate threshold and ventilatory threshold time occurrence during exercise (Aunola & Rusko, 1986; Gaesser & Poole, 1986; Green et al., 1983; Poole & Gaesser, 1985; Simon et al., 1986). A time lag of about 35 s between the AT determined by the NIRS technique and the AT determined by the V-slope method has been identified (Im et al., unpublished data). This lag time may be due to peripheral muscle properties including the biochemical state of different muscle groups and the proportion of slow-twitch fibers (Aunola & Rusko, 1986). cw-NIRS can directly measure the local periphery of the muscle metabolism, whereas the V-slope method is an indirect method of determining AT but may depend

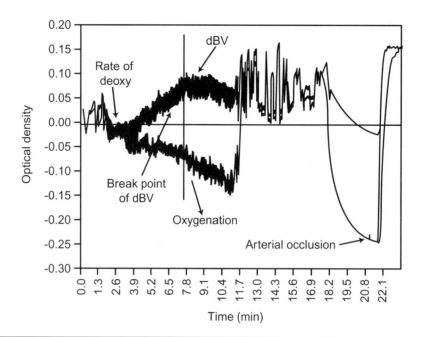

**Figure 10.11**   Im et al. (unpublished data) extended their work examining deoxygenation rate at threshold to include blood volume change. This work demonstrated a linear increase in muscle blood volume during incremental exercise to lactate threshold, at which point blood volume increase reached a plateau (ΔBV) while the rate of deoxygenation increased. This breakpoint in ΔBV suggests that blood delivery either is impeded by intramuscular pressure or has reached a maximal rate.

on a time lag between muscle $\dot{V}O_2$ kinetic and metabolic measurements. This may explain the disparity between AT observed by the cw-NIRS technique and the V-slope method.

**Rate of Deoxygenation and Anaerobic Threshold**   The rate of muscle deoxygenation can be used as an indicator of oxidative metabolism. Hamaoka et al. (1998) found that the initial rate of deoxygenation (deoxy-rate) at the onset of the exercise was positively correlated to the %ST + %FTa fiber type in human skeletal muscle. A positive, strong relationship of 0.87 was found between the rate of deoxy-Hb/Mb and %$\dot{V}O_2$ at AT (Im et al., unpublished data). No relationship was observed, however, between the rate of deoxy-Hb/Mb and maximal $O_2$ consumption, perhaps because of the fitness level and training status of the subjects. The time of which AT occurs may depend on many factors including aerobic enzyme activity and mitochondrial activity, muscle capillary density, and muscle mass; with highly trained athletes, AT or %$\dot{V}O_2$ at AT may be a better indicator of their training status than $\dot{V}O_2$peak is.

### Training Effects Measured Using NIRS

Physical training causes adaptations to both the cardiopulmonary system and the peripheral system. Identifying the relative adaptations and contributions of each of these systems to exercise performance can be difficult and often involves invasive procedures. Current NIRS technology provides a noninvasive method to explore peripheral adaptations in the exercising muscle to training. As with any new technology, or technological application, initial research using NIRS in sport has been primarily descriptive in nature.

Recent work (Costes et al., 2001; Neary et al., 2002) has taken NIRS technology to the next step by evaluating effects of training on the specific muscle used in a particular exercise. Costes et al. examined the training effect on untrained subjects and found that muscle deoxygenation was reduced during submaximal exercise at specific workload intensities after training. These authors suggested that the attenuated $O_2$ desaturation during exercise was due to lowered $O_2$ unloading because of a training-induced reduction in blood lactate and was hence related to the Bohr effect. This study was the first to demonstrate reduced $O_2$ desaturation from training. The authors suggested that this observation was a result of improved muscle oxidative metabolism after training.

Neary et al. (2002) examined the effects of a training program on changes in cardiopulmonary and peripheral responses to an incremental $\dot{V}O_2$max test and a 20 km cycling time trial on a wind-loaded trainer. This study showed an improved $\dot{V}O_2$max after training without significant change in muscle oxygen desaturation during an incremental $\dot{V}O_2$max test, suggesting that the improved $\dot{V}O_2$max was due primarily to cardiopulmonary adaptations. Interestingly, these authors found that $\dot{V}O_2$ during the 20 km time trial did not improve after training in spite of improved performance, but muscle deoxygenation was significantly greater in five of the eight cyclists. Hb/Mb deoxygenation was highly related to time trial performance ($r = -.93$). This paradoxical finding led the authors to conclude that improvements in $\dot{V}O_2$max derived from training were due to increased oxygen transport, while improvements in 20 km time trial performance, an effort intensity below $\dot{V}O_2$ max, resulted from peripheral adaptations that allowed a widening of the $(a-v)O_2$ difference, as indicated by NIRS-measured $O_2$ desaturation in the muscle capillary bed.

Outwardly, these two studies appeared to have contradictory results. When the difference in exercise protocols is examined, however, the results support the concept of improved muscle metabolism; that is, training-induced peripheral adaptations can be seen in submaximal exercise using NIRS. Because the Costes et al. (2001) study evaluated $O_2$ desaturation pre- and posttraining at the same absolute workload intensities, peripheral adaptations from training would be expressed as a reduction in desaturation. In contrast, in the Neary et al. (2002) study, peripheral adaptations allowed for greater oxygen extraction, greater deoxygenation values, and improved power output and subsequent time-trial performance. One could hypothesize that if the trained cyclists in the Neary et al. study performed their posttraining time trials at the same pace as the pretraining time trial, the $O_2$ desaturation would exhibit a similar pattern as that observed in the Costes et al. study. In summary, the results of these two studies demonstrate that NIRS can noninvasively measure training-induced oxidative changes in skeletal muscle.

## Summary

NIRS has been demonstrated to be reliable and effective in evaluating muscle hemodynamics and oxidative metabolism during exercise. The portability and noninvasive nature of NIRS make it a useful tool to use with athletes of all ages. NIRS has shown sensitivity to positional changes during speed skating and alpine skiing that provides information that can be used in designing specific training regimens. Likewise, NIRS provides a noninvasive measurement for lactate threshold determination and has recently been used to evaluate the peripheral changes in the muscle because of training.

The technological advances in recent years and new software packages currently in development will make NIRS an easy-to-use tool for evaluating athletes. Future use of NIRS will undoubtedly increase to the point where coaches will be able to perform routine field evaluations on athletes to improve technique and monitor training.

# eleven

# Anthropometry and Body Composition Measurement

**James E. Graves, PhD**
*Syracuse University*

**Jill A. Kanaley, PhD**
*Syracuse University*

**Linda Garzarella, MS**
*University of Florida at Gainesville*

**Michael L. Pollock, PhD**
*University of Florida at Gainesville*

This chapter describes the more popular techniques used for the determination of body composition. Several aspects of the various methods are discussed—equipment used, methodology, pitfalls associated with the measurement techniques, and accuracy (reliability) and validity. Where appropriate, controversial issues concerning the various protocols and procedures are noted.

Finally, normative data will be presented, and inference about the interpretation of body composition results for general health and fitness and athletic competition will be offered. The following techniques for measuring body composition will be reviewed: hydrostatic (underwater) weighing, air displacement plethysmography, anthropometry, bioelectrical impedance, ultrasound, dual-energy x-ray absorptiometry (DEXA), isotopic dilution, magnetic resonance imaging, and computed tomography.

The measurement of body composition has become a popular and standard practice for many physicians, athletic trainers, allied health professionals, and researchers. Evidence supports the notion that being overweight (having excess body fat) is related to musculoskeletal injury, nonadherence to exercise training, and reduced athletic performance.

Athletic events that require participants to move their bodies quickly and efficiently penalize those who have an accumulation of body fat by reducing their running speed, jumping ability, and overall performance. In addition, numerous health problems are associated with excess body fat, such as hypertension, diabetes mellitus, depression, hyperlipidemia, and coronary heart disease (CHD) (National Heart, Lung, and Blood Institute, 1998). In a 26-year follow-up of participants from the Framingham Heart Study, Hubert, Feinlab, McNamara, and Castelli (1983) showed that obesity in itself is an independent risk factor for mortality from CHD.

The term *overweight* refers to an amount of total body mass (body weight) above what is recommended based upon stature. The Metropolitan Life Insurance tables, using height and weight, have been employed for years as a standard index by health professionals to determine appropriate body weight (Harrison, 1985), but these tables did not consider body frame size. More recently, the body mass index (BMI), which is the ratio of weight to height squared (BMI = $kg/m^2$), became more popular for use in epidemiological research (National Heart, Lung, and Blood Institute, 1998). The problem with the term *overweight* and the use of height–weight or BMI measures is their lack of specificity in describing leanness or fatness. Individuals with a large body weight because of a large muscle mass may appear to be overweight or obese when using BMI or height–weight charts. For example, the BMI indicated that football players were overweight, but body composition as determined by hydrostatic weighing showed that they were not overweight or obese; they were overweight because of having large amounts of fat-free mass (FFM), not fat weight. Obesity refers to the overfat condition that accompanies many comorbidities and includes some or all of the following components of the "obese syndrome," or metabolic syndrome X: glucose intolerance, insulin resistance, dyslipidemia, type 2 diabetes, hypertension, elevated plasma leptin concentrations, increased visceral adipose tissue, and increased risk of coronary heart disease and cancer (McArdle, Katch, & Katch, 2001; National Heart, Lung, and Blood Institute, 1998). Table 11.1 shows the BMI values associated with the overweight or obese condition. Because BMI is not sensitive to the amount of muscle mass or fat mass, various techniques are used to measure body composition; these models differentiate fat from FFM.

## Multicomponent Models

Body composition can be measured at different levels of complexity—atomic, molecular, cellular, tissue-system, and whole body (Wang, Pierson, & Heymsfield, 1992). The most common model used is at the molecular level (figure 11.1). Techniques for estimating body composition use a two-component molecular model, although three- and four-component models are available. To use a three- or four-

component model, it is necessary to combine the various methods of body composition assessment. For example, the body density ($D_b$) that is obtained from hydrodensitometry is adjusted for total body water and for bone mineral. Thus, for these calculations the subject would have to be underwater weighed, have a DEXA scan, and undergo isotope dilution for the measurement of total body water. Using all these methods, the values for bone, fat mass, and total body water are included in an equation that considers each compartment. Heymsfield and colleagues (1990) have developed such equations, and they are useful in populations that we expect to have changes in bone or total body water. Equations for the different models are listed in figure 11.2.

The two-compartment model as proposed by Brozek, Grande, Anderson, and Keys (1963) and Siri (1956) assumes that body composition is made up of fat and fat-free body compartments. The fat-free body (FFM) includes the muscle, bone, and other nonfatty tissues (Wang et al., 1992) (figure 11.1). Although the two-compartment model is well accepted and used extensively in the research and clinical settings, it is not without problems. This system assumes that the composition of fat and FFM is constant for all individuals, that is, that the density of fat is 0.900 g/cc and FFM is 1.100 g/cc. But FFM can vary because subjects vary considerably in density for bone and water. The advantage of the multicompartment model is pronounced in the study of populations other than young healthy adults or the study of those who have deviations in the bone and water compartment. For example, errors involved in determining body composition may occur in children and youth before their age of "chemical maturity" (ages 15 to 18 years, for most) (Lohman, 1986). The FFM is not stable in growing children and youth because water content

**Table 11.1   Body Mass Index and Waist Circumference and the Risk for Disease**

| Classification | Obesity class | BMI (kg/m²) | Disease risk relative to normal weight and waist circumference* | |
|---|---|---|---|---|
| | | | Men ≤ 40 in. (102 cm) | ≥ 40 in. (102 cm) |
| Underweight | | <18.5 | | |
| Normal | | 18.5-24.9 | | |
| Overweight | | 25.0-29.9 | Increased | High |
| Obesity | I | 30.0-34.9 | High | Very high |
| | II | 35.0-39.9 | Very high | Very high |
| Extreme obesity | III | ≥40 | Extremely high | Extremely high |

*Disease risk for type 2 diabetes, hypertension, and cardiovascular disease (National Heart, Lung, and Blood Institute, 1998).

From National Heart, Lung, and Blood Institute, 1998.

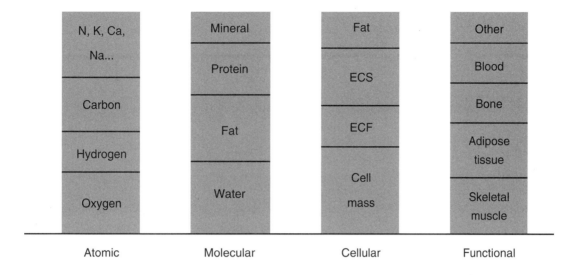

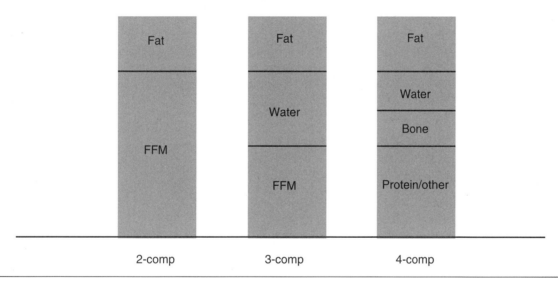

**Figure 11.1** Schematic of the atomic molecular, cellular, and functional models as adapted from Wang, Pierson, and Heymsfield, 1992. In addition, the two-, three-, and four-compartment models are outlined.

ECS = extracellular solids; ECF = extracellular fluid.

Adapted, by permission, by the *American Journal of Clinical Nutrition.* © American Journal of Clinical Nutrition. American society for Clinical Nutrition.

---

2C model—hydrostatic weighing or air displacement plethysmography
%fat = (4.95/$D_b$ − 4.50) × 100          (Siri, 1956)
%fat = (4.57/$D_b$ − 4.142) × 100          (Brozek et al., 1963)
3C model—HW or ADP + isotopic dilution
%fat = [(2.11/$D_b$) − 0.78W − 1.354] × 100     (Siri, 1961)
%fat = proprietary equation supplied by the manufacturer
4C model—HW or ADP + isotopic dilution + DEXA
%fat = [(2.748/D(P + F) − 2.051] [BW − (A + M)]/BW × 100     (Heymsfield et al., 1990)

---

**Figure 11.2** Equations for multicomponent models using values for bone, fat mass, and total body water. D(P+F) = density of protein plus fat compartment mixtures; BW = body weight (kg); A = aqueous mass (kg); M = mineral mass (kg); $D_b$ = body density; W = total body water as a fraction of body weight

decreases and body solids (bone density) increase in concentration until maturity. At the opposite end of the age spectrum, FFM also changes composition in older adults, with decreases in total body water as well as decreases in bone mineral density. The four-component model is now recognized as the gold standard of body composition (Clasey et al., 1999).

The limitation of using the four-compartment model is that measuring each compartment using a different method is expensive and time consuming. Thus, most researchers, athletic trainers, and clinicians do not use this model and instead rely on techniques that use less complex models that make assumptions about the various compartments. Further, the four-compartment model is associated with the error of each method individually; and a propagation of errors occurs when all methods are included in one equation. Wang, Dilmanian, and colleagues (1998), however, noted that several multicomponent molecular models produce fat mass estimates nearly identical to those produced by neutron activation analysis (mean differences in fat mass of 0.0-0.2 kg), a highly precise atomic model.

Researchers, athletic trainers, and clinicians continue to use the two-compartment system extensively because of the technical difficulty of measuring bone mineral (density) and the chemical components of FFM and fat. More recently, the availability of technology such as total-body dual-energy x-ray absorptiometry (DEXA) has made the in vivo measurement of the components of the four-compartment model more feasible. Development of computerized tomography and magnetic resonance imaging has allowed more specific analysis of the fat patterning in the various fat depots. Even so, the cost of such technology limits the use of such techniques to a few researchers.

## Hydrostatic Weighing

Hydrostatic, or underwater, weighing is the most widely used laboratory procedure for measuring body density ($D_b$), often serves as a criterion method for other indirect techniques (McArdle et al., 2001), and is referred to as the gold standard for other body composition measurements. This method uses Archimedes' principle that states a body immersed in a fluid is acted on by a buoyancy force that is evidenced by a loss of weight equal to the weight of the displaced fluid (Behnke & Wilmore, 1974). Because the density of fat is less than that of water, fat contributes to the buoyancy force, as does any air in the lungs or trapped in the body or on the surface of the body. The density of bone and muscle tissue, however, is greater

than that of water and will cause a person to sink. Therefore, a person having more bone and muscle mass and less fat will weigh more in the water and thus have a greater $D_b$, that is, a lower percent body fat and a greater percent FFM.

## Equipment and Methodology

The basic method of hydrostatic weighing used for determining $D_b$ is underwater weighing. Depending on the specific equipment design, hydrostatic weighing can be performed in either a sitting or a kneeling position. For the seated position, which is used most commonly, necessary equipment includes a tank (recommended size $4 \times 4 \times 5$ ft, or $1.2 \times 1.2 \times 1.5$ m), a chair or seat, an autopsy or similar type of scale (15 to 25 g increments) or load cells, a lift system (optional), and a weighted belt. The hydrostatic weighing tank can be made of stainless steel, fiberglass, ceramic tile, Plexiglas, or other material. In a simple system, a chair made from PVC or a combination of PVC and webbing material (figure 11.3) can be suspended from a scale. The lift system, which is a luxury item, requires a 1/4-ton (225 kg) hoist or ratchet system that is mounted to the ceiling. Ideally, the tank should have load cells placed in it. The chair can be placed on the four load cells. This design produces a more sensitive reading than an autopsy scale does, and it allows any of the load cells to detect any movement on the scale (figure 11.3).

In a system with load cells, the subject can kneel on the basket. Use of a scuba belt helps keep the legs of the individual on the basket and prevents overweight subjects from floating off the basket. The chair seat rests on load cells that are directly connected to a recorder or a computer to provide near-instant analysis and feedback.

Observing specific procedures, equipment, calibration, subject instruction, and preparation before underwater weighing is critical to obtaining valid results. Subjects should follow a normal diet, consume a normal amount of fluid, and exercise normally the day before testing (Pollock & Wilmore, 1990). Subjects should abstain from eating 2 h before testing and avoid foods that cause excessive amounts of gas in the gastrointestinal (GI) tract. The underwater weighing measure is greatly affected by factors that influence hydration status, such as menstruation, drugs, physical activity, and saunas (Girandola, Wiswell, & Romero, 1977). Ideally, subjects should wear tight-fitting bathing suits underwater to avoid trapping air.

Before entering the tank, the subject should be asked to void his or her bladder and defecate if nec-

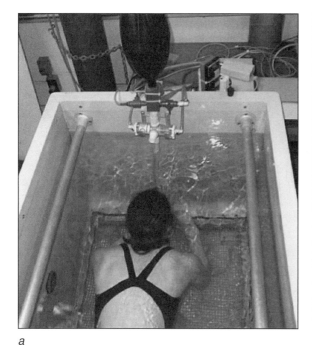

a

b

**Figure 11.3**   An underwater weighing tank used to determine body density. *(a)* Load cells, one under each corner of the basket, will detect any movement on the basket. *(b)* The RV apparatus in the tank. The measurement of RV begins as the subject takes his or her first breath on coming out of the water.

essary. The subject is weighed on land to determine dry weight or weight in air (W) to the nearest 100 g. The subject then climbs into the water tank, stands on the bottom if possible, gets completely wet, and rubs the hands over the entire body to remove any trapped air bubbles from skin, hair, and bathing suit. The subject then sits or kneels on the basket and, if necessary, puts on the weight belt.

Once the subject is seated, the chair height is adjusted so that water level is just below the subject's chin. The subject is instructed to take four or five deep breaths, and start a full expiration while the head is still above the water. When the subject has expelled most of the air in the lungs, he or she bends forward slowly at the waist until the head is completely submerged. The subject continues to exhale air. The subject's hands should be relaxed yet placed on the basket handles to maintain balance and to avoid a rocking motion. The reading from the load cells can be made after the complete exhalation and after the reading stabilizes. If using an autopsy scale, the reading is taken when minimal oscillation of the scale is occurring. After the underwater weight is obtained, the tester taps on the side of the tank to signal the subject to surface. The procedure is repeated 6 to 10 times or more if necessary, because the subject will need several attempts to learn how to perform the technique properly (Katch, 1968). The tester should critique each trial, make the subject comfortable in the water, and give feedback so that the subject can perform the next trial better.

Ideally, the measurement of the residual volume, the volume of air left in the lungs, is measured in the tank. If it is taken simultaneously with each weight, the residual volume can measure any error in total exhalation. In this type of system the subject will have a mouthpiece in the mouth that is connected to a hose for outside air. When the subject submerges, he or she will exhale completely through the hose. The subject will then hold his or her breath for approximately 5 s while the basket reading is taking place. As the subject comes up out of the water and takes his or her first breath, the hose is switched over to a bag with pure oxygen (Organ, Eklund, & Ledbetter, 1994). The subject will then be asked to breathe deeply for four to five breaths from the bag until the nitrogen equilibrates in the bag. The subject then takes the mouthpiece from the mouth.

After the subject's weight levels off after repeated trials and the nitrogen values are consistent for the rebreathing, the procedure is completed. The water temperature should be recorded for each subject, because water density ($D_w$) is temperature dependent (table 11.2). Additionally a tare weight is determined after the subject gets off the basket and sits in a corner of the tank. Tare weight is the combined weight of the subject, chair, mouthpiece, and, when applicable, the weighted belt, taken at the same level at which the underwater weight was observed. Tare weight is then subtracted from the underwater weight to derive the net underwater weight of the subject ($W_w$). Body density ($D_b$) is then calculated using the equation of Goldman and Buskirk (1961):

$$D_b = \frac{W_a}{\dfrac{(W_a - W_w)}{D_w} - (RV + 100\ cc)}$$

where $D_b$ is body density (g/cc), $W_a$ is body weight in air (kg), $W_w$ is body weight in water (kg), $D_w$ is density of water (g/ml), and RV is residual lung volume (ml). One hundred cc is a correction factor that represents the approximate amount of air trapped in the gastrointestinal tract.

Figure 11.4 is a sample data collection form used for determining RV and $D_b$ by underwater weighing. From the data found in this form, trials 5, 6, and 7 were used in the calculation of $D_b$. The subject in figure 11.4 leveled off after trial 4, but additional trials were taken to assure the subject's best effort.

Table 11.2    **Conversion Chart for Determining Water Density ($D_w$) at Various Water Temperatures (W Temp)**

| W temp (°C) | $D_w$ | W temp (°C) | $D_w$ |
|---|---|---|---|
| 23 | 0.997569 | 31 | 0.995372 |
| 24 | 0.997327 | 32 | 0.995057 |
| 25 | 0.997075 | 33 | 0.994734 |
| 26 | 0.996814 | 34 | 0.994403 |
| 27 | 0.996544 | 35 | 0.994063 |
| 28 | 0.996264 | 36 | 0.993716 |
| 29 | 0.995976 | 37 | 0.993360 |
| 30 | 0.995678 | | |

Reprinted from *Exercise and prescription for prevention and rehabilitation*, 2nd edition, M.L. Pollock and J.H. Wilmore, Copyright (1990), with permission from Elsevier.

## Body Composition Analysis: Data Collection Form

Name_____    Date_02-07-03_

Age_19___    Sex_F___    Tester_JEG___    SF_LMG___    UWW____    Group_____

Weight_58.2_(kg) _128.2_(lb)    Height_168.1_(cm) _66.2_(in.)

**Hydrostatic weighing**        **Skinfolds (mm)**

| Trial | Wt (kg) | Trial | Wt (kg) | | |
|---|---|---|---|---|---|
| 1 | 3.475 | 9 | _____ | Chest | 15.5 |
| 2 | 3.500 | 10 | _____ | Axilla | 14.0 |
| 3 | 3.400 | 11 | _____ | Triceps | 20.0 |
| 4 | 3.450 | 12 | _____ | Subscapular | 11.5 |
| 5 | 3.500 | 13 | _____ | Abdominal | 26.5 |
| 6 | 3.500 | 14 | _____ | Suprailiac | 18.5 |
| 7 | 3.500 | 15 | _____ | Thigh | 34.0 |
| 8 | _____ | 16 | _____ | Sum of 7 | 140.0 |
| | | | | Suprapatellar | 27.0 |
| | | | | Medial calf | 19.5 |

$T_i$ (°C) _35.0_    $T_f$ (°C) _35.0_

Average weight = _3.500_ kg

Tare weight = _2.000_ kg

True UWW = _1.500_ kg    % fat (UWW) = _27.4_ %

Body density = _1.037_ g/ml    Fat wt. = _35.1_ lb
                          = _15.9_ kg

**Residual volume**        Fat-free wt. = _93.2_ lb

Trial 1 _0.783_ L    = _42.3_ kg

Trial 2 _0.834_ L    Target % fat = _23.0_ %

Trial 3 _____ L    Target wt. = _121.0_ lb
                          = _54.9_ kg

Average = _0.809_ L

**Summary**

% fat (SF) = _26.1_ %

**Figure 11.4**    Sample of a body composition and pulmonary function form used to determine body density by underwater weighing.

Residual volume, the amount of air remaining in the lungs at the end of a maximal expiration, should be determined when the subject is in the tank. If residual volume cannot be obtained in the tank simultaneously with the underwater weight, then it should be obtained immediately thereafter on land. If this is not possible, an estimation can be used. The following equations, developed by Goldman and Becklace (1959), are recommended:

Men: RV = 0.017 (age in years) + 0.06858 (height in inches) − 3.477

Women: RV = 0.009 (age in years) + 0.08128 (height in inches) − 3.900

Percent fat can then be calculated from $D_b$ from one of the following equations:

Brozek and colleagues (Brozek et al., 1963): % fat = $[(4.57/D_b) − 4.142] \times 100$

Siri (1961): % fat = $[(4.95/D_b) − 4.500] \times 100$

In the calculation of $D_b$ in figure 11.4, RV was measured twice and averaged. For verification, usually more than one RV measure is determined. Measures should be within the reported variation of the technique (usually 50-100 ml).

## Sources of Error

Although it is considered the criterion or gold standard for body composition measurement when developing prediction equations for field methods, the hydrostatic technique is not free from measurement error. One potential source of error comes from the determination of total body volume. To calculate $D_b$ by the hydrostatic method, total body volume must be corrected for the RV and the volume of air trapped in the GI tract. Although GI gas volume can vary, Buskirk (1961) proposed the use of the constant correction value of 100 ml. The constant of 100 ml has been used routinely, but recently Ploutz-Snyder and associates (1999) demonstrated that consumption of a beverage with high carbonation content just before being weighed underwater will alter the measure of body density. This study showed an 11% overestimate of percent body fat compared with measurements made before consumption of a carbonated beverage (Ploutz-Snyder et al., 1999). Thus, the constant of 100 ml can be used but is sensitive to changes in gas volume in the individual.

The error associated with RV can be large and can have a sizable effect on the final calculation of $D_b$ (Pollock & Wilmore, 1990). Residual volume can be estimated from average population values based on age, sex, and height as shown earlier (Goldman & Becklace, 1959) or by an estimated percentage of the vital capacity (approximately 25% to 30%) (Katch & Katch, 1980). Wilmore (1969) compared the use of actual RV and the two estimation techniques for measuring RV in the assessment of body composition by underwater weighing for college-aged males and females. A close agreement (<0.001 g/cc) is found between measures of $D_b$, percent fat, and FFM using the actual measurement of RV and the two techniques used for estimating RV. Thus, estimated RVs can be used satisfactorily for screening purposes and when research is conducted on large populations. Because the RV estimates vary based on age, height, weight, or vital capacity, only actual RV measures are valid for use with individuals who need counseling or in research with small groups of subjects. Katch and Katch (1980) showed that a 600 ml difference in RV could affect relative fat by 8% and FFM by 12 kg. Further, controversy exists whether RV should be measured out of the water or in the water during the actual procedure (figure 11.3). Although 200 to 300 ml differences in RV have been reported in a subject's measurements taken in and out of the water (Behnke & Wilmore, 1974), if care is taken to have the subject exhale fully and use similar body positions for the measurement of RV and underwater weight, the error is thought to be small (Behnke & Wilmore, 1974). For example, a 200 ml error in RV accounts for only a 1% error in percent body fat (table 11.3). This

### Table 11.3 The Effect of Measurement Error on Determination of Body Density ($D_b$) From Underwater Weighing

| Measure | Actual* | Errors** | | |
| --- | --- | --- | --- | --- |
| | | 1 | 2 | 3 |
| RV (L) | 1.200 | 1.400 | 1.700 | 2.200 |
| $D_b$ (g/cc) | 1.0605 | 1.0631 | 1.0669 | 1.0734 |
| Fat (%) | 16.74 | 15.63 | 13.95 | 11.16 |
| $W_1$ (kg) | 4.24 | 4.29 | 4.34 | 4.44 |
| $D_b$ (g/cc) | 1.0605 | 1.0612 | 1.0618 | 1.0631 |
| Fat (%) | 16.74 | 16.46 | 16.18 | 15.62 |
| $W_1$ (kg) | 88.70 | 88.80 | 89.20 | 89.70 |
| $D_b$ (g/cc) | 1.0605 | 1.0605 | 1.0601 | 1.0598 |
| Fat (%) | 16.74 | 16.78 | 16.91 | 17.09 |

*Actual values are from figure 6-27, Pollock and Wilmore (1990).

**Each error for $D_b$ and percent fat is calculated with the other two variables from the actual values.

RV = residual volume; $W_1$ = scale weight in water

Reprinted from *Exercise and prescription for prevention and rehabilitation*, 2nd edition, M.L. Pollock and J.H. Wilmore, Copyright (1990), with permission from Elsevier.

amount of error is acceptable for use in the clinical and research setting. The important factor is to replicate technique consistently so that it will not affect serial analysis of $D_b$. In this case, the error in RV is constant in one direction. Although in most cases the difference between measuring RV in the water at the time of underwater weighing and measuring it out of the water is small (Heyward & Stolarczyk, 1996), the number of trials needed to obtain an accurate $D_b$ can be reduced when measuring RV in the water. When RV is measured in the water, it is not as imperative for subjects to exhale maximally, because RV measurement can detect the difference. Another advantage of measuring RV in the water at the time of weighing is that many subjects feel uncomfortable exhaling maximally in the water.

To address the problem of measuring residual volume in and out of the tank, a six-center research project was conducted. A subject visited each laboratory within a 2-week period for assessment of body composition. The purpose of the visit was to standardize testing procedures and determine interlaboratory variation among measures. All laboratories had experienced investigators and determined body density using the underwater weighing technique. Four of the laboratories measured RV while the subject was out of the water, and four measured it while he was in the water as part of the procedure. Among laboratories, body weight varied from 77.4 to 78.3 kg ($X = 77.9 \pm 0.35$ kg), $D_b$ from 1.0717 to 1.0761 g/cc ($X = 1.0733 \pm 0.00136$ g/cc), the RV from 1.00 to 1.44 L ($X = 1.22 \pm 0.13$ L), and percent fat from 10.0% to 11.9% ($X = 11.2\% \pm 0.6\%$). Variation among laboratories was small, and there was no mean difference in $D_b$ and percent fat between laboratories that measured RV in the water as compared with those that measured it out of the water. Note, however, that the investigators were well trained in the measurement of RV (Lohman, 1992).

The determination of underwater weight is another source of error for the hydrostatic technique. How many trials are necessary to get valid results, and which trial or trials should be used for the calculation of $D_b$? Katch (1968) demonstrated that a learning curve is associated with successive trials of underwater weighing and concluded that 9 to 10 trials were necessary to obtain the most representative underwater weight for an individual. Katch recommended that the average of the last 3 trials be used to calculate $D_b$. Behnke and Wilmore (1974) also suggested performing 10 trials, but they selected the true underwater weight as

1. the highest weight if it is observed more than once;

2. the second-highest weight, if it observed more than once and the first criterion is not met; or

3. the third-highest weight, if neither the first nor the second criterion is attained.

They chose this method of selection to reduce the possibility of underestimating the actual underwater weight on the subject who attained the highest value during the first 5 to 7 trials.

The measurement errors associated with the underwater weighing technique are mainly associated with errors in RV, body weight out of the water ($W_a$), and body weight in the water ($W_w$). For example, table 11.3 shows the effect of errors in RV, $W_a$, and $W_w$ on the determination of $D_b$ from underwater weighing. The hypothetical errors of 200, 500, and 1,000 ml for RV; 50, 100, and 200 g for $W_w$; and 100, 500, and 100 g for $W_a$ were added to the actual calculated $D_b$. As shown in table 11.3, errors can have a dramatic effect on $D_b$ (2% to 5.5% fat). Errors of this magnitude are common when RV is estimated from age and height or vital capacity, and for that reason estimations of RV are not recommended when $D_b$ values are used for research or individual counseling. An error of 50 g or more in underwater scale weight ($W_w$) would be highly unusual; thus, the error associated with an incorrect $W_w$ should be less than 0.5% fat. This is more likely a problem with autopsy scales and with subjects who have difficulty holding their breath long enough for the scale reading to be made. Body weight ($W_a$) can be quite variable depending on time of day, dietary pattern, hydration status, and illness. As previously mentioned, excessive dehydration or hydration caused by such factors as exercise, sauna, diarrhea, medications, or menstrual cycle can have a significant effect on $D_b$ (Girandola et al., 1977; Pollock & Wilmore, 1990). For example, a 2 to 3 kg weight fluctuation could cause a 1.0% change in percent body fat.

A final consideration is the conversion of $D_b$ to percent fat. The theoretical relationship between $D_b$ and fatness can be derived if the fat and the FFM are assumed to have a constant density for all individuals (Lohman, 1981). Percent fat is usually predicted from either the Siri (1961) equation or the Brozek et al. (1963) equation. Siri also proposed a second equation that avoided the assumption of constant water content of the FFM and allowed for correction of hydration. This equation is considered a three-component model (% body fat = $[(2.1176/D_b) - (0.78W) - 1.351] \times 100$, where W is equal to TBW measured by isotopic dilution (Siri, 1961). The equation, however, is still based on the assumption that fat density is 0.900 g/cc and FFM density is 1.100 g/cc (Werdein & Kyle, 1960). The equation of Brozek et al. (1963)

uses the concept of a reference man of a specified $D_b$ and body composition (15.3% fat). Within $D_b$ of 1.09 and 1.03 g/cc, the two formulas agree within 1% fat (Lohman, 1981).

Although universally accepted, the Siri and Brozek equations are not without problems. Both are based on the results of direct compositional analysis of human cadavers, but only a few cadavers were used, and they did not represent a distribution of the normal population (Behnke & Wilmore, 1974; Pollock & Wilmore, 1990). As previously mentioned in this chapter, the density of FFM and fat is quite variable in humans and can cause an error in estimating percent fat from $D_b$ by 2.5% to 3.8% (Lohman, 1981).

Despite the potential sources of error, the hydrostatic technique has been shown to be highly reliable, with Pearson-product correlation coefficients greater than .95 when measurements were made over time intervals ranging from 30 min to a couple days (Keys & Brozek, 1953; Mendez & Lukaski, 1981; Pascale, Grossman, Sloan, & Frankel, 1956; Pollock, Laughbridge, Coleman, Linnerud, & Jackson, 1975). A small standard error of measurement (less than 0.002 g/cc) has also been observed (Jackson, Pollock, Graves, & Mahar, 1988; Oppliger, Looney, & Tipton, 1987).

# Air Displacement Plethysmography

The past decade has seen the emergence of a relatively new technique, also based on the two-component model, that provides an estimate of percent body fat from body density without submerging underwater. This technique, based on air displacement plethysmography (ADP), uses the relationship between pressure and volume to derive body volume ($V_b$) (Wagner, Heyward, & Gibson, 2000). The machine used with this technique consists of two chambers, one for the subject and the other as a reference chamber. After the door is closed and sealed with the subject inside, pressure is slightly increased, resulting in oscillation of a diaphragm that separates the two chambers (Dempster & Aitkens, 1995; McCrory, Gomex, Bernauer, & Mole, 1995). The relationship between pressure and volume at a fixed temperature is used to determine the volume of the subject chamber (Wagner et al., 2000).

## Equipment and Methods

Air displacement plethysmography is currently measured using equipment that has been designed by BODPOD (Life Measurement Instruments, Concord,

California) (figure 11.5). During testing the subject wears a bathing suit and a swim cap. In lieu of a swimsuit, Lycra biking shorts can be worn. The Bod Pod consists of a single fiberglass structure with two internal chambers separated by a moving diaphragm. The diaphragm oscillates back and forth to create small volume changes in the two chambers that are equal in magnitude but opposite in sign (Wagner et al., 2000). These volume shifts result in complementary pressure fluctuations in the two chambers. The pressure and volumes are calculated using Poisson's law ($P_1V_1 = P_2V_2$) (Wagner et al., 2000). After calibration of the chamber, the subject sits quietly in the chamber of the Bod Pod while body volume is measured until two values within 150 ml are obtained. A minimum of two trials are performed on each subject. Subsequently, thoracic gas volume is estimated by having the subject breathe normally for three breathing cycles through a tube connected to the internal system while wearing a nose clip. At the midpoint of an exhalation, the airway tube is momentarily occluded and the subject is signaled to give three small puffs of air into the tube while maintaining a tight seal around the end of the tube, using a sort of a panting maneuver (Dempster & Aitkens, 1995). When the subject does this correctly, the measured thoracic gas volume is used to calculate a corrected body volume (corrected body volume = raw body volume – thoracic gas volume). The subject is usually given two to three trials to complete this maneuver successfully. If the subject cannot comply, an estimated thoracic gas volume can be used for calculation of body fat (McCrory, Mole, Gomez, Dewey, & Bernauer, 1998).

As with hydrostatic weighing, the attire worn during ADP will affect its accuracy and may affect reliability of the percent fat estimation because of the distinction between gas under isothermal or adiabatic conditions (Vescovi, Zimmerman, Miller, & Fernhall, 2002). The isothermal nature of clothing makes it more compressible and will result in a greater pressure change in the front chamber during body volume measurements (Vescovi et al., 2002). A smaller pressure ratio and decreased body volume will result in greater body density and reduced percent fat estimation for a given individual. For example, Vescovi and colleagues noted that wearing a hospital gown significantly affects accuracy, reducing percent fat estimation by approximately 9% compared with the recommended swimsuit. Measurement in the nude, however, does not provide more accuracy than measurement while wearing a swimsuit. Vescovi and colleagues compared a swimsuit, hospital gown, and nude. Reliability of ADP does not appear to be compromised by wearing clothing.

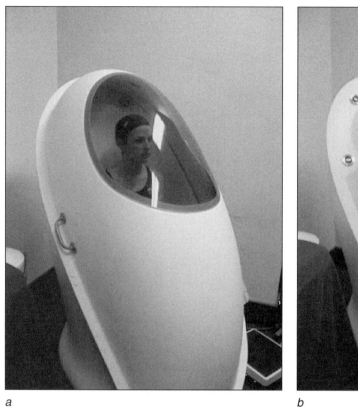
a

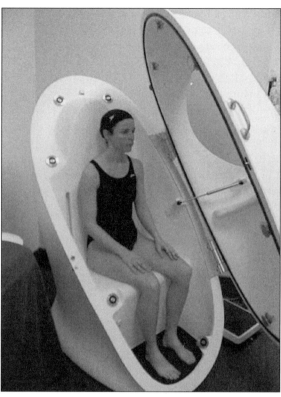
b

**Figure 11.5** Photo of the Bod Pod (Life Measurement Instruments) while the subject is in the chamber *(a)* and when the door is open and the subject chamber is visible *(b)*.

## Advantages and Disadvantages

The advantage of ADP over hydrostatic weighing is that it is noninvasive, comfortable, and quick. ADP does not require the subject to be submerged in water, a condition that makes hydrostatic weight testing impractical in many populations (e.g., children, older adults, burn victims, quadriplegics, and so on). Both methods are sensitive to wearing a tight-fitting suit that will not hold a considerable amount of air. One limitation of ADP is that it is sensitive to temperature and pressure changes in the room; opening of doors or the cycling on and off of the heating system during the testing period can cause problems in measurement. Furthermore, movement, gas and water vapor exchange, and heat generated by a human in a closed space can significantly affect the measurement (Dempster & Aitkens, 1995).

The $D_b$ obtained from ADP varies slightly but consistently from other methods. ADP results in consistently higher values for $D_b$ than hydrostatic weighing (HW) does (Millard-Stafford et al., 2001). The percent fat estimates from $D_b$-ADP were lower compared with $D_b$-HW. Weyers et al. (2002) reported

that percent body fat as estimated by DEXA was slightly but statistically higher than corresponding ADP values. Studying the effect of weight loss on ADP measures, they noted small but consistent differences between measurements from DEXA and ADP. Percent body fat values before and after weight loss were highly correlated between the DEXA and ADP across a relatively wide range of body fat levels in middle-aged women and men (Weyers et al., 2002). DEXA estimates for FM were significantly higher and estimates of FFM significantly lower both before and after weight loss when compared with corresponding ADP values. They concluded that both methods can be used to measure body composition changes with weight loss over time, but that these methods should not be interchanged (Wagner et al., 2000).

## Anthropometry

Anthropometry is the science that deals with the measurement of size, weight, and proportions of the human body. Anthropometric techniques (skin-fold fat, circumference, and diameter measurements) are popular for predicting body composition in the field setting because they are inexpensive to attain, require little space, and are easy to perform with adequate training (Behnke & Wilmore, 1974; Pollock & Wilmore, 1990). Until recently, the techniques used to assess skinfold, circumference, and diameter measures varied and lacked standardization.

In 1985 at the Anthropometric Standardization Conference held in Airlie, Virginia, 50 professionals from the fields of epidemiology, exercise science, human biology, medicine, nutrition, physical anthropology, and physical education met to discuss how to standardize anthropometric procedures. The *Anthropometric Standardization Reference Manual (ASRM)* was a written consensus developed by the many experts attending this conference (Lohman, Roche, & Martorell, 1988). The recommended procedures in the standardization manual will be described in this chapter. Because of the popularity of the generalized equations for the estimating of $D_b$ from skinfold and circumference measures developed by Jackson, Pollock, and Gettman for men (1978) and Jackson, Pollock, and Ward for women (1980), the differences of these techniques from those in the standardization manual will also be outlined.

This section will describe two of the popular techniques used for estimating body composition: skinfold fat and circumferences. Because diameter measures are more related to skeletal size than to body fat, they are not often used to predict body composition. See Lohman et al. (1988) and Pollock and Wilmore (1990) for a more detailed description of diameter measures.

## Skinfold Measurements

The premise behind skinfold measurements is that the skinfold is actually two layers of skin and two layers of fat. This subcutaneous fat depot is assumed to be a reflection of the deeper fat depots in the body, such as the visceral fat depot. To measure this thickness, a caliper that is accurately calibrated and has a constant pressure of 10 g/mm² is recommended (Cataldo & Heyward, 2000; Lohman et al., 1988). Figure 11.6 shows a Harpenden (top) and a Lange (bottom) caliper. When measuring skinfold thickness, site selection and location must be standardized, because small differences in location can cause significant errors in measurement (Lohman et al., 1988). When body composition is predicted from anthropometric measures, the skinfold techniques used to develop the prediction equation should be used to obtain the measures. Depending on which caliper is used, measures are recorded to the nearest

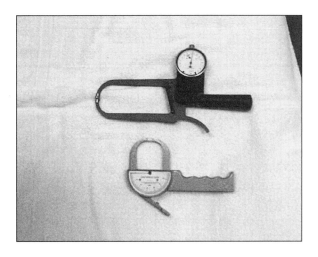

**Figure 11.6**   The Harpenden (top) and Lange (bottom) skinfold calipers.

0.1 cm or 0.5 cm. For standardization purposes, all measurements should be taken from the right side of the body.

### Methodology and Technique

The following description of skinfold measurement techniques is taken from the *Anthropometric Standardization Reference Manual (ASRM)* (Lohman et al., 1988). When different, the techniques of Pollock and Wilmore (1990) will also be presented.

When taking skinfold measurements, the thumb and index finger of one hand are used to palpate the fat–muscle interface at about 1 cm proximal to the site at which the skinfold is to be measured (Lohman et al., 1988). The thumb and index finger should be placed on the skin about 8 cm apart, on a line that is perpendicular to the long axis of the skinfold to be measured. The thicker the fat layer, the greater the distance needed between the thumb and index finger. The thumb and index finger are then drawn toward each other and a fold is grasped firmly between them. The amount of subcutaneous tissue picked up should form a fold that has parallel sides, and the fold should be perpendicular to the surface of the body at the measurement site. With the other hand, the caliper is opened, and the caliper is placed over the skinfold so that the fixed arm of the caliper is positioned on one side of the fold approximately 1 cm adjacent to the first finger. The pressure on the caliper is then released gradually so that the caliper arms come toward each other. The best method is to line up one point of the caliper at one side of the fat fold first and then line up the other point as the caliper is released. The measurement is made while the skinfold is still being held by the other hand,

and the measurement is made about 2 to 4 s after the pressure is released (Lohman et al., 1988). All skinfold sites are measured once, and then the series of measurement sites are repeated two more times until consistency is observed. All skinfolds at a given site should be within 3 mm. If they are not, a fourth measure should be made. The mean of the measures is then calculated.

The following paragraphs describe skinfold measurements for specific sites.

• **Pectoral (chest) skinfold.** The *ASRM* recommends that the same chest skinfold site be used for both males and females (figure 11.7a). Pectoral skinfold thickness is measured using a skinfold with its long axis directed to the nipple. The skinfold is picked up on the anterior axillary fold as high as possible; the thickness is measured 1 cm inferior to this. The measurement is taken while the subject stands with the arms hanging at the sides. According to Pollock and Wilmore (1990), for men the chest skinfold is a diagonal fold one-half the distance between the anterior axillary line and the nipple, and for women, one-third the distance from the anterior axillary line and the equivalent position (figure 11.7b).

• **Midaxillary skinfold.** The subject stands erect with the right arm slightly abducted and flexed at the shoulder joint, while the tester stands facing the subject's right side. The subject may find it more comfortable to rest the arm on the tester's shoulder. The *ASRM* recommends that the midaxillary skinfold be measured at the level of the xiphisternal junction in the midaxillary line, with the skinfold horizontal (figure 11.8). According to Pollock and Wilmore (1990), the location of the skinfold is the same, but a vertical fold is measured.

• **Triceps skinfold.** The *ASRM* recommends that the triceps skinfold be measured "vertically" in the midline of the posterior aspect of the arm, over the triceps muscle. This point is approximately midway between the lateral projection of the acromion process of the scapula and the inferior margin of the olecranon process of the ulna (figure 11.9). The subject is measured standing with the arm hanging loosely and comfortably at his or her side.

• **Subscapular skinfold.** The *ASRM* recommends that the subscapular skinfold be picked up on a diagonal, inclined infero-laterally approximately 45° to the horizontal plane in the natural cleavage lines of

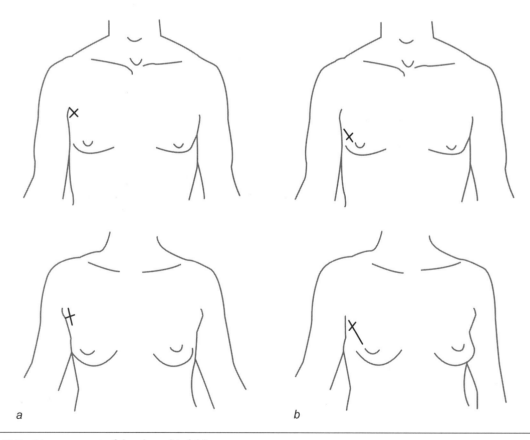

a                    b

**Figure 11.7**    Measurement of the chest skinfold.

*(a)* Adapted from Lohman, Roche, and Martorell 1988; *(b)* Adapted from Pollock and Wilmore 1990.

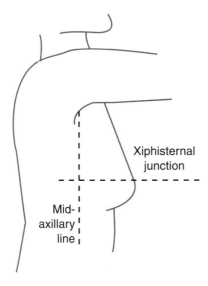

**Figure 11.8**  Illustration of the level of the xiphisternal junction at which the midaxillary skinfold is taken.

the skin. This site is just inferior to the inferior angle of the scapula (figure 11.10*a*). The subject should stand comfortably erect, with the upper extremities relaxed at the sides of the body. To locate the site, the tester palpates the scapula and runs the fingers inferiorly and laterally along its vertebral border until the inferior angle is identified. In obese subjects, gently placing the right arm behind the back "wings" the scapula and helps in locating the site. The skinfold is taken on a diagonal line coming from the vertebral border, 1 to 2 cm from the inferior angle of the scapula (Pollock & Wilmore, 1990) (figure 11.10*b*).

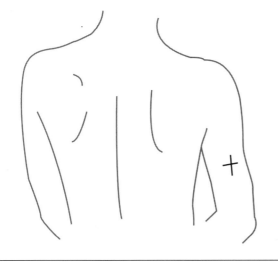

**Figure 11.9**  Landmark for the triceps skinfold. The elbow is bent to help locate the site, but the skinfold is measured with the arm hanging loosely at the subject's side.

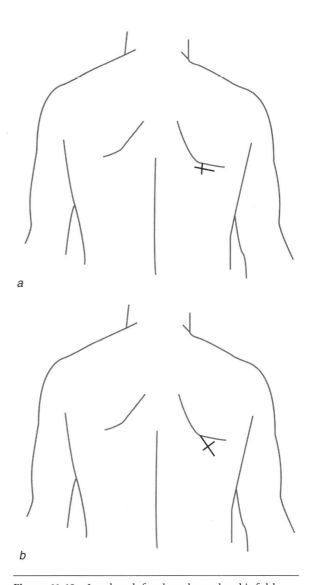

**Figure 11.10**  Landmark for the subcapular skinfold.

(*a*) Adapted from Lohman, Roche, and Martorell 1988; (*b*) Adapted from Pollock and Wilmore 1990.

• **Abdominal skinfold.** The ASRM states that "the abdominal skinfold is taken horizontally at a site which is 3 cm lateral to the midpoint of the umbilicus and 1 cm inferior to it" (figure 11.11*a*). The measure is taken with a horizontal fold. The subject stands and relaxes the abdominal wall musculature as much as possible and breaths normally. According to Pollock and Wilmore (1990), a vertical fold is taken at a lateral distance approximately 2 cm from the umbilicus (figure 11.11*b*).

• **Suprailiac skinfold.** The *ASRM* states that "the suprailiac skinfold is measured in the midaxillary line immediately superior to the iliac crest. A diagonal skinfold is grasped just posterior to the midaxillary line following the natural cleavage lines of the skin"

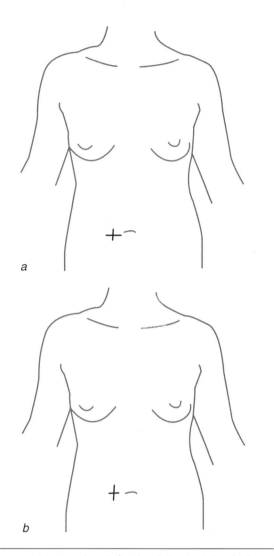

**Figure 11.11** Landmark for the abdominal skinfold.

*(a)* Adapted from Lohman, Roche, and Martorell 1988; *(b)* Adapted from Pollock and Wilmore 1990.

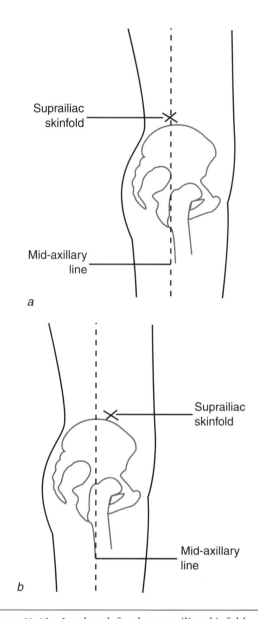

**Figure 11.12** Landmark for the suprailiac skinfold.

*(a)* Adapted from Lohman, Roche, and Martorell 1988; *(b)* Adapted from Pollock and Wilmore 1990.

(figure 11.12*a*). The subject should stand erect with the arms hanging comfortably at his or her sides, but the right arm or hand may need to be moved slightly posterior or be abducted to improve access to the site. According to Pollock and Wilmore (1990), the suprailiac fold is measured as a diagonal fold above the crest of the ilium at the spot where an imaginary line would come down from the anterior axillary line (figure 11.12*b*).

• **Anterior thigh skinfold.** The thigh skinfold site is located in the midline of the anterior aspect of the thigh, midway between the inguinal crease and proximal border of the patella, as recommended by the ASRM. The subject flexes the hip to assist the measurer in locating the inguinal crease (figure 11.13*a*). A vertical skinfold is measured while the

subject stands with body weight shifted to the left leg, the right leg relaxed, and the right knee slightly flexed.

• **Suprapatellar skinfold.** The ASRM states that "the suprapatellar skinfold site is located in the midsagittal plane on the anterior aspect of the thigh, 2 cm proximal to the proximal edge of the patella" (figure 11.13*a*). A vertical skinfold is measured while the subject stands relaxed with body weight shifted to the left leg and the right knee slightly flexed.

• **Medial calf skinfold.** The ASRM recommends that the medial calf fold be measured with the sub-

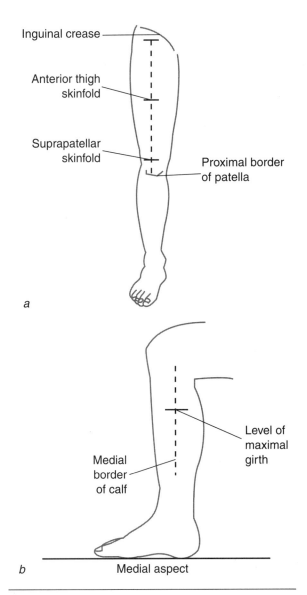

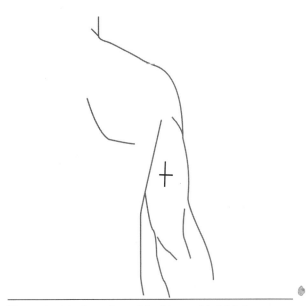

**Figure 11.14**   Location of the biceps skinfold site.

**Figure 11.13**   Location of the *(a)* anterior thigh and suprapatellar skinfold sites and *(b)* the medial calf skinfold site.

ject standing and the right foot placed on a stool or bench so that the knee and hip are flexed to about a 90° angle (figure 11.13*b*). This fold can also be measured with the subject in the sitting position. A vertical fold is measured on the medial aspect of the calf at the level of maximal girth.

• **Biceps skinfold.** The ASRM states that "the biceps skinfold thickness is measured as the thickness of a vertical fold raised on the anterior aspect of the arm, over the belly of the biceps muscle" (figure 11.14). The subject stands with arms hanging relaxed at the sides with the right palm directed anteriorly.

## Sources of Error

Important potential sources of measurement error associated with skinfold measurements are caliper selection and tester reliability, including inter- and intraobserver measurement error and the variance associated with the selection of skinfold site (Lohman et al., 1988). Differences in skinfold fat readings may result from the use of different calipers. The pressure exerted by the caliper has a significant effect on both the skinfold measurement and the consistency with which the measurement is repeated (Edwards, Hammond, Healy, Tanner, & Whitehouse, 1955). Sloan and Shapiro (1972) compared three calipers, the Harpenden, Lange, and MNL, and found that the Harpenden caliper yielded slightly better between-observer results but that no systematic differences occurred between measurements at any site among the calipers. The results of Lohman, Pollock, Slaughter, Brandon, and Boileau (1984) were different, with the Lange calipers consistently overestimating the results of the Harpenden calipers by approximately 1 to 2 mm per site. Others (Gruber, Pollock, Graves, Colvin, & Braith, 1990) found a high correlation between Harpenden and Lange skinfold fat calipers, but the Harpenden values were 10% lower per skinfold site. This finding demonstrated that using the Harpenden rather than the Lange caliper resulted in a 10% difference in the sum of seven skinfolds, which caused a 10% underprediction of $D_b$. Thus, those who estimate $D_b$ and calculate percent fat from skinfold equations must use the same caliper that the investigator used to develop the equation. For example, if the Harpenden caliper is used and the generalized equations by Jackson, Pollock, and Ward (1978, 1980) are used in estimating $D_b$ and percent fat, the 10% difference in the sum of skinfolds measured should be added.

Intratester reliability can be attained rapidly with practice and good instruction (Heyward & Stolarczyk, 1996; Lohman et al., 1988; Roche, Heymsfield, & Lohman, 1996). Test–retest correlations well above .9 have been shown by Pollock et al. (1976, 1975) and Jackson, Pollock, & Gettman (1978) for seven measures (chest, axilla, triceps, subscapula, abdomen, suprailium, and anterior thigh). Researchers have suggested that a tester must measure approximately 50 to 100 persons to attain a high level of competency in taking skinfolds (Katch & Katch, 1980; Pollock & Wilmore, 1990).

The best way to avoid interobserver variability is to have the same investigator take all the measurements. However, in studies that use published regression equations to estimate body composition, and in long-term epidemiological studies in which several different investigators take skinfold measurements, the intertester source of measurement error cannot be avoided (Jackson, Pollock, & Gettman, 1978; Lohman et al., 1984). Lohman et al. (1984) showed significant variation in skinfolds when experienced testers did not practice together or standardize procedures. One 30 min practice session minimized such errors. One study investigated intertester reliability of skinfold measurements and percent fat, and found that the variation among experienced testers who had practiced together was a relatively small source of measurement error (intertester reliability estimates exceeded .93) (Jackson, Pollock, & Gettman, 1978). These results agree with earlier findings reported (Keys & Brozek, 1953; Munro, Joffe, Ward, Syndham, & Fleming, 1966).

The most important factor associated with intertester error is variation in selection of skinfold site (Heyward & Stolarczyk, 1996; Roche et al., 1996). Using four experienced testers Lohman et al. (1984) observed significant variation in the selection of sites, with the variation being greatest at the anterior thigh and suprailium. Measurement error can also be attributed to the size of the skinfold. Research has shown that the standard error among testers was approximately 1 mm for every 10 mm of skinfold fat, that is, 1 mm variation with a skinfold value of 10 mm, 2 mm with 20 mm, 3 mm with 30 mm, and so on (Pollock, Jackson, & Graves, 1986).

# Circumference Measurements

Circumference measurements, used either alone or in combination with skinfold measurements, provide information about body composition, growth, nutritional status, and fat patterning (Behnke & Wilmore, 1974; Jackson & Pollock, 1976; Lohman et al., 1988). Relatively accurate estimates of body composition ($D_b$ and percent fat) have been found with circumference measures (Katch & McArdle, 1973) using abdomen, right thigh, and right forearm circumferences for young women; abdomen, right thigh, and right calf for middle-aged women; right upper arm, abdomen, and right forearm for young men; and buttocks (hip), abdomen, and right forearm for middle-aged men. Refer to McArdle et al. (2001, pp. 813-816) for conversion charts to predict percent fat from the previously mentioned circumferences. The circumference measures before and after intervention programs for weight loss or exercise training are useful in the clinical setting. Recent research has shown that the waist circumference can be used as an indicator of health risks, such that the health risk is greater in normal weight, overweight, and class I obese women with high waist circumference values compared with normal weight, overweight, and class I obese women with normal waist circumference values (Janssen, Katzmarzyk, & Ross, 2002).

## Methodology and Techniques

The following description of circumference measurement technique is taken from the *Anthropometric Standardization Reference Manual* (Lohman et al., 1988). Where different, the techniques of Pollock and Wilmore (1990) will also be presented.

The tape measure should be flexible, nonstretchable (steel), approximately 0.7 cm wide, and easily retractable. Preferably, the tape should be in metric units on one side and inches on the other. Circumferences should be measured with the zero end of the tape held in the left hand (Lohman et al., 1990). In the taking of circumference measurements, the tape is positioned in a horizontal plane or perpendicular to the length of the segment being measured and should be placed directly on the skin. A mirror or assistant should be used to help ensure that the tape is horizontal or not accidentally caught on a piece of clothing. Placement of the tape for each specific measurement is important because inconsistent positioning reduces validity and reliability. Tension on the tape should be snug around the body part but not so tight that it compresses the subcutaneous fat layer. For trunk circumferences (shoulder, chest, waist, and abdomen), measurements should be taken at the end of a normal expiration. According to Pollock and Wilmore (1990), all trunk circumferences are taken at the end of expiration. Measures are recorded to the nearest 0.1 cm. Measures should be taken twice and agree within certain limits. The *ASRM* recommends that intra-intermeasure limits for circumferences should be within 1.0 cm for the shoulder, chest, abdomen, waist, and buttocks; 0.5 cm for the thigh; and 0.2 cm for the calf, ankle, wrist, arm, and forearm. If this standard is not met, the tester should repeat the measure twice.

The following paragraphs describe circumference measurements for specific sites.

• **Shoulder circumference.** The subject stands upright with arms by the sides and body weight evenly distributed. The tape is positioned horizontally at the maximal circumference of the shoulders at the level of the greatest protrusion of the deltoid muscles (figure 11.15).

• **Chest circumference.** The subject stands erect and abducts the arms so that the tape can be placed around the chest at the level of the fourth costosternal joint. The measurement is taken in a horizontal plane at the end of a respiration. After the tape is snugly in place, the arms are lowered to their natural position by the sides and the measurement is taken (figure 11.16).

• **Waist circumference.** The waist circumference measurement should not be made over clothing. The subject stands erect with arms by the sides, feet together, and abdomen relaxed. The tape is placed in a horizontal plane at the level of the narrowest part of the torso as seen from the anterior (front) aspect. The measurement is taken after a normal expiration (figure 11.17).

• **Abdomen circumference.** The abdomen circumference measurement also should not be taken

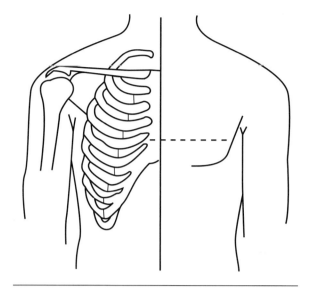

**Figure 11.16**   The level of the fourth costosternal joint for chest circumference measurements.

over clothing. The subject stands erect with arms by the sides, feet together, and abdomen relaxed. The tape is positioned horizontally at the level of the greatest anterior extension of the abdomen. This location is often, but not always, at the level of the umbilicus. Pollock and Wilmore (1990) recommend that the tape be positioned at the level of the umbilicus (figure 11.17). The measurement is taken at the end of a normal expiration.

• **Buttocks (gluteal or hip) circumference.** Ideally, the gluteal circumference measure is taken in a bikini (females) or supporter or swim brief

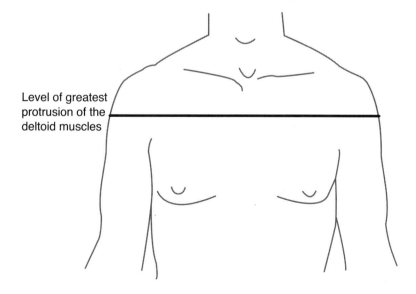

Level of greatest protrusion of the deltoid muscles

**Figure 11.15**   Tape placement for measurement of shoulder circumference.

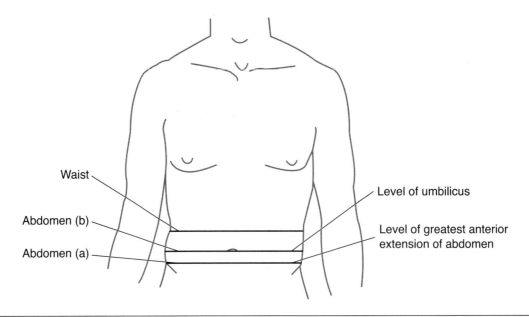

**Figure 11.17**   Tape replacement for measurement of waist circumference and abdominal circumference.
*(a)* Adapted from Lohman, Roche, and Martorell 1988; *(b)* Adapted from Pollock and Wilmore 1990.

(males). The subject stands erect with arms by the sides and feet together. The tape is placed in the horizontal plane at the level of maximum extension of the buttocks (figure 11.18*a*). Pollock and Wilmore (1990) recommend that the gluteal circumference be measured while the buttock muscles are tensed.

• **Thigh circumference (proximal).** With arms by the sides, the subject stands erect, feet approximately

10 cm apart, and weight evenly distributed. The tape is positioned horizontally immediately distal to the gluteal furrow (usually but not always the maximum circumference of the thigh) (figure 11.18*b*). Pollock and Wilmore (1990) recommend that the subject shift weight to the right leg and tense the thigh during this measurement. Midthigh and distal thigh measures are not shown but are described by Lohman et al. (1988).

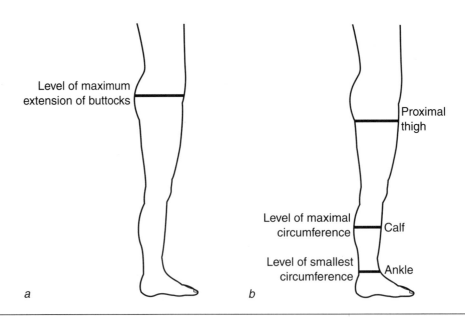

**Figure 11.18**   *(a)* Tape placement for measurement of gluteal circumference. *(b)* Lateral view of locations for thigh, calf, and ankle circumferences.

• **Calf circumference.** The subject either sits on a table so that the leg to be measured hangs freely or stands with the feet about 20 cm apart and weight evenly distributed. The tape is placed horizontally around the calf at the level of maximal circumference in a plane perpendicular to the long axis of the calf (figure 11.18b).

• **Ankle circumference.** The subject stands barefoot with feet slightly separated and weight distributed equally on both feet. The tape is placed at the smallest circumference of the lower leg just proximal to the malleoli in a plane perpendicular to the long axis of the calf (figure 11.18b).

• **Arm (biceps) circumference.** The subject stands erect with the arms hanging freely at the sides and the palms facing the thighs. The tape is positioned horizontally perpendicular to the long axis of the arm at the midpoint of the upper arm (figure 11.19a). To locate the midpoint, the subject's elbow is flexed at 90° with the palm facing superiorly. The measurer stands behind the subject and locates the lateral tip of the acromion by palpating laterally along the superior surface of the spinous process of the scapula. A small mark is made at the identified point. The most distal point of the acromial process is located and marked. A tape is placed so that it passes over the two marks, and the midpoint between them is marked. Pollock and Wilmore (1990) recommend that the measurement be taken at maximal girth of the midarm when flexed to the greatest angle, with the underlying muscles fully contracted (figure 11.19b).

• **Forearm circumference.** The subject stands erect with the arms hanging downward but slightly away from the trunk, with the palms facing anteriorly. The tape is placed around the proximal part of the forearm, perpendicular to its long axis, at the level of maximum circumference (figure 11.20a). Pollock and Wilmore (1990) recommend that the measurement be taken at the largest circumference, with the forearm parallel to the floor, the elbow joint at a 90° angle, the hand clenched and in the supinated position, and the muscles flexed (figure 11.20b).

• **Wrist circumference.** The subject stands erect and flexes the arm at the elbow so that the palm is uppermost and the hand muscles are relaxed. The tape is positioned perpendicular to the long axis of the forearm and in the same plane on the anterior and posterior aspects of the wrist. The tape must be no more than 0.7 cm wide so that it can fit into the medial and lateral depressions at this level (figure 11.20a).

## Prediction Equations for Determining $D_b$ From Anthropometric Measures

In 1951 Brozek and Keys (1951) published the first regression equations predicting $D_b$ using anthropometric techniques. Since then, more than 100 equations have been reported in the literature (Jackson & Pollock, 1982). Early equations were developed from

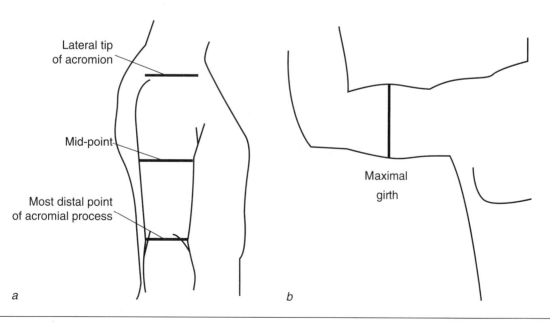

**Figure 11.19** Tape placement for measurement of biceps circumference.

*(a)* Adapted from Lohman, Roche, and Martorell 1988; *(b)* Adapted from Pollock and Wilmore 1990.

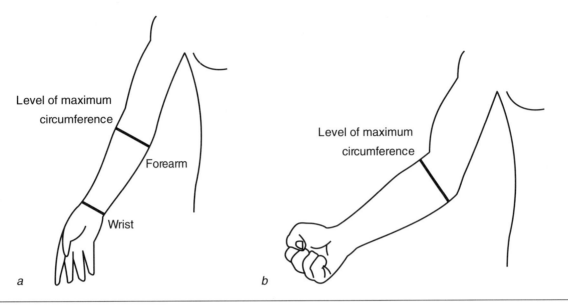

**Figure 11.20**   Tape placement for measurement of forearm and wrist circumferences.

*(a)* Adapted from Lohman, Roche, and Martorell 1988; *(b)* Adapted from Pollock and Wilmore 1990.

various combinations of skinfold fat measurements (Sloan, Burt, & Blyth, 1962; Young, Martin, Tensuan, & Blondin, 1962). In an attempt to produce prediction equations that were more accurate, researchers began adding body circumference and bone diameters as independent variables in the regression model (Katch, 1968; Katch & McArdle, 1973; Pollock et al., 1976, 1975; Sinning, 1978; Wilmore & Behnke, 1970; Young, 1964). These early equations were population specific; that is, they were developed from relatively small, homogeneous samples. Application of these equations to different populations has shown that age, gender, and degree of fatness are important sources of $D_b$ variation (Jackson & Pollock, 1982; Lohman, 1982; Pollock & Wilmore, 1990). Furthermore, these population-specific equations were erroneously based on the assumption that the relationship between hydrostatically determined $D_b$ and subcutaneous fat is linear. Figure 11.21 illustrates that this relationship is actually curvilinear. According to Pollock and Wilmore (1990), population-specific equations predict most accurately at the mean of the population in which the data were collected and the equation developed. As subjects differ from the mean, the standard error of measurement increases significantly (Jackson & Pollock, 1982).

To avoid the limitations of population-specific equations, generalized equations were developed from larger, heterogeneous samples. Unlike population-specific equations, which can be applied only to samples having similar age and physical characteristics, generalized equations can be used with samples varying greatly in age and body fatness (Jackson & Pollock, 1982). The main advantage of the generalized approach is that one equation replaces several without a loss in prediction accuracy (Jackson & Pollock, 1982; Pollock & Wilmore, 1990).

Durnin and Womersley (1974) were the first to use the generalized approach. They studied 272 women, ranging in age from 16 to 68 years, and developed five equations, one for each of the following age groups: 16 to 19, 20 to 29, 30 to 39, 40 to 49, and 50 to 68 years. All equations had a common slope, but the intercepts were adjusted to account for aging. The standard errors of the estimate for these equations ranged from 0.008 g/cc (the 50- to 68-year-old group) to 0.013 g/cc (for the 30- to 39-year-old group). As an extension of the work of Durnin and Womersley, Jackson, Pollock, and Gettman developed generalized equations for men (1978) and Jackson, Pollock, and Ward developed them for women (1980). To account for the potential changes in the ratio of internal to external fat and bone density with age, they added age to the prediction equation as an independent variable. Standard errors of the estimate for percent fat ranged from 3.4% to 3.9% for men and women, respectively.

The studies of Durnin and Womersley (1974) and Jackson, Pollock, and Ward (1978, 1980) realized the curvilinear nature of the data (figure 11.21) and therefore used logarithmic transformations and quadratic forms of various sums of skinfold thickness as independent variables. See table 11.4 for generalized equations for predicting $D_b$ in women and men.

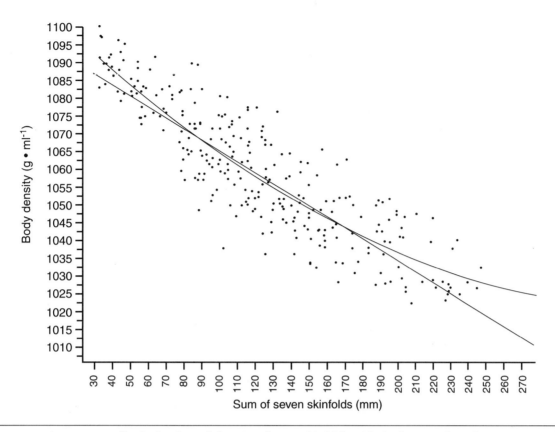

**Figure 11.21**   Scattergram of body density and the sum of seven skinfolds, with the linear and quadratic regression lines for adult men age 18 to 61 years.

Reprinted, by permission, from A.S. Jackson and M.L. Pollock, 1978, "Generalized equations for predicting body density in men," *British Journal of Nutrition* 40: 497-504.

For ease of determination of percentage of body fat, calculated percentages of body fat using age and the sum of chest, abdomen, and thigh skinfolds for men and the sum of triceps, suprailium, and thigh skinfolds for men and women are shown in tables 11.5 and 11.6, respectively. Note that these skinfolds were taken with Lange calipers.

Pollock and Wilmore (1990) note that the use of the quadratic component and age in generalized equations did not increase the correlation substantially over the linear version of the same equations. But the value of the generalized equations over the linear equations is that they minimize large prediction errors that occur at the extremes of the $D_b$ distribution. Several investigators have cross-validated the equations of Jackson and Pollock (Bulbulian, 1984; Jackson et al., 1988; Latin, 1987; Sinning & Wilson, 1984; Smith & Mansfield, 1984; Thorland, Johnson, Tharp, Fagot, & Hammer, 1984) and have shown them to predict $D_b$ accurately for men and women up to 60 years of age and 40% fat. Estimating body composition in the obese population has been difficult with most prediction equations

and methods, which significantly underpredict the values of percent fat (Pollock & Wilmore, 1990). Other methods are generally recommended in the obese population.

When studying obese populations, obtaining accurate skinfold measurements is often difficult for the following reasons: (1) The skinfold may exceed the maximum opening capacity of the caliper; (2) caliper tips may slide on larger skinfolds; and (3) readings tend to decrease with subsequent measurements, which might be associated with edema and repeated compression of the subcutaneous fat (Kuczmarski, Fanelli, & Koch, 1987). To overcome these limitations, Weltman et al. (1988, 1987) developed and cross-validated body composition prediction equations for obese men and women, using height, weight, and circumference measurements as predictor variables. They found that a combination of the mean of two abdominal girths, height, and weight accurately predicted relative fat:

obese males: % fat = 0.31457 (mean Abd) −
0.10969 (Wt) + 10.8336 ($R$ = .54, *SEE* = 2.9%)

Table 11.4    **Generalized Regression Equations for Predicting Body Density ($D_b$) for Adult Women and Men**

| Variables | Regression equation | r | SE ($D_b$) | SE (%F) |
|-----------|--------------------|-----|-----------|---------|
| **Adult women** | | | | |
| $\Sigma$ 7, age | $D_b = 1.0970 - 0.00046971\ (X_1) + 0.00000056\ (X_1)^2 - 0.00012828\ (X_6)$ | .85 | 0.008 | 3.8 |
| $\Sigma$ 3, age | $D_b = 1.0994921 - 0.0009929\ (X_2) + 0.0000023\ (X_2)^2 - 0.0001392\ (X_6)$ | .84 | 0.009 | 3.9 |
| $\Sigma$ 3, age | $D_b = 1.0902369 - 0.0009379\ (X_5) + 0.0000026\ (X_2)^2 - 0.0001087\ (X_6)$ | .84 | 0.009 | 3.9 |
| **Adult men** | | | | |
| $\Sigma$ 7, age | $D_b = 1.11200000 - 0.00043499\ (X_1) + 0.00000055\ (X_1)^2 - 0.00028826\ (X_6)$ | .90 | 0.008 | 3.5 |
| $\Sigma$ 3, age | $D_b = 1.1093800 - 0.0008267\ (X_3) + 0.0000016\ (X_3)^2 - 0.0002574\ (X_6)$ | .91 | 0.008 | 3.4 |
| $\Sigma$ 3, age | $D_b = 1.1125025 - 0.0013125\ (X_4) + 0.0000055\ (X_4)^2 - 0.0002440\ (X_6)$ | .89 | 0.008 | 3.6 |

*Note.* $X_1$ = Sum of seven skinfolds (mm); $X_2$ = sum of triceps, suprailium, and thigh skinfolds (mm); $X_3$ = sum of chest, abdomen, and thigh skinfolds (mm); $X_5$ = sum of chest, triceps, and subscapular skinfolds (mm): $X_5$ = sum of triceps, suprailium, and abdomen skinfolds (mm); $X_6$ = age in years.

Reprinted with permission. © American Society of Contemporary Medicine and Surgery. *Comp Therapy* 1980; 6(9): 12-27.

obese females: % fat = 0.11077 (mean Abd) − 0.17666 (Ht) + 0.14354 (Wt) + 51.03301 ($R = .76$, *SEE* = 2.9%)

(Mean Abd = the average of two circumferences shown as waist and abdomen *(b)* in figure 11.17, Wt = weight, Ht = height.)

# Bioelectric Impedance Analysis

Bioelectric impedance analysis (BIA) is based on the principle that the resistance to an applied electric current is inversely related to the amount of the fat-free mass (FFM) contained within the body. This relationship exists because FFM has greater water and electrolyte content, and therefore greater conductivity, than adipose tissue and bone do. The greater the FFM, the greater the conductivity and the lower the resistance. Relative body fat can be easily calculated from FFM by this formula: % fat = [(weight − FFM)/weight] 100.

The specific physical principles of BIA are somewhat complicated and beyond the scope of this chapter. Baumgartner, Chumlea, and Roche (1990) have written a detailed review of the theory and engineering behind BIA. Most studies using BIA to estimate body composition have been based on the equation $V = p\,L^2/R$, where $V$ represents the volume of the conductor, p is the specific resistivity of the tissue being analyzed, $L$ is the length of the conductor, and $R$ is the observed resistance (Baker, 1989). The volume calculated by the previous equation is assumed to represent total body water (TBW), which is highly correlated with FFM (Hoffer, Meador, & Simpson, 1969).

The application of the equation $V = p\,L^2/R$ to the estimation of body composition in humans has limitations. The assumption that the conductor is homogeneous with respect to composition, shape (cross-sectional area), and current density distribution is not the case in the human body (Rush, Abidskoo, & Fee, 1963). Stature *(S)*, not the actual length of the conductor (although stature and conductor length are highly correlated), is used as a measure of conductor length, and the specific resistivity (p) varies among individuals, based on the amount and distribution of tissues and fluids within the body (Rush et al., 1963). These limitations indicate that $S^2/R$ (the resistive index) does not have the same relationship to TBW or FFM in all individuals. In spite of these limitations, high correlations between $S^2/R$, TBW, and

Table 11.5  **Percent Body Fat Estimation for Men From Age and the Sum of Triceps, Suprailium, and Thigh Skinfolds**

| Sum of skinfolds (mm) | Age to the last year | | | | | | | | |
|---|---|---|---|---|---|---|---|---|---|
| | Under 22 | 23-27 | 28-32 | 33-37 | 38-42 | 43-47 | 48-52 | 53-57 | Over 57 |
| 8-10 | 1.3 | 1.8 | 2.3 | 2.9 | 3.4 | 3.9 | 4.5 | 5.0 | 5.5 |
| 11-13 | 2.2 | 2.8 | 3.3 | 3.9 | 4.4 | 4.9 | 5.5 | 6.0 | 6.5 |
| 14-16 | 3.2 | 3.8 | 4.3 | 4.8 | 5.4 | 5.9 | 6.4 | 7.0 | 7.5 |
| 17-19 | 4.2 | 4.7 | 5.3 | 5.8 | 6.3 | 6.9 | 7.4 | 8.0 | 8.5 |
| 20-22 | 5.1 | 5.7 | 6.2 | 6.8 | 7.3 | 7.9 | 8.4 | 8.9 | 8.5 |
| 23-25 | 6.1 | 6.6 | 7.2 | 7.7 | 8.3 | 8.8 | 9.4 | 9.9 | 9.5 |
| 26-28 | 7.0 | 7.6 | 8.1 | 8.7 | 9.2 | 9.8 | 10.3 | 10.9 | 11.4 |
| 29-31 | 8.0 | 8.5 | 9.1 | 9.6 | 10.2 | 10.7 | 11.3 | 11.8 | 12.4 |
| 32-34 | 8.9 | 9.4 | 10.0 | 10.5 | 11.1 | 11.6 | 12.2 | 12.8 | 13.3 |
| 35-37 | 9.8 | 10.4 | 10.9 | 11.5 | 12.0 | 12.6 | 13.1 | 13.7 | 14.3 |
| 38-40 | 10.7 | 11.3 | 11.8 | 12.4 | 12.9 | 13.5 | 14.1 | 14.6 | 15.2 |
| 41-43 | 11.6 | 12.2 | 12.7 | 13.3 | 13.8 | 14.4 | 15.0 | 15.5 | 16.1 |
| 44-46 | 12.5 | 13.1 | 13.6 | 14.2 | 14.7 | 15.3 | 15.9 | 16.4 | 17.0 |
| 47-49 | 13.4 | 13.9 | 14.5 | 15.1 | 15.6 | 16.2 | 16.8 | 17.3 | 17.9 |
| 50-52 | 14.3 | 14.8 | 15.4 | 15.9 | 16.5 | 17.1 | 17.6 | 18.2 | 18.8 |
| 53-55 | 15.1 | 15.7 | 16.2 | 16.8 | 17.4 | 17.9 | 18.5 | 19.1 | 19.7 |
| 56-58 | 16.0 | 16.5 | 17.1 | 17.7 | 18.2 | 18.8 | 19.4 | 20.0 | 20.5 |
| 59-61 | 16.9 | 17.4 | 17.9 | 18.5 | 19.1 | 19.7 | 20.2 | 20.8 | 21.4 |
| 62-64 | 17.6 | 18.2 | 18.8 | 19.4 | 19.9 | 20.5 | 21.1 | 21.7 | 22.2 |
| 65-67 | 18.5 | 19.0 | 19.6 | 20.2 | 20.8 | 21.3 | 21.9 | 22.5 | 23.1 |
| 68-70 | 19.3 | 19.9 | 20.4 | 21.0 | 21.6 | 22.2 | 22.7 | 23.3 | 23.9 |
| 71-73 | 20.1 | 20.7 | 21.2 | 21.8 | 22.4 | 23.0 | 23.6 | 24.1 | 24.7 |
| 74-76 | 20.9 | 21.5 | 22.0 | 22.6 | 23.2 | 23.8 | 24.4 | 25.0 | 25.5 |
| 77-79 | 21.7 | 22.2 | 22.8 | 23.4 | 24.0 | 24.6 | 25.2 | 25.8 | 26.3 |
| 80-82 | 22.4 | 23.0 | 23.6 | 24.2 | 24.8 | 25.4 | 25.9 | 26.5 | 27.1 |
| 83-85 | 23.2 | 23.8 | 24.4 | 25.0 | 25.5 | 26.1 | 26.7 | 27.3 | 27.9 |
| 86-88 | 24.0 | 24.5 | 25.1 | 25.7 | 26.3 | 26.9 | 27.5 | 28.1 | 28.7 |
| 89-91 | 24.7 | 25.3 | 25.9 | 26.5 | 27.1 | 27.6 | 28.2 | 28.8 | 29.4 |
| 92-94 | 25.4 | 26.0 | 26.6 | 27.2 | 27.8 | 28.4 | 29.0 | 29.6 | 30.2 |
| 95-97 | 26.1 | 26.7 | 27.3 | 27.9 | 28.5 | 29.1 | 29.7 | 30.3 | 30.9 |
| 98-100 | 26.9 | 27.4 | 28.0 | 28.6 | 29.2 | 29.8 | 30.4 | 31.0 | 31.6 |
| 101-103 | 27.5 | 28.1 | 28.7 | 29.3 | 29.9 | 30.5 | 31.1 | 31.7 | 32.3 |
| 104-106 | 28.2 | 28.8 | 29.4 | 30.0 | 30.6 | 31.2 | 31.8 | 32.4 | 33.0 |
| 107-109 | 28.9 | 29.5 | 30.1 | 30.7 | 31.3 | 31.9 | 32.5 | 33.1 | 33.7 |
| 110-112 | 29.6 | 30.2 | 30.8 | 31.4 | 32.0 | 32.6 | 33.2 | 33.8 | 34.4 |
| 113-115 | 30.2 | 30.8 | 31.4 | 32.0 | 32.6 | 33.2 | 33.8 | 34.5 | 35.1 |
| 116-118 | 30.9 | 31.5 | 32.1 | 32.7 | 33.3 | 33.9 | 34.5 | 35.1 | 35.7 |
| 119-121 | 31.5 | 32.1 | 32.7 | 33.3 | 33.9 | 34.5 | 35.1 | 35.7 | 36.4 |
| 122-124 | 32.1 | 32.7 | 33.3 | 33.9 | 34.5 | 35.1 | 35.8 | 36.4 | 37.0 |
| 125-127 | 32.7 | 33.3 | 33.9 | 34.5 | 35.1 | 35.8 | 36.4 | 37.0 | 37.6 |

Reprinted with permission. © American Society of Contemporary Medicine and Surgery. *Comp Therapy* 1980; 6(9): 12-27.

**Table 11.6** **Percent Body Fat Estimation for Women From Age and the Sum of Triceps, Suprailium, and Thigh Skinfolds**

| Sum of skinfolds (mm) | Age to the last year | | | | | | | | |
|---|---|---|---|---|---|---|---|---|---|
| | Under 22 | 23-27 | 28-32 | 33-37 | 38-42 | 43-47 | 48-52 | 53-57 | Over 57 |
| 23-25 | 9.7 | 9.9 | 10.2 | 10.4 | 10.7 | 10.9 | 11.2 | 11.4 | 11.7 |
| 26-28 | 11.0 | 11.2 | 11.5 | 11.7 | 12.0 | 12.3 | 12.5 | 12.7 | 13.0 |
| 29-31 | 12.3 | 12.5 | 12.8 | 13.0 | 13.3 | 13.5 | 13.8 | 14.0 | 14.3 |
| 32-34 | 13.6 | 13.8 | 14.0 | 14.3 | 14.5 | 14.8 | 15.0 | 15.3 | 15.5 |
| 35-37 | 14.8 | 15.0 | 15.3 | 15.5 | 15.8 | 16.0 | 16.3 | 16.5 | 16.8 |
| 38-40 | 16.0 | 16.3 | 16.5 | 16.7 | 17.0 | 17.2 | 17.5 | 17.7 | 18.0 |
| 41-43 | 17.2 | 17.4 | 17.7 | 17.9 | 18.2 | 18.4 | 18.7 | 18.9 | 19.2 |
| 44-46 | 18.3 | 18.6 | 18.8 | 19.1 | 19.3 | 19.6 | 19.8 | 20.1 | 20.3 |
| 47-49 | 19.5 | 19.7 | 20.0 | 20.2 | 20.5 | 20.7 | 21.0 | 21.2 | 21.5 |
| 50-52 | 20.6 | 20.8 | 21.1 | 21.3 | 21.6 | 21.8 | 22.1 | 22.3 | 22.6 |
| 53-55 | 21.7 | 21.9 | 22.1 | 22.4 | 22.6 | 22.9 | 23.1 | 23.4 | 23.6 |
| 56-58 | 22.7 | 23.0 | 23.2 | 23.4 | 23.7 | 23.9 | 24.2 | 24.4 | 24.7 |
| 59-61 | 23.7 | 24.0 | 24.2 | 24.5 | 24.7 | 25.0 | 25.2 | 25.5 | 25.7 |
| 62-64 | 24.7 | 25.0 | 25.2 | 25.5 | 25.7 | 26.0 | 26.7 | 26.4 | 26.7 |
| 65-67 | 25.7 | 25.9 | 26.2 | 26.4 | 26.7 | 26.9 | 27.2 | 27.4 | 27.7 |
| 68-70 | 26.6 | 26.9 | 27.1 | 27.4 | 27.6 | 27.9 | 28.1 | 28.4 | 28.6 |
| 71-73 | 27.5 | 27.8 | 28.0 | 28.3 | 28.5 | 28.8 | 29.0 | 29.3 | 29.5 |
| 74-76 | 28.4 | 28.7 | 28.9 | 29.2 | 29.4 | 29.7 | 29.9 | 30.2 | 30.4 |
| 77-79 | 29.3 | 29.5 | 29.8 | 30.0 | 30.3 | 30.5 | 30.8 | 31.0 | 31.3 |
| 80-82 | 30.1 | 30.4 | 30.6 | 30.9 | 31.3 | 31.4 | 31.6 | 31.9 | 32.1 |
| 83-85 | 30.9 | 31.2 | 31.4 | 31.7 | 31.9 | 32.2 | 32.4 | 32.7 | 32.9 |
| 86-88 | 31.7 | 32.0 | 32.2 | 32.5 | 32.7 | 32.9 | 33.2 | 33.4 | 33.7 |
| 89-91 | 32.5 | 32.7 | 33.0 | 33.2 | 33.5 | 33.7 | 33.9 | 34.2 | 34.4 |
| 92-94 | 33.2 | 33.4 | 33.7 | 33.9 | 34.2 | 34.4 | 34.7 | 34.9 | 35.2 |
| 95-97 | 33.9 | 34.1 | 34.4 | 34.6 | 34.9 | 35.1 | 35.4 | 35.6 | 35.9 |
| 98-100 | 34.6 | 34.8 | 35.1 | 35.3 | 35.5 | 35.8 | 36.0 | 36.3 | 36.5 |
| 101-103 | 35.3 | 35.4 | 35.7 | 35.9 | 36.2 | 36.4 | 36.7 | 36.9 | 37.2 |
| 104-106 | 35.8 | 36.1 | 36.3 | 36.6 | 36.8 | 37.1 | 37.3 | 37.5 | 37.8 |
| 107-109 | 36.4 | 36.7 | 36.9 | 37.1 | 37.4 | 37.6 | 37.9 | 38.1 | 38.4 |
| 110-112 | 37.0 | 37.2 | 37.5 | 37.7 | 38.0 | 38.2 | 38.5 | 38.7 | 38.9 |
| 113-115 | 37.5 | 37.8 | 38.0 | 38.2 | 38.5 | 38.7 | 39.0 | 39.2 | 39.5 |
| 116-118 | 38.0 | 38.3 | 38.5 | 38.8 | 39.0 | 39.3 | 39.5 | 39.7 | 40.0 |
| 119-121 | 38.5 | 38.7 | 39.0 | 39.2 | 39.5 | 39.7 | 40.0 | 40.2 | 40.5 |
| 122-124 | 39.0 | 39.2 | 39.4 | 39.7 | 39.9 | 40.2 | 40.4 | 40.7 | 40.9 |
| 125-127 | 39.4 | 39.6 | 39.9 | 40.1 | 40.4 | 40.6 | 40.9 | 41.1 | 41.4 |
| 128-130 | 39.8 | 40.0 | 40.3 | 40.5 | 40.8 | 41.0 | 41.3 | 41.5 | 41.8 |

FFM indicate that $S^2/R$ can be used as a reasonable predictor of TBW and FFM (Chumlea, Baumgartner, & Roche, 1988; Kushner & Schoeller, 1986; Lukaski, Bolonchuk, Hall, & Siders, 1986; Lukaski, Johnson, Bolonchuk, & Lykken, 1985).

To minimize variability associated with whole-body impedance, several investigators have predicted FFM using impedance measurements from various body segments (Chumlea et al., 1988; Chumlea, Roche, Guo, & Woynarowska, 1987). In some instances the sum of specific segment indices can predict FFM more accurately than $S^2/R$ can (Chumlea & Baumgartner, 1990). This approach is more complicated (less efficient) than making a single whole-body measurement. The segmental approach may be beneficial when accurate measures of stature are difficult to obtain or for estimating body composition in amputees (Vettorazzi, Barillas, Pineda, & Solomons, 1987).

Validation statistics for selected regression equations developed to predict TBW and FFM from $S^2/R$ are illustrated in table 11.7. Standard errors of prediction are ~2.5 L for TBW and range from 2.0 kg to 4.2 kg for FFM. The addition of certain anthropometric variables (skinfolds, circumferences), age, and gender can improve the statistical association between $S^2/R$ and body composition (Guo, Roche, Chumlea, Miles, & Pohlman, 1987). Some typical multivariate pre-diction equations and their associated validation statistics are presented in table 11.8.

## Equipment and Methodology

A variety of BIA analyzers are commercially available. They range in price from several hundred to several thousand dollars. The differences among BIA analyzers are primarily in the amount of computer hardware and software included. A basic low-cost BIA unit will generally display the measured impedance in ohms. More expensive units may calculate and display parameters such as relative body fat and include a printer to generate reports.

The technology required for BIA instrumentation is not complicated, and the more expensive BIA units do not necessarily provide a more accurate measure of impedance. When a BIA analyzer is used that calculates body composition parameters, validation of the algorithms (prediction equations) employed is critical. Reputable manufacturers should be willing to provide the user with the specific equations used and documentation of independently conducted validation studies. A BIA analyzer should have the capability of displaying impedance in ohms. This feature enables the user to choose from a variety of published prediction equations. The best prediction equation for a given situation might not be supplied

Table 11.7   **Selected Validation Statistics for Prediction of Body Composition From Stature$^2$/Resistance (S$^2$/R)**

| Authors | N | Age (yr) | Regression equation | R$^2$ | RMSE |
|---|---|---|---|---|---|
| **TBW** | | | | | |
| Lukaski et al. (1985) | 37 M | 19-42 | 2.03 + 0.63 (S$^2$/R) | .90 | 2.1 L |
| Kushner & Schoeller (1986) | 40 M & W | 32-54 | 0.83 + 0.714 (S$^2$/R) | .94 | 2.5 L |
| **FFM** | | | | | |
| Lukaski et al. (1985) | 37 M | 19-42 | 3.04 + 0.85 (S$^2$/R) | .96 | 2.6 kg |
| Lukaski et al. (1986) | 84 M | 18-50 | 5.21 + 0.83 (S$^2$/R) | .96 | 2.5 kg |
| | 67 W | | 4.92 + 0.82 (S$^2$/R) | .91 | 2.0 kg |
| Chumlea et al. (1988) | 24 boys | 9-17 | −1.23 + 0.92 (S$^2$/R) | .88 | 4.0 kg |
| | 26 girls | 9-17 | −1.38 + 0.96 (S$^2$/R) | .84 | 4.2 kg |
| | 28 M | 18-62 | 3.50 + 0.87 (S$^2$/R) | .81 | 2.9 kg |
| | 44 W | 18-62 | 11.55 + 0.69 (S$^2$/R) | .80 | 2.7 kg |
| Cordain, Whicker, & Johnson (1988) | 30 boys & girls | 9-14 | 6.86 + 0.81 (S$^2$/R) | .69 | 4.1 kg |

M = men; W = women; *RMSE* = root mean squared error.

Reprinted, by permission, from R. Baumgartner, W. Chumlea, and A. Roche, 1990, Bioelectrical impedance for body composition. In *Exercise and Sport Sciences Reviews*, edited by K. Pandolf and J. Holloszy (Baltimore: Williams & Wilkins), 193-224.

Table 11.8   **Selected Equations, Cross-Validated, for Prediction of Body Composition Variables From Impedance, Anthropometric Variables, Age, and Gender**

| Authors | Age (yr) | N | Regression equations | $R^2$ | RMSE |
|---|---|---|---|---|---|
| **TBW** | | | | | |
| Kushner & Schoeller (1986) | 31.8-53.7 | 20 M | $0.396 (x_1) + 0.143 (x_2) + 8.399$ | .95 | 1.7 L |
| | (means for groups differing by obesity and sex) | 20 W | $0.382 (x_1) + 0.105 (x_2) + 8.315$ | .95 | 0.8 L |
| Lukaski & Bolonchuk (1988) | 20-73 | 110 M + W | $0.377 (x_1) + 0.14 (x_2) - 0.08 (x_3) + 2.9 (x_4) + 4.65$ | .97 | 1.5 L |
| **FFM** | | | | | |
| Guo, Roche, & Houtkooper (1989) | 20-73 | 140 M | $-2.93 + 0.646 (x_2) - 0.116 (x_5) - 0.375 (x_6) + 0.475 (x_7) + 0.156 (x_1)$ | .98 | 2.3 kg |
| | | 110 W | $4.34 + 0.682 (x_2) - 0.185 (x_5) - 0.244 (x_8) - 0.202 (x_9) + 0.182 (x_1)$ | .95 | 2.2 kg |
| Graves et al. (1989) | 23.4 ± 2.8 | 46 M | $0.485 (x_1) + 0.338 (x_2) + 5.32$ | .75 | 3.8% |
| | 22.8 ± 3.2 | 46 M | $0.475 (x_1) + 0.295 (x_2) + 5.49$ | .71 | 3.6% |
| **%fat** | | | | | |
| Guo et al. (1987) | 18-30 | 77 M | $1.50 - 0.279 (x_1) + 0.632 (x_8) + 0.346 (x_2)$ | .74 | 3.3% |
| | | 71 M | $-8.48 + 0.434 (x_{10}) + 1.341 (x_{11}) - 0.845 (x_1) + 0.384 (x_2)$ | .81 | 3.2% |

Note. M = men; W = women; $x_1$ = stature$^2$/resistance (cm$^2$/ohms); $x_2$ = weight (kg); $x_3$ = age (yr); $x_4$ = gender (males = 1, females = 0); $x_5$ = lateral calf skinfold (cm); $x_6$ = midaxillary skinfold (cm); $x_7$ = arm circumference (cm); $x_8$ = triceps skinfold (cm); $x_9$ = subscapular skinfold (cm); $x_{10}$ = biceps skinfold (cm); $x_{11}$ = calf circumference (cm); RMSE = root mean squared error.

[a] Cross-validation statistics are for percent fat.

by the manufacturer. In addition, BIA practitioners may want to take advantage of new and improved equations generated by continued research.

One concern when selecting an equation to predict body composition from BIA is the type of instrument on which the equation was developed. Prediction equations are most accurate when used with data collected on the instrument used to develop the equation (Graves, Pollock, Colvin, Van Loan, & Lohman, 1989). Therefore, a prediction equation developed on one type of BIA analyzer should not be used with a different analyzer unless the impedance measure can be corrected to account for the difference.

The methodology employed for whole-body BIA is relatively simple. The subject to be measured lies supine on a nonconductive surface. The legs should be abducted so that the thighs do not touch. The arms should also be abducted slightly so that they do not contact the torso. A pair of electrodes (aluminum foil "spot" electrodes are often used) is placed on the ankle and foot, and a second pair is placed on the wrist and hand (figure 11.22). Because bioelectric impedance is related to the length of the conductor, proper placement of the reference electrodes is critical. The most frequently cited landmarks for the placement of the reference electrodes are the distal condyles of radius and ulna (for the wrist) and the most prominent portions of the malleoli of the tibia and fibula (for the ankle) (Van Loan, 1990). The leading edge (superior linear border) of the reference electrodes should be centered on an imaginary line connecting centers of (bisecting) these bony protrusions. Displacement of the electrodes even slightly from the specified landmarks can result in relatively large changes in observed resistance and error in the estimation of body composition (Elsen, Siu, Pineda, & Solo-

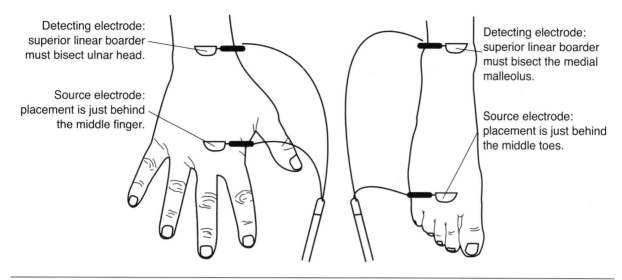

Detecting electrode: superior linear boarder must bisect ulnar head.

Source electrode: placement is just behind the middle finger.

Detecting electrode: superior linear boarder must bisect the medial malleolus.

Source electrode: placement is just behind the middle toes.

**Figure 11.22**   Placement of source and detecting (reference) electrodes (tetrapolar configuration).

mons, 1987; Schell & Gross, 1987). Cleaning the skin where the electrodes will be placed is crucial to ensuring that no interference occurs on the skin. Each pair of electrodes contains a source and a reference (or detecting) electrode. A low-level electrical current (typically 800 μA at a frequency of 50 kHz) is introduced at the source electrodes, and the voltage drop between the reference electrodes is measured as the bioelectric resistance or impedance *(R)*.

As previously mentioned, stature is generally used to represent conductor length. This approximation is reasonable because stature and actual conductor length (the distance between the source and reference electrodes) are highly correlated. The accurate prediction of body composition from BIA therefore requires an accurate measure of stature.

Bioelectric impedance is systematically greater on the left side of the body by 8 to 10 ohms (Graves et al., 1989). Thus, impedance must be measured on the side for which the prediction equation was developed. For the purpose of the standardization, BIA measurements are usually made on the right side of the body. As stated earlier, stature represents conductor length (the distance between the source and reference electrodes); thus, an alteration of conductor length will affect the accuracy of the prediction of body composition. This can create problems with some of the newer BIA equipment that requires the individual either to hold the machine in both hands or to stand on the machine and use both feet as the connecting points.

BIA is not extremely sensitive to age or gender. Although prediction errors of age- and gender-specific regression equations are lower than prediction errors for equations developed on the entire population (Deurenberg, Kusters, & Smit, 1990), the improvement in accuracy following the addition of age and gender to prediction equations is usually small (<2.0 kg of FFM) (Baumgartner et al., 1990). Of more critical importance than age and gender is the hydration status of the subject (Baumgartner et al., 1990). Controlling for factors that can affect total body water is important. Because body tissues tend to dehydrate with age, current prediction equations may not be valid when applied to the elderly.

Standardizing the use of diuretics (Zebatakis et al., 1987) and other drugs that influence hydration status is prudent. BIA measurements should not be taken after strenuous exercise or other conditions that may induce fluid loss (Deurenberg, Weststrate, & van der Kooy, 1989). Although there is no evidence of cyclical changes in bioelectric impedance associated with menstruation (Chumlea et al., 1987), BIA measurements should not be taken during the week before menses to avoid variability that may be associated with water retention.

Although diurnal variations in BIA have been negligible in some studies, reductions in impedance over the course of a day have been reported (Cugini, Salandri, Petrangeli, Capodaglio, & Giovannini, 1996). The change observed by Pollock et al. was twice the magnitude of change seen following exercise-induced dehydration (–9 ohms) in the same study. Diurnal changes in bioelectric impedance may be related to daily shifts in fluid volumes (Chumlea et al., 1987). To get the most accurate results, serial measurements on the same individual should occur at the same time of day. Changes in hydration status can create difficulties when using this equipment in longitudinal studies.

Bioelectric impedance measures are temperature dependent (Caton, Mole, Adams, & Heustis, 1988; Heyward & Stolarczyk, 1996). Temperature-related effects on BIA may be due to changes in fluid volume or distribution associated with changes in temperature. BIA measurements should be made at a comfortable and unchanging ambient temperature. The prediction of body composition from BIA in individuals of different ethnic origins requires use of different and carefully selected prediction equations. See Heyward and Stolarczyk (1996) for a summary of the BIA equations.

## Reliability and Cross-Validation

Bioelectric impedance measurements are highly reliable for both interobserver and intraobserver comparisons (Elsen et al., 1987; Graves et al., 1989; Jackson et al., 1988; Kushner & Schoeller, 1986; Schell & Gross, 1987; Siu, Elsen, Mazariegos, Solomons, & Pineda, 1987). Reliability coefficients are typically $r \geq .990$. Lukaski, Bolonchuk, Johnson, Lykken, and Sandstead (1984) reported $r = .999$ for a single measurement, $r = .995$ for measurements taken 2 h apart, and $r = .99$ for measurements over 5 d (Lukaski et al., 1985). When BIA prediction equations are validated on independently selected populations (cross-validations), results have been variable. For example, cross-validation of the equation developed by Lukaski et al. (1985) for predicting relative fat in men has yielded statistics of $r_{yy}' = .93$, $SEE = 2.7\%$ (Lukaski et al., 1986); $r_{yy}' = .87$, $SEE = 4.8\%$ (Graves et al., 1989); and $r_{yy}' = .71$, $SEE = 6.3\%$ (Jackson et al., 1988). Segal, Gutin, Presta, Wang, and Van Itallie (1985) have also reported a relatively high $SEE$ (6.1%) for predicting relative fat with BIA using a cross-validation experimental design. The relatively high prediction errors reported for the BIA technique have been a primary concern regarding its use as a field method of predicting body composition. The most recent studies, however, have been more favorable and indicate that BIA can predict body composition with the same degree of accuracy as anthropometric techniques (Campos, Chen, & Meguid, 1989; Kushner & Haas, 1988). BIA has been shown to be valid for a variety of populations, including children (Cordain, Whiker, & Johnson, 1988; Deurenberg et al., 1990), athletes (Lukaski, Bolonchuk, Siders, & Hall, 1990), and certain patient populations (Hannan, Cowen, Freeman, & Shapiro, 1990; Shols, Wouters, Soeters, & Westerterp, 1991).

# Ultrasound

During the 1950s, the meat industry was using ultrasound to determine body composition of livestock (Claus, 1957; Dumont, 1957, 1959). In the 1960s, researchers began applying the ultrasound method to the assessment of human body composition (Bullen, Quaade, Oleson, & Lund, 1965; Sloan, 1967). The first human studies used the A-scan mode, which records depth readings of changes in tissue density. Advances in technology led to the development of the B-scan mode, which provides two-dimensional, cross-sectional images of tissue configuration.

Cromwell, Weibell, and Pfeiffer (1980) provided the principles of ultrasound B. Ultrasound is sonic energy at frequencies above the audible range (greater than 20 kHz). Ultrasound exists as a sequence of alternate compressions and rarefactions of a suitable medium (e.g., tissue, bone) and is propagated through that medium at some velocity. Its behavior depends on the frequency (wavelength) of the sonic energy and the density and compliance of the medium through which it travels. Ultrasound can be focused into a beam and therefore obeys the laws of reflection and refraction. Whenever the ultrasound beam passes from one medium to another, a portion of the sonic energy is reflected and the remainder is refracted. The amount of energy depends on the difference in density between the two media and the angle at which the transmitted beam strikes the medium. The greater the difference in media, the greater the amount reflected. Also, the nearer the angle of incidence between the beam and the interface is to 90°, the greater will be the reflected portion. At interfaces of extreme differences in medium, such as between muscle (density = 1.07 g/cc) and bone (density = 1.77 g/cc), almost all the energy will be reflected and practically none will continue through the second medium.

The velocity of sound propagation through a given medium varies with its density and elastic properties as well as the temperature. As a general rule, the greater the density, the greater the velocity. For example, fat has a density of 0.90 g/cc and a velocity of 1,440 m/s, whereas muscle, which has a density of 1.07 g/cc, has a velocity of 1,570 m/s. As ultrasound travels through a material, some of the energy is absorbed and the wave is attenuated a certain amount for each centimeter through which it travels. The amount of attenuation is a function of both the frequency of the ultrasound and the characteristics of the material. Attenuation increases with higher frequencies. Therefore, the higher the frequency, the less distance it can penetrate into the body. For this reason, lower frequencies are used for deeper penetration.

## A-Mode Versus B-Mode Ultrasound

With A-mode ultrasound, a pulse of mechanical vibration is sent out from the sound source (probe) into the subject and is partially reflected by an interface where the specific acoustic impedance of the medium changes (e.g., skin–fat, fat–muscle, and muscle–bone interfaces). The reflected pulse returns along its original path back to the source. The time the echo takes to return gives a measure of the depth of the interface from the surface of the sound source. This information is presented visu-

ally on an oscilloscope screen. Because the trace travels across the screen uniformly with time, the distance along the horizontal trace from the zero point to the vertical deflection is proportional to the time interval for ultrasonic waves to pass from the probe to the reflection interface and back again (Bullen et al., 1965). To convert this distance into actual length, the sound velocity of human fat and muscle must be known (1,440 m/s for fat and 1,570 m/s for muscle). Correlating the individual echoes seen along the line with the surfaces that produced them is often difficult (Newell, 1961). B-mode ultrasound overcomes this difficulty because it allows one to visualize an interface in two dimensions and to see its relation to other interfaces more easily (figure 11.23). This is done by coupling the transducer (probe) to an echo camera that converts the reflected impulses into a picture of the internal tissue structures.

## Equipment and Methodology

Necessary equipment includes a transducer (5 MHz), echo camera, videographic printer, Vernier caliper, and ultrasound transmission gel. The transducer is coupled with the echo camera, which is interfaced to the videographic printer. With B-mode ultrasound, a cross-sectional image of the underlying tissues is produced and printed from the graphic printer. The Vernier caliper is then used to measure the fat layer (distance between skin–fat interface and fat–muscle interface) and the muscle layer (distance between the fat–muscle interface and muscle–bone interface) to the nearest 0.1 mm. Figure 11.24 shows a diagrammatic view of the various interfaces of the arm (biceps) and an actual B-mode ultrasonic view. A small amount of ultrasound transmission gel is applied to the head of the transducer before taking measurement to assure airless contact with the skin.

Subjects should stand with feet comfortably apart while the measurements are taken. The transducer with gel is then placed gently (to avoid compression of tissue) over the desired sites, perpendicular to the bone. The gain knob and distance controls are used to focus the image on the echo camera. When the image is clear, it is frozen on the screen of the echo camera by pressing the freeze button of the videographic printer.

The transducer is placed on the same sites used for skinfolds. The transducer is held horizontally for all sites except the subscapular and medial calf, for which the transducer is held vertically. The main

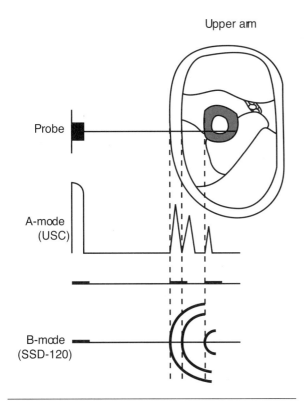

**Figure 11.23** Illustration of results obtained from A-mode and B-mode ultrasound. A-mode records depth readings of changes in tissue density; B-mode produces a two-dimensional, cross-sectional image of underlying tissue configuration.

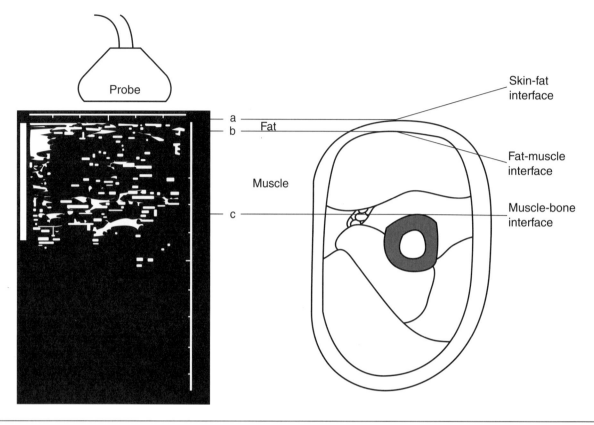

**Figure 11.24**   Thicknesses of the fat (distance between a and b) and muscle (distance between b and c) layers of the biceps as viewed by B-mode ultrasound.

advantages of the ultrasound technique as compared with the traditional caliper method are (1) no tissue compression occurs, (2) palpating the fat–muscle interface is not necessary, (3) larger fat layers can be easily measured, and (4) ultrasound also provides information on muscle thickness.

## Validity and Reliability of Ultrasound

Early studies established the validity of the ultrasound method to estimate subcutaneous fat thickness in humans. Bullen et al. (1965) compared A-mode ultrasonic determinations of subcutaneous fat (abdominal site) to fat thickness measurements obtained by direct needle puncture ($N$ = 13). The correlation between the two methods was $r$ = .97 over a range of 3 to 40 mm of fat, indicating good agreement. Reliability coefficients of correlation between duplicate ultrasonic measurements ($N$ = 100) were also high (triceps, $r$ = .98; subscapular, $r$ = .99; abdominal, $r$ = .99). Booth, Goddard, and Paton (1966) also used the abdominal site to compare the A-mode ultrasound technique with electrical conductivity methods ($N$ = 20). The correlation between these two methods was $r$ = .98 (standard error = ±0.24 mm).

The results of Ishida et al. (1990) also confirmed the high reliability of the ultrasound method. Thirty volunteers (17 men, 13 women) were measured at 14 sites (forearm, biceps, triceps, axilla, subscapular, lumbar, chest [men only], abdomen, suprailiac, quadriceps, hamstrings, suprapatellar, and medial and posterior calf) on 2 separate days. Four identical ultrasound images were printed for each site, and two investigators each measured two of the images. Intratester reliability coefficients ranged from $r$ = .96 to $r$ = .99, and intertester reliability was greater than .90 at 11 of the 14 sites for fat and at 8 of the 11 sites for muscle. The abdomen, lumbar, and hamstrings sites yielded inconsistent between-day correlations for both fat and muscle. Generalizability theory was used to determine the contribution from testers, days, trials, and subjects to the total variance. For the fat and muscle measurements at each site, less than 2% of the variability was due to the effect of testers, days, and trials. Generalizability coefficients of at least $r$ = .92 were obtained for all muscle measurement sites, whereas coefficients for fat measurements exceeded $r = .90$ for all but the axilla site.

The validity of B-mode underlying ultrasound was investigated by comparing ultrasonic fat thicknesses (at the suprailiac and triceps sites) with fat thicknesses obtained by soft-tissue roentgenograms ($N = 37$) (Haymes, Lundegren, Loomis, & Buskirk, 1976). Correlations between fat measurements for these two techniques were considered moderate to high (suprailiac, $r = .78$; triceps, $r = .88$), but were not as high as the correlations obtained previously (Booth et al., 1966; Bullen et al., 1965).

Using direct cadaver analysis, the validity of B-mode ultrasound was evaluated by comparing ultrasonically determined fat and muscle thicknesses of the upper arm, thigh, and abdomen with thicknesses obtained by cadaver analysis (Fukunaga et al., 1989). These three sites were ultrasonically scanned and surgically cut open, and fat and muscle thicknesses were measured directly with slide calipers. The mean fat thickness was $5.0 \pm 0.3$ mm, according to direct cadaver analysis (D-method), compared with $4.8 \pm 0.3$ mm as determined by the ultrasound method (U-method). This corresponds to a D – U/D ratio of 2.7%. The mean muscle thickness was $17.5 \pm 1.3$ mm (D-method), compared with a mean value of $17.0 \pm 1.3$ mm (U-method) (D – U/D ratio = 1.8%). These investigators concluded that there were no significant differences in muscle and fat thickness between D and U methods. Therefore, B-mode ultrasound is a valid technique for measuring subcutaneous fat and muscle thickness.

## Prediction Equations

Few researchers have attempted to predict body composition from ultrasound measurements. The earliest studies used A-mode ultrasound (Borkan, Halts, Cardarelli, & Burrows, 1982; Sloan, 1967) and showed moderate to good prediction estimates of $D_b$ from ultrasonically determined skinfold fat ($R = .81$ and $R = .75$, respectively). Fanelli and Kuczmarski (1984) developed three prediction equations ($N = 124$) for $D_b$ using two B-mode ultrasonic fat measurements: waist and thigh ($R = .81$, SEE = 0.0078 g/cc), triceps and waist ($R = .76$, SEE = 0.0085 g/cc), and triceps and thigh ($R = .76$, SEE = 0.0086 g/cc). Volz and Ostrove (1984) studied college-aged women ($N = 66$) and derived two regression equations for predicting $D_b$ from the log transformation of B-mode ultrasound fat measurements: suprailiac ($R = .74$, SEE = 0.0074 g/cc) and thigh and suprailiac ($R = .78$, SEE = 0.0069 g/cc). Kuczmarski et al. (1987) studied obese males ($n = 13$) and females ($n = 31$) between the ages

of 26 and 69 years. A subject was considered obese if weight-for-height exceeded by at least 20% the upper values listed for a large frame in the 1983 Metropolitan height–weight tables. They developed a regression equation for $D_b$ based on three fat variables: thigh, biceps, and the thigh $\times$ biceps cross-product ($R = .82$, SEE = 0.0095 g/cc).

Garzarella et al. (1991) developed and cross-validated generalized prediction equations for $D_b$ and FFM using the B-mode ultrasound technique. They studied a sample of 254 women who varied greatly in age (range = 18 to 74 years) and percent fat (range = 9.2% to 52.8%). The "best" $D_b$ equation was the quadratic form of the sum of four ($\Sigma4$) fat thicknesses (anterior thigh, abdomen, biceps, and triceps) and age [$D_b = 1.08812 - 0.00085\ (\Sigma4) + 0.000002\ (\Sigma4)^2 - 0.00031\ (\text{age})$]; $R = .93$, SEE = 0.0070 g/cc. The sum of five ($\Sigma5$) muscle thicknesses (anterior forearm, biceps, triceps, abdomen, and anterior thigh) with age, height, and weight yielded the "best" FFM equation [FFM = $- 44.417 + 0.14276\ (\Sigma5) - 0.06426\ (\text{age}) + 0.37284\ (\text{Ht}) + 0.18147\ (\text{Wt})$]; $R = .92$, SEE = 2.4 kg. These later correlations and standard errors are comparable with results obtained from anthropometric prediction equations.

## Sources of Error

Although these data show a reasonably high reliability and validity of the ultrasound technique, certain limitations still exist (Lukaski, 1987). The appropriate signal frequency has not been well defined. A range of 2.5 to 7.5 MHz has been reported in the literature, with the best predictive accuracy associated with the highest frequency. Another difficulty is the need for uniform and constant pressure when applying the probe to the scan site (Haymes et al., 1976). Changes in pressure by probe application can affect the distribution of adipose tissue and prejudice the ultrasonic determination of adipose thickness. Finally, the landmarks used for measurement sites need to be well defined and standardized so that published prediction equations can be used with better accuracy.

## Dual-Energy Projection Methods

Dual-energy projection methods have been used for over a decade to measure bone and soft-tissue composition in vivo (Mazess, Barden, Bisek, & Hanson, 1990). There are two types of dual-energy projection methods. Dual-photon absorptiometry (DPA) is based on the differential attenuation by tissues of transmitted photons at two energy levels (Witt & Mazess, 1978). DPA uses a gadolinium radionuclide ($^{153}$Gd) and has been widely used for the measurement of regional bone mineral density (BMD) and bone mineral content (BMC), particularly of the spine and proximal femur (Mazess et al.,1990).

Dual-energy x-ray absorptiometry (DEXA) is a noninvasive radiologic projection technique. The energy source for DEXA is x-ray rather than the gadolinium used in DPA. This technological change has improved the ability to quantify various parameters of body composition; radiation exposure is minimal, evaluation time is reduced, and precision is improved by enhanced resolution. A Hologic QDR-4500 (Waltham, Massachusetts) is shown in figure 11.25. DPA and DEXA instruments differentiate body weight into three chemical compartments—mineral-free lean soft tissue, fat soft tissue, and bone (Heymsfield & Waki, 1991)—and have the ability to distinguish regional as well as whole-body parameters of body composition. Figure 11.26 illustrates BMD results of total body scan and the body composition measurements. DEXA correlates well with hydrostatic weighing and multicompartment models and is a precise measure of body composition (Clasey et al., 1997; Kohrt, 1998; Salamone et al., 2000; Wang et al., 1998).

Unlike hydrostatic densitometry, DEXA is not limited by the assumptions associated with the two-compartment constant-density model. Tissue densities are measured directly and are differentiated. Because the bone compartment is measured, as well as the lean and fat compartments, it is considered a three-compartment model. But it does not separate out a water measurement; thus, this technique has limited utility in a population that would have changes in water. Both hydrostatic weighing and DEXA assume a constant hydration state for the FFM of all individuals (73.2%). Although good agreement in percent fat estimates is found by hydrodensitometry and DEXA (Hologic QDR-1000, Waltham, Massachusetts) in young people, in older individuals discrepancies between methods were reported (Snead, Birge, & Kohrt, 1993). Potential problems with assessing percent fat with DEXA in older adults were noted when they manipulated the FM by placing lard over the center and peripheral regions of the body. DEXA correctly assessed the additional fat mass when it was positioned over the lower body (DEXA identified 96% of the additional fat mass) but correctly identified only 55% of the exogenous fat when it was overlying the trunk region. Soft-tissue evaluation is possible only in those pixels that do not contain bone mineral, and this circumstance may have resulted in the inaccurate assessment of soft tissue (Snead et al., 1993). Thus, in those individuals with increased abdominal adipose tissue, DEXA may underestimate percent fat. Similarly, findings have been reported with a Lunar DPX-1 (Milliken, Going, & Lohman, 1996). Exogenous fat mass was underestimated at both the thigh and the abdomen, with a greater underestimation at the abdomen (Milliken et al., 1996). Many suggest that DEXA should be the criterion measure for body composition, but as mentioned earlier, it has limitations. These limitations include that DEXA depends on geometric models, that constant hydration in fat-free soft tissue is assumed, and that not all DEXA machines give the same results (Lantz, Samuelson, Bratteby, Mallmin, & Sjostrom, 1999; Wang et al., 1998). The equations in the DEXA machines need to be refined using the multicompartment models (Wang et al., 1998).

Reliability and validity of DPA (Heymsfield, Wang, & Aulet, 1990; Peppler & Mazess, 1981) and DEXA (Haarbo, Gotfredsen, Hassager, & Christiansen, 1991; Mazess et al., 1990) are well established. Precision errors are less than 3% for fat and 1.1 kg and 30 g for muscle and bone, respectively (Haarbo et al., 1991). Correlations between body composition parameters as measured by DPA and other methods such as hydrodensitometry and neutron activation analysis are typically greater than $r = .90$ (Heymsfield et al., 1989; Heymsfield, Wang, Heshka, Kehayias, &

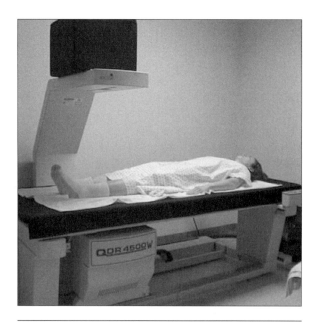

**Figure 11.25** Hologic QDR 4500 (Waltham, Massachusetts) bone density scanner.

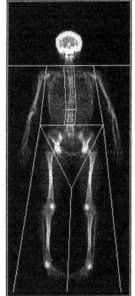

## MEDICAL IMAGING CENTER

C03220316      Sat Mar 22 11:31 2003

Name:
Comment:
I.D.:                        Sex:      F
S.S.#:        – –           Ethnic:   W
ZIP Code:                   Height: '    "
Operator:           CR      Weight:
BirthDate: 02/15/78         Age:      25
Physician:
Image not for diagnostic use

TOTAL BMC and BMD CV is < 1.0%
C.F.    1.029    1.009    1.000

| Region | Area (cm2) | BMC (grams) | BMD (gms/cm2) |
|---|---|---|---|
| L Arm | 150.23 | 114.73 | 0.764 |
| R Arm | 164.67 | 130.95 | 0.795 |
| L Ribs | 109.65 | 76.71 | 0.700 |
| R Ribs | 113.94 | 72.68 | 0.638 |
| T Spine | 119.79 | 114.01 | 0.952 |
| L Spine | 39.41 | 50.57 | 1.283 |
| Pelvis | 236.85 | 298.17 | 1.259 |
| L Leg | 291.87 | 357.45 | 1.225 |
| R Leg | 298.90 | 350.73 | 1.173 |
| SubTot | 1525.31 | 1566.01 | 1.027 |
| Head | 237.24 | 569.10 | 2.399 |
| TOTAL | 1762.55 | 2135.10 | 1.211 |

oMar 22 11:41 2003   [318 x 150]
Hologic QDR-4500W (S/N 47730)
Whole Body V8.26a:5

## HOLOGIC

## MEDICAL IMAGING CENTER

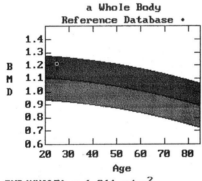

a Whole Body
Reference Database •

BMD(WHOLE) = 1.211 g/cm$^2$

| T(20.0) | Z |
|---|---|
| +1.26 110% | +1.34 111% |

C03220316      Sat Mar 22 11:31 2003

Name:
Comment:
I.D.:                        Sex:      F
S.S.#:        – –           Ethnic:   W
ZIP Code:                   Height: '    "
Operator:           CR      Weight:
BirthDate: 02/15/78         Age:      25
Physician:

• Age and sex matched
T = peak BMD matched
Z = age matched          PS      10/25/91

## HOLOGIC

**Figure 11.26** Results of a total body scan on a 25-year-old female. The results also include her body composition measures.

Reprinted, by permission, from Hologic, Inc.

Pierson, 1989; Lichtman, Heymsfield, & Kehayias, 1990). In a recent study that evaluated the interday reliability of 13 bone density parameters measured by DEXA, correlation coefficients were $r = .90$ to $r = .99$, and *SEE* ranged from 0.01 to 0.08 $g/cm^2$, which represented less than 4% of mean density values (Tucci et al., 1991).

Unfortunately, dual-energy projection methods are expensive (cost can exceed $75,000) and in many states require trained radiology personnel. These considerations limit the applicability of DEXA and DPA to clinical and research laboratory settings.

# Isotopic Dilution

Water makes up 40 to 60% of body weight (Schoeller, 1996) and is associated predominantly with fat-free tissue (FFM). Using isotopic dilution techniques, total body water (TBW) can be measured to provide an estimate of FFM and consequently percent body fat (Pace & Rathbun, 1945). According to the dilution principle (Edelman, Olney, & James, 1952), the concentration of a compound in a solvent depends on the volume of the solvent and the amount of compound added to it. Thus if the concentration and amount of the compound (isotope) are known, the solvent (body water) can be calculated (Wagner et al., 2000). The isotopic dilution methods to be discussed provide information on total body water, but are not useful to measure changes in intracellular and extracellular water.

## Equipment and Methodology

For determination of TBW, the volume of a compartment is defined as the ratio of the dose of a tracer, administered orally or intravenously, to its concentration in the body compartment after a short equilibrium period. For selection of an isotopic tracer, it must distribute only in body water and evenly throughout it, be nonmetabolizable, and be nontoxic (Schoeller, 1991). Several tracers meet this specification with tritiated water ($^3H_2O$) and $D_2O$ being the most popular (Sheng & Huggins, 1979); $D_2O$ is most commonly used in children because it is not radioactive. Samples of $D_2O$, however, are more difficult to process than are those with $^3H_2O$, and it is more expensive. A tracer dose of labeled water ($^3H_2O$, $D_2O$, or $^{18}O$) is used, and two body fluid samples are collected, one predose and the second after an equilibration time of ~2 to 3 h (Ellis, 2000). The method of analysis of the sample depends on the choice of tracer: radioactive β-counting for tritium, mass spectroscopy for $^{18}O$, and infrared absorption, gas chromatography, or mass spectroscopy for deuterium. Just before dosing with an isotope, a fluid sample (blood, saliva, or urine) is collected to determine the background levels, and a second sample is collected after an equilibration period (Ellis, 2000). For example, several researchers (Khaled, Lukaski, & Watkins, 1987; Lukaski et al., 1985) have used the following protocol: a 10 ml saliva sample to determine baseline isotope levels, ingestion of 10 g of deuterium oxide ($D_2O$) mixed in 300 ml of deionized water, a 4 h equilibration period, and then a second 10 ml saliva sample.

## Limitations

Isotopic dilution has a few limitations. Research has shown that 2 to 4% of the hydrogen tracers exchange with nonaqueous hydrogen, but only ~1% of the $^{18}O$ tracer (Schoeller, 1991, 1996). For each of these tracers, the estimated error for a TBW measurement is typically <1 kg. This uncertainty in TBW translates to an error of 10% (~1.4 kg) for the absolute fat mass, which translates to 2% of the percent fat estimate in reference man (Ellis, 2000). Technical errors in estimating TBW from isotope dilution can occur, but careful methodology can minimize those errors. Variations occur in the physiological fluid measured (saliva, urine, blood, or respiratory water), the equilibration period, water changes during equilibration, and the method for measuring the isotopic enrichment following equilibration (Roche et al., 1996).

TBW is similar when the isotope concentration is derived from different body fluids, but the equilibration times are not the same for the different compartments. In saliva, equilibration of the isotopic tracer is complete in 3 h (Schoeller, 1991); however, a 4 h equilibration is preferred (Kushner & Schoeller, 1986; Schoeller, 1991), and the equilibration time given for serum or urine tracer concentrations is 6 h.

The largest potential error appears to occur in calculating the isotopic dilution space (Wagner, Heyward, & Gibson, 2000). The isotopes will take up a larger space than the TBW because the hydrogens in $D_2O$ and $^3H_2O$ also exchange with proteins and carbohydrates in the body. Indications are that an overestimation of 0.5% to 2.0% occurs in measuring TBW from isotope dilution, using the following assumptions:

1. Protein constitutes 15% of body weight; 6% of protein is hydrogen, and 15% of its hydrogen exchanges readily.

2. Carbohydrate constitutes 0.5% of body weight; 6% of the carbohydrate is hydrogen, and 35% of the hydrogen atoms exchange readily.

3. Fat has no rapidly exchanging hydrogen atoms.

Thus, the TBW value obtained by isotope dilution should be decreased by 0.5% to 2.0% to correct for hydrogen exchange (Schoeller, 1991). More recently Schoeller (1996) estimated that a 4% correction factor is needed when using $D_2O$ as the isotopic tracer. For each of these tracers, the estimated error for a TBW measurement is typically <1 kg. This uncertainty in TBW translates to an error of 10% (~1.4 kg) for the absolute fat mass, which translates to 2% of the percent body fat estimate in reference man (Ellis, 2000).

The test–retest reliability in TBW measures obtained from isotope dilution is excellent. In one study, a 14 d interval separated the tests, with some participants at energy balance ($n = 5$, SEM= 0.009 L) and others at negative ($n = 6$, SEM = 0.006 L) or positive energy balance ($n = 6$, SEM = 0.011 L) (Schoeller, 1991). Average TBW changed by only 1% during the 12 d interval for the participants in energy balance (Wagner et al., 2000).

Although the precision from isotope dilution for estimating TBW is good, caution should be used when estimating percent fat from TBW (Pace & Rathbun, 1945). The mean hydration fraction of the FFM is 73.2%, and the following equation was developed: %BF = 100 – %TBW/0.732. Because of the large interindividual variation in the proportion of water in the FFM, the use of this method to estimate %BF has been questioned. Even with no technical error in measuring TBW, a biological variability of 2% in the water content of the FFM occurs, which would correspond to an uncertainty in percent body fat of 3.6% (Siri, 1961). The percentage of water in the FFM ranges from 70 to 76% in most species (Sheng & Huggins, 1979). Thus the equation (Pace & Rathbun, 1945) for the estimation of fat needs to be used cautiously. Even with changes in intra- and extracellular water ratio, this hydration ratio of TBW/FFM remains firm (Wang et al., 1998). The isotopic dilution technique, however, requires a laborious and costly analysis, thus limiting its use in many situations when it would be valuable to know the changes in TBW.

# Magnetic Resonance Imaging and Computed Tomography

As stated earlier, body composition can be divided into a five-level model: atomic, molecular, cellular, tissue system, and whole body (Heymsfield et al., 1990). On the tissue level, computerized axial tomography (CT) and magnetic resonance imaging (MRI) techniques can be used to assess subcutaneous and visceral adipose tissue volumes. The advantage of CT and MRI over other methods is that they offer direct visualization of images depicting skeletal muscle or abdominal adipose tissue cross sectional area. These methods measure precisely and reliably the amount of deep abdominal adipose tissue and provide an indication of regional fat distribution. These images can be used individually or combined with mathematical reconstruction algorithms to estimate the mass of fat stores, individual muscle groups, or total-body skeletal muscle mass (Heymsfield, Ross, Wang, & Frager, 1997). Assessment of the fat depots is important because they are associated with metabolic complications that are risk factors for cardiovascular disease.

# Equipment and Methodology: CT Scans

The CT image is produced from an x-ray beam emitted from a source rotating around the patient. The x-ray passes through the patient. Using a series of detectors, the exit transmission intensity can be monitored (Plourde, 1997). The x-ray beam creates cross-sectional slices about 10 mm thick through the patient. The transmission at any angle can be used to calculate the average attenuation coefficient along the length of the x-ray beam measure. The CT image represents a two-dimensional map of pixels, a picture element, corresponding to a three-dimensional section composed of an equal number of voxels, or volume elements (Goodpaster, Thaete, & Kelley, 2000). Voxels, which are similar to pixels, have the same area but also include slice thickness dimension. In the two-dimensional CT image, each pixel has a specific number that corresponds to a specific location within the patient (Goodpaster et al., 2000) and the numerical value of each pixel within the matrix corresponds to a specific level of gray within the image. These values are attenuation coefficients and are reported in Hounsfield units (HU). Tissues can be differentiated in vivo based on their attenuation characteristics. The attenuation characteristics of a tissue depend on the density and electrons per unit mass (Bushberg, Seibert, Leidholdt, & Boone, 1994). The ratios of electrons to mass for carbon, nitrogen, and oxygen are 0.5, whereas the ratio for hydrogen is 1.0 (Bushberg et al., 1994). Thus, the higher proportion of hydrogen in adipose tissue separates it particularly well from other tissues (Goodpaster et al., 2000). The different attenuation characteristics of fat and muscle are what allow them to be discerned from each other. The attenuation values for fat are in the negative range (–190 to –30 HU), whereas muscle mass has a positive attenuation (0 to 100 HU), allowing for differentiation of the tissue. Bone has a very high attenuation on CT and can also be separated out (Goodpaster et al., 2000). At the same time, low attenuation values within muscle reflect increased lipid content (Goodpaster et al., 2000). Thus, CT is appropriate for quantifying separate adipose tissue depots and whole-body composition (Van der Kooy & Siedell, 1993), permitting direct visualization of internal adipose tissue compartments. These internal compartments are not accessible by the methods discussed previously.

## Equipment and Methodology: MR image

The MR image is created by the interaction between nuclei of hydrogen atoms. These atoms behave like small magnets because they have a nonzero magnetic moment, which is orientated randomly (Ross, 1996). This characteristic tends to cancel their magnetic effects. The magnet provides an external magnetic field along the body axis, and gradient coils add a smaller identification field (Plourde, 1997). Inside an MR magnet, the strength of the magnetic field is 10,000 times stronger than outside, resulting in the protons aligning themselves within the subject longitudinally to the external magnetic field (Ross, 1996). With the hydrogen protons aligned in a known direction, a pulsed radio frequency (RF) is applied to the body tissues, causing the hydrogen protons to absorb energy. The radio frequency generated by these fields provides the force necessary to rotate the nuclear spins away from the direction of the external magnetic field. When the RF field is turned off, the nuclear spins press back toward the direction of their original positions, and in the process the release of energy emits a radio frequency signal (Kelsey, 1993). These signals can be combined to form an image. With variations in the radio frequency pulse sequence, the image can be made up of predominantly T1 or T2 information (Kelsey, 1993). T1, or longitudinal relaxation time, is the time required for the protons to return to the original position. Images developed using T1 information are heavily weighted toward proton density, whereas T2 produces images that provide information on tissue differences (Kelsey, 1993). The T1 for fat is much shorter compared with the T1 for protons in water.

The T1 signal is used to develop the MR images by computer. This signal can be varied by changing the parameters of the time to repeat (TR) and the time to echo (TE) of the RF pulse. By manipulating the TR and TE, the RF pulse sequence varies (Ross, 1996). Spin echo is one such sequence that varies the TR parameter to exploit the difference in T1 relaxation times of adipose tissue, short T1, and nonadipose tissue, long T1. This variation provides good tissue contrast required for high-quality MR images. Spin echo is the most commonly used pulse sequence because of its inherent tissue contrast properties, robust signal strength, and relative insensitivity to artifacts that affect other MR sequences. The principal drawback of spin echo is the time required for the preferred routines. Fast spin echo pulse sequences can reduce time considerably compared with conventional spin echo (Ross, 1996). Inversion recovery is another RF pulse sequence that uses a different T1 time. The spin echo sequence is used more frequently because it more quickly obtains the MR image with good tissue contrast compared with using the inversion recovery sequence.

The development of MRI technology over the past decade has greatly reduced the time required to obtain an image and improved the quality of the image. Now that the MRI pulse sequence can be obtained quickly, subjects can hold their breath to prevent distortion of the image (Ross, 1996). This method substantially reduces the potential negative effects that respiratory motion may have on image quality in the abdominal region. Respiratory gaiting, which times the MR imaging to the breathing frequency, is another technique that is used.

## Scan Acquisition

MRI requires a cylindrical magnet with an internal diameter large enough to enclose the human body, but the space is often not large enough for morbidly obese subjects. For the measurement of abdominal fat, scans can be taken from the pubic symphysis to the top of the diaphragm and done as a series of consecutive slices or one slice can be taken at the intervertebral disk between the fourth and the fifth lumbar vertebrae. Research studies have also used CT and MRI to measure the muscle component of a limb. Typically the midthigh area is measured, but other regions have been studied. This measurement quantifies not only the amount of muscle and subcutaneous adipose tissue but also the adipose tissue interspersed in and around the muscle (Goodpaster, Thaete, Simoneau, & Kelley, 1997; Mitsiopoulos et al., 1998). Using these techniques to measure the muscle component of a limb excluding the visible adipose tissue interspersed in and around muscle has provided more precise measurement of the lean tissue of the body.

Typically, tissue area within the CT scan or MR image is calculated by segmentation techniques. By selecting regions of interest on the scan, fat and lean tissue can be quantified. In an abdominal scan, the total abdominal fat on the scan is usually calculated first (figure 11.27). The region of interest, for example, visceral fat or the intra-abdominal depot, is traced by a mouse pointer, and this area is calculated. To calculate subcutaneous fat, the visceral depot is subtracted from the total abdominal fat. The area of each region is determined by multiplying the number of pixels in the highlighted region by their known area (Abate, Burns, Peshock, Garg, & Grundy, 1994). If a single image is used the area values are computed, but if multiple images are used the volume values are derived.

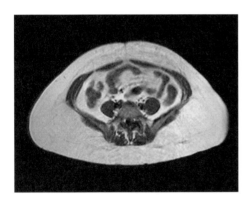

**Figure 11.27** Slice from a magnetic resonance image. The top is the stomach, and the bottom of the image is the back and spine.

Validation of MRI has been primarily conducted using animal models. Fowler, Fuller, Glasbey, Cameron, & Foster (1992) compared MRI adipose tissue measurement with those obtained by dissection of lean and obese pigs and observed that the distribution of MRI adipose tissue correlated strongly with adipose tissue distribution by dissection ($r = .98$) and had a mean square error of 2.1%. Using a rat model, Ross, Léger, Guardo, DeGuise, & Pike (1991) reported that whole-carcass chemically extracted lipid was highly correlated with MRI adipose tissue mass ($r = .97$, $P < .01$) and the standard error of estimate was 10.5%. Using three human cadavers, Abate et al. (1994) compared MRI measures of abdominal subcutaneous and visceral AT with those obtained by direct weighing of the same adipose tissue compartments. The mean difference between the two compartments was 6%. Overall MRI measures of adipose tissue compared well with cadaver and animal data with an error of 2 to 10% (Ross, 1996).

Recent work by Tang, Vasselli, Wu, Boozer, and Gallagher (2002) has reported on the use of MRI to measure various soft tissues and organ volumes in Spraque-Dawley rats in vivo. MRI was highly sensitive to organ and tissue changes with the exception of the heart, providing a technique to study changes in organs with aging, growth, and disease progression. MRI estimated the weight of the brain, kidney, and spleen with high accuracy ($r < .90$), but overestimated intra-abdominal adipose tissue, skeletal mass, and liver volumes. MRI estimated heart and lung weight with the lowest precision because of motion artifacts. Hence, these investigators are the first to show that MRI can be used to measure organ mass (Tang et al., 2002).

The most accurate MRI method for measuring total body fat uses contiguous slices covering the entirety of a volunteer's body (Thomas et al., 1998). This method demands considerable scanning and analytical time. At the other extreme is the use of a single slice at a predetermined level to extrapolate total fat content (Abate, Garg, Coleman, Grundy, & Peshock, 1997), thereby reducing scanning and analysis time at some cost in accuracy. Thomas et al. (1998) have shown that measurement uncertainty increases as slices are removed from a contiguous data set. This uncertainty highlights the possible inaccuracies in using single-slice data sets. Ross, Léger, Morris, DeGuise, and Guardo (1992) found that although a single-slice analysis produced significant correlation with total body fat, it accounted for only 81% of the variance in the group as a whole. Using single-slice data may be valid for providing a rough estimate of internal fat, but potentially important information and subtle differences will be compromised. Typically, multiple slices are acquired at one time.

Mitsiopoulos et al. (1998) demonstrated that both CT and MRI images of cadaver sections provided accurate estimates of subcutaneous fat, interstitial adipose, and adipose tissue-free skeletal muscle. The intraobserver correlation of MRI adipose tissue-free skeletal muscle for duplicate measurements in vivo was 0.99 ($SEE = 8.7$ cm$^2$ (2.9%), $P < .001$). MRI was correlated with cadaver values of adipose tissue-free skeletal muscle with an $r = .99$ ($SEE = 2.9$ cm$^2$, $P < .001$), as was CT with similar estimates of adipose tissue-free skeletal muscle ($r = .99$, $SEE = 3.8$ cm$^2$, $P < .001$).

## Comparison Among Methods

The only direct method of evaluating body composition is chemical digestion and subsequent analysis of the tissues. Obviously, this approach is not practical in humans. Computer imaging, dual-energy projection, and radioisotope techniques are highly accurate but expensive and require specially trained personnel. For these reasons, hydrostatic densitometry and air displacement plethysmography are the most commonly used methods of body composition analysis.

Relative fat values derived from hydrostatic weighing have been compared with relative fat from DPA and DEXA in several studies (Haarbo et al., 1991; Verlooy, Dequeker, Geusens, Nijs, & Goris, 1991; Wang, Heymsfield, Aulet, Thornton, & Pierson, 1989). Wang et al. (1989) found the reliability of DPA ($r = .97$) to be similar to that of hydrostatic weighing ($r = .95$). Relative fat determined by hydrostatic weighing was related to DPA at $r = .87$ and $SEE = 3.4\%$. Haarbo et al. (1991) found a correlation of $r = .97$ between hydrostatically determined percent fat and percent fat derived from DEXA, but the $SEE$ (5.6%) was somewhat higher than that observed by Wang et al. (1989). Verlooy et al. (1991) reported an $r = .96$ and $SEE = 1.68\%$ for a comparison of relative fat values obtained by hydrostatic weighing and DEXA.

Hydrostatic densitometry requires relatively expensive, specialized equipment and is time consuming. Thus, it is not a suitable field method of measuring body composition. Anthropometry and BIA are the most common field techniques used for predicting body composition. Both are relatively simple to perform and can be completed in less than 10 min.

Because skinfolds are more highly correlated with body fat than are other anthropometric measures (circumferences, skeletal diameters), prediction equations for body composition from anthropometry generally include skinfolds (Jackson & Pollock, 1976). As with BIA, the addition of one or more circumference measures to skinfold prediction equations can slightly improve accuracy (Jackson et al., 1980). Anthropometric and BIA techniques for predicting body composition have typically employed hydrostatic densitometry as the criterion method during the development of prediction equations. Thus, the 2.5% error associated with hydrostatic densitometry is inherent in the prediction of relative fat from anthropometry and BIA. An error of ±2.5% is the best possible degree of accuracy one could expect from a prediction technique based on hydrostatic weighing. The accuracy reported for anthropometry and BIA generally ranges from ±3% to 6% fat overall and from ±3% to 4% fat for the most accurate equations. Both BIA and anthropometry tend to overpredict in very lean individuals and underpredict in obese individuals when equations

developed from normal populations are employed. For comparative purposes, the accuracy of predicting relative fat from anthropometric and BIA techniques is listed with age, height, weight, and BMI in table 11.9 (Pollock, Schmidt, & Jackson, 1980).

Table 11.9   **Correlation and Standard Error of Predicting Relative Fat From Selected Variables**

|  | Women | | Men | |
|---|---|---|---|---|
| Variable | r | SEE (%) | r | SEE (%) |
| Age | −.35 | 6.7 | −.38 | 7.4 |
| Height | −.08 | 7.2 | .01 | 8.0 |
| Weight | −.63 | 5.6 | −.62 | 6.3 |
| BMI | −.70 | 5.1 | −.69 | 5.8 |
| Σ 7 skinfolds | −.85 | 3.8 | −.88 | 3.8 |
| BIA | .71 | 3.6 | .75 | 3.8 |

Note: BMI = body mass index: $Wt(kg)/Ht^2(cm^2)$

Reprinted with permission. ©American Society of Contemporary Medicine and Surgery. *Comp Therapy 1980*; 6 (9): 12-27.

Population-specific equations are often invalid when applied to persons who do not fit the characteristics of the population from which they were developed. Therefore, selection of the prediction equation is critical for both BIA and anthropometric prediction techniques. Generalized equations that predict body composition with an acceptable degree of accuracy over a wide range of individuals have been developed for skinfold (Jackson, Pollock, & Gettman, 1978; Jackson et al., 1980) and BIA techniques (Deurenberg et al., 1990; Segal, Van Loan, Fitzgerald, Hogdon, & Van Itallie, 1988). Comparisons of body composition derived from anthropometric and bioelectric impedance methods have been made (Campos et al., 1989; Kushner & Haas, 1988; Shols et al., 1991). Heyward and Stolarczyk (1996) have summarized many of the population-specific equations for hydrostatic weighing, BIA, and skinfolds. Campos et al. (1989) found that $D_b$, total body water, relative fat, and FFM could be predicted by both anthropometry (weight, height, and two skinfold measures) and BIA with similar accuracy.

Kushner and Haas (1988) also concluded that there is excellent agreement between the estimation of FFM by BIA and skinfold anthropometry. Schols et al. (1991), however, found BIA to be more accurate than skinfold thicknesses for predicting FFM in 32 pulmonary patients. Fuller and Elia (1989) studied 24 normal men ($n = 14$) and women ($n = 10$) and found a small advantage for predicting $D_b$ from BIA when compared with the skinfold method described by Durnin and Womersley (1974). Anthropometry is probably the least expensive method of predicting body composition. Additional advantages of anthropometric measurements over BIA include the fact that the measurements provide regional information on body composition and are often of value by themselves for comparative purposes and tracking changes over time. Whole-body resistance in ohms from BIA means little to most individuals by itself.

Predicting body composition from ultrasound measurements is similar in theory to the skinfold technique. When compared with anthropometric and bioelectrical impedance (BIA) equations, the ultrasound technique is equally reliable and valid. But the ultrasound technique has an important advantage over the other methods in that it provides useful information on muscle thickness. This information would be valuable in studying the effects of a resistance-training program on body composition. Circumference measurements alone cannot determine whether a change in circumference resulted from an alteration in the amount of fat, the amount of muscle, or both. The ultrasound technique, however, can make this distinction. Furthermore, unlike the BIA method, which gives information only on changes in total body composition, the ultrasound technique can identify the specific sites at which the changes occurred in both fat and muscle. A major disadvantage of ultrasound compared with anthropometry and BIA is cost (approximately $30,000).

The advantage of BIA over anthropometry and ultrasound is that only a single measurement, largely unaffected by interinvestigator variability, is required to predict body composition. Because hydration status can greatly influence BIA measures, however, careful standardization of factors influencing hydration is important, and BIA does not provide a good measure of changes in body composition over time. Neither anthropometry nor BIA is a suitable technique for obtaining criterion measures of body composition for research purposes.

Many reports indicate that DEXA overestimates percent body fat and fat mass and underestimates FFM compared with ADP (Lockner, Heyward, Baumgartner, & Jenkins, 2000; Wagner & Heyward, 1999). Finding consistency with these earlier studies, Weyers et al. (2002) observed that DEXA and ADP measured similar absolute changes in percent body fat, fat mass, and FFM after weight loss in women and men. DEXA estimates of percent body fat, however, were slightly but statistically higher than corresponding ADP values.

Studies have been conducted to determine whether estimates of body water from BIA could be substituted for estimates of total body water from isotopic dilution in estimating body fatness from multicompartment models (Evans, Arngrimsson, & Cureton, 2001). Despite a strong relation between $TBW_{BIA}$ and $TBW_{D2O}$, estimates of percent body fat from three- and four-component models using $TBW_{BIA}$ were considerably less accurate than estimates from the same models using $TBW_{D2O}$ and were not more accurate than estimates from body density alone using a two-component model. The accuracy of estimating $TBW_{D2O}$ from single-frequency BIA at 50 Hz in this study ($r = .94$, $SEE = 2.4$ L) was similar to that of earlier reports (Evans et al., 2001). Errors in estimating body fat from multicomponent models that included $TBW_{BIA}$ were larger than those using body density alone. Apparently, $TBW_{BIA}$ introduces new error when used in equations validated using isotopic dilution methods. Any theoretical advantage of taking into account variation in body water when predicting percent body fat from body density is lost when $TBW_{BIA}$ is included (Evans et al., 2001). These data indicate that despite the strong association between $TBW_{D2O}$ and $TBW_{BIA}$, the magnitude of individual differences between $TBW_{BIA}$ and $TBW_{D2O}$ are such that it is not advantageous to substitute $TBW_{BIA}$ for $TBW_{D2O}$ in estimating body composition from multicomponent models.

Both MRI and CT scans provide valuable information concerning the intra-abdominal fat depots that cannot be detected with a DEXA scan. CT is advantageous because it is not as expensive as MRI, but it exposes the subject to high doses of ionizing radiation. MRI does not expose the subject to ionizing radiation, allowing multiple images to be obtained. The high cost and the priority assigned for clinical use make access to MRI systems difficult. With MRI, research can examine both whole-body and regional measurements of adipose tissue and lean tissue, which makes it well suited for describing the effects of various nutritional perturbations on adipose and lean tissue (Tang et al., 2002). The multiple scans that can be obtained with MRI allow the quantification of total abdominal fat. Intra-abdominal adipose tissue volume and L2-L3, L3-L4, or L4-L5 interspaces were good predictors of total intra-abdominal adipose tissue ($r < 0.87$, $P < .0001$ for each slice) (Jensen, Kanaley, Reed, & Sheedy, 1995).

Because the cost and availability of MRI and CT scanning may be prohibitive for some researchers, several studies have tried to differentiate visceral from subcutaneous abdominal fat mass by combining anthropometric parameters and DEXA. Tothill, Han, Avenell, McNeill, and Reid (1996) reported that the Norland DEXA versus MRI overestimates total abdominal FM by 50 to 125 g/cm depending on the distance from the sternal notch. An association of total abdominal fat mass determined by Lunar DEXA with waist-to-hip ratio and trunk skinfolds best predicted visceral fat mass by CT in postmenopausal women ($r^2 = 0.91$) (Svendsen, Hassage, Bergmann, & Christiansen, 1993), whereas others have not confirmed the accuracy of this combination and have suggested instead the use of a Lunar DEXA and a single-slice CT (Jensen et al., 1995). Recently, Park, Heymsfield, and Gallagher (2002) found that DEXA regions of interests are highly correlated with total VAT, as well as MRI-derived VAT area at L4-5 in nonobese men. Multiple regression analysis revealed that 86% of the variance in total VAT was predicted by VAT area at L4-5, DEXA region of interest L2-L4, and the waist-to-hip ratio (Park et al., 2002). Bertin, Marcus, Ruiz, Eschard, and Leutenegger (2000), however, reported that the combination of abdominal transverse diameters measured by Hologic DEXA with the sagittal diameter at the umbilical level in 71 overweight subjects correlated strongly with visceral abdominal fat mass measured by CT ($r = .94$ for women and $r = .88$ for men). The determination of the accuracy of DEXA for measuring limb muscle mass is at its very beginning.

In a recent study on older individuals, DEXA (Hologic QDR-1500) was found to underestimate total abdominal fat by about 10% compared with CT slice, particularly in those people with less abdominal fat (Snijder et al., 2002). Visceral fat by a CT was similarly associated with DEXA subregion (white and black men and women respectively: $r = .66, .78, .79$, and $.65$) as it was with sagittal diameter ($r = .74, .70, .84, .68$). Snijder et al. (2002) concluded that DEXA was a good alternative to CT for predicting total abdominal fat but was not superior to sagittal diameter in predicting visceral fat.

Many researchers try to use waist circumference or waist-to-hip ratio (WHR) as an estimate of abdominal fat. Two individuals with a similar waist circumference or WHR may have very different visceral and subcutaneous fat content. WHR has been shown to be a poor anthropometric estimator of visceral adipose tissue (Kekes-Szabo et al., 1994). Waist circumference alone is a more useful indictor of visceral adiposity, and the prediction of visceral fat varies with gender and age (Lemieux, Prud'homme, Bouchard, Tremblay, & Despres, 1996). Gender differences in the ability of anthropometric measures to predict visceral

fat have also been shown (Van der Kooy & Siedell, 1993). When compared with MRI scans, sagittal diameter and sagittal transverse diameter are better than waist circumferences and WHR when assessing abdominal fat in men, but not in women. Care must be taken when using waist circumferences for the estimation of visceral adipose tissue because of the large interindividual variation in subcutaneous and visceral adipose tissue.

When compared with hydrostatic weighing, the estimates of body density and percent body fat using ADP were highly correlated with the corresponding measures (Demerath et al., 2002). The mean differences in the estimates between ADP and hydrostatic weighing were small in magnitude but were statistically significant in adults. At the individual level the differences were large and ranged from ~ –9% to +7% body fat. These differences were even greater in children. ADP overestimated percent body fat at all levels of fat in adults, but it overestimated body fat in lean children and underestimated body fat in fatter children, compared with hydrostatic weighing (Demerath et al., 2002). Millard-Stafford et al. (2001) found that in young adults, ADP results in consistently higher values for $D_b$ than HW does. They also noted that because of relatively small bias and low individual error, $D_b$-AP is an acceptable substitute for $D_b$-HW when estimating percent fat with a four-component model in young adults, but they should not be used interchangeably in a given study. These findings support previous reports that DEXA overestimates percent body fat and FM and underestimates FFM compared with ADP (Sardinha, Lohman, Teixeira, Guedes, & Going, 1998; Yee et al., 2001). In one study in which males and females ($n = 47$) were studied together, the percent body fat was not different between hydrostatic weighing and ADP, but in the men percent body fat was underestimated and in the women it was overestimated. Thus the sex effect was significant (Braggi et al., 1999).

## Normative Data

Body composition has been evaluated extensively in humans, and normative data are available by age and for a variety of athletic populations (Wilmore & Costill, 1988). For younger women, normative values for percent fat range from 22% to 29%, for younger men from 12% to 15%, for older women from 25% to 34%, and for older men from 18% to 27% (McArdle et al., 2001). Recommended levels of relative fat are 15% for men and 23% for women, based on the Behnke and Wilmore (1974) models of the reference man and woman. Although relative fat increases as a function of age, no known advantages result from gaining fat beyond recommended levels. In fact, increased levels of body fat are associated with an increased risk of cardiovascular and metabolic disease (McArdle et al., 2001). Thus, maintaining a healthy level of relative fat with a combination of sensible diet and physical activity throughout life is advantageous.

## Summary

Measurement of body composition has become a popular practice for health professionals, researchers, and physical trainers. Because of the known health consequences due to increases in body fat, there is considerable interest by both researchers and the lay public to get an estimate of percent body fat. This chapter has discussed the numerous methods available for measuring percent body fat. Frequently, the method selected is a function of equipment availability and expense. Although some methods perform better in some populations than others, it is necessary to remember the limitation inherent to each method.

Traditionally, hydrostatic weighing is considered the "gold standard"; however it is limited because it is only a two-component model. Other methods based on the two-component model include air displacement plethsmography, anthropometry, bioelectric impedance, isotopic dilution, and ultrasound. Dual-energy x-ray absorptiometry is not limited by the assumptions of a two-component model, because the bone compartment is also measured; however, the water compartment is not estimated. Although all of the body composition methods have their limitations, the multicomponent model is currently the most accurate estimate of body fat. To utilize a three- or four-component model, it is necessary to combine the various methods of body composition assessment. The combination most frequently used is hydrodensitometry (UWW), total body water ($^3H_2O$) and measurement of bone mineral (DEXA).

More recently, CT and MRI techniques have been developed to assess abdominal subcutaneous and visceral adipose tissue volume. The advantage of these methods over other methods is the direct visualization of images depicting the various compartments. These techniques are also used to look at the cross-sectional area of skeletal muscle.

# twelve

# Static Techniques for the Evaluation of Joint Range of Motion and Muscle Length

Peter J. Maud, PhD

*New Mexico State University*

Kate M. Kerr, PhD, MCSP

*University of Nottingham, England*

The term *flexibility* is commonly used to describe the range of motion about a joint or about a series of joints. Flexibility may also imply freedom to move (Alter, 1996) or the ability to move a part, or parts, of the body through a wide range of purposeful movements at the required speed (Galley & Forster, 1982). Flexibility may be measured by both static and dynamic means, the former requiring relatively simple, inexpensive measurement tools, and the latter more sophisticated equipment, ranging from video analysis to expensive and complex optoelectronic motion analysis systems. These methods allow examination of flexibility during dynamic skilled movement; they probably are the preferred methods for analyzing the contribution of range of motion in the assessment of athletic ability. Description of these methods, however, is beyond the scope of this chapter.

This chapter focuses on the measurement of static flexibility. Assessment of static flexibility may involve the measurement of both range of joint motion and range of muscle length. The terms *range of joint motion* and *range of muscle length* have specific meanings (Kendall, McCreary, & Provance, 1993). Range of joint motion refers to the number of degrees through which a joint is capable of moving, whereas range of muscle length (also usually expressed in degrees) refers to the length of individual muscles or groups of muscles. In situations in which muscles pass over only one joint, the measurement of range of joint motion and range of muscle length will be the same. When a muscle or group of muscles passes over two or more joints, however, the total range of muscle length will not permit total range of motion at the joints over which it passes simultaneously. In this situation, the normal range of muscle length will be less than the total (summed) range of the joints over which the muscle passes. Thus, if a muscle or muscle group passes over two joints, the muscle must be allowed to be slack over one of the joints when measuring joint range of motion of the other joint.

The purpose of this chapter is to present gross range-of-motion tests for the upper and lower extremities and trunk and to outline accepted methods of measuring joint range of motion and muscle length using static evaluation techniques. Indirect methods of assessment that use linear displacement, rather than angular displacement, are presented as gross range-of-motion tests, and direct measures for using the inclinometer and the goniometer are described in detail.

# Rationale for Measurement of Flexibility

Although flexibility is generally considered one of the five components of physical fitness, its exact contribution to general health is even less clearly defined than its importance to athletic performance. Within the realm of sport, high degrees of flexibility in specific joints are desirable for enhanced performance in many quantitative and qualitative athletic activities. Although it seems apparent that tight muscles may predispose an athlete to muscle strains and tendon injuries and that highly extensible muscles, in the absence of coordination and stability, may lead to joint dislocations and ligamentous sprains, few prospective studies verify this phenomenon (Cowan, Jones, Tomlinson, Robinson, & Polly, 1988; Ekstrand & Gillquist, 1983). The existence of an optimal joint range appears at least to be sport specific and variable on an individual basis. Because flexibility, as defined by joint range of motion, is affected by joint congruence, tendons, ligaments, fascia, joint capsules, and adipose tissue in addition to muscle, a direct relationship between flexibility and injury becomes difficult to establish. The arrangement of muscles in an agonist–antagonist relationship also suggests that alterations in the relationship between strength and flexibility will result in agonist–antagonist imbalance. Excessively long muscles are usually weak and allow adaptive shortening of opposing muscles; muscles that are too short are usually strong and maintain opposing muscles in a lengthened position (Kendall et al., 1993). This type of imbalance might be thought to predispose the athlete to injury, but research results do not consistently support this contention either (Knapik, Bauman, Jones, Harris, & Vaughan, 1991; Knapik, Jones, Bauman, & Harris, 1992). The reader may wish to consult an excellent review article by Knapik et al. (1992) on the contribution of flexibility to the prevention and prediction of athletic injury.

Given this difficulty in quantifying the flexibility necessary for optimal performance and injury prevention, what then is the purpose of flexibility testing for general fitness? Does flexibility become more important in injury prevention as one becomes older? Can people who engage in sporadic recreational activities afford to be more or less flexible than the competitive athlete who constantly stresses his or her musculo-skeletal system? One partial answer to these questions is that flexibility testing is important because it establishes a baseline of range of motion before intervention. This baseline data can be compared with normative values for exercise prescription or kept for future reference when retesting the individual after an intervening exercise program or during rehabilitation following injury. Therefore, the method of measurement chosen should fit the purpose. For example, if an employee in a work-related fitness program decides to begin a weight-training program to improve groundstrokes in tennis, flexibility testing might be directed toward gross assessment of range of motion of the trunk and extremities. Follow-up measurements of specific joints that are found to be restricted or that will be specifically placed in demand would also be indicated. But if the same employee complained of a lack of power in his or her serve, detailed shoulder range-of-motion tests, in addition to strength testing, would be incorporated into the assessment.

# Methods of Measurement

One of the major problems of flexibility measurement is that flexibility is not a general characteristic but is highly specific to individual joints (Corbin & Noble, 1980; Dickinson, 1968; Golding, Myers, & Sinning, 1989). Therefore, comprehensive flexibility assessment would involve measurement of all joints through a variety of planes, an extremely time-consuming task. As stated earlier, specific joints need to be isolated for assessment based on their contribution to health fitness or athletic performance. Methods of measurement include direct measurement techniques, which involve angular measures, and indirect measurement techniques, which involve linear measures. Goniometry is the most common technique used to measure single-joint systems, and, because it involves a direct measure of joint angle, it can generate normative values that can be used for intersubject comparisons. Indirect measurement techniques are commonly used when several joints or body segments are involved, but because linear measures may be influenced by body segment dimensions, it is not always possible to use these measures for intersubject comparisons. Measurement of muscle length may employ either direct (goniometric) or indirect (linear) measures.

Indirect methods are appropriate for large-scale screening or for documentation of change resulting from specific flexibility exercise training programs. Indirect methods have the advantage of ease and speed of measurement and low cost of equipment. But when accurate scientific data are required, particularly when comparing between individuals or between groups, or when assessing range of motion to determine whether impairment is present within a joint, direct methods of assessment are essential.

## Techniques for Direct Measurement of Joint Range of Motion

Several types of instruments are commonly recognized as appropriate for the measurement of range of motion. These include laboratory-based 3D systems, such as CODA, ELITE, and VICON optoelectronic measurement systems, which use reference systems and markers to estimate body segment orientation in space and define joint movement, and electrogoniometry. Clinical and field measurement tools tend to be less sophisticated and less expensive, and, if used within a standardized protocol, can provide accurate and reliable measures of static range of motion. The typical universal goniometer, which is available in different lengths for measuring both small and large joints, is the most commonly used measurement tool; others include the OB Goniometer ("Myrin") (Clarkson & Gilewich, 1989), the flexometer, and the clinical goniometer, or inclinometer (American Medical Association, 1988), as shown in figure 12.1.

Other methods of measuring static-joint range of motion include still photography and radiography, both of which have the advantage of providing a permanent record of the measurement. Both methods, however, particularly the latter, are relatively expensive, and radiography has the added disadvantage of exposure to radiation.

### Universal Goniometer

The most common device for measuring angles or range of motion is the universal goniometer. A goniometer is essentially a 180° or 360° protractor with two long arms. The stationary arm is structurally a part of the body of the goniometer. The moving arm is attached to the body by a rivet that allows it to move freely. The body of the goniometer, referred to as the fulcrum in this text, can be a semicircle or full-circle protractor that is marked in either 1°, 2°, 5°, or 10° increments. Goniometers can be made of clear plastic or metal, and they come in a variety of sizes to accommodate different joint sizes. Hubley-Kozey (1990) has described some of the limitations of using the standard goniometer. The primary problem is that its reliability is questionable because of

**Figure 12.1**   A clinical goniometer, or inclinometer.

the difficulty of identifying the axis of motion and position of the goniometer arms at certain joints. Several studies have established the reliability of goniometric measurements (Gogia, Braatz, Rose, & Norton, 1987; Mitchell, Millar, & Sturrock, 1975; Elveru, Rothstein, and Lamb, 1988; Watkins, Riddle, Lamb, & Personius, 1991). Certain skills, however, are required to maintain the reliability and validity of the instrument. Good palpation skills and knowledge of anatomy are necessary to identify the bony landmarks for alignment of the arms and fulcrum of the goniometer. Proper stabilization of the subject's body to prevent excess movement, and thus extra range of motion, is also imperative. For reliable measurements, range of motion should be measured at the same time of day, by the same assessor, using the same measuring tool, using the same subject position, and using a standardized protocol (Clarkson & Gilewich, 1989). The goniometer must be read at eye level to prevent errors in reading the scale. If this procedure is followed, reliability of repeated measures by the same assessor should be within 3 to 4° for both upper- and lower-extremity measurements. If different assessors carry out repeated measures using a standardized protocol, measurements should be within 5° for the upper extremity and within 6° for the lower extremity (Clarkson & Gilewich, 1989).

Goniometric alignment refers to the positioning of the arms of the goniometer in relation to the proximal and distal segments of the joint being evaluated. Accurate alignment relies on the use of bony anatomical landmarks to visualize the joint segments, so the subject must be appropriately dressed to ensure that these landmarks are visible to the measurer. Generally, the positioning of the arms of the goniometer using bony landmarks ensures content validity for the measurements, and this, in conjunction with the use of recommended testing positions, increases the accuracy and reliability of goniometric measurements (Norkin & White, 1995).

The stationary arm is usually aligned parallel to the longitudinal axis of the proximal segment of the joint to be measured, and the moving arm is aligned parallel to the longitudinal axis of the distal segment of the joint. The anatomical landmarks provide reference points to ensure that the alignment of the arms is accurate. Some sources emphasize that the fulcrum of the goniometer should be positioned accurately over the axis of motion of the joint. As the axis of motion changes during the movement, the position of the fulcrum should be adjusted accordingly. However, accurate and careful positioning of the arms of the goniometer along the longitudinal axes of the segments of the joint will ensure that the fulcrum lies over the axis of the joint (Clarkson & Gilewich, 1989). Therefore, accurate alignment of the arms of the goniometer with the proximal and distal joint segments should be emphasized more than placement of the fulcrum over the approximate axis of motion (Norkin & White, 1995). The most effective way to achieve this alignment is to position the joint segments in the required position (neutral, or the limit of range) and *then* align the stationary arm of the goniometer with the proximal joint segment and the moving arm with the distal joint segment. The fulcrum will then lie approximately over the axis of motion of the joint, and the appropriate reading may be taken. This method is simpler than attempting to hold the arms of the goniometer in place as the joint moves through range.

Figures 12.2 and 12.3 show a typical large goniometer being used to measure internal rotation of the shoulder.

The American Academy of Orthopaedic Surgeons (1994) has produced a useful text that gives average ranges of motion and describes methods of measurement for joints, using the standard goniometer. Norkin and White (1995) present detailed descriptions of positioning and stabilization for goniometry, as well as norms. But considerable variations are apparent in published "normal"

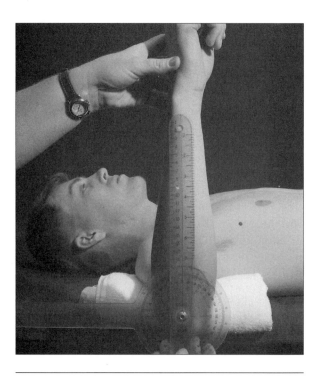

**Figure 12.2**  The initial position for measurement of internal rotation. Note that correct alignment of the goniometer arms is critical for accurate baseline measurement.

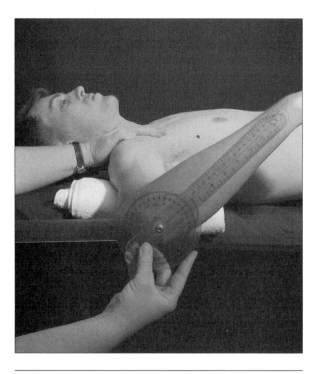

**Figure 12.3**  Final position for measurement of internal rotation using the goniometer. Note that stabilization of the shoulder is necessary to prevent the shoulder from rolling up from the table and giving a falsely greater reading of internal rotation.

range-of-motion values from different sources. For example, wrist flexion ranges from 60° to 90° and hip extension from 10° to 50° (Rothstein, 1985). These variations may be explained by different measurement techniques, variation in starting position, whether the range of motion was active or passive, and other factors such as age, sex, and activity profile (Rothstein, 1985).

The American Medical Association's text *Guides to the Evaluation of Permanent Impairment, Third Edition* (1988) give descriptions of methods of measurement for specific joints as well, but it also provides a chart or table of ranges of motion where some degree of impairment is purported to exist. For example, if a subject is capable of 60° of motion, the table might indicate that this represents an 11% impairment to total movement of the joint. Table 12.1 shows average range of motion for selected joints, excluding the spine, according to the American Academy of Orthopaedic Surgeons text (1994), and a range of motion below the average indicated by the American Medical Association (1988) as having the minimum percent of impairment. For example, if the average range of motion is listed as 80%, then the text might indicate that a minimal degree of impairment occurs if the subject can only manage 70° of movement. Measured ranges of motion at or below the level designated as comprising minimal impairment could serve as a guide to indicate undesirably low levels of range of motion. Both sides should be evaluated to determine whether the restricted joint movement is general or specific to only one side.

The goniometer is not generally advocated for measurement of range of motion of the spine (American Medical Association, 1988), although some sources have described goniometric methods (American Medical Association, 1984; Norkin & White, 1995). Rather, use of the clinical goniometer or inclinometer is preferred. Mayer (1990) suggests that using an inclinometer for the measurement of range of motion in the spine provides the only valid technique to differentiate between true lumbar motion as compared with hip motion. But Williams, Binkley, Bloch, Goldsmith, and Minuk (1993) claim that the modified-modified Schober technique is more reliable than the double inclinometer method.

## Inclinometer

The AMA text (1988) uses the term *inclinometer* to describe what has otherwise been called a clinical goniometer or a gravity-and-compass goniometer (Hubley-Kozey, 1990). Although a number of varia-

### Table 12.1   Normal Range of Motion and Range at Which Impairment Exists for Selected Joints

| Joint | Normal range (mean values) | Range at which impairment exists |
|---|---|---|
| Shoulder | | |
| Flexion | 167 | 160 |
| Extension | 62 | 40 |
| Abduction | 184 | 160 |
| Internal rotation | 69 | 60[a] |
| External rotation | 104 | 60[a] |
| Elbow | | |
| Flexion and extension | 141 | 130 |
| Forearm | | |
| Supination | 81 | 60 |
| Pronation | 75 | 70 |
| Wrist | | |
| Flexion | 75 | 50 |
| Extension | 74 | 50 |
| Ulnar deviation | 35 | 25 |
| Radial deviation | 21 | 15 |
| Hip | | |
| Flexion | 121 | 90 |
| Extension | 20 | 12 |
| Adduction | 27 | 10 |
| Abduction | 41 | 30 |
| Internal rotation | 44 | 30[a] |
| External rotation | 44 | 40[a] |
| Knee | | |
| Flexion and extension | 143 | 140[a] |
| Ankle | | |
| Plantarflexion | 56 | 30 |
| Dorsiflexion | 13 | 10 |
| Inversion | 37 | 20 |
| Eversion | 21 | 10 |

*Note.* All data reported in degrees.

[a]Method of measurement description differs between the two texts. Values given are relative to the methods described in the chapter and are similar to those described in the AMA text.

Data from M.J. Alter 1996 and American Medical Association 1984.

tions are available, each consists of a 360° dial that can be rotated on a mount and either a weighted pointer ("Plurimeter-V") or a fluid contained within the circumference of the dial (MIE). The dial and the pointer or liquid operate freely and independently of each other so that the dial can be rotated on its mount while gravity determines the position of the liquid or pointer. To perform a measurement, the mount is positioned with its free edge on the body segment in the neutral position and the dial is rotated until the edge of the fluid or the pointer registers zero. The segment is then put through its range of motion, and the position of the edge of the fluid or pointer on the dial is noted. The neutral position of the body part, however, does not always coincide with either the vertical or horizontal position (e.g., the contours of the lumbar spine). In these situations, the dial should be rotated so that its zero coincides with either the horizontal or vertical plane, and the initial (neutral) and final (end of range) positions are noted.

Because the inclinometer can be used in place of either the clinical goniometer or the OB goniometer, no further description of movement at joints to be measured will be given except for selected measurements of the spine. The spinal motion measurements are as described in the AMA text (1988). Although methods were described for using either a single inclinometer or two instruments, only the preferred, two-instrument method of evaluation will be described (except for the measurement of cervical rotation, which requires only one inclinometer). Table 12.2, adapted from the AMA text, gives the range of motion below which impairment may be considered to exist for selected measurements of motion of the spine.

Table 12.2    **Range at Which Impairment May Exist for Selected Measures of Spine Motion**

| Joint | Range at which impairment may exist |
| --- | --- |
| Neck | |
| Flexion | <40 |
| Extension | <50 |
| Right lateral flexion | <30 |
| Left lateral flexion | <30 |
| Right rotation | <60 |
| Left rotation | <60 |
| Thoracic region | |
| Flexion and extension | <30 |
| Right rotation | <20 |
| Left rotation | <20 |
| Lumbosacral region | |
| Right lateral flexion | <20 |
| Left lateral flexion | <20 |

*Note:* All data are reported in degrees.

Data from American Medical Association 1984.

## Techniques for Indirect Measurement of Joint Range of Motion

Indirect measures of range of motion may involve measurement of the change in distance between one body segment and another, or between a body segment and a fixed object such as the supporting bench or seat, or the floor or wall. The measurement may be of the shortest distance between two points (i.e., a straight line, as in fingertips to floor measurement), or it may be an increase or decrease in curvature (as in measures of flexion and extension of the trunk). As with placement of the universal goniometer, accurate identification of bony landmarks is essential to provide the points between which the measurement will be taken.

Although few published studies have addressed the validity and reliability of linear measurements, both intertester and intratester reliability have been show to be good (Leighton, 1964a, 1964b; Newton & Waddell, 1991). A linear measure is obtained by using a tape measure (Clarkson & Gilewich, 1989; Norkin & White, 1995) or a rigid rule. But because these measurements may be influenced by individual anthropometric variables, no published normative values are available for them; thus the linear measures can be used only for intrasubject comparisons. Functional or positional descriptors may also be used to reflect normal range (Hoppenfield, 1976; Kendall et al., 1993).

## Pretest Activity

Regardless of the method chosen for flexibility testing, a standardized procedure for warm-up before testing should be established. Leighton (1966) advocated no warm-up before measurement, but as Hubley-Kozey (1990) has noted, a warm-up probably should be included for the sake of safety. Because warm-up and muscle stretching before measurement may affect test results, protocols should be formulated to specify both the type and duration of warm-up and muscle-stretching techniques to be used. Three measurements are usually required at each joint, with the greatest angle measured recorded as the degrees of motion for that particular joint. But Hubley-Kozey has suggested that the mean of three measurements should be recorded. If, as is normally advocated, a warm-up precedes the evaluation, then it would be more appropriate to use the mean of the three measures.

# Trunk Range of Motion

This section will include gross measures of range of motion, which encompass the hip and trunk, measures involving the trunk alone, and measures of range of motion in specific parts of the trunk.

## Back Flexibility and Hamstring Length

Probably the most common example of indirect measurement is the sit-and-reach test used to evaluate hamstring and low-back flexibility. A lack of flexibility in these areas alone, or when accompanied by relative weakness in the abdominal muscles, has frequently been cited as a possible cause of low-back pain syndrome (American Alliance for Health, Physical Education, Recreation and Dance, 1984; Blair, Falls, & Pate, 1983). But little clinical evidence supports this claim, and Nachemson (1990) did not specifically mention lack of flexibility as a contributor to low-back pain in his review of the relationship between low-back pain, exercise, and physical fitness. Studies by Jackson and Baker (1986) and Jackson and Langford (1989), however, have found that although the sit-and-reach test appeared to be a valid measure of flexibility in the hamstring area, it was not necessarily a valid measure of low-back flexibility. For both young and adult females, the sit-and-reach test correlated moderately well with passive hamstring flexibility, r = .64 (Jackson & Langford, 1989) and r = .70 (Jackson & Langford, 1989), respectively, but poorly with low-back flexibility, r = .28 (Jackson & Baker, 1986) and r = .12 (Jackson & Langford, 1989). On the other hand, for adult males, the sit-and-reach test correlation was r = .89 for hamstring and r = .59

for low-back flexibility. Kippers and Parker (1987) evaluated a similar test, the standing toe-touch test, and found that although this test was positively related to trunk and hip flexion, it correlated very poorly with vertebral flexion in males.

A more valid description of the sit-and-reach test would be that it assesses the flexibility of the posterior muscles—hamstrings and the lower middle and upper paraspinals and calf muscles. Tightness in any of these muscles can limit the subject's ability to reach forward. But one or more of the other muscles being tested may compensate for the tightness, thereby resulting in a "normal" sit-and-reach test score. Figures 12.4 through 12.6 depict three subjects. Subject A has normal flexibility of the back and hamstrings, and the back has a nice, rounded curve. Subject B has reasonable hamstring flexibility but limited back flexibility, resulting in a lower sit-and-reach score. Subject C has poor hamstring flexibility, giving rise to an even lower sit-and-reach test score.

These examples explain why correlations between low-back and hamstring flexibility may be low. Therefore, when used in general health and fitness screening, the sit-and-reach test should be evaluated both quantitatively and qualitatively. If indicated, the qualitative aspect of the test would lead the tester to perform more specific tests for low-back or hamstring flexibility. For example, for subject B in figure 12.5, the tester would specifically test low-back flexibility using an indirect method such as the modified-modified Schober technique or a direct method such as the double inclinometer technique. Both techniques are described later in this chapter. Differing limb and torso lengths also have an effect on the indirect measurement of flexibility. Anthropometric variables have been assumed to have little effect on sit-and-reach test results (American Alliance for

**Figure 12.4**   Flexibility characteristics shown during the sit-and-reach test, subject A.

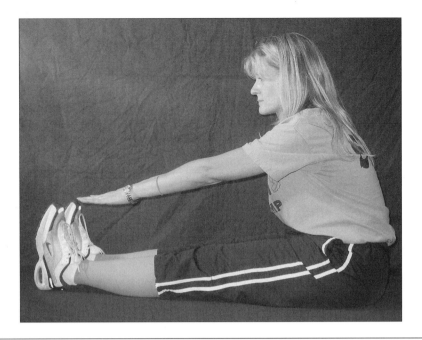

**Figure 12.5**   Flexibility characteristics shown during the sit-and reach test, subject B.

**Figure 12.6**   Flexibility characteristics shown during the sit-and-reach test, subject C.

Health, Physical Education, Recreation and Dance, 1984), although Broer and Galles (1958) found that, with extreme body types, test results were affected. Individuals with long trunk and arm measurements but short legs were at an advantage, whereas those with short trunk and arm measurements and long legs were at a disadvantage. Wear (1963) reported similar findings. To eliminate problems associated with differing arm and leg lengths, Hoeger (1991)

developed a modified sit-and-reach test that standardized the finger-to-box distance to allow for proportional differences between arms and legs. Hoeger and Hopkins (1990) demonstrated that the modified procedure eliminates the problem of disproportional limb lengths. Therefore, when a sit-and-reach test protocol is deemed appropriate, the modified procedure of Hoeger and Hopkins should be used. (The next paragraph gives a description of the test.) Variations in head position have also been found to have a significant effect on sit-and-reach test results (Smith & Miller, 1985). Although these effects are small, they emphasize the importance of exact test description and adherence to protocol. Despite these problems, the sit-and-reach test is extensively used in the assessment of flexibility in both adult and child populations, and extensive norms have been developed (Minister of State, Fitness and Amateur Sports, 1987; Golding et al., 1989; Nachemson, 1990; Summary of findings, 1985; Youth fitness in 1985, 1985). Because several different protocols are available and each uses a different measuring scale, a specific test protocol must be adhered to and the appropriate norms for that specific test must be identified. Johnson (1972) describes and gives normative data for his version of the sit-and-reach test and eight additional indirect measures. These additional tests include a bridge-up test to measure hyperextension of the spine, front and side splits, shoulder and wrist elevation, trunk and neck extension measured from the prone position, ankle plantar and dorsiflexion tests, and shoulder rotation. All these tests use the fleximeasure, which basically consists of a yardstick with an adjustable ruler attached at an angle of 90° by a special sliding aluminum box. This device is designed to facilitate and increase the accuracy and reproducibility of measurement.

In the modified sit-and-reach test developed by Hoeger (1987), the subject sits on the floor with buttocks, shoulders, and head in contact with the wall. The legs are extended, with the knees straight and the soles of the feet placed against a box approximately 12 in. (30 cm) high. The hands are placed one on top of the other, with neither set of fingers extending beyond the other. A yardstick is placed on top of the box with the zero end toward the subject. The subject reaches forward as far as possible without allowing either the head or shoulders to come away from the wall, and the yardstick is positioned and held so that the zero end touches the extended fingers. The yardstick must now be held firmly in place until completion of the test. The subject leans forward slowly, allowing the head and shoulders to move away from the wall and the fingers to slide along the top of the yardstick. The subject makes three slow, forward movements, and on the third forward motion the subject leans as far forward as possible, holding this position for a minimum of 2 s. A reading is taken of the distance moved by the fingers along the yardstick. Two separate trials are made, and the mean of the two is recorded as the sit-and-reach score. Table 12.3 gives percentile ranks for the test (Hoeger, 2005).

Table 12.3   **Percentile Ranks for the Modified Sit-and-Reach Test**

| Percentile Rank | Age Category—Men | | | |
| | ≤18 In. | 19-35 In. | 36-49 In. | ≥50 In. |
| --- | --- | --- | --- | --- |
| 99 | 20.8 | 20.1 | 18.9 | 16.2 |
| 95 | 19.6 | 18.9 | 18.2 | 15.8 |
| 90 | 18.2 | 17.2 | 16.1 | 15.0 |
| 80 | 17.8 | 17.0 | 14.6 | 13.3 |
| 70 | 16.0 | 15.8 | 13.9 | 12.3 |
| 60 | 15.2 | 15.0 | 13.4 | 11.5 |
| 50 | 14.5 | 14.4 | 12.6 | 10.2 |
| 40 | 14.0 | 13.5 | 11.6 | 9.7 |
| 30 | 13.4 | 13.0 | 10.8 | 9.3 |
| 20 | 11.8 | 11.6 | 9.9 | 8.8 |
| 10 | 9.5 | 9.2 | 8.3 | 7.8 |
| 05 | 8.4 | 7.9 | 7.0 | 7.2 |
| 01 | 7.2 | 7.0 | 5.1 | 4.0 |

| Percentile Rank | Age Category—Women | | | |
| | ≤18 In. | 19-35 In. | 36-49 In. | ≥50 In. |
| --- | --- | --- | --- | --- |
| 99 | 22.6 | 21.0 | 19.8 | 17.2 |
| 95 | 19.5 | 19.3 | 19.2 | 15.7 |
| 90 | 18.7 | 17.9 | 17.4 | 15.0 |
| 80 | 17.8 | 16.7 | 16.2 | 14.2 |
| 70 | 16.5 | 16.2 | 15.2 | 13.6 |
| 60 | 16.0 | 15.8 | 14.5 | 12.3 |
| 50 | 15.2 | 14.8 | 13.5 | 11.1 |
| 40 | 14.5 | 14.5 | 12.8 | 10.1 |
| 30 | 13.7 | 13.7 | 12.2 | 9.2 |
| 20 | 12.6 | 12.6 | 11.0 | 8.3 |
| 10 | 11.4 | 10.1 | 9.7 | 7.5 |
| 05 | 9.4 | 8.1 | 8.5 | 3.7 |
| 01 | 6.5 | 2.6 | 2.0 | 1.5 |

## Lumbar Flexion and Extension

The modified-modified Schober test, also known as a skin distraction test, was first described by Schober (1937) as a means of measuring lumbar flexion. The test was later modified by Macrae and

Wright (1969) and again by Van Adrichem and van der Korst (1973) to enhance its reliability and ease of application. The technique as modified by Macrae and Wright requires that the lumbosacral junction be identified and skin marks made 5 cm below and 10 cm above the lumbosacral junction. A tape measure is placed on the subject's spine between the two skin marks, and the subject flexes forward from a standing position, keeping the knees straight. The new distance between these two skin marks is recorded, to the nearest millimeter, as lumbar flexion (Mellin, 1986). Lumbar extension is measured similarly, with the subject bending backward as far as possible, with the hands on the buttocks. The new distance between the superior and inferior skin marks is recorded, to the nearest millimeter, as lumbar extension. Measurement of lumbar flexion and extension with an inclinometer is described later.

The difficulty in reliability arises from problems with determining the lumbosacral junction. Van Adrichem and van der Korst (1973), therefore, modified this technique by using the posterior superior iliac spines (PSIS) as a reference point for the tape measure. Their modified-modified Schober technique requires that an ink mark be placed on the midline of the lumbar spine between the subject's PSIS. Another mark is made 15 cm above the mark at the PSIS. The tape measure is aligned between the two skin marks with the zero at the inferior skin mark as shown in figure 12.7. The subject flexes forward

or extends backward, as previously described, and the difference between the initial length and the new flexed or extended length is recorded. Figure 12.8 shows the flexed position.

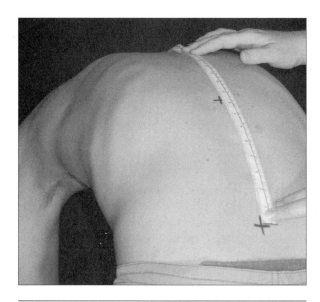

**Figure 12.8**   The modified-modified Schober test: Van Adrichem and van der Korst variation, flexed position.

### Inclinometer

• **Starting position.** The subject stands with knees straight, weight evenly distributed, and hands resting on hips. The T12 spinous process is located, and a skin mark is made. A second mark is made at the midpoint of the sacrum. Both inclinometers are aligned in the sagittal plane, one over the spinous process of the T12 and the other over the midpoint of the sacrum. With the trunk in neutral position, the inclinometers are set to zero (figure 12.9).

• **Movement.** The subject flexes maximally at the hips (figure 12.10), the inclinometers are read, and the reading from the midpoint of the sacrum is subtracted from the one obtained at T12 (figure 12.9). The resultant difference is recorded as maximal lumbar flexion. *Note. The inclinometer method allows differentiation between the thoracic and the lumbosacral regions of the vertebral column. The subject then returns to the neutral position, where both inclinometers are at the zero setting. The subject then extends backward as far as possible, and the inclinometers are again read (figure 12.11). The reading from the midpoint of the sacrum is subtracted from the one obtained at T12, and the difference is recorded as the maximal lumbar extension.

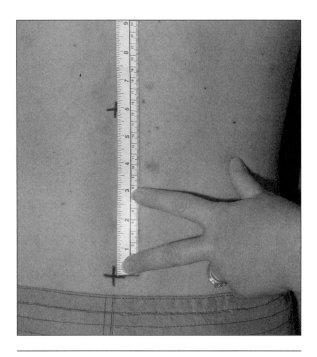

**Figure 12.7**   The modified-modified Schober test: Van Adrichem and van der Korst variation, neutral position.

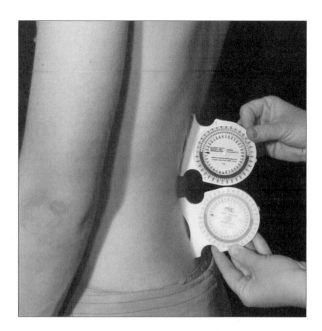

**Figure 12.9** Use of two inclinometers for the measurement of lumbosacral flexion and extension. The photo shows the neutral or starting position.

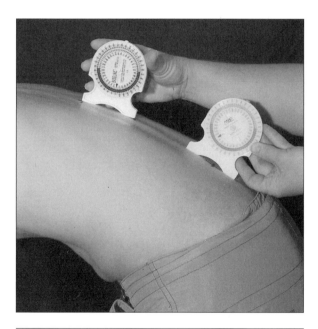

**Figure 12.10** Measurement of lumbosacral flexion. This photo shows maximal flexion.

• **Note.** According to the AMA text (1988), "Perceived lumbar flexion is actually a compound movement of both the lumbar and the hips (measured at the sacrum), in which hip flexion normally accounts for at least 50% of total flexion" (p. 89). For this reason the AMA developed criteria to validate the lumbar flexion test, which necessitates the

measurement of a straight-leg raise of the tightest leg. The lumbar flexion test is invalid if the joint range of motion for the straight-leg raise exceeds the total of sacral flexion and extension by 10° or more. For measurement of the straight-leg raise, the subject lies supine with the knees of both legs extended. The inclinometer is placed on the tibial spine of one leg and set at zero, and the investigator raises the leg. At the position of maximal hip flexion, the inclinometer is read. The leg not being tested remains flat on the bench. The subject keeps the knees of both legs straight throughout the test.

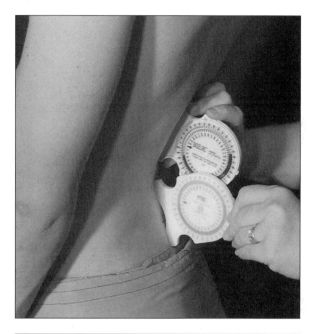

**Figure 12.11** Measurement of lumbosacral extension. This photo shows maximal extension.

## Trunk Flexion

Trunk flexion may be assessed in standing by simply asking the subject to bend forward. The linear measurement of the distance between fingertips and the floor has demonstrated good intertester reliability (ICC = .98) (Newton & Waddell, 1991). This movement involves hip flexion along with trunk flexion, however, and to assess trunk range of motion, one must ignore the contribution from hip flexion. With normal trunk flexion, the lumbar spine will negate its normal lordosis to flatten, and the thoracic spine will increase its forward concavity and should describe a smooth, continuous curve (figure 12.12). A combination of normal trunk flexion and normal hamstring length will permit an adult to touch fingertips to toes (Kendall et al., 1993).

**Figure 12.12**   Linear measurement of trunk flexion.

To assess flexion of the trunk by eliminating hip flexion, the subject should lie supine and rest on forearms with elbows at 90° and arms close to the body. With normal or good range of trunk flexion, the subject can flex the trunk comfortably, with the pelvis flat on the bench (figure 12.13).

## Trunk Extension

Trunk extension can be assessed in the standing position by asking the subject to lean backward (figure 12.14). This movement also involves hip extension.

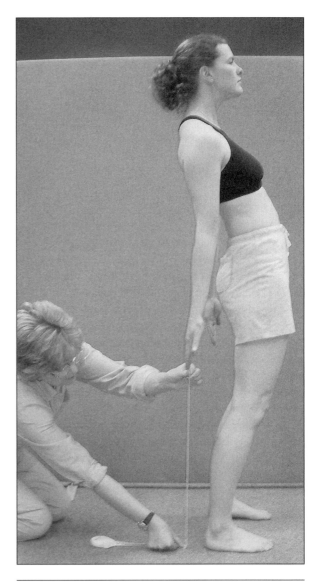

**Figure 12.14**   Linear measure of trunk extension.

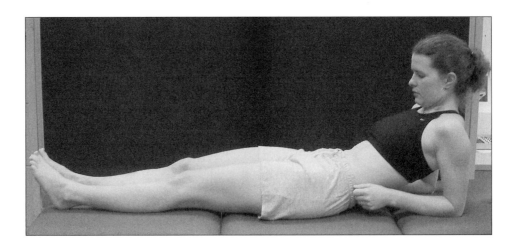

**Figure 12.13**   Trunk flexion with hip flexion eliminated.

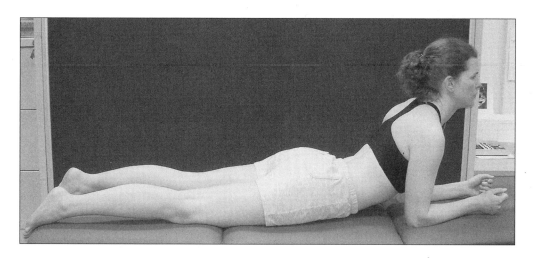

**Figure 12.15** Trunk extension with hip extension eliminated.

With normal trunk extension, the lumbar lordosis will increase, and the forward concavity of the thoracic spine will flatten (Kendall et al., 1993).

To assess extension of the trunk by eliminating hip extension, the subject should lie prone and rest on forearms with elbows at 90° and arms close to the body. With normal or good range of trunk extension, the subject can extend the trunk comfortably, with the pelvis flat on the bench (i.e., anterior superior iliac spines in contact with the bench) (figure 12.15).

## Trunk, Lateral Flexion

Lateral flexion of the trunk can be measured with an inclinometer, with a goniometer, or linearly. Descriptions of all three procedures follow.

### Inclinometer

- **Starting position.** The subject stands with knees straight, weight evenly distributed, and hands resting on hips. The T12 spinous process is located and a skin mark is made. A second mark is made at the midpoint of the sacrum. Both inclinometers are aligned in the coronal plane, one over the spinous process of the T12 and the other over the midpoint of the sacrum. With the trunk in neutral position, the inclinometers are set to zero.

- **Movement.** The subject flexes maximally to the right lateral side, and the inclinometer readings are recorded. The sacral reading is subtracted from the T12 reading, and the difference is recorded as maximum right lateral flexion. The subject then returns to the neutral position, with the inclinometers at the zero position, and then flexes maximally to the left.

The inclinometer readings are taken, and the sacral reading is subtracted from the T12 reading. The difference is recorded as the maximum left lateral flexion.

### Goniometer

- **Starting position.** The subject stands with knees straight, weight evenly distributed, and feet hip distance apart. The examiner stands behind the subject and centers the fulcrum of the goniometer over the spinous process of S1. The moveable arm of the goniometer hangs so that it is perpendicular to the floor, and the fixed arm is aligned with the spinous process of the C7 (Norkin & White, 1995).

- **Movement.** The subject flexes laterally to the right, without flexion or extension in the sagittal plane. The examiner realigns the fixed arm of the goniometer with the spine of C7.

- **Observation.** The subject should perform the movement strictly in the coronal plane and avoid flexion, extension, and rotation of the trunk or lateral tilting of the pelvis (Clarkson & Gilewich, 1989).

### Linear

- **Starting position.** The subject stands with knees straight, weight evenly distributed, and feet hip distance apart. The arms should hang vertically at the side, with elbows extended.

- **Movement.** The subject flexes laterally to the right, without flexion or extension in the sagittal plane, and stretches the hand down the lateral side of the lower limb. The examiner measures and records the distance between the tip of the middle finger and the floor (Clarkson & Gilewich, 1989) (figure 12.16).

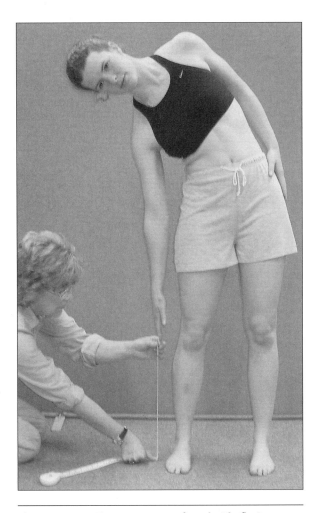

**Figure 12.16**　Linear measure of trunk side flexion.

- **Observation.** The subject should perform the movement strictly in the coronal plane and avoid flexion, extension, and rotation of the trunk or lateral tilting of the pelvis.

# Neck Range of Motion

This section will include gross measures of range of motion, which encompass flexion, extension, lateral flexion, and rotation about the neck joints. Inclinometer and linear measurements will be described for each.

### Inclinometer (Flexion and Extension)

- **Starting position.** The subject sits in a chair with the thoracic and lumbar spine in contact with the back of the chair and the head in the neutral position. Two inclinometers are used and oriented in the sagittal plane. One is placed over the T1 spinous process (the location is previously marked on the skin), and the other is placed over the occiput of the head. The inclinometers are then set to the zero position.

- **Movement**
  - Flexion: The subject flexes the neck maximally, and the reading is taken from both inclinometers. The reading from T1 is subtracted from the reading over the occiput, with the difference recorded as degrees of cervical flexion.
  - Extension: The subject then returns the head to the neutral position so that both inclinometers return to the zero reading. The subject then extends the neck maximally. Readings are again taken from both inclinometers, and the difference between the two readings is recorded as degrees of cervical extension.

- **Observation.** If, during extension, the positioning of the inclinometers causes them to contact each other and prevent maximum extension, then the T1 inclinometer should be moved laterally to avoid contact and the location should be marked. Care should be taken to ensure that when the inclinometer is moved, it is still maintained in the sagittal plane.

### Linear (Flexion and Extension)

- **Starting position.** The subject sits on a chair with a back support. The head and neck are in the anatomical position. The back of the chair provides support for the thoracic and lumbar spine. Stabilization by the measurer is not practical in this measurement. The subject is instructed to relax and maintain the shoulders in the neutral position and to maintain contact with the back of the chair to prevent thoracic and lumbar spine movement. The distance between the chin and the sternal notch is measured (figure 12.17).

- **Movement**
  - Flexion: The subject flexes the neck forward to the limit of motion. A tape measure is used to measure the distance between the tip of the chin and the suprasternal notch. Full range of motion is achieved when the subject is able to touch the chin to the chest (figure 12.18). The linear measure reflects the decreased range of motion.
  - Extension: The same reference points are used. The subject extends the neck backward to the limit of motion. A measure is taken in the neutral position and in the fully extended position (figure 12.19). The difference between the two measures reflects the range of neck extension. Normal range of extension allows the subject to look at the ceiling directly above his or her head.

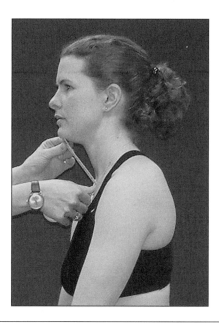

**Figure 12.17**   Linear measure of neck in the neutral position.

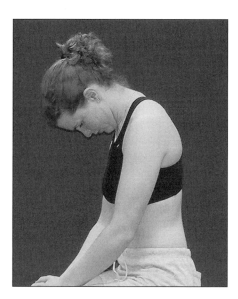

**Figure 12.18**   Full neck flexion. The chin approximates the suprasternal notch.

• **Observation.** In all these measurements the shoulders must remain in the neutral position and not be elevated or protracted (Clarkson & Gilewich, 1989).

*Inclinometer (Lateral Flexion)*

• **Starting position.** The starting position is the same as that used for flexion and extension except that the inclinometers are aligned in the coronal plane.

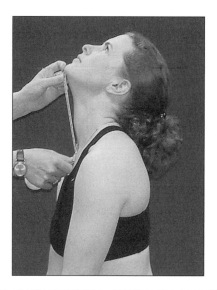

**Figure 12.19**   Linear measure of neck extension.

• **Movement.** The subject maximally flexes the neck laterally to the left, and readings are taken from both inclinometers. The difference between the two readings is recorded as maximal cervical left lateral flexion. The subject then returns the head to the neutral position so that both inclinometers return to the zero setting. The subject then laterally flexes the neck maximally to the right. The inclinometers are again read, with the difference between the two readings recorded as maximal cervical right lateral flexion.

• **Observation.** Subject should keep shoulders level and avoid side flexion of the trunk.

*Linear (Lateral Flexion)*

• **Starting position.** The subject sits on a chair with a back support. The head and neck are in the anatomical position. The back of the chair provides support for the thoracic and lumbar spine. The subject is instructed to relax and maintain the shoulders in the neutral position and to maintain contact with the back of the chair to prevent thoracic and lumbar spine movement.

• **Movement.** The subject flexes the neck to the side (without rotation) to the limit of motion. A tape measure is used to measure the distance between the mastoid process of the skull and the acromion process. A measure is taken in the neutral, or anatomical, position and at the limit of range of lateral flexion (figure 12.20). Normal range is indicated when the head can tilt laterally approximately 45° (Hoppenfield, 1976).

• **Observation.** Subject should keep the shoulders level and avoid neck flexion, extension, or rotation.

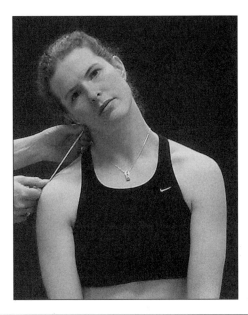

**Figure 12.20**   Linear measure of neck side flexion.

*Inclinometer (Rotation)*

- **Starting position.** The subject lies supine on a flat table. (The shoulders of the subject should be exposed so that any excessive shoulder rotation may be noted and avoided.) The single inclinometer required is placed in the coronal plane on top of the head with the base "near the back of the head approximately in line with the cervico-occipital junction" (Blair et al., 1983, p. 83). The inclinometer is set to the zero position, with the head of the subject perpendicular to the table.

- **Movement.** The subject rotates the head maximally to the right, and the inclinometer reading is recorded as maximal cervical right rotation. The subject returns the head to the neutral position so that the inclinometer returns to the zero setting, and then maximally rotates the head to the left. The inclinometer reading is recorded as maximal cervical left rotation.

- **Observation.** The subject should avoid rotation of the trunk.

*Linear (Rotation)*

- **Starting position.** The subject sits on a chair with a back support. The head and neck are in the anatomical position. The back of the chair provides support for the thoracic and lumbar spine. The subject is instructed to relax, maintain the shoulders in the neutral position, and maintain contact with the back of the chair to prevent thoracic and lumbar spine movement.

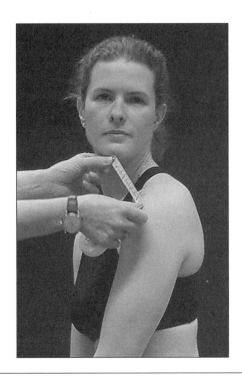

**Figure 12.21**   Linear measure of neck rotation.

- **Movement.** The subject rotates the head in the horizontal plane (without flexing or extending) to the limit of the range. A tape measure is used to measure the distance between the chin and the acromion process. A measure is taken in the neutral, or anatomical, position and at the limit of range of rotation (figure 12.21). Normal range of rotation should allow the subject to bring the chin almost in line with the shoulder (Hoppenfield, 1976).

- **Observation.** The subject should avoid neck flexion and extension.

## Upper Limb Range of Motion

This section will include gross measures of range of motion, which encompass flexion and extension, adduction and abduction, and rotation about the shoulder joint. This section will also cover elbow flexion and extension, forearm pronation and supination, wrist flexion and extension, and wrist ulnar and radial deviation. All measures will be described using a goniometer.

*Shoulder, Flexion and Extension*

- **Starting position.** The subject may stand or lie supine with the arms by the side of the body and palms against the sides of the thighs. This is the neutral, or zero, position. The stationary arm of the goniometer is aligned with the midaxillary line of the thorax. The fulcrum is close to the acromion process. The moving arm is aligned with the lateral midline of the humerus, using the lateral epicondyle of the humerus as a bony landmark, or reference point. For measurement of extension, the subject may stand or lie prone.

- **Movement**
  - Flexion is recorded as the maximal forward and upward movement of the arm from the neutral position.
  - Extension is recorded as the maximal backward and upward movement of the arm from zero.
- **Observation**. The subject should not be allowed to arch the back.

### Shoulder, Adduction and Abduction

- **Starting position**. The neutral position is with the subject either supine or sitting with the arms at the side. Hands are supinated. The stationary arm of the goniometer is placed just lateral to the anterior chest wall, parallel to the midline of the body. The fulcrum is placed over the anterior aspect of the acromion process. The moving arm is aligned with the midline of the humerus, using the medial epicondyle as a reference.
- **Movement**
  - Abduction is recorded as the maximal movement of the arm in the coronal plane from neutral.
  - Adduction is recorded as the return of the shoulder from full abduction.
- **Observation**. The subject should not be allowed to lean laterally.

### Shoulder, Rotation

- **Starting position**. The neutral position is with the subject lying supine, the arm abducted to 90°, and the elbow flexed to 90° with the forearm upright and perpendicular to the ground. The stationary arm of the goniometer is placed perpendicular to the ground but pointing downward. The fulcrum is placed over the olecranon, between the medial and lateral epicondyles of the humerus. The moving arm is aligned with the lateral surface of the ulna, using the ulnar styloid process as the reference point. (This starting position is shown in figure 12.2.)
- **Movement**. External rotation is recorded as the maximal movement of the forearm from the upright, neutral position to a position of backward

rotation. Internal rotation is recorded as the maximal movement of the forearm toward a forwardly rotated position. (This is illustrated in figure 12.3.)

- **Observation**. During internal rotation, the subject should not be allowed to rotate the scapula forward up off the table or arch the back during external rotation.

### Elbow, Flexion and Extension

- **Starting position**. In the neutral position, the subject is either lying supine with the arm at the side of the body and the hand supinated or sitting with the arm flexed to 90° with the elbow joint fully extended. The stationary arm of the goniometer is aligned with the midline, lateral surface of the humerus, using the acromion process as a reference point. The fulcrum is placed over the lateral epicondyle of the humerus. The moving arm is aligned with the midline, lateral surface of the radius, using the radial styloid process as a reference point.
- **Movement**
  - Flexion is recorded as the maximal movement of the forearm from neutral toward the humerus.
  - Extension is the return of the forearm from the fully flexed position to neutral. If the subject cannot reach neutral position, a negative value (e.g. –10°) should be recorded.
- **Note**. Because hyperextension is sometimes present at the elbow, the subject should be positioned so that the surface supporting the arm does not limit full extension. A towel roll might be needed just proximal to the elbow. Neutral is 0°, so hyperextension would be recorded as a negative number.

### Forearm, Pronation and Supination

- **Starting position**. The neutral position is with the subject seated, the arm by the side of the body, and the elbow flexed to 90° with the thumb pointing upward (the palm is facing in). The stationary arm of the goniometer is aligned in the coronal plane with the lateral midline of the humerus for pronation and with the medial midline for supination. The fulcrum is placed lateral to the ulnar styloid process for pronation and medial to the radial styloid process for supination. The moving arm is placed across the flat part of the wrist just proximal to the ulnar styloid process on the dorsal surface for pronation and on the palmar surface for supination.
- **Movement**
  - Pronation is recorded as the movement of the forearm from the neutral position to a

position in which the palm of the hand is facing the floor.

- Supination is recorded as the movement of the forearm to a position in which the palm of the hand is facing upward.

- **Observation.** The subject should not be allowed to rotate the humerus during measurement. Be particularly aware of this substitution when measuring pronation, because the subject might tend to allow the elbow to come away from the body.

### Wrist, Flexion and Extension

- **Starting position.** The neutral position is with the subject seated next to a table, arm abducted to 90° and elbow flexed to 90°, with the forearm resting on the table. The wrist and hand extend just beyond the edge of the table, with the palm facing downward. The stationary arm of the goniometer is aligned with the lateral midline of the ulna using the olecranon process as a reference point. The fulcrum is placed on the lateral surface of the wrist, just distal to the ulnar styloid process. The moving arm is aligned with the lateral midline of the fifth metacarpal.

  - **Movement**
    - Wrist flexion is recorded as the movement of the wrist from the neutral position to a position of full flexion, with the palm of the hand directed down and toward the table.
    - Wrist extension is recorded as the movement from the neutral position to full extension, with the palm of the hand facing forward.

- **Observation.** The forearm must remain in contact with the surface of the table at all times.

### Wrist, Ulnar and Radial Deviation

- **Starting position.** The neutral position is the same as for wrist flexion and extension, but the wrist and hand may be supported on the table. The stationary arm of the goniometer is aligned with the dorsal midline of the forearm, using the lateral epicondyle of the humerus for a reference point. The fulcrum is placed over the dorsal midline of the wrist. The moving arm is aligned with the dorsal midline of the third metacarpal.

  - **Movement**
    - Ulnar deviation is recorded as the movement from neutral to a position in which the hand is maximally directed toward the outside.
    - Radial deviation is recorded as the movement from neutral to a position in which the hand is maximally directed toward the inside.

- **Stabilization.** The subject should not be allowed to pronate or supinate the forearm during measurement.

## Lower Limb Range of Motion

This section will include gross measures of range of motion, which encompass flexion and extension, adduction and abduction, and rotation about the hip joint. This section will also cover knee flexion and extension, ankle plantar flexion and dorsiflexion, and ankle inversion and eversion. All measures will be described using a goniometer.

### Hip, Flexion and Extension

- **Starting position.** The neutral position is with the subject prone, with the hip in 0° abduction or adduction and 0° internal or external rotation. The stationary arm of the goniometer is aligned with the lateral midline of the pelvis. The fulcrum is placed over the greater trochanter. The moving arm is aligned with the lateral midline of the femur, using the lateral femoral epicondyle as a reference point. For hip flexion, the same goniometric alignments are used, except that the subject lies supine.

  - **Movement**
    - Hip extension is recorded as the maximal upward movement of the femur from the neutral position (prone).
    - Hip flexion is recorded as the maximal movement of the femur toward the chest from the neutral position.

- **Observation.** The subject must not be allowed to bring the pelvis up off the table or arch or flex the lumbar spine. For hip flexion, the leg not being tested may be flexed at the hip and knee, with the foot placed flat on the table to help stabilize the pelvis.

### Hip, Adduction and Abduction

- **Starting position.** The neutral position is with the subject supine and the hip in 0° internal or external rotation and 0° flexion or extension. Both legs are positioned in midline relative to the torso. The stationary arm of the goniometer is aligned across the pelvis, "connecting" the left and right anterior superior iliac spines. The fulcrum is placed over the anterior superior iliac spine of the leg being measured. The moving arm is aligned with the anterior midline of the femur, using the midline of the patella for reference. For hip adduction, the same goniometric placements are used, except that the subject's other leg is abducted so that it does not interfere with the movement of the leg being tested. The neutral position is still with the leg to be measured in midline.

- **Movement**
  - Hip abduction is recorded as the maximal movement of the thigh away from the midline of the body.
  - Hip adduction is recorded as the maximal movement of the thigh across the midline.
- **Observation.** The subject should not be allowed to rotate the pelvis or internally or externally rotate the thigh.

### Hip, Rotation

- **Starting position.** The neutral position is with the subject seated (hips at 90° flexion and 0° abduction or adduction, and knees at 90° flexion). A small towel roll is placed beneath the distal thigh of the leg to be tested; this ensures that the thigh is parallel to the surface upon which the subject is seated. The stationary arm of the goniometer is held downward, perpendicular to the floor. The fulcrum is placed over the anterior surface of the patella. The moving arm is aligned with the midline of the tibia, using the crest of the tibia and the midpoint of the two malleoli as reference points.
- **Movement.** Hip external rotation is recorded as the maximal movement of the thigh from the neutral position to a position in which the thigh is rolled outward. Hip internal rotation is recorded as the movement of the thigh from neutral to a position of maximal inward rotation.
- **Observation.** The subject should not be allowed to let the pelvis rotate up off the table or chair or to lean to one side, especially during hip internal rotation.

### Knee, Flexion and Extension

- **Starting position.** The neutral position is with the subject supine, with the knee in full (or as near full as possible) extension. The stationary arm of the goniometer is aligned with the lateral midline of the thigh, using the greater trochanter as a reference point. The fulcrum is placed over the lateral epicondyle of the femur. The moving arm is aligned with the lateral midline of the fibula, using the lateral malleolus as a reference point.
- **Movement**
  - Knee flexion is recorded as the movement of the lower leg from the neutral position to a position in which the lower leg and heel are maximally drawn toward the buttocks.

  - Extension is measured with the knee in the neutral (or maximally extended) position. If the knee cannot extend fully, a negative value (e.g. -10°) is recorded; if the knee reaches full extension, but no further, a value of 0° is recorded. If knee extension reaches beyond neutral, a value of (e.g. 10°) hyperextension is recorded.
- **Observation.** The subject should not be allowed to rotate the thigh internally or externally. Hyperextension is frequently encountered in the knee and is recorded as a negative number (see elbow extension).

### Ankle, Plantar Flexion and Dorsiflexion

- **Starting position.** The neutral position is with the patient seated, knee flexed to 90° and ankle in 0° inversion or eversion. The plantar surface of the foot is placed parallel to the ground. The knee is flexed to eliminate tightness of the gastrocnemius muscle as a limit to true ankle dorsiflexion. But dorsiflexion is often measured with the knee at 0° and 90° flexion to determine if in fact the gastrocnemius is tight. The stationary arm is aligned with the lateral midline of the fibula, using the fibular head as a reference point. The fulcrum is placed over the lateral malleolus. The moving arm is aligned parallel with the lateral aspect of the fifth metatarsal.
- **Movement**
  - Ankle plantar flexion is recorded as the maximal movement of the foot downward from the neutral position.
  - Ankle dorsiflexion is recorded as the maximal movement of the foot upward from the neutral position.
- **Observation.** The subject must not be allowed to invert or evert the foot.

### Ankle, Inversion and Eversion

- **Starting position.** The neutral position for inversion and eversion is the same as for dorsiflexion and plantarflexion. The stationary arm of the goniometer is aligned with the anterior midline of the tibia, using the crest of the tibia as a reference point. The fulcrum is placed over the anterior ankle joint, using the midpoint of the malleoli as a reference point. The moving arm is aligned with the anterior surface of the second metatarsal.
- **Movement**
  - Ankle inversion is recorded as the maximal movement of the foot inward from neutral.

– Ankle eversion is recorded as the maximal outward movement of the foot.

• **Observation.** The foot should not be allowed to move into extreme plantar flexion during inversion or into extreme dorsiflexion during eversion.

# Muscle Length Tests

Muscle length tests are performed to determine whether the range of muscle length is normal, restricted, or excessive. Reduced muscle length is characterized by the inability of the muscle to be stretched to its normal length. Normal length may be determined by an angular measurement or by the position attained with respect to a supporting surface.

In the following sections we will describe muscle length tests for glenohumeral and scapular muscles as well as for lower limb and pelvic muscles.

### Latissimus Dorsi, Teres Major, and Rhomboids

• **Starting position.** The subject lies on an unpadded bench in the supine position, with arms at the sides of the body, elbows extended, and palms facing downward. The knees are bent with feet flat on the bench, and the low back is flat on the bench.

• **Movement.** Both arms are raised overhead upward and backward into flexion, keeping the arms close to the sides of the head. With normal length, the subject should be able to rest the arms flat on the bench, close to the sides of the head (figure 12.22).

• **Shortness.** The subject is unable to get the arms to the level of the table. Measurement of the limitation may be either by the angle between the humerus and the bench or by the distance between the lateral epicondyle of the humerus and the bench.

• **Observation.** The examiner should note whether the low back arches from the bench; this movement occurs in compensation for short shoulder flexors.

### Pectoralis Major

• **Starting position.** The subject lies on an unpadded bench in the supine position, with arms at the sides of the body, elbows extended, and palms facing upward. The knees are bent with feet flat on the bench, and the low back is flat on the bench.

• **Movement**

– Upper or clavicular fibers: The arm is placed in horizontal abduction, with the shoulder in lateral rotation and the elbow in extension. With normal length, the subject should be able to rest the arm flat on the bench, without rotation of the trunk (figure 12.23).

– Lower or sternal fibers: The arm is placed in 135° of abduction, with the shoulder in lateral rotation and the elbow in extension. With normal length, the subject should be able to rest the arm flat on the bench, without arching the low back (figure 12.24).

• **Shortness**

– Upper or clavicular fibers: The subject is unable to get the arm to the level of the bench. Measurement of the limitation may be either by the angle between the humerus and the bench or by the distance between the lateral epicondyle of the humerus and the bench.

– Lower or sternal fibers: The subject is unable to get the arm to the level of the bench. Measurement of the limitation may be either by the angle between the humerus and the

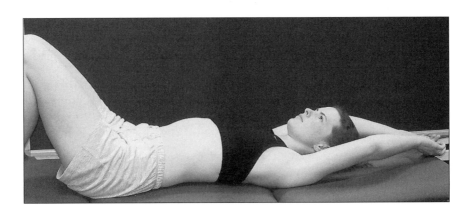

**Figure 12.22**   Measurement of length of latissimus dorsi, teres major, and rhomboids.

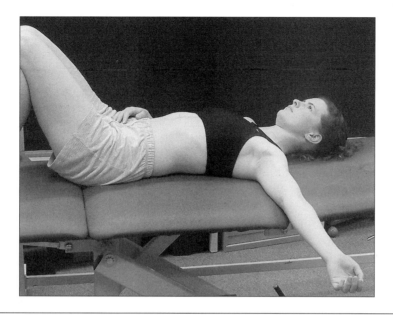

**Figure 12.23**    Measurement of the length of the upper (clavicular) fibers of pectoralis major.

bench or by the distance between the lateral epicondyle of the humerus and the bench.

- **Observation**
    - Upper or clavicular fibers: The examiner should note whether the subject rotates the trunk to the side of the testing arm.
    - Lower or sternal fibers: The examiner should note whether the subject arches the low back or rotates the trunk to the side of the testing arm.

*Pectoralis Minor*

- **Starting position.** The subject lies on an unpadded bench in the supine position, with arms at the sides of the body, elbows extended, and palms facing downward. The knees are bent with feet flat on the bench, and the low back is flat on the bench.

- **Movement.** The arms remain by the sides, and the examiner stands above the subject's head and observes the position of the shoulder girdle. With normal length, the subject should be able to rest the shoulder girdle flat on the bench, without rotation of the trunk to the test side (figure 12.25).

- **Shortness.** The subject is unable to rest the shoulder girdle on the bench. The limitation may be recorded by measuring the distance between the posterior border of the acromion and the bench (figure 12.25).

- **Observation.** The examiner should note whether the subject rotates the trunk to the test side.

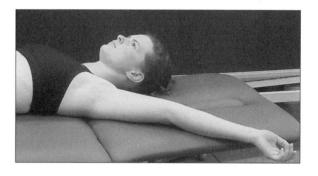

**Figure 12.24**    Measurement of the length of the lower (sternal) fibers of pectoralis major.

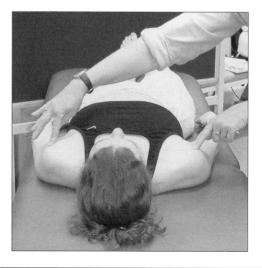

**Figure 12.25**    Measurement of the length of pectoralis minor (showing shortness on the left).

### Shoulder Medial Rotators

- **Starting position.** The subject lies on an unpadded bench in the supine position, with the arms in 90° of abduction, elbows in 90° of flexion, and forearms in midprone and perpendicular to the bench. The knees are bent with feet flat on the bench, and the low back is flat on the bench.

- **Movement.** The arms remain in 90° of abduction, and the forearm moves back toward the bench, beside the head. With normal length, the subject should be able to rest the forearm flat on the bench, without arching the trunk (i.e., 90° lateral rotation) (figure 12.26).

- **Shortness.** The subject is unable to rest the forearm on the bench. The limitation may be recorded by measuring either the angle between the forearm and the bench or the distance between the ulnar styloid process and the bench.

- **Observation.** The examiner should note whether the subject arches the trunk.

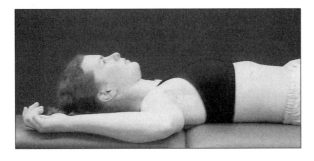

**Figure 12.26**  Measurement of the length of the medial rotators of the shoulder.

### Shoulder Lateral Rotators

- **Starting position.** The subject lies on an unpadded bench in the supine position, with the arms in 90° of abduction, elbows in 90° of flexion, and forearms in midprone and perpendicular to the bench. The knees are bent with feet flat on the bench, and the low back is flat on the bench.

- **Movement.** The arms remain in 90° of abduction, and the forearm moves forward or down toward the bench, beside the trunk, while the examiner places a hand on the anterior aspect of the humeral head to stabilize the shoulder girdle. With normal length, the forearm should make a 20° angle with the bench (i.e., 70° medial rotation) (figure 12.27).

- **Shortness.** The subject is unable to achieve the correct position of the forearm. The limitation may be recorded by measuring either the angle between

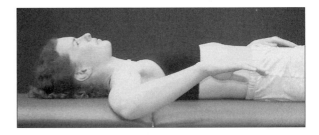

**Figure 12.27**  Measurement of the length of the lateral rotators of the shoulder.

the forearm and the bench or the distance between the ulnar styloid process and the bench.

- **Observation.** The examiner should note whether the subject protracts the shoulder girdle or flexes the thoracic spine.

### Hip Flexors

- **Starting position.** The subject sits at the end of an unpadded bench, with lower limbs supported to the level of approximately midthigh. The examiner stands on the side opposite that being tested and helps the subject lie down on the bench, placing one arm behind the subject's back and the other behind the knee of the nontested limb. The examiner draws the knee of the nontested limb toward the subject's chest.

- **Movement.** The subject places his or her hands behind the thigh of the nontested limb and draws it up toward the chest, to the extent that the lumbar spine is flattened (but not flexed) against the bench. The opposite (tested) limb is allowed to drop toward the bench. With normal length, the thigh of the tested limb should rest on the bench, and the leg should hang free of the bench, at a knee angle of 80° (figure 12.28).

- **Shortness.** The subject is unable to achieve the correct position of the thigh or leg. If the shortness is in iliopsoas alone, the thigh will not rest on the bench but the knee angle of 80° will be achieved. If the shortness is in rectus femoris or tensor fascia lata but not in iliopsoas, the thigh will rest on the bench but the knee flexion angle will be less than 80°. The limitation may be recorded by measuring the angle between the thigh and the bench (iliopsoas) or the knee flexion angle (rectus femoris or tensor fascia lata).

- **Note.** When testing length of rectus femoris when the length of iliopsoas is normal, the examiner should make sure that the thigh of the test limb is resting on the bench before measuring knee flexion angle.

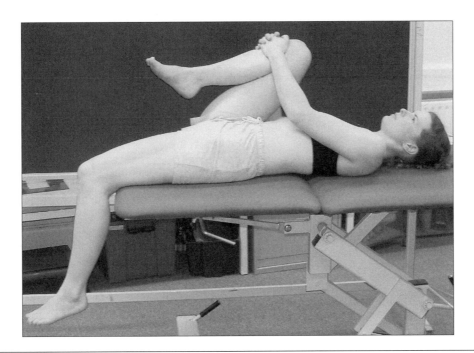

**Figure 12.28**   Measurement of the length of the hip flexors.

• **Observation.** The examiner should note whether the subject pulls the nontest knee too far toward the chest, because this action will produce a posterior tilt of the pelvis and an apparent (not true) hip flexor shortness.

*Sartorius*

• **Starting position.** The subject sits at the end of an unpadded bench, with lower limbs supported to the level of approximately midthigh. The examiner stands on the side opposite that being tested and helps the subject lie down on the bench, placing one arm behind the subject's back and the other behind the knee of the nontested limb. The examiner draws the knee of the nontested limb toward the subject's chest.

• **Movement.** The subject places his or her hands behind the thigh of the nontested limb and draws it up toward the chest, to the extent that the lumbar spine is flattened (but not flexed) against the bench. The opposite (tested) limb is allowed to drop toward the bench. With normal length, the thigh of the tested limb should rest on the bench and the leg should hang free of the bench, at a knee angle of 80°.

• **Shortness.** The subject is unable to achieve the correct position of the thigh or leg. The thigh may not rest on the bench, it may abduct or rotate externally, and the knee may flex more than 80°. A combination of three or more of these factors indicates tightness of sartorius (Kendall et al., 1993).

• **Observation.** The examiner should note whether the subject pulls the nontest knee too far toward the chest, because this action will produce a posterior tilt of the pelvis and an apparent (not true) shortness.

*Hamstrings*

• **Starting position.** The subject lies on an unpadded bench in the supine position, with arms at the sides of the body, elbows extended, and palms facing downward. The lower limbs are extended flat on the bench, and the sacrum is flat on the bench. The examiner stands on the side opposite the test limb.

• **Movement.** The examiner places one hand on the anterior aspect of the nontest thigh to ensure that it remains in contact with the bench. The other hand is placed under the calf of the test leg. The examiner raises the test limb, keeping the knee in extension, until the subject feels some discomfort, usually behind the knee. The foot and ankle should be relaxed to negate any influence of tension in gastrocnemius on knee position. If, in the absence of gastrocnemius tension, the knee starts to flex, the examiner should lower the limb slightly until the maximum amount of hip flexion concurrent with full knee extension is achieved. With normal length, hip flexion of approximately 80° should be achieved (figure 12.29).

• **Shortness.** The subject is unable to achieve the correct position of 80° of hip flexion.

**Figure 12.29**   Measurement of the length of hamstrings.

- **Observation.** The examiner should note whether the low back is hyperextended (which will make normal length hamstrings appear short) or flexed (which will make short hamstrings appear normal).

### Tensor Fascia Lata or Iliotibial Band

- **Starting position.** The subject lies on an unpadded bench on the opposite side to that being tested, with the hip and knee of the nontest side flexed to flatten the low back and stabilized against anterior pelvic tilt. The examiner stands behind the subject, places one hand on the iliac crest of the test side, and pushes upward to stabilize the pelvis and keep the lateral trunk in contact with the bench.

- **Movement.** The examiner supports the test leg and ensures that it is in line with the trunk and in neutral rotation. The leg is then allowed to drop toward the bench (figure 12.30). A drop of 10° is considered normal.

- **Shortness.** The test leg fails to drop when the pelvis is stabilized.

- **Observation.** A negative test result in the presence of shortness will occur if there is an anterior tilt of the pelvis, a lateral tilt downward on the test side, or a medial rotation of the hip. A positive test result in the presence of normal length will occur if there is a lateral tilt of the pelvis upward on the test side.

### Plantar Flexors

- **Starting position**
  - Soleus: The subject lies prone on a firm bench, with the knee and hip of the test limb flexed to approximately 90°.
  - Gastrocnemius: The subject lies prone on a firm bench, with the knee and hip of the test limb in extension and the foot over the end of the bench.

- **Movement**
  - Soleus: The examiner supports the test leg, and the subject dorsiflexes the foot and ankle. With normal length, the foot can be dorsiflexed 20°.
  - Gastrocnemius: The examiner stabilizes the test leg at the calf to prevent knee flexion, and the subject dorsiflexes the foot. With normal length, the foot can be dorsiflexed 10°.

- **Shortness.** The subject is unable to achieve the normal range of 20° with the knee flexed (soleus) or 10° with the knee extended (gastrocnemius).

- **Observation.** For the gastrocnemius test, the knee must remain in full extension.

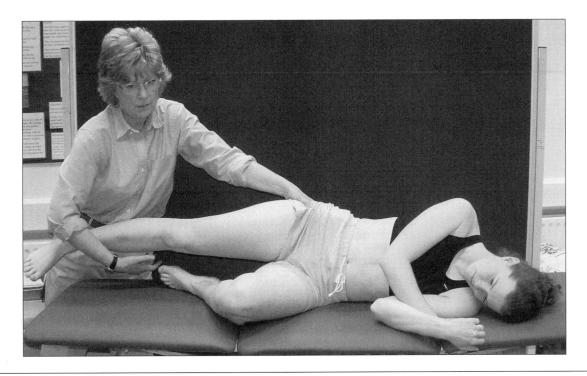

**Figure 12.30**   Measurement of the length of tensor fascia lata or iliotibial band.

## Summary

Assessment of flexibility is complicated by the fact that it is not a general characteristic but specific to the individual numerous joints that comprise the human body and the specific planes of measurement. It is further complicated by being a function not only of joint range of motion, but also of muscle length. Flexibility requirements are specific to the individual and his or her needs, whether they are for overall optimal health, prevention of injury, or for enhanced athletic performance where range of motion in specific joints may play an important part in improved performance in both qualitative and quantitative athletic events. Methods for determination of this parameter have been described using both direct measures, where joint range of motion is usually measured by use of the goniometer or inclinometer and recorded in degrees of joint range of motion, or by indirect means involving linear measurement. The ultimate decision as to which method is to be used depends, for example, upon the purpose of the evaluation, the location, the number of subjects to be tested, availability of equipment, and the expertise of the person performing the evaluation.

# thirteen

# Field Testing of Athletes

**Carl Foster, PhD**
*University of Wisconsin at La Crosse*
**Jack T. Daniels, PhD**
*State University of New York at Cortland*
**Jos J. deKoning, PhD**
*Vrije Universiteit at Amsterdam*
**Holly M. Cotter, MS**
*University of Wisconsin at La Crosse*

Few things are more useful in bridging the gap between the sport scientist and athletes than thoughtful research in the field under conditions that represent what athletes experience in training and competition. Laboratories can be intimidating places for athletes. The laboratory is the domain of the scientist, where controlled protocol rules. Test circumstances and methods of producing work are unfamiliar to most athletes. In the pool, on the track, on the ice, or in a lake is where athletes are in their milieu and feel in charge. In those environments, such unfamiliar technologies as heart rate monitors, blood lactate analyzers, gas collection devices, and even invasive procedures like muscle biopsies become much more meaningful and acceptable to athletes. Because the session occurs in the athlete's environment, he or she has less sense of missing some aspect of regular training by participating in a testing program. Another advantage of field testing is that more investigators can participate in research. Although portable state-of-the-art technology has been developed and is becoming more widely used, simple questions can be asked in the field setting with remarkable ease. You can acquire a basic exercise laboratory for the field study of athletes for less than $10,000, so research is possible even if you do not have the capital for a mass spectrophotometer. Finally, field testing makes investigators more honest. In the fully equipped laboratory we can become too comfortable with the "friendliness" of our computerized hardware and software. Sometimes we forget the calculations necessary to perform studies. More important, we often forget the assumptions involved in the study. When errors occur in the normal laboratory environment, we often do not recognize them. A visit to the field makes us appreciate where we are coming from and what we are measuring, and forces us to think about how it is to be measured.

Field testing has unique problems—the valve that does not work, the hose that was forgotten, the spilled chemical that cannot be replaced by simply stepping over to the next room. It takes some planning to keep from forgetting the relevant tools. Electrical outlets might not be the type that you need. Environmental conditions can be unpredictable (or harder to measure). Even the right kind of water might not be available. Most of these problems are easy to solve with a little planning before you go into the field.

You should keep in mind why you are doing your study and communicate honestly with athletes who are participating. If the purpose of a study is to test a hypothesis, then the athletes should know that they are research subjects, that the results might not be available to them immediately, and that they might benefit only in the longer term. They need to understand that in this case they are like laboratory animals that you recruit for normal experiments. They should know that you are there to serve the needs of the experiment and that they are agreeing to volunteer. On the other hand, if science is a desired by-product of a service function that you are providing to the athlete and coach, you need to remember that the short-term needs of the athlete and coach are far more important than answering potential experimental questions. Particularly as you work with more accomplished athletes, remember that their competitive lifetimes are short and that they are unlikely to tolerate lack of understanding and consideration on your part.

# General Principles for the Field Laboratory

You will be working in an environment that is likely as alien to you as the laboratory is to the athlete. Carefully consider how you plan your studies so that you can accomplish your goals. We will discuss specific techniques later, but certain general principles are always valid.

- **Practice.** If it is at all feasible, do a pilot or practice study using the experimental protocol that you plan to use in the field. Often ideas that work well on paper or in the lab do not work well in the field. If at all feasible the primary investigator should function as a preliminary subject. Having the shoe on the other foot often provides a valuable perspective.

- **Timing.** Provide considerably more time per test than you would in the lab. Assume that you will spend half your day fixing things that break because you are moving them about. Allow for the unexpected in your scheduling so that athletes do not spend half the day waiting around to be studied.

- **The nature of athletes.** Understand that most athletes, particularly young athletes, are social creatures. They tend to show up in groups, and you will not have the peace and quiet of the laboratory to which you are accustomed. If controlled conditions, privacy, or confidentiality are important, you must specifically plan for them. This new "laboratory" is likely to be noisy, often filled with music that you may be too old to appreciate.

- **Schedules.** Rely as little as possible on external groups for scheduling. If you cannot meet with coaches and athletes before the testing session to set up a master schedule, assume that whatever can go wrong with scheduling will go wrong. Plan breaks into the schedule so that you can catch up after disruptions. Expect to spend considerable time sitting and waiting. Insofar as you can control scheduling,

pick a time of day when conditions are likely to be favorable for your testing. For outdoor testing, early morning and early evening weather is usually more predictable than midday weather, when wind and cloud cover are more variable. For winter sports, however, the temperature may be tolerable only during midday. Wind and cloud cover may be the smaller problems compared with cold and darkness.

- **Planning and packing.** Moving your lab to the field can be confusing because you must take everything out of its normal place. Prepare a detailed list of facilities, equipment, and conditions that you need for testing and what each subject should provide. Be specific about each subject's needs (shoes, warm-ups, equipment, and so on). Use a checklist when you pack your equipment; if you forget something, you cannot go to the next room to retrieve it.

- **Athlete requirements.** Spell out any dietary or training requirements or constraints that you expect the athletes or coaches to observe. Your expectations can present a particular problem at training camps, where athletes typically work harder than they do in normal training and may be more tired than usual. Your most important function may be to insist on a rest day for the integrity of your testing. An additional advantage of this rest day is that the athletes may better adapt to the coach's schedule.

- **Finances.** Be clear about expenses for the entire project. Do you expect to pay subjects? If so, you must be sure that it is legal for each athlete to accept money. Who pays the subjects' room and board and transportation? Who pays your expenses? Answers to these questions should be clear. All other things being equal, it is probably better to pay the athletes' expenses yourself rather than requiring them to await reimbursement. At the same time you should recognize that many elite athletes are comparatively wealthy. If they are expecting first-class travel and accommodations and your budget does not provide for such treatment, you may have a problem unless the details are initially clarified.

- **Pilot testing.** Run through several complete pilot tests yourself or with dummy subjects to identify and address problems before involving athletes in any real tests. Also, work through how you plan to give feedback to coaches and athletes. You have much less time for analysis than you might prefer, and producing "pretty" results may be difficult. If you have thought through your feedback information, quick and dirty results now are often preferable to elegant conclusions 6 weeks later. Program your computer before you get to the field. Preprinted forms to convey results can also be useful.

- **Securing equipment.** Consider your security needs for equipment that you must leave at the test site when you are not actually conducting tests. You may misplace items in field situations simply because the testing area is different from your home laboratory. In your practice trials and in setting up your field-testing area, aim for consistency to reduce loss of equipment. The other security issue relates to how airport security personnel will view your equipment. In an environment of greatly increased security consciousness, trying to explain a respiratory gas analyzer or blood analyzer to airport security can be daunting. On the other hand, putting your state-of-the-art analyzer into baggage claim can also be frightening. Talk to airport security people well before the day of your departure.

- **Transportation.** Know in advance how long it will take to transport equipment, samples, subjects, and so on from the test site to other facilities such as dorms, showers, or toilets. Have sufficient transportation available to make trips back and forth. Make sure that the transport you organize to pack your equipment into is adequate to hold everything that you need to move.

- **Waste disposal.** Determine how you will dispose of any hazardous materials that you will use. Bloodstained gauze, needles, and lances are all hazardous. Although bloodstained gauze is not a biohazard and can normally be disposed of in regular trash, doing so may create a public relations problem with venue managers who may view all blood-related items as dangerous. In our experience it is better to save and transport this form of waste back to your home laboratory. Make sure that you have containers in which hazardous materials can be safely stored and transported back to your home laboratory or other proper disposal areas. If you plan to use a local disposal site such as a hospital, make your arrangements before you show up with a trunk full of containers of hazardous materials.

- **Permission.** Performing research in the field usually means getting permission to use facilities of one kind or another (e.g., a swimming pool, running track, laboratory, or classroom that can be used as a field lab). Although your peers may be collegial at professional meetings, they will rarely be enthusiastic about lending you their lab for a week. They have their own laboratory agendas. Work out details with local authorities well ahead of time. If this is not feasible, show up at the testing venue a day early and be prepared to be creative. We have used many sites as our "lab"—the basement of a college dorm, local tracks, college classrooms, our dormitory room, and group shower rooms.

# Specific Field-Test Conditions and Considerations

In order to get optimal benefit out of field testing, there are several issues that need to be considered, simply because in some environments, mistakes are easy to make and can ruin the outcome of the studies. This may be thought of as an attempt to add some of the consistency that one expects to have in the laboratory. Recognizing that this is not possible in many field situations, one still must try to eliminate as many extraneous variables as possible.

## Testing on or in Water

Equipment must remain functional in high humidity and when wet. Many investigators rely on tape to prepare and repair equipment, but most tapes do not work well when moist or wet. Duct tape is an exception and will remain secure underwater if it is applied before the surface or tape first becomes wet.

Take precautions to make all pieces of equipment secure or at least retrievable if dropped or mishandled around water. Losing a nose clip during a test is always frustrating, but it can be a real problem if the nose clip goes to the bottom of a lake where it cannot be retrieved. Securing the nose clip to the subject's headgear or a larger piece of equipment with dental floss can help save time or even the whole study. Even large items like gas collection valves can profitably be attached to yourself or the boat with dental floss.

Any apparatus used for inspiration or expiration of air during exercise while in or under the water must be airtight and capable of withstanding whatever pressure to which it is subjected. Any water leaking into any part of the subject's air-supply tubing can cause considerable concern, especially for the subject who is not comfortable being tested by performing intense exercise in the water. Besides being reliable and watertight, any equipment worn by subjects in the water should be designed to offer limited resistance to moving through the water and must not obstruct the subject's vision.

Equipment worn in the water should be easily removable by the subject or anyone else in a position to help during an emergency. Equipment worn by subjects should be constructed so that it neither weighs down the subject nor adds to the subject's buoyancy. Performance must remain realistic.

Data-recording forms invariably get wet when used around water. Use heavy-grade paper and cover it with a clear plastic cover. Ballpoint pens work better than felt pens or pencils on damp paper. For extreme circumstances, you might use a non-water-soluble grease pen on a plastic slate (but you cannot write much down because you have to use large letters, and you may need alcohol swabs to clean the slate).

You must take special precautions to prevent liquid samples, such as blood or urine, from being diluted with water. This problem can easily occur with the finger-stick blood samples unless particular care is taken to dry the skin surface before each sampling. Even after the subject's hand has been dried, water can drip from the hair or other parts of the body and rewet the hand. Many small washcloths may be more practical than a few towels, which will become soaked quickly anyway. Have a large cloth bag for disposal.

If tests on or in the water are designed to assess the subject's reaction to performing at a variety of speeds, you must be able to measure or control that speed of movement accurately. You should make clear to the subjects exactly when, during the test, speed is required, how much speed is required, and how that speed can be controlled. Will there be pace lights or some sound to follow, or will pace be given in terms of time per measured distance covered? If pace lights are not available, giving time at each pool length can be adequate, but in open water the situation is different. Paddlers and rowers must have buoys or a clearly visible line along which they travel to help them control speed. If a line is used, it must be taut and firmly anchored at each end so that movement of the line is not counted as movement through the water. Provide markers of varied designs and colors at relatively short intervals for more accurate control of speed. As a rule of thumb, the athlete should have some feedback regarding pace at least every 20 to 30 s, with rapid reporting so that they can adjust pace.

## Testing in the Cold

Everything can freeze. Blood samples can freeze in capillary tubes while being taken from the finger. The term *hyperemic fingertip* is close to meaningless in a cold environment. Breathing valves can freeze between or even during tests. Batteries, heart rate monitors, and analyzers can suddenly give much less power than they do when warm. Do everything you can to protect your equipment from the cold and to keep its exposure of short duration. Insulated containers can protect equipment and keep samples from freezing.

Subjects can freeze, so do all you can to keep them warm. Provide ways that they can stay reasonably comfortable when they are waiting their turns, during recovery intervals, and between tests. Provide warming tents, shelters, and extra cloth-ing. Construct minienvironments for taking blood samples and keeping subjects comfortable between tests. If prolonged testing might produce sweat and wet clothes, all subjects should have several changes of clothing.

Investigators can freeze. The subjects are usually exercising, which can help keep them warm. They may show up at the venue, do their thing, and then leave. Meanwhile, you are waiting for the next subject and not moving much. Think of the weather as being much colder than it is, because you're going to be standing out in it for quite a while. If you are not accustomed to cold weather, you will have a long day, and your judgment and good humor are going to be mightily tested. Wear multiple layers of clothing, heavy boots, good gloves or mittens, and something warm on your head—a hat *and* a hood. Plan for breaks that have as their only purpose warming up the investigators. Access to hot beverages is sometimes the only thing that makes experiments in the cold tolerable.

## Testing in the Heat

In a warm environment, heat can accumulate, and the temperature can change significantly over a short time. If control over heat exposure is important, allow considerable time for testing because you may be limited to a narrow temperature window each day.

Like water, sweat can contaminate blood samples. Make sure that the subject's skin is well dried before blood sampling. (This precaution can be important even in cold environments, where mittens or hats can make fingertips or earlobes sweat.) The lactate concentration in sweat is usually much higher than it is in blood, and sweat contamination may spuriously change lactate or other metabolite concentrations.

Maintain a source of shade near the testing site. Shade allows control of radiant exposure to the subject, provides a location for cooling, and may be critical to the use of some instruments, such as dry gas meters. When measuring gas volume in warm (or any other) environments, control for consistency of the environment for a few moments before measurement. Exposure of instrumentation, or even meteorological balloons, to direct sunlight can confound results. If you are not sure that an adequate source of shade will be present at your testing site, take one with you—an extra large beach umbrella, for instance (with a method for securing it against the wind).

Know and allow for the temperature limits of your instrumentation. In a cold environment, keep instruments in a warming house or your hotel room. In a warm environment, even if it is shady, the limits of enzymatically driven instruments may be exceeded even when you are relatively comfortable. Portable lactate and glucose analyzers may not remain stable beyond about 30 °C. Likewise, having a portable cooling device for storing temperature-sensitive samples is important. For more prolonged storage, you could rent a portable refrigerator or freezer from an appliance store.

Heat stress and resultant physiological responses are dose dependent, so you should have a way to assess the magnitude of stress. A thermometer, a sling psychrometer, and a scale for measuring changes in body weight should be used frequently.

## Tests at Altitude

The single greatest need for altitude studies is for a reliable source of uncorrected (for sea level) "station" barometric pressure. Portable barometers are available from the same scientific supply house that provided your main laboratory barometer. They are usually restricted to relatively low (<1,500 m) or relatively high (>1,500 m) altitudes. Technological advances have made wristwatches with built-in barometers feasible. If you work in a situation in which knowledge of air density is important, such devices are worthwhile investments. Before use in the field, they should be calibrated against a standard mercury barometer.

On arrival at altitude, it is particularly important to unpack all equipment and supplies carefully but promptly. If you are coming from sea level or an altitude different from the one where testing will take place, the change in barometric pressure will affect the pressure in storage vessels, sealed containers, and analyzers. Get all supply vessels and equipment "acclimatized" to the new pressure as soon as possible. Pay particular attention to the hydration status of your subjects. Because of the low relative humidity and strong radiant exposure at altitude, fluid losses may be large and not obvious. Dehydration may influence your results almost as much as the reduced barometric pressure, particularly during the first few days at altitude.

Beyond an altitude of about 10,000 ft (3,000 m) be alert for changes in mental status of both your subjects and yourself. In the unacclamatized individual, rapid ascent to altitudes of this magnitude can impair mental functioning. If your procedures are not rote, consider taking along supplemental $O_2$.

Because of the dryness of the air during winter at altitude, you may have a considerable problem with static electricity. Your laboratory equipment, particularly computerized equipment, may not function as well as usual. Portable room humidifiers can help you maintain an environment suitable for contemporary electrical equipment and make you more comfortable during sleep as well.

## Marking of Collection Venue and Method for Pacing Subjects

For tracks and pools, standard course marking is usually evident. For roads or supplemental marking, take along a long metric tape measure and brightly colored tape for marking the site (alternatively, take cones or buoys that can be easily placed and removed). Make sure that you have permission to use the venue and that you are not violating local ordinances.

Pacing usually requires auditory or visual feedback to subjects at frequent intervals (every 200 m for cycling or skating, every 100 m for running, every 25 m for swimming). Blowing a whistle when the subject should be at the marker or giving splits against a known split schedule is effective. For running studies, you may be able to have an investigator on a pacing cycle. Make sure that the subject does not draft the pacing investigator. If the subjects are moving rapidly, having two investigators with headphones (available at electronics store) is helpful. One investigator takes the split and radios the second, who then provides feedback to the subject. You may wish to use hand signals to indicate that a subject is below (hand low), on (hand level), or over (hand high) the desired pace. If money is no object, pacing lights synchronized to the desired pace are state of the art. GPS systems that are linked to HR monitors are beginning to appear in the marketplace. Although these do not work well indoors and may not account for changes in elevation, they may still be useful for measuring pace in high-speed events (e.g., cycling) in which the investigator may not be able to ride fast enough to pace the athlete, or in open-water situations. As a rule, the more frequent the feedback to the subject, the more even the pace.

# Measurement of Hemodynamics in the Field

During the last 25 years the development of radiotelemetry pulse monitors has greatly changed the field evaluation of athletes. The newest systems are being developed with integrated GPS systems that allow calculation of pace with reasonably good resolution of velocity.

Beginning about 1960, tape recorders attached to ECG leads were employed to record ambulatory ECGs. Holter monitors today are fairly advanced and of reasonably small size. In most cases two ECG leads can be recorded with enough fidelity to make diagnoses of exercise-induced ischemia and dysrhythmias. These systems can essentially record every heartbeat for periods exceeding 24 h. Rapid scanning systems can retrieve virtually any information regarding heart rate or ECG. These systems depend on a hard-wire attachment from the subject to the recorder and a recorder large enough to accommodate a cassette tape recorder, which renders them somewhat impractical for many athletic settings.

Approximately 10 years ago, small-scale radio transmitters designed to transmit over a distance of less than 5 ft (1.5 m) to a recorder worn by the subject radically changed the capabilities of field evaluation of heart rate. With current microchip technology, several hours of averaged heart rate data can be recorded for essentially instantaneous retrieval to a laptop computer. These systems have taken much of the work out of measuring heart rate in the field. Simply start the watch, punch the event marker at appropriate times to indicate correlation with events occurring during data collection (beginning and end of intervals, collection of lactate, completion of laps, and so on), and retrieve the data later at your convenience. These systems are accurate and easy to use. In aquatic environments, or wherever the receiver is likely to get wet, a frozen-food baggie can provide protection without compromising function. Given the limited transmission range, interference from one subject to another is usually not a problem if the subjects are reasonably aware of the need to avoid close contact. Newer models are tuned from receiver to transmitter to minimize the likelihood of cross talk.

Blood pressure can be measured in the field with about the same ease as the recording of ambulatory heart rates for clinical purposes. The monitors are reasonably compact, and the microchip technology for the automated detection of Karotkov sounds is improving. For highly competitive or strenuous athletic pursuits, the system is not yet suitable, but good progress is being made in its development. Portable systems for measuring pulse oximetry are also available.

Also available now are portable gamma scintillation cameras suitable for measuring left-ventricular ejection fraction and relative changes in left-ventricular volume during ambulatory activities and moderate exercise. Although these systems are currently designed for clinical needs, possibilities for field research are clearly available at the other end of the exercise spectrum.

## Measurement of Blood Lactate

Blood collection in the field is easy today. Capillary blood samples may be obtained easily from a fingertip puncture. In cold environments, athletes should wear warm gloves to optimize blood flow to the hand. Even so, some of the assumptions about the degree of hyperemia in the fingertip are probably violated in this setting. With the use of gloves, or in warm environments, care must be taken to remove sweat from the sampling site. Of course, given the risk of blood-borne infections, the investigator should observe universal precautions.

The development of the enzyme electrode system for measuring blood lactate (or other blood-borne constituents) has also greatly changed the possibilities for field evaluation of the athlete. Enzymatic laboratory chemistry methods have been available for years, as has the possibility of securing deproteinized blood samples in the field. But the time required to analyze samples made serious field studies of large numbers of athletes less than feasible. The enzyme electrode system allowed the development of portable analyzers that could be used with the simple injection of a blood sample or a blood sample–buffer mixture. These systems are currently small enough to fit into a small backpack or under an airplane seat. Deproteinized buffered samples are usually stable for at least 24 h, and longer in refrigerated conditions. In many cases, portable analyzers depend on a source of distilled or deionized water for mixing reagents. In the field, an adequate source of distilled water is available at most grocery stores. Portable blood lactate analyzers based on dry chemistry have developed to the point where their accuracy is comparable to bench enzyme electrode systems.

## Other Useful Items for Your Traveling Laboratory

Doing field research is something like being a good scout—your motto has to be "Be Prepared." You have to be prepared to pretend that you are McGyver and can assemble a mass spectrophotometer from baling wire, duct tape, and three mirrors using only your Swiss Army knife (which you cannot carry on the airplane any more), while suspended by one foot below a ski lift rigged to explode, during a blizzard. Many of the factors that you take for granted in your home laboratory cannot be assumed in the field. Prepare a traveling box with support equipment. This box might include the following items:

- Electrical surge protector with multiple plugs (if outside the United States and Canada, a transformer to convert power sources of U.S.-made instruments)
- Electrical adapter to allow grounded plugs to be attached to non-three-prong sockets
- Electrical extension cords
- Extra batteries for all battery-powered equipment, such as lactate analyzers and heart rate monitors; chargers for rechargeable batteries
- Flashlights
- Blank computer disks, computer paper, extra printer cartridge
- Clerical supplies for organizing data such as preformatted data collection forms, graph paper, three-hole punch, staple remover, paper clips, whiteout, tabbed organizer pages, three-ring binder, ruler, plastic page protectors, multicolored pens and magic markers, scissors, small knife (Swiss Army knives are perfect), rubber bands, string, and so on
- Metric tape measure
- Marking tape of several colors
- Various other types of tape (plastic, adhesive, duct)
- Small toolkit including regular and Phillips screwdrivers, Allen and regular wrenches of several sizes, pliers (regular and needle-nose)

You can often rent portable spectrophotometers from local clinical laboratories at reasonable cost. These devices should allow analysis of at least some blood samples in the field. Make sure that your analysis mode is kit oriented so that you are not mixing a lot of chemicals in the field.

You can usually rent portable refrigerators, freezers, dehumidifiers, humidifiers, and such from appliance stores. The availability of dry ice is unpredictable, particularly in small towns. Call ahead to make sure that you can get it in the quantity you need. Be careful about storing dry ice in your car or hotel room. Waking up in the middle of the night hyperventilating because your room is full of $CO_2$ is no fun.

Last, bring at least two or three mirrors. Why? Because a good bit of your magic will need to be done with mirrors. If the protocol gets broken or the data do not present in the way that you anticipated and planned, a little creative thought and careful use of your magical mirrors will help you make the best of a bad situation. For us this is critical, because we are ever mindful that most good science depends on a creative response to serendipity (or disaster).

## Summary

Field testing, if carefully performed, can be much better than laboratory testing. It can challenge athletes in their own environment and provide coaches with data that are much more relevant to the practical questions they have. However, careful planning and attention to detail are as important in the field as in the laboratory.

# references

## Preface

Gollnick, P.D., & Matoba, H. (1984). The muscle fiber composition of skeletal muscle as a predictor of athletic success. *American Journal of Sports Medicine, 12*(3), 212-216.

## Chapter 1

American College of Cardiology, and American Heart Association. (1986). Guidelines for exercise testing. *Circulation, 74*, 653A-667A.

American College of Sports Medicine. (2000). *ACSM's guidelines for exercise testing and prescription.* Philadelphia: Lippincott Williams & Wilkins.

Åstrand, P.-O., & Rodahl, K. (1986). *Textbook of work physiology.* New York: McGraw-Hill.

Australian Sports Commission. (2000). *Physiological tests for elite athletes.* C.J. Gore (Ed.) Champaign, IL: Human Kinetics.

Barach, J.H. (1919). The energy index (S.D.R.) of the cardiovascular system. *Archives of Internal Medicine, 24*(5), 509.

Bar-Or, O. (1987). The Wingate anaerobic test. An update on methodology, reliability and validity. *Sports Medicine, 4*, 381-394.

Bar-Or, O., Dotan, R., Inbar, O., Rothstein, A., Karlsson, J., & Tesch, P. (1980). Anaerobic capacity and muscle fiber type distribution in man. *International Journal of Sports Medicine, 1*, 82-85.

Beunen, G., & Borms, J. (1990). Kinanthropometry: Roots, developments and future. *Journal of Sports Sciences, 8*, 1-15.

Brace, D.K. (1927). *Measuring motor ability.* New York: Barnes.

Brouha, L. (1943). The step test: A simple method of measuring physical fitness for muscular work in young men. *Research Quarterly, 14*, 31-36.

Canadian Society for Exercise Physiology. (1994). *PAR-Q and you.* Gloucester, Ontario: Canadian Society for Exercise Physiology.

Carlson, J.S., & Cera, M.A. (1984). Cardiorespiratory, muscular strength and anthropometric characteristics of elite Australian junior male and female tennis players. *Australian Journal of Science and Medicine in Sport, 16*(4), 7-13.

Clarke, H.H. (1966). *Muscular strength and endurance in man.* Englewood Cliffs, NJ: Prentice Hall.

Cozens, F.W. (1929). *The measurement of general athletic ability in college men.* Eugene, OR: University of Oregon Press.

Crampton, C.W. (1913). Blood ptosis; A test of vasomotor efficiency. *New York Medical Journal, 98*, 916-918.

Cureton, T.K. (1947). *Physical fitness appraisal and guidance.* St. Louis: Mosby.

Fleishman, E.A. (1964.) *The structure and measurement of physical fitness.* Englewood Cliffs, NJ: Prentice Hall.

Fleishman, E.A. (1975). Toward a taxonomy of human performance. *American Psychologist, 30*, 1127-1149.

Foster, W.L. (1914). A test of physical efficiency. *American Physical Education Review, 19*, 632.

Fox, E., Robinson, S., & Wiegman, D. (1969). Metabolic energy sources during continuous and interval running. *Journal of Applied Physiology, 27*, 174-178.

Gabbett, T.J. (2002). Physiological characteristics of junior and senior rugby league players. *British Journal of Sports Medicine, 36*(5), 334-339.

Gollnick, P.D., & Matoba, H. (1984). The muscle fiber composition of skeletal muscle as a predictor of athletic success. *American Journal of Sports Medicine, 12*(3), 212-216.

Guidetti, L., Musulin, A., & Baldari, C. (2002). Physiological factors in middleweight boxing performance. *Journal of Sports Medicine and Physical Fitness, 42*(3), 309-314.

Hitchcock, E., & Seelye, H.H. (1893). *An anthropometric manual.* Amherst, MA: Carpenter & Morehouse.

Johnson, G.B. (1932). Physical skills test for sectioning classes into homogeneous units. *Research Quarterly, 3*, 128-136.

Kellogg, J.H. (1896). *The value of strength tests in the prescription of exercise* (Vol. 2). Battle Creek, MI: Modern Medicine Library.

Kreider, R.B., Fry, A.C., & O'Toole, M.L. (Eds.). (1998). *Overtraining in sport.* Champaign, IL: Human Kinetics.

MacCurdy, H.L. (1933). *A test for measuring the physical capacity of secondary school boys.* Yonkers, NY: Author.

Margaria, R., Aghemo, P., & Rovelli, E. (1966). Measurement of muscular power (anaerobic) in man. *Journal of Applied Physiology, 21*, 1662-1664.

Maud, P.J., & Longmuir, G.E. (1983). A survey of health-fitness evaluation centers. *Public Health Reports, 98*, 30-34.

Maud, P.J., & Shultz, B.B. (1984). The U.S. national rugby team: A physiological and anthropometric assessment. *Physician and Sportsmedicine, 12*(9), 86-99.

McCloy, C.H. (1937). An analytical study of the stunt type tests as a measure of motor educability. *Research Quarterly, 8,* 46-55.

McCloy, C.H. (1939). *Tests and measurements in health and physical education.* New York: Croft.

McCurdy, J.H., & Larson, L.A. (1939). *The physiology of exercise.* Philadelphia: Lea & Febiger.

*Mosby's medical and nursing dictionary.* (1986). St. Louis: Mosby.

National Alpine Staff. (1990). *United States Ski Team training manual.* Park City, UT: United States Ski Team.

Noakes, T.D. (1988). Implications of exercise testing for prediction of athletic performance: A contemporary perspective. *Medicine and Science in Sports and Exercise, 20*(4), 319-330.

Parkin, D. (1982). Fitness appraisal in Australian football. *Sports Coach, 6*(1), 40-43.

Parr, R.B., Hoover, R.H., Wilmore, J.H., Bachman, D., & Kerlan, R.K. (1978). Professional basketball players: Athletic profiles. *Physician and Sportsmedicine, 6,* 77-84.

Rannou, F., Prioux, J., Zouhal, H., Gratas-Delamarche, A., & Delamarche, P. (2001). Physiological profile of handball players. *Journal of Sports Medicine and Physical Fitness, 41*(3), 349-353.

Rate, R., & Pyke, F. (1978). Testing and training of women field hockey players. *Sports Coach, 2*(2), 14-17.

Raven, P., Gettman, L., Pollock, M.L., & Cooper, K.H. (1976). A physiological evaluation of professional soccer players. *British Journal of Sports Medicine, 10,* 209-216.

Rogers, F.R. (1927). *Tests and measurement programs in the redirection of physical education.* New York: Columbia University Bureau of Publications.

Salmoni, A.W., Guay, M., & Sidney, K. (1988). Skill analysis in racquetball. *Perceptual and Motor Skills, 67,* 208-210.

Sargent, D.A. (1921). The physical test of man. *American Physical Education Review, 25,* 188-194.

Schneider, E.C. (1920). A cardiovascular rating as a measure of physical fitness and efficiency. *Journal of the American Medical Association, 74*(5), 1506-1507.

Seaver, J.W. (1890). *Anthropometry in physical education.* New Haven, CT: Tuttle, Morehouse and Taylor.

Sheldon W.H., Stevens, S.S., & Tucker, W.B. (1940). *The varieties of human physique.* New York: Harper and Brothers.

Shephard, R.J. (1978). Aerobic versus anaerobic training for success in various athletic events. *Canadian Journal of Applied Sport Sciences, 3,* 9-15.

Skinner, J.S., & Morgan, D.W. (1984). Aspects of anaerobic performance. In *Limits of human performance* (American Academy of Physical Education Paper No. 18, pp. 31-44). Champaign, IL: Human Kinetics.

Smith, L.K. (1988). Health appraisal. In S.N. Blair, P. Painter, R.R. Pate, L.K. Smith, & C.B. Taylor (Eds.), *Resource manual for guidelines for exercise testing and prescription* (pp. 155-169). Philadelphia: Lea & Febiger.

Steinhaus, A.H. (1936). Health and physical fitness from the standpoint of the physiologist. *Journal of Health and Physical Education, 7*(4), 224.

Thomas, D.Q., Seegmiller, J.G., Cook, T.L., & Young, B.A. (2004). Physiologic profile of the fitness status of collegiate cheerleaders. *Journal of Strength and Conditioning Research, 18*(2), 252-254.

Tuttle, W.W. (1931). The use of the pulse rate test for rating physical efficiency. *Research Quarterly, 11*(2), 5.

Willgoose, C.E. (1961). *Evaluation in health education and physical education.* New York: McGraw-Hill.

Wilmore, J.H., Parr, R.B., Haskell, W.L., Costill, D.L., Milburn, L.J., & Kerlan, R.K. (1976). Football pros' strengths—and CV weakness—charted. *Physician and Sportsmedicine, 4,* 44-54.

Withers, R.T. (1978). Physiological responses of international female lacrosse players to pre-season conditioning. *Medicine and Science in Sports, 10*(4), 238-242.

# Chapter 2

Åstrand, P.-O. (1952). *Experimental studies of physical work capacity in relation to sex and age.* Copenhagen: Munksgaard.

Balke, B., & Ware, R.W. (1959). An experimental study of "physical fitness" of Air Force personnel. *U.S. Armed Forces Medical Journal, 10,* 675-688.

Beaver, W.L., Wasserman, K., & Whipp, B.J. (1973). On-line computer analysis and breath-by-breath graphical display of exercise function tests. *Journal of Applied Physiology, 34,* 128-132.

Bruce, R.A., Kusumi, F., & Hosmer, D. (1973). Maximal oxygen intake and nomographic assessment of functional aerobic impairment in cardiovascular disease. *American Heart Journal, 85,* 546-562.

Buchfuhrer, M.J., Hansen, J.E., Robinson, T.E., Sue, D.Y., Wasserman, K., & Whipp, B.J. (1983). Optimizing the exercise protocol for cardiopulmonary assessment. *Journal of Applied Physiology, 55,* 558-564.

Cumming, G.R., & Borysyk, L.M. (1972). Criteria for maximum oxygen uptake in men over 40 in a population survey. *Medicine and Science in Sports, 14,* 18-22.

Cumming, G.R., & Friesen, W. (1967). Bicycle ergometer measurement of maximal oxygen uptake in children. *Canadian Journal of Physiology and Pharmacology, 45,* 937-946.

Cunningham, D.A., Van Waterschoot, B.M., Paterson, D.H., Lefcoe, M., & Sangal, S.P. (1977). Reliability and reproducibility of maximal oxygen uptake measurement in children. *Medicine and Science in Sports and Exercise, 9,* 104-108.

Davis, J.A., & Kasch, F.W. (1975). Aerobic and anaerobic differences between maximal running and cycling in middle–aged males. *Australian Journal of Sports Medicine, 7,* 81-84.

Davis, J.A., & Lamarra, N. (1984). A turbine device for accurate volume measurement during exercise [Abstract]. *Aviation, Space, and Environmental Medicine, 55,* 472.

Davis, J.A., Storer, T.W., Caiozzo, V.J., & Pham, P.H. (2002). Lower reference limit for maximal oxygen uptake in men and women. *Clinical Physiology and Functional Imaging, 22,* 332-338; Erratum (2003) in *Clinical Physiology and Functional Imaging, 23,* 62.

Ellestad, M.H. (1986). *Stress testing—principles and practice* (3rd ed.). Philadelphia: Davis.

Fay, L., Londeree, B.R., LaFontaine, T.P., & Volek, M.R. (1989). Physiological parameters related to distance running performance in female athletes. *Medicine and Science in Sports and Exercise, 21,* 319-324.

Freedson, P., Kline, G., Porcari, J., Hintermeister, R., McCarron, R., Ross, J., et al. (1986). Criteria for defining $\dot{V}O_2max$: a new approach to an old problem [Abstract]. *Medicine and Science in Sports and Exercise, 18,* S36.

Hermansen, L., & Saltin, B. (1969). Oxygen uptake during maximal treadmill and bicycle exercise. *Journal of Applied Physiology, 26,* 31-37.

Hughson, R.L., Northey, D.R., Xing, H.C., Dietrich, B.H., & Cochrane, J.E. (1991). Alignment of ventilation and gas fraction for breath-by-breath respiratory gas exchange calculations in exercise. *Computers and Biomedical Research, 24,* 118-128.

Jones, N.L. (1984). Evaluation of microprocessor controlled exercise testing system. *Journal of Applied Physiology, 57,* 1312-1318.

Jones, N.L., Makrides, L., Hitchcock, C., Chypchar, T., & McCartney, N. (1985). Normal standards for an incremental progressive cycle ergometer test. *American Review of Respiratory Disease, 131,* 700-708.

Macfarlane, D.J. (2001). Automated metabolic gas analysis systems. *Sports Medicine, 31,* 841-861.

McArdle, W.D., Katch, F.I., & Pechar, G.S. (1973). Comparison of continuous and discontinuous treadmill and bicycle tests for max $\dot{V}O_2$. *Medicine and Science in Sports, 5,* 156-160.

Maksud, M.G., & Coutts, K.D. (1971). Comparison of a continuous and discontinuous graded treadmill test for maximal oxygen uptake. *Medicine and Science in Sports, 3,* 63-65.

Mitchell, J.H., Sproule, B.J., & Chapman, C.B. (1958). The physiological meaning of the maximal oxygen uptake test. *Journal of Clinical Investigation, 37,* 538-547.

Noakes, T.D. (1988). Implications of exercise testing for prediction of athletic performance: a contemporary perspective. *Medicine and Science in Sports and Exercise, 20,* 319-330.

Proctor, D.N., & Beck, K.C. (1996). Delay time adjustments to minimize errors in breath-by-breath measurements of $\dot{V}O_2$ during exercise. *Journal of Applied Physiology, 81,* 2495-2499.

Saltin, B., Blomquist, G., Mitchell, J.H., Johnson, R.L., Wildenthal, K., & Chapman, C.B. (1968). Response to exercise after bed rest and after training. *Circulation (Suppl. VII),* 1-78.

Thoden, J.S. (1991). Testing aerobic power. In J.D. MacDougall, H.A. Wenger, & H.J. Green (Eds.), *Physiological testing of the high-performance athlete* (2nd ed.) (pp.107-173). Champaign IL: Human Kinetics.

Wasserman, K., & Whipp, B.J. (1975). Exercise physiology in health and disease. *American Review of Respiratory Disease, 112,* 219-249.

Whipp, B.J., Davis, J.A., Torres, F., & Wasserman, K. (1981). A test to determine the parameters of aerobic function during exercise. *Journal of Applied Physiology, 5,* 217-221.

Wilmore, J.H., & Costill, D.L. (1974). Semiautomated systems approach to the assessment of oxygen uptake during exercise. *Journal of Applied Physiology, 36,* 618-620.

Wilmore J.H., Davis, J.A., & Norton, A.C. (1976). An automated system for assessing metabolic and respiratory function during exercise. *Journal of Applied Physiology, 40,* 619-624.

# Chapter 3

Ainsworth, B.E., Richardson, M.T., Jacobs, D.R, Jr., Leon, A.S., and & Sternfeld, B. (1999). Accuracy of recall of occupational physical activity by questionnaire. *Journal of Clinical Epidemiology, 52,* 219-27.

Åstrand, P.O., & Rodahl, K. (1986). *Textbook of work physiology: Physiological bases of exercise.* New York: McGraw-Hill.

Åstrand, P.O., & Ryhming, I.A. (1954). Nomogram for calculation of aerobic capacity (physical fitness) from pulse rate during submaximal work. *Journal of Applied Physiology, 7,* 218-221.

Balke, B. (1963). Report 63-6. Federal Aviation Agency, Aeromedical Research Division Civil. Aeromedicine Research Institute: Oklahoma City, OK.

Berthouze, S.E., Minaire, P.M., Castells, J., Busso, T., Vico, L., and & Lacour, J.R. (1995). Relationship between mean habitual daily energy expenditure and maximal oxygen uptake. *Medicine and Science in Sports and Exercise, 27,* 1170-1179.

Billat, V., & Koralsztein, J.P. (1996). Significance of the velocity at $\dot{V}O_2max$ and time to exhaustion at this velocity. *Sports Medicine, 22,* 90-108.

Billat, L.V. (2001a). Interval training for performance: A scientific and empirical practice. Special recommendations for middle- and long-distance running. Part I: Aerobic interval training. *Sports Medicine, 31,* 13-31.

Billat, L.V. (2001b). Interval training for performance: A scientific and empirical practice. Special recommendations for middle- and long-distance running. Part II: Anaerobic interval training. *Sports Medicine, 31,* 75-90.

Billat, V., Bernard, O., Pinoteau, J., Petit, B., & Koralsztein, J.P. (1994). Time to exhaustion at $\dot{V}O_2$max and lactate steady state velocity in sub elite long-distance runners. *Archives Internationales de Physiologie de Biochimie et de Biophysique,*102, (4) 215-219.

Billat, V., Faina, M., Sardella, F., Marini, C., Fanton, F., Lupo S., et al. (1996). A comparison of time to exhaustion at $\dot{V}O_2$max in elite cyclists, kayak paddlers, swimmers and runners. *Ergonomics, 39,* 267-277.

Billat, V., Renoux, J.C., Pinoteau, J., Petit, B., & Koralsztein, J.P. (1995). Hypoxemia and exhaustion time to maximal aerobic speed in long-distance runners. *Canadian Journal of Applied Physiology, 20,* 102-11.

Billat, V., Renoux, J.C., Pinoteau, J., Petit, B., & Koralsztein, J.P. (1994a). Reproducibility of running time to exhaustion at $\dot{V}O_2$max in subelite runners. *Medicine and Science in Sports and Exercise, 26,* 254-257.

Billat, V., Renoux, J.C., Pinoteau, J., Petit, B., & Koralsztein, J.P. (1994b). Times to exhaustion at 100% of velocity at $\dot{V}O_2$max and modelling of the time-limit/velocity relationship in elite long-distance runners. *European Journal of Applied Physiology and Occupational Physiology, 69,* 271-273.

Billat, V., Sirvent, P., Py, G., Koralsztein, J.P., & Mercier, J. (2003). The concept of maximal lactate steady state: A bridge between biochemistry, physiology and sport science. *Sports Medicine, 33,* 407-426.

Billat, V.L., Slawinski, J., Danel, M., & Koralsztein, J.P. (2001). Effect of free versus constant pace on performance and oxygen kinetics in running. *Medicine and Science in Sports and Exercise, 33,* 2082-2088.

Billat, V.L., Blondel, N., & Berthoin, S. (1999). Determination of the velocity associated with the longest time to exhaustion at maximal oxygen uptake. *European Journal of Applied Physiology and Occupational Physiology, 80,* 159-161.

Billat, V.L., Flechet, B., Petit, B., Muriaux, G., & Koralsztein, J.P. (1999). Interval training at $\dot{V}O_2$max: Effects on aerobic performance and overtraining markers. *Medicine and Science in Sports and Exercise, 31,* 156-63.

Billat, V.L., Morton, R.H., Blondel, N., Berthoin, S., Bocquet, V., Koralsztein, J.P., et al. (2000). Oxygen kinetics and modelling of time to exhaustion whilst running at various velocities at maximal oxygen uptake. *European Journal of Applied Physiology, 82,* 178-187.

Billat, V.L., Richard, R., Lonsdorfer, E., & Lonsdorfer, J. (2001). Stroke volume increases in all-out severe cycling exercise in moderate trained subjects. *Medicine and Science in Sports and Exercise, 33 (Suppl.),* S18.

Bodner, M.E., & Rhodes, E.C. (2000). A review of the concept of the heart rate deflection point. *Sports Medicine, 30,* 31-46.

Borg, G. (1982). Ratings of perceived exertion and heart rates during short-term cycle exercise and their use in a new cycling strength test. *International Journal of Sports Medicine, 3,* 153-158.

Boutcher, S.H., McLaren, P.F., Cotton, Y., & Boutcher, Y. (2003). Stroke volume response to incremental submaximal exercise in aerobically trained, active, and sedentary men. *Canadian Journal of Applied Physiology, 28,* 12-26.

Brickley, G., Doust, J., & Williams, C.A. (2002). Physiological responses during exercise to exhaustion at critical power. *European Journal of Applied Physiology, 88,* 146-151.

Brooke, J.D., Hamley, E.J., & Thomason, H. (1968). Relationship of heart rate to physical work. *Journal of Physiology—London, 197,* 61P-63P.

Camus, G., Juchmes, J., Thys, H., & Fossion, A. (1988). Relation between endurance time and maximal oxygen consumption during supramaximal running. *Journal de Physiologie, 83,* 26-31.

Capelli, C., Rosa, G., Butti, F., Ferretti, G., Veicsteinas, A., & di Prampero, P.E. (1993). Energy cost and efficiency of riding aerodynamic bicycles. *European Journal of Applied Physiology and Occupational Physiology, 67,* 144-149.

Conconi, F., Ferrari, M., Ziglio, P.G., Droghetti, P., & Codeca, L. (1982). Determination of the anaerobic threshold by a noninvasive field test in runners. *Journal of Applied Physiology, 5,* 869-73.

Cooper, K.H. (1968). A means of assessing maximal oxygen intake. Correlation between field and treadmill testing. *Journal of the American Medical Association, 203,* 201-204.

Dabonneville, M., Berthon, P., Vaslin, P., & Fellmann, N. (2003). The 5 min running field test: Test and retest reliability on trained men and women. *European Journal of Applied Physiology, 88,* 353-360.

Daniels, J., Scardina, N., Hayes J., & Folet, P. (1984). Elite and subelite female middle- and long-distance runners. In D.M. Landers (Ed). *Sport and elite performers,* Vol. 3 (pp. 57-72). Proceedings of the 1984 Olympic Scientific Congress, July 19-23, Oregon. Champaign, IL: Human Kinetics.

di Prampero, P.E. (1986). The energy cost of human locomotion on land and in water. *International Journal of Sports Medicine, 7,* 55-72.

di Prampero, P.E., Cortili, G., Mognoni, P., & Saibene, F. (1979). Equation of motion of a cyclist. *Journal of Applied Physiology, 47,* 201-206.

Dolgener, F.A., Hensley, L.D., Marsh, J.J., & Fjelstul, J.K. (1994). Validation of the Rockport Fitness Walking Test in college males and females. *Research Quarterly for Exercise and Sport, 65,* 152-158.

Ekblom, B., & Hermansen, L. (1968). Cardiac output in athletes. *Journal of Applied Physiology, 25,* 619-625.

Gaesser, G.A., & Poole, D.C. (1996). The slow component of oxygen uptake kinetics in humans. *Exercise and Sport Sciences Reviews, 24,* 35-71.

Gore, C.J., Booth, M.L., Bauman, A., & Owen, N. (1999). Utility of pwc75% as an estimate of aerobic power in epidemiological and population-based studies. *Medicine and Science in Sports and Exercise, 31,* 348-51.

Higgs, S.L. (1973). Maximal oxygen intake and maximal work performance of active college women. *Research Quarterly, 44,* 125-131.

Hill, A.V. (1927). *Muscular movement in man: The factors governing speed and recovery from fatigue.* New York: McGraw-Hill.

Hofmann, P., Pokan, R., von Duvillard, S.P., Seibert, F.J., Zweiker, R., & Schmid, P. (1997). Heart rate performance curve during incremental cycle ergometer exercise in healthy young male subjects. *Medicine and Science in Sports and Exercise, 29,* 762-768.

Horvath, S.M., & Michael, E.D., Jr. (1970). Responses of young women to gradually increasing and constant load maximal exercise. *Medicine and Science in Sports and Exercise, 2,* 128-131.

Jackson, A.S., Blair, S.N., Mahar, M.T., Wier, L.T., Ross, R.M., & Stuteville, J.E. (1990). Prediction of functional aerobic capacity without exercise testing. *Medicine and Science in Sports and Exercise, 22,* 863-70.

Javorka, M., Zila, I., Balharek, T., & Javorka, K. (2003). On- and off-responses of heart rate to exercise—relations to heart rate variability. *Clinical Physiology and Functional Imaging, 23,* 1-8.

Jeukendrup, A.E., Hesselink, M.K., Kuipers, H., & Keizer, H.A. (1996). The Conconi test. *International Journal of Sports Medicine, 17,* 509-519.

Kuipers, H., Verstappen, F.T., Keizer, H.A., Geurten, P., & van Kranenburg, G. (1985). Variability of aerobic performance in the laboratory and its physiologic correlates. *International Journal of Sports Medicine, 6,* 197-201.

Kukkonen-Harjula, K., Laukkanen, R., Vuori, I., Oja, P., Pasanen, M., Nenonen, A., et al. (1998). Effects of walking training on health-related fitness in healthy middle-aged adults—a randomized controlled study. *Scandinavian Journal of Medicine and Science in Sports, 8,* 236-242.

Laukkanen, R.M., Kukkonen-Harjula, T.K., Oja, P., Pasanen, M.E., & Vuori, I.M. (2000). Prediction of change in maximal aerobic power by the 2-km walk test after walking training in middle-aged adults. *International Journal of Sports Medicine, 21,* 113-116.

Laukkanen, R., Oja, P., Pasanen, M., & Vuori, I. (1992). Validity of a two kilometre walking test for estimating maximal aerobic power in overweight adults. *International Journal of Obesity and Related Metabolic Disorders, 16,* 263-268.

LaVoie, N.F., & Mercer, T.J. (1987). Incremental and constant load determination of $\dot{V}O_2max$ and maximal constant load performance time. *Canadian Journal of Sport Sciences, 12,* 229-232.

Leger, L., & Boucher, R. (1980). An indirect continuous running multistage field test: The Universite de Montreal track test. *Canadian Journal of Applied Sport Sciences, 5,* 77-84.

Leger, L., & Mercier, D. (1984). Gross energy cost of horizontal treadmill and track running. *Sports Medicine, 1,* 270-277.

Leger, L.A., & Lambert, J.A. (1982). A maximal multistage 20-m shuttle run test to predict $\dot{V}O_2max$. *European Journal of Applied Physiology and Occupational Physiology, 49,* 1-12.

Matthews, C.E., Heil, D.P., Freedson, P.S., & Pastides, H. (1999). Classification of cardiorespiratory fitness without exercise testing. *Medicine and Science in Sports and Exercise, 31,* 486-93.

Monod, H., & Scherrer, J. (1965). The work capacity of synergy muscular groups. *Ergonomics, 8,* 329-338.

Morgan, D.W., Baldini, F.D., Martin, P.E., & Kohrt, W.M. (1989). Ten kilometer performance and predicted velocity at $\dot{V}O_2max$ among well-trained male runners. *Medicine and Science in Sports and Exercise, 21,* 78-83.

Morton, R.H., Green, S., Bishop, D., & Jenkins, D.G. (1997). Ramp and constant power trials produce equivalent critical power estimates. *Medicine and Science in Sports and Exercise, 29,* 833-836.

Perrey, S., Grappe, F., Girard, A., Bringard, A., Groslambert, A., Bertucci, W., et al. (2003). Physiological and metabolic responses of triathletes to a simulated 30-min time-trial in cycling at self-selected intensity. *International Journal of Sports Medicine, 24,* 138-143.

Pringle, J.S., & Jones, A.M. (2002). Maximal lactate steady state, critical power and EMG during cycling. *European Journal of Applied Physiology, 88* (3), 214-226.

Pugh, L.G. (1970). Oxygen intake in track and treadmill running with observations on the effect of air resistance. *Journal of Physiology—London, 207,* 823-835.

Riley, M., Wasserman, K., Fu, P.C., & Cooper, C.B. (1996). Muscle substrate utilization from alveolar gas exchange in trained cyclists. *European Journal of Applied Physiology and Occupational Physiology, 72,* 341-348.

Ross, R.M., & Jackson, A.S. (1990). *Exercise concepts & computer applications.* Carmel, IN: Benchmark Press.

Safrit, M.J., Glaucia Costa, M., Hooper, L.M., Patterson, P., & Ehlert, S.A. (1988). The validity generalization of distance run tests. *Canadian Journal of Sport Sciences, 13,* 188-96.

Seiler, K.S., & Kjerland, G.Ø. (In press). The polarized training model: An optimal distribution of training intensity? *Medicine and Science in Sports and Exercise.*

Shephard, R.J. (1984). Tests of maximum oxygen intake. A critical review. *Sports Medicine, 1,* 99-124.

Shephard, R.J., & Sidney, K.H. (1978). Exercise and aging. *Exercise and Sport Sciences Reviews, 6,* 1-57.

Smith, C.G., & Jones, A.M. (2001). The relationship between critical velocity, maximal lactate steady-state velocity and lactate turnpoint velocity in runners. *European Journal of Applied Physiology, 85,* 19-26.

Snyder, A.C., Jeukendrup, A.E., Hesselink, M.K., Kuipers, H., & Foster, C. (1993). A physiological/psychological indicator of over-reaching during intensive training. *International Journal of Sports Medicine, 14,* 29-32.

Taylor, H.L., Buskirk, E., & Henschel, A. (1955). Maximal oxygen intake as an objective measure of cardiorespiratory performance. *Journal of Applied Physiology, 8,* 73-80.

Tong, T.K., Fu, F.H., & Chow, B.C. (2001). Reliability of a 5-min running field test and its accuracy in V̇O₂max evaluation. *Journal of Sports Medicine and Physical Fitness, 41*, 318-323.

Vachon, J.A., Bassett, D.R., Jr., & Clarke, S. (1999). Validity of the heart rate deflection point as a predictor of lactate threshold during running. *Journal of Applied Physiology, 87*, 452-459.

Vainamo, K., NIssila, S., Makikallio, T., and Tulppo, M. (1996). Artificial neural networks for aerobic fitness approximation. *Proceedings of the International Neural Network ICNN*, 1939-1949.

Volkov, N.I., Shirkovets, E.A., & Borilkevich, V.E. (1975). Assessment of aerobic and anaerobic capacity of athletes in treadmill running tests. *European Journal of Applied Physiology and Occupational Physiology, 34*, 121-130.

Wasserman, K., Hanson, J., Sue, D.Y., Whipp, B.J., & Casaburi, R. (1999). *Principles of exercise testing and interpretation*, 3rd ed. Philadelphia: Lippincott Williams & Wilkins.

Whipp, B.J. (1994). The slow component of O₂ uptake kinetics during heavy exercise. *Medicine and Science in Sports and Exercise, 26*, 1319-1326.

Whipp, B.J., & Wasserman, K. (1972). Oxygen uptake kinetics for various intensities of constant-load work. *Journal of Applied Physiology, 33*, 351-356.

Williford, H.N., Scharff-Olson, M., Wang, N., Blessing, D.L., Smith, F.H., & Duey, W.J. (1996). Cross validation of non-exercise predictions of V̇O₂peak in women. *Medicine and Science in Sports and Exercise, 28*, 926-930.

Zoladz, J.A., Duda, K., & Majerczak, J. (1998a). Oxygen uptake does not increase linearly at high power outputs during incremental exercise test in humans. *European Journal of Applied Physiology, 77*, 445-51.

Zoladz, J.A., Duda, K., & Majerczak, J. (1998b). Oxygen uptake-power output relationship during incremental exercise tests in humans. *Journal of Physiology—London, 511P*, 8P.

Zoladz, J.A., Szkutnik, Z., Majerczak, J., & Duda, K. (1999). Change point in V̇CO₂ during incremental exercise test: A new method for assessment of human exercise tolerance. *Acta Physiologica Scandinavica, 167*, 49-56.

## Chapter 4

Airaksinen, K., Ikaheimo, M.J., Linnaluoto, M.K., Neimela, M., & Takkunen, J.T. (1987). Impaired vagal heart rate control in coronary heart disease. *British Heart Journal, 58*, 592-597.

Akselrod, S., Gordon, D., Ubel, F.A., Shannon, D.C., Barger, A.C., & Cohen, R.J. (1981). Power spectrum analysis of heart rate fluctuations: A quantitative probe of beat to beat cardiovascular control. *Science, 230*, 200-222.

Akselrod, S. (1985). Hemodynamic regulation: Investigation by spectral analysis. *American Journal of Physiology, 249*, H867-H875.

Anrep, G.V., Pascual, W., & Rossler, R. (1936). Respiratory variations of the heart rate. I. The reflex mechanism of the respiratory arrhythmia. *Proceedings of the Royal Society, Series B, 119*, 191-217.

Appel, M.L, Berger, R.D., Saul J.P., Smith J.M., & Cohen R.J. (1989). Beat to beat variability in cardiovascular variables: Noise or music? *Journal of the American College of Cardiology, 14, 5*, 1139-1148.

Armstrong, L.E., & VanHeest, J.L. (2002). The unknown mechanism of the overtraining syndrome: Clues from depression and psychoneuroimmunology. *Sports Medicine, 32*(3), 185-209.

Azuaje, F., Dubitzky, W., Wu, X., Lopes, P., Black, N., Adamson, K., et al. (1997). A neural network approach to coronary heart disease risk assessment based on short-term measurement of RR intervals. *Proceedings Computers in Cardiology, 24*, 53-56, Sweden.

Azuaje, F., Dubitzky, W., Lopes, P., Black, N., Adamson, K., Wu, X., et al. (1999). Predicting coronary disease based on short-term electrocardiogram patterns: A neural network approach. In *Artificial Intelligence in Medicine*. Amsterdam: Elsevier.

Baselli, G., et al. (1985). Heart rate and arterial blood pressure variability signals for the evaluation of normal and pathological subjects. IEEE Computers in Cardiology Conference, Linkoping, Sweden.

Berntson, G.G., Caciopo, J.T., & Quigley, R.S. (1993). Respiratory sinus arrhythmia: Autonomic origins, physiological mechanisms, and psychophysiological implications. *Psychophysiology, 30*, 183-196.

Bigger, J.T., Jr., Fleiss, J.L., Rolnitzky, L.M., & Steinman, R.C. (1993). Frequency domain measures of heart period variability to assess risk late after myocardial infarction. *Journal of the American College of Cardiology, 21*(3), 729-736.

Billman, G.L., Schwarz, P.J., & Stone, H.L. (1982). Baroreflex control of heart rate: A predictor of sudden death. *Circulation, 66*, 874-80.

Brown, T.E., Beightol, L.A., Koh, J., & Eckberg, D.L. (1993). Important influence of respiration on human R-R interval power spectra is largely ignored. *Journal of Applied Physiology, 75* (5), 2310-2317.

Budgett R. (1998). Fatigue and underperformance in athletes: The overtraining syndrome. *British Journal of Sports Medicine, 32*(2), 107-10.

Casadei, B., Cochrane, S., Johnston, J., Conway, J., & Sleight, P. (1995). Pitfalls in the interpretation of spectral analysis of the heart rate variability during exercise in humans. *Acta Physiologica Scandinavia, 153*(2), 125-31.

Casolo, G.C., Stroder, P., Signorini, C., Calzolari, F., Zucchini, M., Balli, E., et al. (1992). Heart rate variability during the acute phase of myocardial infarction. *Circulation, 85*(6), 2073-2079.

Cerutti, M., Baselli, G., Bianchi, A., Mainardi, L.T., Signorini, M.G., & Malliani, A. (1994). Cardiovascular variability signals: From signal processing to modelling complex physiological interactions. *Automedica, 16*, 45-69.

Clark, J.W. (1978). The origin of potentials. In J.G. Webster (Ed.), *Medical instrumentation, application and designs.* Boston: Houghton Mifflin.

Davies, C.T.M., & Neilson, J.M.M. (1967). Sinus arrhythmia in man at rest. *Journal of Applied Physiology, 22*(5), 947-955.

De Meersman, R.E. (1992). Respiratory sinus arrhythmia alteration following training in endurance athletes. *European Journal of Applied Physiology, 64,* 434-36.

De Meersman, R.E. (1993a). Aging as a modulator of respiratory sinus arrhythmia. *Journal of Gerontology, 48*(2), B74-B78.

De Meersman, R.E. (1993b). Heart rate variability and aerobic fitness. *American Heart Journal, 125*(3), 726-731.

Dixon, E.M., Kamath, M.V., McCartney, N., & Fallen, E.L. (1992). Neural regulation of heart rate variability in endurance trained and sedentary controls. *Cardiovascular Research, 26,* 713-719.

Dubitzky, W., Lopes, P., White, J.A., Anderson, J., Dempsey, G., Hughes, J.G., et al. (1996). A holistic approach to coronary heart disease risk assessment using case-based reasoning. Second International Conference on Neural Networks and Expert Systems in Medicine and Health Care, Plymouth, England, August 28-30.

Eckberg, D.L. (1983). Human sinus arrhythmia as an index of vagal cardiac outflow. *Journal of Applied Physiology, 54,* 961-966.

Eckberg, D.L. (1997). Sympathovagal balance: A critical appraisal. *Circulation, 96*(9), 3224-3232.

Farrell, T.G., Bashir, Y., Cripps, T., Malik, M., Poloniecki, J., Bennett, E.D., et al. (1991). Risk stratification for arrhythmic events in post-infarction patients based on heart rate variability, ambulatory electrocardiographic variables and the signal-averaged electrocardiogram. *Journal of the American College of Cardiology, 18*(3), 687-697.

Fouad, F.M., Tarazi, R.C., Ferrario, C.M., Fighaly, S., & Alicandri, C. (1984). Assessment of parasympathetic control of heart rate by a non-invasive method. *American Journal of Physiology, 246* (Heart Circulation Physiology 15), H838-H842.

Fozzard, H.A. (1979). Electrophysiology of the heart: The effects of ischaemia. In A.M. Katz & A. Selwyn, *Handbook of physiology. Section 2: The cardiovascular system. Vol. 1. The heart* (pp. 335-356). Bethesda, Maryland: American Physiology Society.

Frick, M.H., Elovainio, R.O., & Somer, T. (1967). The mechanism of bradycardia evoked by physical training. *Cardiologia, 51,* 46-54.

Furlan, R., Piazza, S., Dell'Orto, S., Gentile, E., Cerutti, S., Pagani, M., et al. (1993). Early and late effects of exercise and athletic training on neural mechanisms controlling heart rate. *Cardiovascular Research, 27*(3), 482-488.

Goldberger, A.L., & West, B.J. (1987). Fractals in physiology and medicine. *Yale Journal Biology Medicine, 60*(5), 421-435.

Goldberger, A.L. (1991). Is the normal heartbeat chaotic or homeostatic? *News in Physiological Science, 6,* 87-91.

Goldberger, A.L. (1996). Non-linear dynamics for clinicians. Chaos theory, fractals, and complexity at the bedside. *Lancet, 347*(9011),1312-4.

Goldsmith, R.L., Bigger, J.T., Steinman, R.C., & Fleiss, J.L. (1992). Comparison of 24 hour parasympathetic activity in endurance trained and untrained young men. *Journal of the American College of Cardiology, 20,* 552-558.

Goldstein, S. (1989). Mechanisms and prevention of sudden death in coronary heart disease. *Journal of Clinical Pharmacology, 29*(5), 413-417.

Green, H. (1990). *The autonomic nervous system and exercise.* London: Chapman and Hall.

Grossman, P., Karemaker, J., & Wieling, W. (1991). Prediction of tonic parasympathetic cardiac control using respiratory sinus arrhythmia: The need for respiratory control. *Psychophysiology, 28*(2), 201-216.

Guyton, A.C. 1977. *Basic human physiology.* Philadelphia: Saunders.

Hayano, J., Mukai, S., Sakakibara, M., Okada, A., Takata, K., & Fujinami, T. (1994a). Effects of respiratory interval on vagal modulation of heart rate. *American Journal of Physiology, 267,* H33-H40.

Hayano, J., Sakakibara, Y., Yamada, M., Ohte, N., Fujinami, T., Yokoyama, K., et al. (1990). Decreased magnitude of heart rate spectral components in coronary artery disease. Its relation to angiographic severity. *Circulation, 81*(4), 1217-1224.

Hayano, J., Yamada, A., Mukai, S., Sakakibara, Y., Yamada, M., Ohte, N., et al. (1991). Severity of atherosclerosis correlates with the respiratory component of heart rate variability. *American Heart Journal, 121*(4), 1070-79.

Hayano J., Taylor, J.A., Mukai, S., Okada, A., Watanabe, Y., Takata, K., et al. (1994b). Assessment of frequency shifts in R-R interval variability and respiration with complex demodulation. *Journal of Applied Physiology, 77*(6), 2879-2888.

Hedelin, R., Wiklund, U., Bjerle, P., et al. (2000). Cardiac autonomic imbalance in an overtrained athlete. *Medicine and Science in Sports and Exercise, 32*(9), 1531-3.

Hedelin, R., Bjerle, P., & Henriksson-Larsén, K. (2001). Heart rate variability in athletes: Relationship with central and peripheral performance. *Medicine and Science in Sports and Exercise, 33*(8), 1394-8.

Hirsch, J.A., & Bishop, B. (1981). Respiratory sinus arrhythmia in humans: How breathing pattern modulates heart rate. *American Journal of Physiology, 241*(4), H620-H629.

Hrushesky, W.J.M. (1991). Quantitative respiratory sinus arrhythmia analysis: A simple non-invasive, reimbursable measure of cardiac wellness and dysfunction. *Annals of the New York Academy of Sciences, 618*(Feb), 67-101.

Janssen, M.J., Swenne, C.A., de Bie, J., Rompelman, O., & van Bemmel, J.H. (1993). Methods in heart rate variability analysis: Which tachogram should we choose? *Computer Methods and Programs in Biomedicine, 41*(1), 1-8.

Jennings, J.R., & McKnight, J.D. (1994). Inferring vagal tone from heart rate variability. *Psychosomatic Medicine, 56*(3), 194-196.

Kamath, M.V., Fallen, E.L., & McKelvie, R. (1991). Effects of steady state exercise on the power spectrum of heart rate variability. *Medicine and Science in Sports and Exercise, 23*(4), 428-34.

Kannel, W.B., Plehn, J.F., & Cupples, L.A. (1988). Cardiac failure and sudden death in the Framingham Study. *American Heart Journal, 115*(4), 869-875.

Kenney, W.L. (1985). Parasympathetic control of resting heart rate: Relationship to aerobic power. *Medicine and Science in Sports and Exercise, 17*(4), 451-455.

Kleiger, R.E., Miller, J.P., Bigger, J.T., Jr., & Moss, A.J. (1987). Decreased heart rate variability and its association with increased mortality after acute myocardial infarction. *American Journal of Cardiology, 59*(4), 256-262.

Kyrozi, E., Maounis, T., Chiladakis, I., Vassilikos, V., Manolis, A., & Cokkinos, D. (1995). Short-term reproducibility of the time and frequency domain parameters of heart rate variability. *Computers in Cardiology IEEE conference,* 481-484.

Lehmann, M., Foster, C., Dickhuth, H.H., et al. (1998). Autonomic imbalance hypothesis and overtraining syndrome. *Medicine and Science in Sports and Exercise, 30*(7), 1140-5.

Liao, D., Cai, J., Rosamond, W.D., Barnes, R.W., Hutchinson, R.G., Whitsel, E.A., et al. (1997). Cardiac autonomic function and incident coronary heart disease: A population-based case-cohort study. The ARIC study. Atherosclerosis Risk in Communities Study. *American Journal of Epidemiology, 145*(8), 696-706.

Lombardi, F. (2002). Clinical implications of present physiological understanding of HRV components. *Cardiac Electrophysiology Review, 6*(3), 245-9.

Makikallio, T.H., Tapanainen, J.M., Tulppo, M.P., & Huikuri, H.V. (2002). Clinical applicability of heart rate variability analysis by methods based on nonlinear dynamics. *Cardiac Electrophysiology Review, 6*(3), 250-5.

Malliani, A., Pagani, M., & Lombardi, F. (1986). Positive feedback reflexes. In A. Zanchetti & R.C. Tarazi, *Handbook of hypertension, Vol. 8, Pathophysiology of hypertension* (pp. 69-81). Amsterdam: Elsevier.

Malliani, A., Pagani, M., Lombardi, F., & Cerutti, S. (1991). Cardiovascular neural regulation explored in the frequency domain. *Circulation, 84*(2), 482-492.

Mehlsen, J., Pagh, K., Nielsen, J.S., Sestoft, L., & Nielson, S.L. (1987). Heart rate response to breathing: Dependency upon breathing pattern. *Clinical Physiology, 7,* 115-124.

Molgaard, H., Sorensen, K.E., & Bjerregaard, P. (1991). Circadian variation and influence of risk factors on heart rate variability in healthy subjects. *American Journal of Cardiology, 68*(8), 777-784.

Monti, A., Médigue, C., & Mangin, L. (2002). Instantaneous parameter estimation in cardiovascular time series by harmonic and time-frequency analysis. *IEEE Transactions on Biomedical Engineering, 49*(12), 1547-1556.

Myers, G.A., Martin, G.J., Magid, N.M., Barnett, P.S., Schaad, J.W., Weiss, J.S., et al. (1986). Power spectral analysis of heart rate variability in sudden death: Comparison to other methods. *IEEE Transactions on Biomedical Engineering, 33,* 1149-56.

O'Brien, I.A., O'Hare, P., & Corrall, R.J. (1986). Heart rate variability in healthy subjects: Effect of age and the derivation of normal ranges for tests of autonomic function. *British Heart Journal, 55*(4), 348-354.

Odemuyiwa, O., Malik, M., Farrell, T., Bashir, Y., Poloniecki, J., & Camm, J. (1991). Comparison of the predictive characteristics of heart rate variability index and left ventricular ejection fraction for all-cause mortality, arrhythmic events and sudden death after acute myocardial infarction. *American Journal of Cardiology, 68*(5), 434-439.

Pagani, M., Lombardi, F., Guzetti, S., Rimolda, O., Farlan, R., Pizzinelli, P., et al. (1986). Power spectral analysis of heart rate and arterial pressure variabilities as a marker of sympathovagal interaction in man and conscious dog. *Circulation Research, 59,* 178-193.

Park, R.C., & Crawford, M.N. (1985). Heart of the athlete. *Current Problems in Cardiology, 5,* 6-73.

Penaz, J., Roukenz, J., & Van Der Waal, H.J. (1968). Spectral analysis of some spontaneous rhythms in the circulation. In H. Drischel & N. Tiedt (Eds.), *Biokybernetik, Bd. I.* Leipzig: Karl Marx University, 233-241.

Pichot, V., Busso, T., Roche, F., Garet, M., Costes, F., Duverney, D., et al. (2002). Autonomic adaptations to intensive and overload training periods: A laboratory study. *Medicine and Science in Sports and Exercise, 34*(10),1660-6.

Pichot, V., Roche, F., Gaspoz, J.M., et al. (2000). Relation between heart rate variability and training load in middle-distance runners. *Medicine and Science in Sports and Exercise, 32*(10), 1729-36.

Pincus, S.M. (1991). Approximate entropy as a measure of system complexity. *Proceedings of the National Academy of Science (USA), 88*(6), 2297-301.

Pincus, S.M. (1995). Approximate entropy (ApEn) as a complexity measure. *Chaos, 5*(1), 110-117.

Pitzalis, M.V., Mastropasqua, F., Massari, F., Forleo, C., Di Maggio, M., Passantino, A., et al. (1996). Short- and long-term reproducibility of time and frequency domain heart rate variability measurements in normal subjects. *Cardiovascular Research, 32*(2), 226-33.

Pomeranz, B., Macauley, R.J.B., Caudill, M.A., et al. (1985). Assessment of autonomic function in humans by heart rate spectral analysis. *American Journal of Physiology (Heart Circulation Physiology), 248,* H151-H153.

Portier, H., Louisy, F., Laude, D., et al. (2001). Intense endurance training on heart rate and blood pressure variability in runners. *Medicine and Science in Sports and Exercise, 33*(7), 1120-5.

Puig, J., Fretas, J., Carvalho, M.T., Puga, N., Ramos, J., Fernandes, P., et al. (1993). Spectral analysis of heart rate variability in athletes. *Journal of Sports Medicine and Physical Fitness, 33,* 44 48.

Rich, M.W., Saini, J.S., Kleiger, R.E., Carney, R.M., teVelde, A., Freedland, K.E. (1988). Correlation of heart rate variability with clinical and angiographic variables and late mortality after coronary angiography. *American Journal of Cardiology, 62*(10), 714-717.

Rinoli, T., & Porges, S.W. (1997). Inferential and descriptive influences on measures of respiratory sinus arrhythmia: Sampling rate, R-wave trigger accuracy, and variance estimates. *Psychophysiology, 34,* 613-621.

Sacknoff, D.M., Gleim, G.B., Stachenfeld, N., & Coplan, N.L. (1994). Effect of athletic training on heart rate variability. *American Heart Journal, 127,* 1275-1278.

Sayers, B. (1973). Analysis of heart rate variability. *Ergonomics, 16,* 17.

Seals, D.R., & Chase, P.B. (1989). Influence of physical training on heart rate variability and baroreflex circulatory control. *Journal of Applied Physiology, 66*(4), 1886-95.

Shin, K., Minamitani, H., Onishi, S., Yamazaki, H., & Lee, M. (1995). The power spectral analysis of heart rate variability in athletes during dynamic exercise—part I. *Clinical Cardiology, 18*(10), 583-6.

Singer, D.H., Martin, G.J., Magid, N., Weiss, J.S., Schaad, J.W., Kehoe, R., et al. (1988). Low heart rate variability and sudden cardiac death. *Journal of Electrocardiology, 21*(Suppl.), S46-S55.

Smith, M.L., Hudson, D.L., Graitzer, H.M., & Raven, P.B. (1989). Exercise training bradycardia: The role of autonomic balance. *Medicine and Science in Sports and Exercise, 21*(1), 40-44.

Somers, V.K., Conway, J., Johnston, J., & Sleight, P. (1991). Effects of endurance training on baroreflex sensitivity and blood pressure in borderline hypertension. *Lancet, 337*(8754), 1363-1368.

Somers, V.K., Conway, J., & Sleight, P. (1986). The effect of physical training on home blood pressure measurement in normal subjects. *Journal of Hypertension, 4*(Suppl. 6), S657-58.

Task Force of the European Society of Cardiology & the North American Society of Pacing and Electrophysiology. (1996). Heart rate variability: Standards of measurement, physiological interpretation and clinical use. *Circulation, 93*(5),1043-1065.

Tulppo, M.P., Mäkikallio, T.H., Takala, T.E.S., Seppänen, T., & Huikuri, H.V. (1996). Quantitative beat-to-beat analysis of heart rate dynamics during exercise. *American Journal of Physiology, 271,* H244-H252.

Uusitalo, A.L., Uusitalo, A.J., & Rusko, H.K. (2000). Heart rate and blood pressure variability during heavy training and overtraining in the female athlete. *International Journal of Sports Medicine, 21*(1), 45-53.

Uusitalo, A.L.T. (2001). Overtraining: Making a different diagnosis and implementing targeted treatment. *Physician and Sports Medicine, 29*(5), 35-50.

Vaishnav, S., Stevenson, R., Marchant, B., Lagi, K., Ranjadayalan, K., & Timmis, A.D. (1994). Relation between heart rate variability early after acute myocardial infarction and long-term mortality. *American Journal of Cardiology, 73*(9), 653-657.

Van Capelle, F.L. (1987). Electrophysiology of the heart. In R.I. Kitney & O. Rompelman (Eds.), *The beat-by-beat investigation of cardiovascular function: Measurement, analysis and applications* (pp. 3-26). Oxford: Clarendon Press.

Vassalle, M. (1976). *Cardiac physiology for the clinician.* New York: New York Academic Press.

Verlinde, D., Beckers, F., Ramaekers, D., & Aubert, A.E. (2001). Wavelet decomposition analysis of heart rate variability in aerobic athletes. *Autonomic Neuroscience, 90*(1-2), 138-41.

Webb, S.W., Adgey, A.A., & Pantridge, J.F. (1972). Autonomic disturbance at onset of acute myocardial infarction. *British Medical Journal, 3,* 818, 89-92.

White, J.A., Semple, E., & Norum, R.A. (1990). Sinus arrhythmia as an estimate of maximal aerobic power in man. *Journal of Physiology, 432,* 41P.

Wolf, M.W., Varigos, G.A., Hunt, D., & Sloman, J.G. (1978). Sinus arrhythmia in acute myocardial infarction. *Medical Journal of Australia, 2,* 52-53.

Yamamoto, Y., Hughson, R.L., & Peterson, J.C. (1991). Autonomic control of heart rate during exercise studied by heart rate variability spectral analysis. *Journal of Applied Physiology, 71*(3), 1136-42.

# Chapter 5

Allen, W.K., Seals, D.R., Hurley, B.F., Ehsani, A.A., & Hagberg, J.M. (1985). Lactate threshold and distance running performance in young and older endurance athletes. *Journal of Applied Physiology, 58,* 1281-1284.

Aunola, S., & Rusko, H. (1984). Reproducibility of aerobic and anaerobic thresholds in 20-50 year old men. *European Journal of Applied Physiology, 53,* 260-266.

Bang, O. (1936). The lactate content of the blood during and after muscular exercise in men. *Skandivica Archives of Physiology Supplement, 10,* 51-82.

Beaver, W.L., Wasserman, K., & Whipp, B.J. (1985). Improved detection of lactate threshold during exercise using a log-log transformation. *Journal of Applied Physiology, 59,* 1936-1940.

Beaver, W.L., Wasserman, K., & Whipp, B.J. (1986). A new method for detecting the anaerobic threshold by gas exchange. *Journal of Applied Physiology, 60,* 2020-2027.

Beneke, R. (1995). Anaerobic threshold, individual anaerobic threshold, and maximal lactate steady state in rowing. *Medicine and Science in Sports and Exercise, 6,* 863-867.

Beneke, R., Hutler, M., & Leithauser, R. (2000). Maximal lactate-steady-state independent of performance. *Medicine and Science in Sports and Exercise, 6,* 1135-1139.

Beneke, R., & von Duvillard, S.P. (1996). Determination of maximal lactate steady state response in selected sports events. *Medicine and Science in Sports and Exercise, 2,* 241-246.

Bishop, D., Jenkins, D., McEniery, M., & Carey, M. (2000). Relationship between plasma lactate parameters and muscle characteristics in female cyclists. *Medicine and Science in Sports and Exercise, 3,* 1088-1093.

Bishop, P.A., Smith, J.F., Kime, J.C., Mayo, J.M., & Murphy, M. (1992). Influence of blood handling techniques on lactic acid concentrations. *International Journal of Sports Medicine, 13,* 56-59.

Bishop, P.A., Smith, J.F., Kime, J.C., Mayo, J.M., & Tin, Y.H. (1992). Comparison of a manual and an automated enzymatic technique for detecting blood lactate concentrations. *International Journal of Sports Medicine, 13,* 36-39.

Bodner, M., & Rhodes, E. (2000). A review of the concept of the heart rate deflection point. *Sports Medicine, 1,* 31-46.

Brooks, G.A. (1985). Anaerobic threshold: Review of the concept and directions for future research. *Medicine and Science in Sports and Exercise, 17,* 22-31.

Bunc, V., Heller, J., Leso, J., Sprynarova, S., & Zdanowicz, S. (1987). Ventilatory threshold in various groups of highly trained athletes. *International Journal of Sports Medicine, 8,* 275-280.

Caiozzo, V.J., Davis, J.A., Ellis, J.F., Azua, J.L., Vandagriff, R., Prietto, C.A., et al. (1982). A comparison of gas exchange indices used to detect the anaerobic threshold. *Journal of Applied Physiology, 53,* 1184-1189.

Cellini, M., Vitiello, P., Nagliati, A., Ziglio, P.G., Martinelli, S., Ballarin, E., et al. (1986). Noninvasive determination of the anaerobic threshold in swimming. *International Journal of Sports Medicine, 7,* 347-351.

Cheng, B., Kuipers, H., Snyder, A.C., Keizer, H.A., Jeukendrup, A., & Hesselink, M. (1992). A new approach for the determination of ventilatory and lactate thresholds. *International Journal of Sports Medicine, 13,* 518-522.

Christansen J., Douglas, C.G., & Haldand, J.S. (1914). The absorption and dissociation of carbon dioxide of human blood. *Journal of Physiology (London), 48,* 244-271.

Coen, B., Schwarz, L., Urhausen, A., & Kinnderman, W. (1991). Control of training in middle and long distance running by means of the individual anaerobic threshold. *International Journal of Sports Medicine, 12,* 519-524.

Coen, B., Urhausen, A., & Kindermann, W. (2001). Individual anaerobic threshold: Methodology aspects of its assessment in running. *International Journal of Sports Medicine, 22,* 8-16.

Conconi, F., Ferrari, M., Ziglio, P.G., Droghetti, P., & Codeca, L. (1982). Determination of the anaerobic threshold by noninvasive field test I runners. *Journal of Applied Physiology, 52,* 869-873.

Costill, D.L., Branam, G., Eddy, D., & Sparks, K. (1971). Determinants of marathon running success. *International Zeitschrift for Angewande Physiologie, 29,* 249-252.

Costill, D.L., Thomason, H., & Roberts, E. (1973). Fractional utilization of the aerobic capacity during distance running. *Medicine and Science in Sports. 5,* 248-252.

Coyle, E.F., Coggan, A.R., Hopper, M.K., & Walters, T.J. (1988). Determinants of endurance in well-trained cyclist. *Journal of Applied Physiology, 64,* 2622-2630.

Daniels, J.T., Yarbrough, R.A., & Foster, C. (1978). Changes in $\dot{V}O_2$max and running performance with training. *European Journal of Applied Physiology, 39,* 249-254.

Davis, J.A. (1985). Anaerobic threshold: Review of the concept and directions for future research. *Medicine and Science in Sports and Exercise, 17,* 6-18.

Davis, J.A., Caiozzo, V.J., Lamarra, N., Ellis, J.F., Vandagriff, R., Prietto, C.A., et al. (1983). Does the gas exchange anaerobic threshold occur at a fixed blood lactate concentration of 2 or 4 mM. *International Journal of Sports Medicine, 4,* 89-93.

Davis, J.A., Frank, M.H., Whipp, B.J., & Wasserman, K. (1979). Anaerobic threshold by endurance training in middle aged men. *Journal of Applied Physiology, 46,* 1039-1046.

Davis, J.A., Vodak, P., Wilmore, J.H., Vodak, J., & Kurtz, P. (1976). Anaerobic threshold and maximal aerobic power for three modes of exercise. *Journal of Applied Physiology, 41,* 544-550.

Davis, J.A., Whipp, B.J., Lamarra, N., Huntsman, D.J., Frank, M.H., & Wasserman, K. (1982). Effect of ramp slope on demonstration of aerobic parameters from the ramp exercise test. *Medicine and Science in Sports and Exercise, 14,* 339-344.

Davis, J.A., Whipp, B.J., & Wasserman, K. (1980). The relation of ventilation to metabolic rate during moderate exercise in man. *Journal of Applied Physiology, 44,* 97-108.

Denis, C., Fouquet, R., Poty, P., Geyssant, A., & Lacour, J.R. (1982). Effect of 40 weeks of endurance training on the anaerobic threshold. *International Journal of Sports Medicine, 3,* 208-214.

Donovan, C.M., & Brooks, G.A. (1983). Endurance training affects lactate clearance, not lactate production. *American Journal of Physiology, 244,* E83-E92.

Dotan, R., Rotstein, A., & Grodjinovsky, A. (1989). Effect of training load on OBLA determination. *International Journal of Sports Medicine, 10,* 346-351.

Droghetti, P., Borsetto, C., Casoni, I., Cellini, M., Ferrari, M., Paolini, A.R., et al. (1985). Noninvasive determination of the anaerobic threshold in canoeing, cross-country skiing, cycling, roller and ice skating, rowing and walking. *European Journal of Applied Physiology, 53,* 299-303.

Dwyer, J., & Bybee, R. (1983). Heart rate indices of the anaerobic threshold. *Medicine and Science in Sports and Exercise, 15,* 72-76.

Farrell, P.A., Wilmore, J.H., Coyle, E.F., Billing, J.E., & Costill, D.L. (1979). Plasma lactate accumulation and distance running performance. *Medicine and Science in Sports and Exercise, 11,* 338-344.

Fell, J.W., Rayfield, J.M., Gulbin, J.P., & Gaffney, P.T. (1998). Evaluation of the Accusport lactate analyser. *International Journal of Sports Medicine, 19,* 199-204.

Fernandez-Garcia, B.J., Perez-Landaluce, J., Rodriguez-Alonso, M., & Terrados N. (2000). Intensity of exercise during road race pro-cycling competition. *Medicine and Science in Sports and Exercise, 32,* 1002-1006.

Fohrenbach, R., Mader, A., & Hollman, W. (1987). Determination of endurance capacity and prediction of exercise intensities for training and competition in marathon runners. *International Journal of Sports Medicine, 8,* 11-18.

Forster, H.V., Dempsey, J.A., Thomson, J., Vidruk, E., & DoPico, G.A. (1972). Estimation of arterial $PO_2$, $PCO_2$, pH and lactate from arterialized venous blood. *Journal of Applied Physiology, 32,* 134-137.

Foster, C., Cohen, J., Donovan, K., Gastrau, P., Killian, P.J., Schrager, M., et al. (1993). Fixed time versus fixed distance protocols for the blood lactate profile in athletes. *International Journal of Sports Medicine, 14,* 264-268.

Foster, C., Fitzgerald, D., & Spatz, P. (1999). Stability of the blood lactate-heart rate relationship in competitive athletes. *Medicine and Science in Sports and Exercise, 4,* 578-582.

Foster, C., Pollock, M.L., Farrell, P.A., Maksud, M.G., Anholm, J.D., & Hare, J. (1982). Training responses of speed skaters during a competitive season. *Research Quarterly for Exercise and Sport, 53,* 243-246.

Foster, C., Spatz, P., & Georgakopoulos, N. (1999). Left ventricular function in relation to the heart rate performance curve. *Clinical Exercise Physiology, 1,* 29-32.

Foster, C., Snyder, A.C., Thompson, N.N., & Kuettal, K. (1988). Normalization of the blood lactate profile in athletes. *International Journal of Sports Medicine, 9,* 198-200.

Foxdal, P., Sjodin, A., Ostman, B., & Sjodin, B. (1991). The effect of different blood sampling studies and analyses on the relationship between exercise intensity and 4.0 mmol/L blood lactate concentration. *European Journal of Applied Physiology, 63,* 52-54.

Foxdal, P., Sjodin, A., & Sjodin, B. (1996). Comparison of blood lactate concentrations obtained during incremental and constant intensity exercise. *International Journal of Sports Medicine, 17,* 360-365.

Foxdal, P., Sjodin, S., Sjodin, A., & Ostman, B. (1994). The validity and accuracy of blood lactate measurements for prediction of maximal endurance running capacity. *International Journal of Sports Medicine, 15,* 89-95.

Francis, K.T., McClatchey, P.R., Sumsion, J.R., & Hansen, D.E. (1989). The relationship between anaerobic threshold and heart rate linearity during cycle ergometry. *European Journal of Applied Physiology, 59,* 273-277.

Freund, H., Ouono-Euguelle, S., Heitz, A., Marbach, J., Ott, C., & Gartner, M. (1989). Effect of exercise duration on lactate kinetics after short muscular exercise. *European Journal of Applied Physiology, 58,* 534-542.

Freund, H., Ouono-Euguelle, S., Heitz, A., Marbach, J., Ott, C., Gartner, M., et al. (1990). Comparative lactate kinetics after short and prolonged submaximal exercise. *International Journal of Sports Medicine, 11,* 284-288.

Freund, H., & Zouloumian, P. (1981a). Lactate after exercise in man: Evolution kinetics in arterial blood. *European Journal of Applied Physiology, 46,* 121-133.

Freund, H., & Zouloumian, P. (1981b). Lactate after exercise in man: Physiological observations and model predictions. *European Journal of Applied Physiology, 46,* 161-176.

Gaesser, G.A., Poole, D.C., & Gardner, B.P. (1984). Dissociation between $\dot{V}O_2$max and ventilatory threshold responses to endurance training. *European Journal of Applied Physiology, 53,* 242-247.

Gladden, L.B., Yates, J.W., Stremel, R.W., & Stamford, B.A. (1985). Gas exchange and lactate anaerobic thresholds: Inter- and intra-evaluator agreement. *Journal of Applied Physiology, 58,* 2082-2089.

Green, H.J., Hughson, R.L., Orr, G.W., & Ranney, D.A. (1983). Anaerobic threshold, blood lactate and muscle metabolites in progressive exercise. *Journal of Applied Physiology, 54,* 1032-1038.

Hagberg, J.M., & Coyle, E.F. (1983). Physiological determinants of endurance performance as studied in competitive racewalkers. *Medicine and Science in Sports and Exercise, 15,* 287-289.

Heck, H., Mader, A., Hess, G., Mucke, S., Muller, R., & Hollmann, W. (1985). Justification of the 4 mmol/L lactate threshold. *International Journal of Sports Medicine, 6,* 117-130.

Heitcamp, H.Ch., Holdt, M., & Scheib, M. (1991). The reproducibility of the 4 mmol/L lactate threshold in trained and untrained women. *International Journal of Sports Medicine, 12,* 363-368.

Hill, A.V., Long, C.N.H., & Lupton, H. (1924). Muscular exercise, lactic acid and the supply and utilization of oxygen: Part VI. The oxygen debt at the end of exercise. *Proceedings of the Royal Society of London, 97,* 127-137.

Hofman, P., Pokan, R., Preidler, K., Leitner, H., Szolar, D., Eber, B., et al. (1994). Relationship between heart rate threshold, lactate turn point and myocardial function. *International Journal of Sports Medicine, 15,* 232-237.

Hofman, P., Pokan, R., von Duvillard, S.P., Seibert, F.J., Azeiker, R., & Schmid, P. (1997). Heart rate performance curve during incremental cycle ergometer exercise in healthy young male subjects. *Medicine and Science in Sports and Exercise, 6,* 762-768.

Hollmann, W. (1985). Historical remarks on the development of the aerobic-anaerobic threshold up to 1966. *International Journal of Sports Medicine, 6,* 109-116.

Hollmann, W., Rost, R., Liesen, H., Dufaux, B., Heck, H., & Mader, A. (1981). Assessment of different forms of physical activity with respect to preventive and rehabilitative cardiology. *International Journal of Sports Medicine, 2,* 67-80.

Hoogeveen, A.R., & Schep, G. (1997). The plasma lactate response to exercise and endurance performance: Relationship in elite triathletes. *International Journal of Sports Medicine, 18,* 526-530.

Hoogeveen, A.R., Schep, G., & Hoogsteen, J. (1999). The ventilatory threshold, heart rate, and endurance performance: Relationships in elite cyclists. *International Journal of Sports Medicine, 20,* 114-117.

Hughes, E.F., Turner, S.C., & Brooks, G.A. (1982). Effects of glycogen depletion and pedaling speed on anaerobic threshold. *Journal of Applied Physiology, 52,* 1598-1607.

Hughson, R.L., & Green, H.J. (1982). Blood acid-base and lactate relationships studied by ramp work tests. *Medicine and Science in Sports, 14,* 297-302.

Hurley, B.F., Hagberg, J.M., Allen, W.K., Seals, D.R., Young, J.C., Cuddihee, R.W., et al. (1984). Effect of training on blood lactate levels during submaximal exercise. *Journal of Applied Physiology, 56,* 1260-1264.

Ivy, J.L., Costill, D.L., van Handel, P.J., Essig, D.A., & Lower, R.W. (1981). Alteration in the lactate threshold with changes in substrate availability. *International Journal of Sports Medicine, 2,* 139-142.

Iwaoka, K., Fuchi, T., Higuchi, M., & Kobayashi, S. (1988). Blood lactate accumulation during exercise in older endurance runners. *International Journal of Sports Medicine, 9,* 253-256.

Iwaoka, K., Fuchi, T., Atomi, Y., & Miyashita, M. (1988). Lactate, respiratory compensation thresholds and distance running performance in runners of both sexes. *International Journal of Sports Medicine, 9,* 306-309.

Jacobs, I. (1981). Lactate, muscle glycogen and exercise performance in man. *Acta Physiologica Scandinavica (Suppl.) 495,* 1-35.

Jacobs, I. (1986). Blood lactate: Implications for training and sports performance. *Sports Medicine, 3,* 10-25.

Janssen, P.G.J.M. (1987). *Training, lactate, pulse-rate.* Oulu, Finland: Polar Electro Oy.

Jones, A., & Doust, J. (1998). The validity of the lactate minimum test for determination of the maximal lactate steady state. *Medicine and Science in Sports and Exercise, 8,* 1304-1313.

Jorfeldt, L., Juhein-Dannfeldt, A., & Karlsson, J. (1978). Lactate release in relation to tissue lactate in human skeletal muscle during exercise. *Journal of Applied Physiology, 44,* 350-352.

Karlsson, J. (1971). Lactate in working muscles after prolonged exercise. *Acta Physiologica Scandinavica, 82,* 123-130.

Karlsson, J., Diamant, B., & Saltin, B. (1971). Muscle metabolites during submaximal and maximal exercise in man. *Scandinavian Journal of Clinical and Laboratory Investigation, 26,* 385-394.

Karlsson, J., & Jacobs, I. (1982). Onset of blood lactate accumulation during muscular exercise as a threshold concept: Theoretical considerations. *International Journal of Sports Medicine, 3,* 190-201.

Karlsson, J., Jacobs, I., Sjodin, B., Tesch, P., Kaiser, P., Sahl, O., et al. (1983). Semiautomatic blood lactate assay: Experiences from an exercise laboratory. *International Journal of Sports Medicine, 4,* 52-55.

Karlsson, J., Nordesjo, L.O., Jorfeldt, L., & Saltin, B. (1972). Muscle lactate, ATP and CP levels during exercise after physical training in man. *Journal of Applied Physiology, 29,* 598-602.

Karlsson, J., & Saltin, B. (1970). Lactate, ATP and CP in working muscles during exhaustive exercise in man. *Journal of Applied Physiology, 29,* 598-602.

Kindermann, W., Simon, G., & Keul, J. (1979). The significance of the aerobic-anaerobic transition for the determination of work load intensities during endurance training. *European Journal of Applied Physiology, 42,* 25-34.

Kumagai, S., Tanaka, K., Matsuura, Y., Matsuzaka, A., Hirakoba, K., & Asano, K. (1982). Relationships of the anaerobic threshold with the 5 km, 10 km, and 10 mile races. *European Journal of Applied Physiology, 49,* 13-23.

LaFontaine, T.P., Londeree, B.R., & Spath, W.K. (1981). The maximal steady state versus selected running events. *Medicine and Science in Sports and Exercise, 13,* 190-192.

Lehmann, M., Berg, A., Kapp, R., Wessinghage, T., & Keul, J. (1983). Correlations between laboratory testing and distance running performance in marathoners of similar performance ability. *International Journal of Sports Medicine, 4,* 226-230.

Ljunggren, G., Ceci, R., & Karlsson, J. (1987). Prolonged exercise at a constant load on a bicycle ergometer: Ratings of perceived exertion and leg aches and pain as well as measurements of blood lactate accumulation and heart rate. *International Journal of Sports Medicine, 8,* 109-116.

Londeree, B.R., & Ames, S.A. (1975). Maximal steady state versus state of conditioning. *European Journal of Applied Physiology, 34,* 269-278.

Lucia, A., Hoyos, J., Carvajal, A., & Chicharro, J.L. (1999). Heart rate response to professional road cycling: The Tour De France. *International Journal of Sports Medicine, 20,* 167-172.

Lucia, A., Hoyos, J., Perez, M., & Chicharro, J.L. (2000). Heart rate and performance parameters in elite cyclists: A longitudinal study. *Medicine and Science in Sports and Exercise, 32,* 1777-1782.

Maassen, N., & Busse, M.W. (1989). The relationship between lactic acid and work load: A measure for endurance capacity or an indicator of carbohydrate deficiency? *European Journal of Applied Physiology, 58,* 728-737.

MacRae, H.S.H., Dennis, S.C., Bosch, A.N., & Noakes, T.D. (1992). Effects of training on lactate production and removal during progressive exercise in humans. *Journal of Applied Physiology, 72,* 1649-1656.

Mader, A., & Heck, H. (1986). A theory of the metabolic origin of "anaerobic threshold." *International Journal of Sports Medicine, 7,* 45-65.

Martin, L., & Whyte, G.P. (2000). Comparison of critical swimming velocity and velocity at lactate threshold in elite triathletes. *International Journal of Sports Medicine, 21,* 366-368.

Mazzeo, R.S., Brooks, G.A., Schoeller, D.A., & Budinger, T.F. (1986). Disposal of blood lactate in human during rest and exercise. *Journal of Applied Physiology, 60,* 232-242.

McLellan, T.M. (1985). Ventilatory and plasma lactate response with different exercise protocols: A comparison of methods. *International Journal of Sports Medicine, 6,* 30-35.

McLellan, T.M., & Cheung, K.S.Y. (1992). A comparative evaluation of the individual anaerobic threshold and the critical power. *Medicine and Science in Sports and Exercise, 24,* 543-550.

McLellan, T.M., Cheung, K.S.Y., & Jacobs, I. (1990). Incremental test protocol, recovery mode and the individual anaerobic threshold. *International Journal of Applied Physiology, 12,* 190-195.

McLellan, T.M., & Gass, G.C. (1989a). Metabolic and cardiorespiratory responses relative to the anaerobic threshold. *Medicine and Science in Sports and Exercise, 21,* 191-198.

McLellan, T.M., & Gass, G.C. (1989b). The relationship between the ventilation and lactate thresholds following normal, low and high carbohydrate diets. *European Journal of Applied Physiology, 58,* 568-576.

McLellan, T.M., & Skinner, J.S. (1981). The use of the aerobic threshold as a basis for training. *Canadian Journal of Applied Sport Sciences, 6,* 197-201.

McLellan, T.M., & Skinner, J.S. (1982). Blood lactate removal during active recovery related to the aerobic threshold. *International Journal of Sports Medicine, 3,* 224-229.

McNaughton, L.R., Thompson, D., Philips, G., Backx, K., & Crickmore, L. (2002). A comparison of the Lactate Pro, Accusport, Analox GM7 and Kodak Ektachem lactate analysers in normal, hot and humid conditions. *International Journal of Sports Medicine, 23,* 130-135.

Myburgh, K., Viljoen, A., & Tereblanche, S. (2001). Plasma lactate concentrations for self-selected maximal effort lasting 1 h. *Medicine and Science in Sports and Exercise, 1,* 152-156.

Mujika, I., & Padilla, S. (2001). Physiological performance characteristics of male professional road cyclist. *Sports Medicine, 31,* 479-487.

Nagel, F., Robinson, D., Howley, E., Daniels, J., Baptista, G., & Stoedefalk, K. (1970). Lactate accumulation during running at submaximal aerobic demands. *Medicine and Science in Sports, 2,* 182-186.

Nemoto, L., Iwaoka, I., Funato, K., Yoshioka, N., & Miyashita, M. (1988). Aerobic threshold, anaerobic threshold and maximal oxygen uptake of Japanese speed skaters. *International Journal of Sports Medicine, 9,* 433-437.

Olbrecht, J., Madsen, Ø., Mader, A., Liesen, A., & Hollmann, W. (1985). Relationship between swimming velocity and lactic concentration during continuous and intermittent training exercises. *International Journal of Sports Medicine, 6,* 74-77.

Orok, C.J., Hughson, R.L., Green, H.J., & Thompson, J.A. (1989). Blood lactate responses in incremental exercise as predictors of constant load performance. *European Journal of Applied Physiology, 59,* 262-267.

Owles, W.H. (1930). Alterations in the lactic acid content of the blood as a result of light exercise and associated changes in the $CO_2$ combining power of the blood and in the alveolar $CO_2$ pressure. *Journal of Physiology, 69,* 214-237.

Oyono-Euguelle, S., Gartner, M., Marbach, J., Heitz, A., Ott, C., & Freund, H. (1989). Comparison of arterial and venous blood lactate kinetics after short exercise. *International Journal of Sports Medicine, 10,* 16-24.

Peronnet, F., Thibault, G., Rhodes, E.C., & McKenzie, D.C. (1987). Correlation between ventilatory threshold and endurance capability in marathon runners. *Medicine and Science in Sports and Exercise, 19,* 610-615.

Pokan, R., Hofman, P., Preidler, K., Leitner, H., Dusleag, J., Eber, et al. (1993). Correlation between inflection of heart rate/work performance curve and myocardial function in exhausting cycle ergometer exercise. *European Journal of Applied Physiology, 67,* 385-388.

Poole, D.C., & Gaesser, G.A. (1985). Response of ventilatory and lactate thresholds to continuous and interval training. *Journal of Applied Physiology, 28,* 1115-1121.

Poole, D.C., Ward, S.A., & Whipp, B.J. (1990). The effects of training on the metabolic and respiratory profile of high-intensity cycle ergometer exercise. *European Journal of Applied Physiology, 59,* 421-429.

Reybrouck, T., Ghesquiere, J., Cattaert, A., Fagard, R., & Amery, A. (1983). Ventilatory threshold during short and long term exercise. *Journal of Applied Physiology, 55,* 1694-1700.

Ribeiro, J.P., Fielding, R.A., Hughes, V., Black, A., Bochese, M.A., & Knuttgen, H.G. (1985). Heart rate break point may coincide with the anaerobic but not the aerobic threshold. *International Journal of Sports Medicine, 6,* 220-224.

Rieu, M., Miladi, J., Ferry, A., & Duvallet, A. (1989). Blood lactate during submaximal exercise: Comparison between intermittent incremental exercises and isolated exercises. *Journal of Applied Physiology, 59,* 73-79.

Robergs, R.A., Chwalbinska-Moneta, J., Mitchell, J.B., Pascoe, D.D., Houmard, J., & Costill, D.L. (1990). Blood lactate threshold differences between arterialized and venous blood. *International Journal of Sports Medicine, 11,* 446-451.

Rodriguez, F.A., Banquells, M., Pons, V., Dropnic, F., & Galilea, P.A. (1992). A comparative study of blood lactate analytic methods. *International Journal of Sports Medicine, 13,* 462-466.

Rusko, H., Luhdtanen, P., Rahkila, P., Viitasalo, J., Rehunen, S., & Harkonen, M. (1986). Muscle metabolism, blood lactate and oxygen uptake in steady state exercise at aerobic and anaerobic thresholds. *European Journal of Applied Physiology, 55,* 181-186.

Rusko, H., Rahkila, P., & Karvinen, E. (1980). Anaerobic threshold, skeletal muscle enzymes and fiber composition in young female cross country skiers. *Acta Physiologica Scandinavica, 108,* 263-268.

Saltin, B., Hartley, L.H., Kilbom, A., & Åstrand, I. (1969). Physical training in sedentary middle aged and older men II: Oxygen uptake, heart rate and blood lactate concentration at submaximal and maximal exercises. *Scandinavian Journal of Clinical and Laboratory Investigation, 24,* 323-334.

Schnable, A., Kindermann, W., Schmitt, W.M., Biro, G., & Stegmann, H. (1982). Hormonal and metabolic consequences of prolonged running at the individual anaerobic threshold. *International Journal of Sports Medicine, 3,* 163-168.

Simon, J., Young, J.L., Gutin, B., Blood, D.K., & Case, R.B. (1983). Lactate accumulation relative to the anaerobic and respiratory compensation thresholds. *Journal of Applied Physiology, 54,* 13-17.

Sjodin, B., Jacobs, I., & Karlsson, J. (1981a). Onset of blood lactate accumulation and marathon running performance. *International Journal of Sports Medicine, 2,* 23-26.

Sjodin, B., Jacobs, I., & Karlsson, J. (1981b). Onset of blood lactate accumulation and enzyme activities in m vastus lateralis in man. *International Journal of Sports Medicine, 2,* 166-170.

Sjodin, B., Jacobs, I., & Svedenhag, J. (1982). Changes in onset of blood lactate accumulation (OBLA) and muscle enzymes after training at OBLA. *European Journal of Applied Physiology, 49,* 45-57.

Sjodin, B., & Svedenhag, J. (1985). Applied physiology of marathon running. *Sports Medicine, 2,* 83-99.

Skinner, J.S., & McLellan, T.M. (1980). The transition from aerobic to anaerobic metabolism. *Research Quarterly for Exercise and Sport, 51,* 234-248.

Smekal, G., Pokan, P., von Duvillard, S.P., Baron, R., Tschan, H., & Bachi, N. (2000). Comparison of laboratory and "on-court" endurance testing in tennis. *International Journal of Sports Medicine, 21,* 242-249.

Snyder, A.C., Woulfe, T.J., Welsh, R., & Foster, C. (1994). A simplified approach to estimating the maximal lactate steady state. *International Journal of Sports Medicine, 15,* 27-31.

Stegmann, H., & Kindermann, W. (1982). Comparison of prolonged exercise tests at the individual anaerobic threshold and the fixed anaerobic threshold of 4 mmol/L lactate. *International Journal of Sports Medicine, 3,* 105-110.

Stegmann, H., Kindermann, W., & Schnabel, A. (1981). Lactate kinetics and individual anaerobic threshold. *International Journal of Sports Medicine, 2,* 160-165.

Svedenhag, J., & Sjodin, B. (1984). Maximal and submaximal oxygen uptakes and blood lactate levels in elite male middle and long distance runners. *International Journal of Sports Medicine, 5,* 255-261.

Svedenhag, J., & Sjodin, B. (1985). Physiological characteristics of elite male runners in and off season. *Canadian Journal of Applied Physiology, 10,* 127-133.

Swensen, T., Harnish, L., Beitman, L., & Keller, B. (1999). Noninvasive estimation of the maximal lactate steady state in trained cyclist. *Medicine and Exercise in Sports and Exercise, 5,* 742-746.

Tanaka, K., Watanabe, H., Konishi, Y., Mitsuzono, R., Sumida, S., Tanaka, S., et al. (1986). Longitudinal associations between anaerobic threshold and distance running performance. *European Journal of Applied Physiology, 55,* 248-252.

Tesch, P.A., Daniels, W.L., & Sharp, D.S. (1982). Lactate accumulation in muscle and blood during submaximal exercise. *Acta Physiologica Scandinavica, 114,* 441-446.

Tesch, P.A., Sharp, D.S., & Daniels, W.L. (1981). Influence of fiber type composition and capillary density on onset of blood lactate accumulation. *International Journal of Sports Medicine, 2,* 252-255.

Thorland, W., Podolin, D.A., & Mazzeo, R.S. (1994). Coincidence of lactate threshold and HR-power output threshold under varied nutritional state. *International Journal of Sports Medicine, 15,* 301-304.

Urhausen, A., Coen, B., & Kindermann, W. (2000). *Exercise and Sport Science.* Philadelphia: Lippincott Williams & Wilkins.

Wasserman, K., Hansen, J.E., Sue, D.Y., & Whipp, B.J. (1987). *Principles of exercise testing and interpretation.* Philadelphia: Lea & Febiger.

Wasserman, K., & McIlroy, M.B. (1964). Detecting the threshold of anaerobic metabolism. *American Journal of Cardiology, 14,* 844-852.

Weltman, A., Seip, R.L., Snead, D., Weltman, J.Y., Haskvitz, E.M., Evans, W.S., et al. (1992). Exercise training at and above the lactate threshold in previously untrained women. *International Journal of Sports Medicine, 13,* 257-263.

Weltman, A., Snead, D., Seip, R.L., Schurrer, R., Levin, S., Rutt, R., et al. (1987). Prediction of lactate threshold and fixed blood lactate concentrations from 3200m running performance in male runners. *International Journal of Sports Medicine, 8,* 401-406.

Withers, R.T., Sherman, W.M., Clark, D.G., Esselbach, P.C., Nolan, S.R., Mackay, M.K., et al. (1991). Muscle metabolism during 30, 60, and 90 s of maximal cycling on an air-braked ergometer. *European Journal of Applied Physiology, 63,* 354-362.

Yamamoto, Y., Miyashita, M., Hughson, R.L., Tamura, S., Shinohara, M., & Mutoh, Y. (1991). The ventilatory threshold gives maximal lactate steady state. *European Journal of Applied Physiology, 63,* 55-59.

Yeh, M.P., Gardner, R.M., Adams, T.S., Yanowitz, F.G., & Crapo, R.O. (1983). Anaerobic threshold: Problems of determination and validation. *Journal of Applied Physiology, 55,* 1178-1186.

Yoshida, T. (1984). Effect of dietary modifications on lactate threshold and onset of blood lactate accumulation during incremental exercise. *European Journal of Applied Physiology, 53,* 200-205.

Yoshida, T. (1984). Effect of exercise duration during incremental exercise on the determination of anaerobic threshold and the onset of blood lactate accumulation. *European Journal of Applied Physiology, 53,* 196-199.

Yoshida, T., Suda, Y., & Takeuchi, N. (1982). Endurance training regimen based upon arterial blood lactate: Effects on anaerobic threshold. *European Journal of Applied Physiology, 49,* 223-230.

Yoshida, T., Takeuchi, N., & Suda, Y. (1982). Arterial versus venous blood lactate increase in the forearm during incremental bicycle exercise. *European Journal of Applied Physiology, 50,* 87-93.

## Chapter 6

Åstrand, P.O., Hultman, E., Juhlin-Danfelt, A., & Reynold, G. (1986). Disposal of lactate during and after strenuous exercise in humans. *Journal of Applied Physiology, 61,* 338-343.

Bangsbo, J., Gollnick, P.D., Graham, T.E., Juel, C., Kiens, B., Mizuno, M., et al. (1990). Anaerobic energy production and O$_2$ deficit-debt relationship during exhaustion exercise in humans. *Journal of Physiology, 442,* 539-559.

Bassett, D.R., & Howley, E.T. (1997). Maximal oxygen uptake: "Classical" versus "contemporary" viewpoints. *Medicine and Science in Sports and Exercise, 29,* 591-603.

Bar-Or, O. (1987). The Wingate anaerobic test: An update on methodology, reliability and validity. *Sports Medicine, 4,* 381-394.

Bosco, C., Luhtanen, P., & Komi, P. (1983). A simple method for measurement of mechanical power in jumping. *European Journal of Applied Physiology, 50,* 273-282.

Bouchard, C., Taylor, A.W., Simoneau, J.A., & Dulac, S. (1991). Testing anaerobic power and capacity. In MacDougall J.D., H.A. Wenger , and H.J. Green, *Physiological Testing of the High Performance Athlete* (2nd ed.) (pp. 175-222). Champaign, IL: Human Kinetics.

Boulay, M.R., Lortie, G., Simoneau, J.A., Hamel, P., Lablanc, C., & Bouchard, C. (1985). Specificity of aerobic and anaerobic work capacities and powers. *International Journal of Sports Medicine, 6,* 325-328.

Brooks, G.A. (1971). Temperature, skeletal muscle mitochondrial functions and oxygen debt. *American Journal of Physiology, 220,* 1053-1059.

Brooks, G.A. (1973). Glycogen synthesis and metabolism of lactic acid after exercise. *American Journal of Applied Physiology, 224,* 1162-1166.

Brooks, G.A. (1991). Current concepts in lactate exchange. *Medicine and Science in Sports and Exercise, 23,* 895-906.

Buck, D., & McNaughton, L. (1999). Maximal accumulated oxygen deficit must be calculated using 10-min time periods. *Medicine and Science in Sports and Exercise, 31,* 1346-1349.

Camus, G., & Thys, H. (1991). An evaluation of the maximal anaerobic capacity in man. *International Journal of Applied Physiology, 12,* 349-355.

Chicharro, J.L., Lopez-Mojares, L.M., Lucia, A., Pérez, M., Alvarez J., Labanda P., et al. (1998). Overtraining parameters in special military units. *Aviation Space Environmental Medicine, 69*(6), 562-568.

Dotan, R., & Bar-Or, O. (1983). Load optimization from the Wingate anaerobic test. *European Journal of Applied Physiology, 51,* 409-417.

Evans, J.A., & Quinney, H.A. (1981). Determination of resistance settings for anaerobic power testing. *Canadian Journal of Applied Sports Science, 6,* 53-56.

Fabian, N.M., Adams, K.J., Durham, M.P., Kipp, R.L., Berning, J.M., Swank, A.M., et al. (2001). Comparison of power production between the Bosco and Wingate 30-second power tests. *Medicine and Science in Sports and Exercise,* (Vol. 33 No. 5 Supplement, Wednesday, May 30).

Foley, M.J., McDonald, K.S., Green, M.A., Schrager, M., Snyder, A.C., & Foster, C. (1991). Comparison of methods for estimation of anaerobic capacity. *Medicine and Science in Sports and Exercise, 23,* S34.

Foster, C.J., deKoning, F. Hettinga, J., Lampen, K., La Chair, C., Dodge, M., et al. (2003). Pattern of energy expenditure during simulated competition. *Medicine and Science in Sports and Exercise.*

Foster, C., Kuettal, K., & Thompson, N.N. (1989). Estimation of anaerobic capacity. *Medicine and Science in Sports and Exercise, 21,* S27.

Fox, E., & Mathews, D. (1974). *Interval training: Conditioning for sports and general fitness.* Philadelphia: Saunders.

Gastin, P., Lawson, D., Hargreaves, M., Carey, M., & Fairweather, I. (1991). Variable resistance loadings in anaerobic power testing. *International Journal of Sports Medicine, 6,* 513-518.

Gastin, P.B. (1994). Quantification of anaerobic capacity. *Scandinavian Journal of Medicine and Science in Sports, 4,* 91-112.

Gastin, P.B., Costill, D.L., Larson, D.L., Krzeminski, K., & McConell, G.K. (1995). Accumulated oxygen deficit during supramaximal all-out and constant intensity exercise. *Medicine and Science in Sports and Exercise, 27,* 255-263.

Green, S., Dawson, B., Goodman, C., & Carey, M. (1996). Anaerobic ATP production and accumulated O$_2$ deficit on cyclists. *Medicine and Science in Sports and Exercise, 28,* 315-321.

Hermansen, L. (1969). Anaerobic energy release. *Medicine and Science in Sports, 1,* 32-38.

Hill, A.V. (1924). Muscular exercise, lactic acid and the supply and utilization of oxygen. *Proceedings of the Royal Society, 96*, 438-455.

Hill, D.W., & Smith, J.C. (1989). Oxygen uptake during the Wingate anaerobic test. *Canadian Journal of Sports Sciences, 14*, 122-125.

Inbar, O., Bar-Or, O., & Skinner, J.S. (1996). *The Wingate anaerobic test*. Champaign, IL: Human Kinetics.

Jacobs, I., Tesch, P.A., Bar-Or, O., Karlsson, J., & Dotan, R. (1983). Lactate in human skeletal muscle after 10 and 30 s of supramaximal exercise. *Journal of Applied Physiology, 55*, 365-367.

Jones, N.L., McCartney, N., & McComas, A.J. (Eds.). (1986). *Human muscle power*. Champaign, IL: Human Kinetics.

Karlsson, J. (1971). Lactate and phosphagen concentrations in working muscle of men. *Acta Physiologica Scandinavica*, S358.

Katz, A.L., Snell, P., & Stray-Gundersen, J. (1989). A combined protocol for running economy, $\dot{V}O_2$max and anaerobic capacity. *Medicine and Science in Sports and Exercise, 21*, S10.

Krogh, A., & Lindhard, J. (1913). The regulation of respiration and circulation during the initial stages of muscular work. *Journal of Physiology (London), 47*, 112-136.

Margaria, R., Aghemo, P., & Rovelli, E. (1966). Measurement of muscular power (anaerobic) in man. *Journal of Applied Physiology, 21*, 1662-1664.

Maud, P.J., & Atwood, A.K. Unpublished data.

Maud, P.J., & Shultz, B. (1989). Norms for the Wingate anaerobic test with comparison to another similar test. *Research Quarterly for Exercise and Sport, 60*(2), 144-151.

Maud, P.J., & Shultz, B.B. (1986). Relationships between, and normative data for, three performance measures of anaerobic power. In T. Reilly, J. Watkins, & J. Borms (Eds.), *Kinanthropometry III: Proceedings of the VIII Commonwealth and International Conference on Sport, Physical Education, Dance, Recreation and Health* (pp. 284-298). London. London: E. & F.N. Spon Ltd.

Medbø, J.L., & Burgers, S. (1990). Effect of training on the anaerobic capacity. *Medicine and Science in Sports and Exercise, 22*, 501-507.

Medbø, J.L., Mohn, A.C., Tabata, I., Bahr, R., Vaage, O., & Sejersted, O.M. (1988). Anaerobic capacity determined by maximal accumulated $O_2$ deficit. *Journal of Applied Physiology, 64*, 50-60.

Medbø, J.L., & Tabata, I. (1989). Relative importance of aerobic and anaerobic energy release during shortlasting, exhaustion bicycle exercise. *Journal of Applied Physiology, 67*, 1881-1886.

Mitchell, J.H., Sproule, B.J., & Chapman, C.B. (1958). The physiological meaning of the maximal oxygen intake test. *Journal of Clinical Investigation, 37*, 538-547.

Noakes, T.D. (1997). Maximal oxygen uptake: "Classical" versus "contemporary" viewpoints: A rebuttal. *Medicine and Science in Sports and Exercise, 29*, 571-590.

Nummela, A., Alberts, M., Rijntjes, R.P., Luhtanen, P., & Rusko, R. (1996). Reliability and variability of the maximal anaerobic running test. *Medicine and Science in Sports and Exercise*, S97-S102.

Nummela, A., Anderson, N., Hakkinen, K., & Rusko, H. (1996). Effects of inclination on the results of the maximal anaerobic running test. *Medicine and Science in Sports and Exercise*, S103-S108.

Nummela, A., Mero, A., & Rusko, H. (1996). Effects of sprint training on anaerobic performance characteristics determined by the MART. *Medicine and Science in Sports and Exercise*, S114-S119.

Nummela, A., Mero, A., Stray-Gundersen, J., & Rusko, H. (1996). Important determinants of anaerobic running performance in male athletes and non-athletes. *Medicine and Science in Sports and Exercise*, S91-S96.

Patton, J.F., Murphy, M.M., & Fredrick, F.A. (1985). Maximal power outputs during the Wingate anaerobic test. *International Journal of Sports Medicine, 6*, 82-85.

Rusko, H., Nummela, A., & Mero, A. (1993). A new method for the evaluation of anaerobic running power in athletes. *European Journal of Applied Physiology, 66*, 97-101.

Scott, C.B., & Bogdonfly, G.M. (1998). Aerobic and anaerobic energy expenditure during exhaustive ramp exercise. *International Journal of Sports Medicine, 19*, 277-280.

Serresse, O., Lortie, G., Bouchard, C., & Boulay, M.R. (1988). Estimation of the contribution of various energy systems during maximal work of short duration. *International Journal of Sports Medicine, 9*, 456-460.

Serresse, O., Simoneau, J.A., Bouchard, C., & Boulay, M.R. (1991). Aerobic energy contribution during maximal work outputs in 90 s determined with various ergocycle workloads. *International Journal of Sports Medicine, 12*, 543-547.

Simoneau, W.N., Lortie, G., Boulay, M.R., & Bouchard, C. (1983). Tests of anaerobic alactacid and lactacid capacities: Description and reliability. *Canadian Journal of Applied Sports Science, 8*, 266-270.

Spencer, M.R., & Gastin, P.B. (2001). Energy system contribution during 200- to 1500-m running in highly trained athletes. *Medicine and Science in Sports and Exercise, 33*, 157-162.

Stainsby, W.N., & Brooks, G.A. (1990). Control of lactic acid metabolism in contracting muscle and during exercise. In K.B. Pandolf (Ed.), *Exercise and sports science reviews* (Vol. 18) (pp. 29-64). Baltimore: Williams & Wilkins.

St. Clair Gibson, A., Lambert, M.I., & Noakes, T.D. (2001). Neural control of force output during maximal and submaximal exercise. *Sports Medicine, 31*, 157-162.

Szogy, A., & Cherebetiu, G. (1974). A one minute bicycle ergometer test for determination of anaerobic capacity. *European Journal of Applied Physiology, 33*, 171-176.

Tossavainen, M., Nummela, A., Paavolainen, L., Mero, A., & Rusko, H. (1996). Comparison of two maximal anaerobic cycling tests. *Medicine and Science in Sports and Exercise*, S120-S124.

Wasserman, K., Hansen, J.E., Sue, D.Y., & Whipp, B.J. (1987). *Principles of exercise testing and interpretation.* Philadelphia: Lea & Febiger.

Withers, R.T., Sherman, W.M., Clark, D.G., Esselbach, P.C., Nolan, S.R., Mackay, M.H., et al. (1991). Muscle metabolism during 30, 60 and 90 s of maximal cycling on an air braked ergometer. *European Journal of Applied Physiology, 63,* 354-362.

Withers, T.T., & Telford R.D. (1987). The determination of maximal anaerobic power and capacity. In G. Gass (Ed.), *Physiological guidelines for the assessment of the elite athlete* (pp. 105-124). Canberra, Australia: Australian Sports Commission.

## Chapter 7

Abate, F. (Ed.) (1996). *The Oxford dictionary and thesaurus, American edition.* New York: Oxford University Press.

Atha, J. (1981). Strengthening muscle. In D. Miller (Ed.), *Exercise and sports sciences reviews* (Vol. 9) (pp. 1-73). Philadelphia: Franklin Institute.

Bar-Or, O., Dotan, R., Inbar, O., Rotstein, A., & Tesch, P. (1980). Anaerobic capacity and muscle fiber type distribution in non-athletes and in athletes. *International Journal of Sports Medicine, 1,* 82-85.

Chaffin, D.B., Chaffin, D., & Andersson, G.B. (1999). *Occupational biomechanics* (3rd ed.). New York: Wiley.

Daly, D., and Cavanagh, P. (1976). Asymmetry in bicycle ergometer pedalling. *Medicine and Science in Sports, 8*(3), 204-208.

Davies, C. (1971). Human power output in exercise of short duration in relation to body size and composition. *Ergonomics, 14*(2), 245-256.

Davies, C., & Rennie, R. (1968). Human power output. *Nature, 217,* 770-771.

Davies, C., & Young, K. (1983). Effect of temperature on the contractile properties and muscle power of triceps surae in humans. *Journal of Applied Physiology, 55*(1), 191-195.

Diefenderfer, J., & Holton, B.E. (1993). *Principles of electronic instrumentation* (3rd ed.). Philadelphia: Saunders.

Driss, T., Vandewalle, H., Quievre, J., Miller, C., & Monod, H. (2001). Effects of external loading on power output in a squat jump on a force platform: a comparison between strength and power athletes and sedentary individuals. *Journal of Sports Science, 19*(2), 99-105.

Enoka, R. (2002). *Neuromechanics of human movement* (3rd ed.) Champaign, IL: Human Kinetics.

Fisher, A.G., & Jensen, C.R. (1990). *Scientific basis of athletic conditioning* (3rd ed.). Philadelphia: Lippincott Williams & Wilkins.

Fox, E., & Mathews, D. (1974). *Interval training: Conditioning for sports and general fitness.* Philadelphia: Saunders.

Fox, E., & Mathews, D. (1981). *The physiological basis of physical education and athletics* (3rd ed.). Philadelphia: Saunders.

Frykman, P., Harman, E., & Kraemer, W. (1987). The effects of all -running, all-lifting, and mixed training on power output at two different speeds. *Journal of Applied Sports Science Research, 9*(4), 59.

Frykman, P., Harman, E., Rosenstein, M., & Rosenstein, R. (1989). A computerized instrumentation system for weight machines. *Journal of Applied Sports Science Research, 3*(3), 77.

Garhammer, J. (1989). Weight lifting and training. In C. Vaughn (ed.), *Biomechanics of sport* (pp. 169-211). Boca Raton, FL: CRC Press.

Harman, E. (1989). An axial force transducer for a weight-lifting bar. *Proceedings of the American Society of Biomechanics 13th Annual Meeting* (pp. 216-217). Burlington, VT: American Society of Biomechanics in cooperation with the University of Vermont.

Harman, E., Frykman, P., & Kraemer, W. (1986). Maximal cycling force and power at 40 and 100 RPM. *National Strength and Conditioning Association Journal, 8*(4), 71.

Harman, E., Knuttgen, H., & Frykman, P. (1987). Automated data collection and processing for a cycle ergometer. *Journal of Applied Physiology, 62*(2), 831-836.

Harman, E., Rosenstein, M., Frykman, P., & Rosenstein, R. (1990). The effects of arms and countermovement on vertical jumping. *Medicine and Science in Sports and Exercise, 22,* 825-833.

Harman, E., Rosenstein, M., Frykman, P., Rosenstein, R., & Kraemer, W. (1991). Estimation of human power output from maximal vertical jump and body mass. *Journal of Applied Sports Science Research, 5,* 116-120.

Hay, J. (1978). *The biomechanics of sports techniques.* Englewood Cliffs, NJ: Prentice-Hall.

Hull, M., & Davis, R. (1981). Measurement of pedal loading in bicycling: Instrumentation. *Journal of Biomechanics, 14*(12), 843-856.

Kirkendall, D., Gruber, J., & Johnson, R. (1987). *Measurement and evaluation for physical educators* (2nd ed). Champaign, IL: Human Kinetics.

Knuttgen, H., & Kraemer, W. (1987). Terminology and measurement in exercise performance. *Journal of Applied Sport Science Research, 1*(1), 1-10.

Komi, P., & Hakkinen, K. (1988). Strength and power. In A. Dirix, H. Knuttgen, & K. Tittel (Eds.), *The Olympic book of sports medicine* (Vol. 1, pp. 181-193). Boston: Blackwell Scientific.

Lakomy, H. (1984). An ergometer for measuring the power generated during sprinting. *Journal of Physiology, 354,* 33P.

Lakomy, H. (1986). Measurement of work and power output using friction-loaded cycle ergometers. *Ergonomics, 29*(4), 509-517.

Lakomy, H. (1987). Measurement of human power output in high intensity exercise. In B. Van Gheluwe & J. Atha (Eds.), *Current research in sports biomechanics* (pp.46-57). Basel, Switzerland: Karger.

Lide, D.R. (Ed.) (2003). *CRC handbook of chemistry and physics* (84th ed.). Boca Raton, FL: CRC Press.

Margaria, R., Aghemo, P., & Rovelli, E. (1966). Measurement of muscular power (anaerobic) in man. *Journal of Applied Physiology, 21*(5), 1662-1664.

McCartney, N., Spriet, L., Heigenhauser, G., Kowalchuk, J., Sutton, J., & Jones, N. (1986). Muscle power and metabolism in maximal intermittent exercise. *Journal of Applied Physiology, 60*(4), 1164-1169.

Meriam, J.L., & Kraige, L.G. (2002). *Engineering mechanics: Dynamics* (5th ed). New York: Wiley.

Newton, R.U., Kraemer, W.J., & Hakkinen, K. (1999). Effects of ballistic training on preseason preparation of elite volleyball players. *Medicine and Science in Sports and Exercise, 31*(2), 323-30.

Patterson, R., & Moreno, M. (1990). Bicycle pedalling forces as a function of pedalling rate and power output. *Medicine and Science in Sports and Exercise, 22*(4), 512-516.

Perrine, J. (1986). The biophysics of maximal muscle power outputs: methods and problems of measurement. In L. Jones, N. McCartney, & A. McComas (Eds.), *Human Muscle Power* (pp. 15-22). Champaign, IL: Human Kinetics.

Perrine, J., & Edgerton, V. (1978). Muscle force-velocity and power-velocity relationships under isokinetic loading. *Medicine and Science in Sports, 10*, 159-166.

Press, W.H., Teukolsky, S.A., Vetterling, W.T., & Flannery, B.P. (Eds.) (2002). *Numerical recipes in C++: The art of scientific computing* (2nd ed). New York: Cambridge University Press.

Rosenstein, M., Harman, E., Frykman, P., & Johnson, M. (1989). An inexpensive method for measurement of torque and power output during accelerative movement. *Journal of Applied Sports Science Research, 3*(3), 77.

Sargeant, A., & Davies, C. (1977). Forces applied to cranks of a bicycle ergometer during one- and two-leg cycling. *Journal of Applied Physiology, 42*(4), 514-518.

Sayers, S.P., Harackiewicz, D.V., Harman, E.A., Frykman, P.N., & Rosenstein, M.T. (1999). Cross-validation of three jump power equations. *Medicine and Science in Sports and Exercise, 31*(4), 572-577.

Taylor, B.N. (Ed.) (2002). *International System of Units (SI): 2001 edition.* Collingdale, PA: DIANE.

Winter, D. (1978). Calculation and interpretation of the mechanical energy of movement. *Exercise and Sports Sciences Reviews, 6*, 183-201.

Winter, D. (1979). A new definition of mechanical work done in human movement. *Journal of Applied Physiology, 46*, 79-83.

Winter, D.A. (1990). *Biomechanics and motor control of human movement* (2nd ed.). New York: Wiley.

## Chapter 8

Abadie, B.R., Altorfer, G.L., & Schuler, P.B. (1999). Does a regression equation to predict maximal strength in untrained lifters remain valid when the subjects are technique trained? *Journal of Strength and Conditioning Research, 13*, 259-263.

Abdo, J.S. (1985). Weight training percentage table. *National Strength and Conditioning Association Journal, 7*, 50-51.

Adams, K.J., Swank, A.M., Barnard, K.L., Berning, J.M., & Sevene-Adams, P.G. (2000). Safety of maximal power, strength, and endurance testing in older African American women. *Journal of Strength and Conditioning Research, 14*, 254-260.

Akebi, T., Saeki, S., Hieda, H., & Goto, H. (1998). Factors affecting the variability of the torque curves at isokinetic trunk strength testing. *Archives of Physical Medicine and Rehabilitation, 79*, 33-35.

Alderman, R.B., & Banfield, T.J. (1969). Reliability estimation in the measurement of strength. *Research Quarterly, 40*, 448-455.

Allerheiligen, B., Arce, J.H., Arthur, M., Chu, D., Lilja, L., Semenick, D., et al. (1983). Coaches roundtable: Testing for football. *National Strength and Conditioning Association Journal, 5*, 12-19, 62-68.

Altug, Z., Altug, T., & Altug, A. (1987). A test selection guide for assessing and evaluating athletes. *National Strength and Conditioning Association Journal, 9*, 62-66.

American College of Sports Medicine. (1998). Position stand: The recommended quantity and quality of exercise for developing and maintaining cardiorespiratory and muscular fitness, and flexibility in healthy adults. *Medicine and Science in Sports and Exercise, 30*, 975-991.

American College of Sports Medicine. (2002). Position stand: Progression models in resistance training for healthy adults. *Medicine and Science in Sports and Exercise, 34*, 364-380.

Amusa, L.O., & Obajuluwa, V.A. (1986). Static versus dynamic training programs for muscular strength using the knee-extensors in healthy young men. *Journal of Orthopedic Sport, Medicine and Physical Therapy, 8*, 243-247.

Anderson, T., & Kearney, J.T. (1982). Effects of three resistance training programs on muscular strength and absolute and relative endurance. *Research Quarterly, 53*, 1-7.

Appen, L., & Duncan, P.W. (1986). Strength relationship of the knee musculature: Effect of gravity and sport. *Journal of Orthopedic Sport, Medicine and Physical Therapy, 7*, 232-235.

Armstrong, L.E. (1988). The impact of hyperthermia and hypohydration on circulation, strength, endurance and health. *Journal of Applied Sport Science Research, 2*, 60-65.

Atha, J. (1981). Strengthening muscle. *Exercise and Sport Science Reviews* (D.I. Miller, ed.), 1-74.

Austin, D., Roll, F., Kreis, E.J., Palmieri, J., & Lander, J. (1987). Roundtable: Breathing during weight training. *National Strength and Conditioning Association Journal, 9*, 17-25.

Aydin, T., Yildiz, Y., Yildiz, C., & Kalyon, T.A. (2001). The stretch-shortening cycle of the internal rotators muscle group measured by isokinetic dynamometry. *Journal of Sports Medicine and Physical Fitness, 41,* 371-379.

Baker, D., Wilson, G., & Carlyon, B. (1994). Generality versus specificity: A comparison of dynamic and isometric measures of strength and speed-strength. *European Journal of Applied Physiology, 68,* 350-355.

Barnard, K.L., Adams, K.J., Swank, A.M., Mann, E., & Denny, D.M. (1999). Injuries and muscle soreness during the one-repetition maximum assessment in a cardiac rehabilitation population. *Journal of Cardiopulmonary Rehabilitation, 19,* 52-58.

Batterham, A., & George, K. (1997). Allometric modeling does not determine a dimensionless power function ratio for maximal muscular function. *Journal of Applied Physiology, 83,* 2158-2166.

Behm, D.G., Reardon, G., Fitzgerald, J., & Drinkwater, E. (2002). The effects of 5, 10, and 20 repetition maximums on the recovery of voluntary and evoked contractile properties. *Journal of Strength and Conditioning Research, 16,* 209-218.

Belanger, A.Y., & McComas, A.J. (1989). Contractile properties of human skeletal muscle in childhood and adolescence. *European Journal of Applied Physiology, 58,* 563-567.

Bemben, M.G., Clasey, J.L., & Massey, B.H. (1990). The effect of the rate of muscle contraction on the force time curve parameters of male and female subjects. *Research Quarterly, 61,* 96-99.

Berger, R. (1962). Effect of varied weight training programs on strength. *Research Quarterly, 33,* 168-181.

Berger, R.A. (1967). Determination of a method to predict 1-RM chin and dip from repetitive chins and dips. *Research Quarterly, 38,* 330-335.

Biddle, S.J. (1986). Personal beliefs and mental preparation in strength and muscular endurance tasks: A review. *Physical Education Review, 8,* 90-103.

Bishop, P., Cureton, K., & Collins, M. (1987). Sex difference in muscular strength in equally-trained men and women. *Ergonomics, 30,* 675-687.

Blazevich, A.J., Gill, N., & Newton, R.U. (2002). Reliability and validity of two isometric squat tests. *Journal of Strength and Conditioning Research, 16,* 298-304.

Brown, E.W., & Kimball, R.G. (1983). Medical history associated with adolescent powerlifting. *Pediatrics, 72,* 630-644.

Brown, L., Whitehurst, M., Findley, B., Gilbert, P., Groo, D., & Jimenez, J. (1998). The effect of repetitions and gender on acceleration range of motion during knee extension on an isokinetic device. *Journal of Strength and Conditioning Research, 12,* 222-225.

Brown, L.E., & Whitehurst, M. (2000). Load range. In L.E. Brown (Ed.), *Isokinetics in human performance* (pp. 97-121). Champaign, IL: Human Kinetics.

Brown, R.D., & Harrison, J.M. (1986). The effects of a strength training program on the strength and self-concept of two female age groups. *Research Quarterly, 57,* 315-320.

Brownstein, B.A., Lamb, R.L., & Mangine, R.E. (1985). Quadriceps torque and integrated electromyography. *Journal of Orthopedic Sport, Medicine and Physical Therapy, 6,* 309-314.

Brzycki, M. (1993). Strength testing: Predicting a one-rep max from reps-to-fatigue. *JOHPERD, 64,* 88-90.

Caiozzo, V.J., Perrine, J.J., & Edgerton, V.R. (1981). Training induced alterations of the invivo force velocity relationship of human muscle. *Journal of Applied Physiology, 51,* 750-754.

Cappaert, T.A. (1999). Review: Time of day effect on athletic performance: An update. *Journal of Strength and Conditioning Research, 13,* 412-421.

Capranica, L., Battent, M., Demarie, S., & Figura, F. (1998). Reliability of isokinetic knee extension and flexion strength testing in elderly women. *Journal of Sports Medicine and Physical Fitness, 38,* 169-176.

Challis, J.H. (1999). Methodological report: The appropriate scaling of weightlifting performance. *Journal of Strength and Conditioning Research, 13,* 367-371.

Chapman, P.P., Whitehead, J.R., & Binkert, R.H. (1998). The 225-lb reps-to-fatigue test as a submaximal estimate of 1-RM bench press performance in college football players. *Journal of Strength and Conditioning Research, 12,* 258-261.

Chilibeck, P.D., Calder, A.W., Sale, D.G., & Webber, C.E. (1998). A comparison of strength and muscle mass increases during resistance training in young women. *European Journal of Applied Physiology, 77,* 170-175.

Chmelar, R.D., Shultz, B.B., Ruhling, R.O., Fitt, S.S., & Johnson, M.B. (1988). Isokinetic characteristics of the knee in female, professional and university ballet and modern dancers. *Journal of Orthopedic Sport, Medicine and Physical Therapy, 9,* 410-418.

Christensen, C.S. (1975). Relative strength in males and females. *Athletic Training, 10,* 189-192.

Cisar, C.J., Johnson, G.O., Fry, A.C., Housh, T.J., Hughes, R.A., Ryan, A.J., et al. (1987). Preseason body composition, build and strength as predictors of high school wrestling success. *Journal of Applied Sport Science Research, 1,* 66-70.

Cisar, C.J., Johnson, G.O., Fry, A.C., & Ryan, A.J. (1987). Assessment of preseason muscular strength as a basis for specific conditioning. *Journal of Applied Sport Science Research, 1,* 60.

Clark, H.M. (1967). *Application of measurement to health and physical education.* Englewood Cliffs, NJ: Prentice Hall.

Clarke, D.M. (1986). Sex differences in strength and fatigability. *Research Quarterly, 57,* 144-149.

Conroy, B., Stanley, D., Fry, A., & Kraemer, W.J. (1984). A comparison of isokinetic protocols. *Journal of Applied Sport Science Research, 3,* 72.

Cook, E.E., Gray, V.L., Savinar-Nogue, E., & Medeiros, J. (1987). Shoulder antagonistic strength ratios: A comparison between college-level baseball pitchers and non-pitchers. *Journal of Orthopedic Sport, Medicine and Physical Therapy, 8*, 451-461.

Coyle, E.F., Feiring, D.C., Rotkis, T.C., Cote, R.W., Roby, F.B., Lee, W., et al. (1981). Specificity of power improvements through slow and fast isokinetic training. *Journal of Applied Physiology, 51*, 1437-1442.

Cresswell, A.G., Blake, P.L., & Thorstensson, A. (1994). The effect of an abdominal muscle training program on intra-abdominal pressure. *Scandinavian Journal of Rehabilitation and Medicine, 26*, 79-86.

Cummings, B., & Finn, K.J. (1998). Estimation of a one repetition maximum bench press for untrained women. *Journal of Strength and Conditioning Research, 12*, 262-265.

Davies, G.J., Heiderscheit, B., & Brinks, K. (2000). Test interpretation. In L.E. Brown (Ed.), *Isokinetics in human performance* (pp. 3-24). Champaign, IL: Human Kinetics.

DiBrezzo, R., & Fort, I.L. (1987). Strength norms for the knee in women 25 years and older. *Journal of Applied Sport Science Research, 1*, 45-47.

Doan, B.K., Newton, R.U., Marsit, J.L., Triplett-McBride, N.T., Koziris, L.P., Fry, A.C., et al. (2002). Effects of increased loading on bench press 1RM. *Journal of Strength and Conditioning Research, 16*, 9-13.

Dooman, C.S., & Vanderburgh, P.M. (2000). Allometric modeling of the bench press and squat: Who is the strongest regardless of body mass? *Journal of Strength and Conditioning Research, 14*, 32-36.

Dudley, G.A., & Djamil, R. (1985). Incompatibility of endurance and strength training modes of exercise. *Journal of Applied Physiology, 59*, 1446-1451.

Dudley, G.A., Tesch, P.A., Miller, B.J., & Buchanan, P. (1991). Importance of eccentric actions in performance adaptations to resistance training. *Aviation Space and Environmental Medicine, 62*, 543-550.

Dvir, Z. (1997). Differentiation of submaximal from maximal trunk extension effort: An isokinetic study using a new testing protocol. *Spine, 22*, 2672-2676.

Dvir, Z., & David, G. (1996). Suboptimal muscular performance: Measuring isokinetic strength of knee extensors with new testing protocol. *Archives of Physical Medicine and Rehabilitation, 77*, 578-581.

Dvir, Z., & Keating, J. (2001). Reproducibility and validity of a new test protocol for measuring isokinetic trunk extension strength. *Clinical Biomechanics (Bristol, Avon), 16*, 627-630.

Earle, R.W., & Baechle, T.R. (2000). Resistance training and spotting techniques. In T.R. Baechle & R.W. Earle (Eds.), *Essentials of strength training and conditioning* (2nd ed.) (pp. 343-389). Champaign, IL: Human Kinetics.

Edgerton, V.R., & Perrine, J.J. (1978). Muscle force-velocity and power-velocity relationships under isokinetic loading. *Medicine and Science in Sports and Exercise, 10*, 159-166.

Emery, C.A., Maitland, M.E., & Meeuwisse, W.H. (1999). Test-retest reliability of isokinetic hip adductor and flexor muscle strength. *Clinical Journal of Sports Medicine, 9*, 79-85.

Enoka, R.M. (1988). Muscle strength and its development—new perspectives. *Sports Medicine, 6*, 146-168.

Epler, M., Nawoczenski, D., & Englehardt, T. (1988). Comparison of the Cybex II standard shin adaptor versus the Johnson anti-sheer device in torque generation. *Journal of Orthopedic Sport, Medicine and Physical Therapy, 9*, 284-286.

Epley, B. (1985). *Dynamic Strength Training for the Athlete.* Lincoln, NE: William C. Brown.

Faigenbaum, A.D., Skrinar, G.S., Cesare, W.F., Kraemer, W.J., & Thomas, H.E. (1990). Physiologic and symptomatic responses of cardiac patients to resistance exercise. *Archives of Physical Medicine and Rehabilitation, 71*, 395-398.

Faigenbaum, A.D., Westcott, W.L., Loud, R.L., & Long, C. (1999). The effects of different resistance training protocols on muscular strength and endurance development in children. *Pediatrics, 104*, 1-7.

Fees, M., Decker, T., Snyder-Mackler, L., & Axe, M.J. (1998). Upper extremity weight-training modifications for the injured athlete: A clinical perspective. *American Journal of Sports Medicine, 26*, 732-742.

Fleck, S.J., & Kraemer, W.J. (1997). *Designing resistance training programs* (2nd ed). Champaign, IL: Human Kinetics.

Fleck, S.J., & Kraemer, W.J. (1988). Resistance training: Basic principles. *Physician and Sportsmedicine, 16*, 160-171.

Fleming, L.K. (1985). Accommodation capabilities of Nautilus weight machines to human strength curves. *National Strength and Conditioning Association Journal, 7*, 68.

Francis, K., & Hoobler, T. (1987). Comparison of peak torque values of the knee flexor and extensor muscle groups using the Cybex II and Lido 2.0 isokinetic dynamometers. *Journal of Orthopedic Sport, Medicine and Physical Therapy, 8*, 480-483.

Fry, A. (1985). Weight room safety. *National Strength and Conditioning Association Journal, 7*, 32-33.

Fry, A.C., Bibi, K.W., Eyford, T., & Kraemer, W.J. (1990). Stature variables as discriminators of foot contact during the squat exercise in untrained females. *Journal of Applied Sport Science Research, 9*, 23-32.

Fry, A.C., & Kraemer, W.J. (1991). Physical performance characteristics of American collegiate football players. *Journal of Applied Sport Science Research, 5*, 126-138.

Fry, A.C., Kraemer, W.J., Weseman, C.A., Conroy, B.P., Gordon, S.E., Hoffman, J.R., et al. (1991). The effects of an off-season strength and conditioning program on starters and non-starters in women's intercollegiate volleyball. *Journal of Applied Sport Science Research, 5*, 174-181.

Fry, A.C., & Powell, D.R. (1987a). A comparison of isokinetic and isometric muscle balance characteristics. *Journal of Applied Sport Science Research, 1*, 59.

Fry, A.C., & Powell, D.R. (1987b). Hamstring/quadriceps parity with three different modes of weight training. *Journal of Sports Medicine and Physical Fitness, 27,* 362-367.

Fry, A.C., Powell, D.R., & Kraemer, W.J. (1992). Validity of isokinetic and isometric testing modalities for assessing short-term resistance exercise strength gains. *Journal of Sport Rehabilitation, 1,* 275-283.

Fry, A.C., Schmidt, R.J., Johnson, G.O., Tharp, G.D., & Kraemer, W.J. (1993). Recovery heart rate and blood pressure responses to a graded exercise test and heavy resistance exercise. *Isokinetics and Exercise Science, 3,* 74-84.

Fry, A.C., Webber, J.M., Weiss, L.W., Fry, M.D., & Li, Y. (2000). Impaired performances with excessive high-intensity free-weight training. *Journal of Strength and Conditioning Research, 14,* 54-61.

Gaines, J.M., & Talbot, L.A. (1999). Isokinetic strength testing in research and practice. *Biology and Research in Nursing, 1,* 57-64.

Giorgi, A., Wilson, G.J., Weatherby, R.P., & Murphy, A.J. (1998). Functional isometric weight training: Its effects on the development of muscular function and the endocrine system over an 8-week training period. *Journal of Strength and Conditioning Research, 12,* 18-25.

Goertzen, M., Schoppe, K., Lange, G., & Schulitz, K.P. (1989). Injuries and damage caused by excess stress in body building and power lifting. *Sportverletzung Sportschaden, 3,* 32.

Going, S.B., Massey, B.H., Hoshizaki, T.B., & Lohman, T.G. (1987). Maximal voluntary static force production characteristics of skeletal muscle in children 8-11 years of age. *Research Quarterly, 58,* 115-123.

Graves, J.E., & James, R.J. (1990). Concurrent augmented feedback and isometric force generation during familiar and unfamiliar muscle movements. *Research Quarterly, 61,* 75-79.

Gregor, R.J., Edgerton, V.R., Perrine, J.J., Capion, D.S., & DeBus, C. (1979). Torque-velocity relationships and muscle fiber composition in elite female athletes. *Journal of Applied Physiology, 47,* 388-392.

Guy, J.A., & Micheli, L.J. (2001). Strength training for children and adolescents. *Journal of the American Academy of Orthopedic Surgeons, 9,* 29-36.

Hageman, P.A., Gillaspie, D.M., & Hill, L.D. (1988). Effects of speed and limb dominance on eccentric and concentric isokinetic testing of the knee. *Journal of Orthopedic Sport, Medicine and Physical Therapy, 10,* 59.

Häkkinen, K. (1989). Neuromuscular and hormonal adaptations during strength and power training. *Journal of Sports Medicine and Physical Fitness, 29,* 9-26.

Häkkinen, K., Kauhanen, H., & Komi, P.V. (1987). Aerobic, anaerobic, assistant exercise, and weightlifting performance capacities in elite weightlifters. *Journal of Sports Medicine and Physical Fitness, 27,* 240-246.

Häkkinen, K., Pakarinen, A., Alen, M., Kauhanen, H., & Komi, P.V. (1988). Neuromuscular and hormonal responses in elite athletes to two successive strength training sessions in one day. *European Journal of Applied Physiology, 57,* 133-139.

Häkkinen, K., Alen, M., Kallinen, M., Newton, R.U., & Kraemer, W.J. (2000). Neuromuscular adaptation during prolonged strength training, detraining and re-strength-training in middle-aged and elderly people. *European Journal of Applied Physiology, 83,* 51-62.

Harman, E., Sharp, M., Manikowski, R., Frykman, P., & Rosenstein, R. (1987). Analysis of a muscle strength data base. *Journal of Applied Sport Science Research, 2,* 54.

Harman, E.A., Frykman, P.N., Clagett, E.R., & Kraemer, W.J. (1988). Intra-abdominal and intra-thoracic pressures during lifting and jumping. *Medicine and Science in Sports and Exercise, 20,* 195-201.

Hemba, G. (1985). Hamstring parity. *National Strength and Conditioning Association Journal, 7,* 30-31.

Heyward, V.H., Johannes-Ellis, S.M., & Romer, J.F. (1986). Gender differences in strength. *Research Quarterly, 57,* 154-159.

Hickson, R.C., Hidaka, K., Foster, C., Falduto, M.T., & Chatterton, R.T. (1994). Successive time courses of strength development and steroid hormone responses to heavy-resistance training. *Journal of Applied Physiology, 76,* 663-670.

Hill, D.W., Collins, M.A., Cureton, K.J., & DeMello, J.J. (1989). Blood pressure response after weight training exercise. *Journal of Applied Sport Science Research, 3,* 44-47.

Hoeger, W.K., Barette, S.L., Hale, D.F., & Hopkins, D.R. (1987). Relationship between repetitions and selected percentages of one repetition maximum. *Journal of Applied Sport Science Research, 1,* 11-13.

Hoeger, W.K., Barette, S.L., Hale, D.F., & Hopkins, D.R. (1990). Relationship between repetitions and selected percentages of one repetition maximum. *Journal of Applied Sport Science Research, 1,* 11-13.

Hoffman, J.R., Fry, A.C., Howard, R., Maresh, C.M., & Kraemer, W.J. (1991). Strength, speed, and endurance changes during the course of a Division I basketball season. *Journal of Applied Sport Science Research, 5,* 144-149.

Hoffman, J.R., Maresh, C.M., & Armstrong, L.E. (1992). Isokinetic and dynamic constant resistance strength testing: Implications for sport. *Physical Therapy Practice, 2,* 42-53.

Hortobagyi, T.T., LaChance, P.F., & Katch, F.I. (1987). Prediction of maximum isokinetic force, power and isotonic velocity and power from maximum force and power measured during hydraulic bench press exercise. *Journal of Applied Sport Science Research, 1,* 58.

Housh, T.J., Johnson, G.O., Hughes, R.A., Cisar, C.J., & Thorland, W.G. (1988). Yearly changes in the body composition and muscular strength of high school wrestlers. *Research Quarterly, 59,* 240-243.

Housh, T.J., Johnson, G.O., Hughes, R.A., Housh, D.J., Hughes, R.J., Fry, A.C., et al. (1989). Isokinetic strength and body composition of high school wrestlers across age. *Medicine and Science in Sports and Exercise, 21,* 105-109.

Housh, T.J., Johnson, G.O., Marty, L., Eischen, G., Eischen, C., & Housh, D.J. (1988). Isokinetic leg extension and extension strength of university football players. *Journal of Orthopedic Sport, Medicine and Physical Therapy, 9*, 365-369.

Housh, D.J., Housh, T.J., Johnson, G.O., & Chu, W.K. (1992). Hypertrophic response to unilateral concentric isokinetic resistance training. *Journal of Applied Physiology, 73*, 65-70.

Huegli, R., Richardson, T., Graffis, K., Kroll, B., & Epley, B. (1989). Roundtable: Safe facility design and standards for safe equipment. *National Strength and Conditioning Association Journal, 11*, 14-27.

Hunter, G.R., & Culpepper, M.I. (1988). Knee extension torque joint position relationships following isotonic fixed resistance and hydraulic resistance training. *Athletic Training, 23*, 16-20.

Hunter, G.R., McGuirk, J., Mitrano, N., Pearman, P., Thomas, B., & Arrington, R. (1984). The effects of a weight training belt on blood pressure during exercise. *Journal of Applied Sport Science Research, 3*, 13-18.

Hunter, G.R., Wetzstein, C.J., McLafferty, C.L., Zuckerman, P.A., Landers, K.A., & Bamman, M.M. (2001). High-resistance versus variable-resistance training in older adults. *Medicine and Science in Sports and Exercise, 33*, 1759-1764.

Hurley, B.F., Redmond, R.A., Pratley, R.E., Treuth, M.S., Rogers, M.A., & Goldberg, A.P. (1995). Effects of strength training on muscle hypertrophy and muscle cell disruption in older men. *International Journal of Sports Medicine, 16*, 378-384.

Ikai, M., & Steinhaus, A.H. (1961). Some factors modifying the expression of human strength. *Journal of Applied Physiology, 16*, 157-163.

Jaric, S., Ugarkovic, D., & Kukolj, M. (2002). Evaluation of methods for normalizing muscle strength in elite and young athletes. *Journal of Sports Medicine and Physical Fitness, 42*, 141-151.

Jenkins, W.L., Thackaberry, M., & Killian, C. (1984). Speed-specific isokinetic testing. *Journal of Orthopedic Sport, Medicine and Physical Therapy, 6*, 181-183.

Kanehisa, H., & Miyashita, M. (1983). Specificity of velocity in strength training. *European Journal of Applied Physiology, 52*, 104-106.

Kellis, E., & Baltzopoulos, V. (1995). Isokinetic eccentric exercise. *Sports Medicine, 19*, 202-222.

Keogh, J.W.L., Wilson, G.J., & Weatherby, R.P. (1999). A cross-sectional comparison of different resistance training techniques in the bench press. *Journal of Strength and Conditioning Research, 13*, 247-258.

Kim, H.J., & Kramer, J.F. (1997). Effectiveness of visual feedback during isokinetic exercise. *Journal of Sports Medicine and Physical Therapy, 26*, 318-323.

Kindig, L.E., Soares, P.L., Wisenbaker, J.M., & Mrvos, S.R. (1984). Standard scores for women's weight training. *Physician and Sportsmedicine, 12*, 67-74.

Klopfer, D.A., & Greij, S.D. (1988). Examining quadriceps/hamstrings performance at high velocity isokinetics in untrained subjects. *Journal of Orthopedic Sport, Medicine and Physical Therapy, 10*, 18-22.

Knapik, J.J., Mawdsley, R.H., & Ramos, M.U. (1983). Angular specificity and test mode specificity of isometric and isokinetic strength training. *Journal of Orthopedics and Sports Physical Therapy, 5*, 58-65.

Knuttgen, H.G., & Kraemer, W.J. (1987). Terminology and measurement in exercise performance. *Journal of Applied Sport Science Research, 1*, 1-10.

Knutzen, K.M., Brilla, L.R., & Caine, D. (1999). Validity of 1RM prediction equations for older adults. *Journal of Strength and Conditioning Research, 13*, 242-246.

Kraemer, W.J. (1983). Measurement of strength. In A. Weltman & C.G. Spain (Eds.), *Proceedings of the White House Symposium on physical fitness and sports medicine* (pp. 35-36). Washington, DC: President's Council on Physical Fitness and Sports.

Kraemer, W.J. (1990). Physiological and cellular effects of exercise training. In W.B. Leadbetter, J.A. Buckwalter, & S.L. Gordon (Eds.), *Sports-induced inflammation* (pp. 659-676). Park Ridge, IL: American Academy of Orthopedic Surgeons.

Kraemer, W.J., Deschenes, M.R., & Fleck, S.J. (1988). Physiological adaptations to resistance exercise: Implications for athletic conditioning. *Sports Medicine, 6*, 246-256.

Kraemer, W.J., & Fleck, S.J. (1993). *Strength training for young athletes.* Champaign, IL: Human Kinetics.

Kraemer, W.J., Fry, A.C., Frykman, P.N., Conroy, B., & Hoffman, J. (1989). Resistance training and youth. *Pediatric Exercise Science, 1*, 336-350.

Kraemer, W.J., Gordon, S.E., Fleck, S.J., Marchitelli, L.J., Mello, R., Dziados, J.E., et al. (1991). Endogenous anabolic hormonal and growth factor responses to heavy resistance exercise in males and females. *International Journal of Sports Medicine, 12*, 228-235.

Kraemer, W.J., & Koziris, L.P. (1992). Muscle strength training: Techniques and considerations. *Physical Therapy Practice, 2*, 54-68.

Kraemer, W.J., Marchitelli, L., McCurry, D., Mello, R., Dziados, J.E., Harman, E., et al. (1990). Hormonal and growth factor responses to heavy resistance exercise. *Journal of Applied Physiology, 69*, 1442-1450.

Kraemer, W.J., Noble, B.J., Clark, M.J., & Culver, B.W. (1987). Physiologic responses to heavy-resistance exercise with very short rest periods. *International Journal of Sports Medicine, 8*, 247-252.

Kraemer, W.J., Fleck, S.J., & Evans, W.J. (1996). Strength and power training: Physiological mechanisms of adaptation. In J.O. Holloszy (Ed.), *Exercise and sport science reviews* (pp. 363-397). Philadelphia: Williams and Wilkins.

Kraemer, W.J., Fleck, S.J., Maresh, C.M., Ratamess, N.A., Gordon, S.E., Goetz, K.L., et al. (1999). Acute hormonal responses to a single bout of heavy resistance exercise in trained power lifters and untrained men. *Canadian Journal of Applied Physiology, 24*, 524-537.

Kraemer, W.J., Ratamess, N., Fry, A.C., Triplett-McBride, T., Koziris, L.P., Bauer, J.A., et al. (2000). Influence of resistance training volume and periodization on physiological and performance adaptations in college women tennis players. *American Journal of Sports Medicine, 28*, 626-633.

Kraemer, W.J., & Ratamess, N.A. (2000). Physiology of resistance training: Current issues. *Orthopaedic Physical Therapy Clinics of North America: Exercise Technologies, 9*, 467-513.

Kraemer, W.J., Ratamess, N.A., & French, D.N. (2002). Resistance training for health and performance. *Current Sports Medicine Reports, 1*, 165-171.

Kulig, K., Andrews, J.G., & Hay, J.G. (1984). Human strength curves. *Exercise and Sports Science Reviews, 12*, 417-466.

LaChance, P.F., Gabriel, D.A., Hortobagyi, T.T., & Katch, F.I. (1987). Muscular peak torque during fast and slow uni- and bi-directional concentric hydraulic resistance exercise. *Journal of Applied Sport Science Research, 1*, 59.

LaChance, P.F., Katch, F.I., Mistry, D.J., & Hortobagyi, T.T. (1988). Day-to-day reliability during high and low resistance bi-directional hydraulic exercise. *Journal of Applied Sport Science Research, 2*, 57.

Lander, J. (1985). Maximum based on reps. *National Strength and Conditioning Association Journal, 6*, 60-61.

Lander, J.E., Simonton, R.L., & Giacobbe, J.K.F. (1990). The effectiveness of weight-belts during the squat exercise. *Medicine and Science in Sports and Exercise, 22*, 117-126.

Legg, S.J., & Pateman, C.M. (1984). A physiological study of the repetitive lifting capabilities of healthy young males. *Ergonomics, 27*, 259-272.

LeSuer, D.A., McCormick, J.H., Mayhew, J.L., Wasserstein, R.L., & Arnold, M.D. (1997). The accuracy of prediction equations for estimating 1-RM performance in the bench press, squat, and deadlift. *Journal of Strength and Conditioning Research, 11*, 211-213.

Leveritt, M., & Abernethy, P.J. (1999). Effects of carbohydrate restriction on strength performance. *Journal of Strength and Conditioning Research, 13*, 52-57.

Lewis, C.L., & Spitler, D.L. (1989). Effect of tibial rotation on measures of strength and endurance of the knee. *Journal of Applied Sport Science Research, 3*, 19-22.

Lundeen, W.A., Nicolau, G.Y., Lakatua, D.J., Sackett-Lundeen, L., Petrescu, E., & Haus, E. (1990). Circadian periodicity of performance in athletic students. In D.K. Hayes, J.E. Pauly, & R.J. Reiter (Eds.), *Chronobiology: Its role in clinical medicine, general biology and agriculture*. New York, NY: Wiley-Liss.

MacDougall, J.D., Tuxen, D., Sale, D.G., Moroz, J.R., & Sutton, J.R. (1985). Arterial blood pressure response to heavy resistance exercise. *Journal of Applied Physiology, 58*, 785-790.

Mayhew, J.L., Ball, T.E., Arnold, M.D., & Bowen, J.C. (1992). Relative muscular endurance performance as a predictor of bench press strength in college men and women. *Journal of Applied Sport Science Research, 6*, 200-206.

Mayhew, J.L., Piper, F.C., & Ware, J.S. (1993). Anthropometric correlates with strength performance among resistance trained athletes. *Journal of Sports Medicine and Physical Fitness, 33*, 159-165.

Mayhew, J.L., Prinster, J.L., Ware, J.S., Zimmer, D.L., Arabas, J.R., & Bemben, M.G. (1995). Muscular endurance repetitions to predict bench press strength in men of different training levels. *Journal of Sports Medicine and Physical Fitness, 35*, 108-113.

Mayhew, J.L., Ware, J.S., Bemben, M.G., Wilt, B., Ward, T.E., Farris, B., et al. (1999). The NFL-225 test as a measure of bench press strength in college football players. *Journal of Strength and Conditioning Research, 13*, 130-134.

Mazur, L.J., Yetman, R.J., & Risser, W.L. (1993). Weight-training injuries: Common injuries and preventative methods. *Sports Medicine, 16*, 57-63.

Mazzetti, S.A., Kraemer, W.J., Volek, J.S., Duncan, N.D., Ratamess, N.A., Gomez, A.L., et al. (2000). The influence of direct supervision of resistance training on strength performance. *Medicine and Science in Sports and Exercise, 32*, 1175-1184.

McBride, J.M., Triplett-McBride, T., Davie, A., & Newton, R.U. (2002). The effect of heavy- vs. light-load jump squats on the development of strength, power, and speed. *Journal of Strength and Conditioning Research, 16*, 75-82.

Meese, G.B., Schiefer, R.E., Kustner, P., Kok, R., & Lewis, M.I. (1986). Subjective comfort vote and air temperature as predictors of performance in factory workers. *European Journal of Applied Physiology, 55*, 195-197.

Moffroid, M., & Whipple, R.H. (1970). Specificity of speed of exercise. *Physical Therapy, 50*, 1692-1700.

Mookerjee, S., & Ratamess, N.A. (1999). Comparison of strength differences and joint action durations between full and partial range-of-motion bench press exercise. *Journal of Strength and Conditioning Research, 13*, 76-81.

Morales, J., & Sobonya, S. (1996). Use of submaximal repetition tests for predicting 1-RM strength in class athletes. *Journal of Strength and Conditioning Research, 10*, 186-189.

Morrissey, M.C. (1987). The relationship between peak torque and work of the quadriceps and hamstrings after meniscectomy. *Journal of Orthopedic Sport, Medicine and Physical Therapy, 8*, 405-408.

Morrissey, M.C., & Brewster, C.E. (1986). Hamstring weakness after surgery for anterior cruciate injury. *Journal of Orthopedic Sport, Medicine and Physical Therapy, 7*, 310-312.

Murphy, A.J., & Wilson, G.J. (1996). Poor correlations between isometric tests and dynamic performance: Relationship to muscle activation. *European Journal of Applied Physiology, 73*, 353-357.

Nagle, F.J., Seals, D.R., & Hanson, P. (1988). Time to fatigue during isometric exercise using different muscle masses. *International Journal of Sports Medicine, 9*, 313-315.

National Strength and Conditioning Association. (1985). Position paper on prepubescent strength training. *National Strength and Conditioning Association Journal, 7*, 27-31.

Nosse, L.J. (1982). Assessment of selected reports on the strength relationship of the knee musculature. *Journal of Orthopedic Sport, Medicine and Physical Therapy, 4*, 78-85.

Nosse, L.J., & Hunter, G.R. (1985). Free weights: A review supporting their use in training and rehabilitation. *Athletic Training, 20*, 206-209.

Nunn, K.D., & Mayhew, J.L. (1988). Comparison of three methods of assessing strength imbalances at the knee. *Journal of Orthopedic Sport, Medicine and Physical Therapy, 10*, 134-138.

O'Conner, B., Simmons, J., & O'Shea, P. (1989). *Weight training today*. St. Paul: West.

Osternig, L.R. (1986). Isokinetic dynamometry: Implications for muscle testing and rehabilitation. In K.B. Pandolf (Ed.), *Exercise and Sport Science Reviews, 14*, 45-104.

Osternig, L.R., Sawhill, J.A., Bates, B.J., & Hamill, J. (1983). Function of limb speed on torque patterns of antagonistic muscles. In H. Matsui & K. Kobayashi (Eds.), *Biomechanics VIII-A* (pp. 251-257). Champaign, IL: Human Kinetics.

Osternig, L.R. (2000). Assessing human performance. In L.E. Brown (Ed.), *Isokinetics in human performance* (pp. 77-96). Champaign, IL: Human Kinetics.

Parcell, A.C., Sawyer, R.D., Tricoli, V.A., & Chinevere, T.D. (2002). Minimum rest period for strength recovery during a common isokinetic testing protocol. *Medicine and Science in Sports and Exercise, 34*, 1018-1022.

Patterson, P., Sherman, J., Hitzelberger, L., & Nichols, J. (1996). Test-retest reliability of selected Life circuit machines. *Journal of Strength and Conditioning Research, 10*, 246-249.

Patteson, M.E., Nelson, S.G., & Duncan, P.W. (1984). Effects of stabilizing the non-tested lower extremity during isokinetic evaluation of the quadriceps and hamstrings. *Journal of Orthopedic Sport, Medicine and Physical Therapy, 6*, 18-20.

Ploutz-Snyder, L.L., & Giamis, E.L. (2001). Orientation and familiarization to 1RM strength testing in old and young women. *Journal of Strength and Conditioning Research, 15*, 519-523.

Poulmedis, P. (1985). Isokinetics maximal torque power of Greek elite soccer players. *Journal of Orthopedic Sport, Medicine and Physical Therapy, 6*, 293-295.

Rahmani, A., Viale, F., Dalleau, G., & Lacour, J.R. (2001). Force/velocity and power/velocity relationships in squat exercise. *European Journal of Applied Physiology, 84*, 227-232.

Rankin, J.M., & Thompson, C.B. (1983). Isokinetic evaluation of quadriceps and hamstring function: Normative data concerning body weight and sport. *Athletic Training, 18*, 110-113.

Raske, A., & Norlin, R. (2002). Injury incidence and prevalence among elite weight and power lifters. *American Journal of Sports Medicine, 30*, 248-256.

Ratamess, N.A., Kraemer, W.J., Volek, J.S., Rubin, M.R., Gómez, A.L., French, D.N., et al. (2003). The effects of amino acid supplementation on muscular performance during resistance training overreaching: Evidence of an effective overreaching protocol. *Journal of Strength and Conditioning Research, 17*, 250-258.

Rhea, M.R., Ball, S.D., Phillips, W.T., & Burkett, L.N. (2002). A comparison of linear and daily undulating periodized programs with equated volume and intensity for strength. *Journal of Strength and Conditioning Research, 16*, 250-255.

Rhea, M.R., Landers, D.M., Alvar, B.A., & Arent, S.M. (2003). The effects of competition and the presence of an audience on weight-lifting performance. *Journal of Strength and Conditioning Research, 17*, 303-306.

Rhea, M.R., Alvar, B.A., Burkett, L.N., & Ball, S.D. (2003). A meta-analysis to determine the dose-response for strength development. *Medicine and Science in Sports and Exercise, 35*, 456-464.

Rizzardo, M., Bay, G., & Wessel, J. (1988). Eccentric and concentric torque and power of the knee extensors. *Canadian Journal of Sport Sciences, 13*, 166-169.

Rose, K., & Ball, T.E. (1992). A field test for predicting maximum bench press lift of college women. *Journal of Applied Sport Science Research, 6*, 103-106.

Rutherford, O.M., Greig, C.A., Sargeant, A.J., & Jones, D.A. (1986). Strength training and power output transference effects in the human quadriceps muscle. *Journal of Sports Sciences, 4*, 101-107.

Rutherford, O.M., & Jones, D.A. (1986). The role of learning and coordination in strength training. *European Journal of Applied Physiology, 55*, 100-105.

Safran, M.R., Garrett, W.E., Seaber, A.V., Glisson, R.R., & Ribbeck, B.M. (1988). The role of warm-up in muscular injury prevention. *American Journal of Sports Medicine, 16*, 123.

Sailors, M., & Berg, K. (1987). Comparison of responses to weight training in pubescent boys and men. *Journal of Sports Medicine and Physical Fitness, 27*, 30-38.

Sale D., & Delman, A. (1983). Fatigability in young men and women during weight lifting exercise. *Medicine and Science in Sports and Exercise, 15*, 146.

Sale, D.G. (1988). Neural adaptation to resistance training. *Medicine and Science in Sports and Exercise, 20*(Suppl.), S135-S145.

Sale, D.G., & MacDougall, J.D. (1979). Effect of strength training upon motoneuron excitability of man. *Medicine and Science in Sports, 11*, 77.

Sale, D.G., Fleck, S.J., & Kraemer, W.J. (1988). Testing strength and power. In D. MacDougall, H.A. Wenger, &

H.J. Green (Eds.), *Physiological testing of the elite athlete* (2nd. ed.) (pp. 21-106). Ithaca, NY: Movement.

Sanborn, K., Boros, R., Hruby, J., Schilling, B., O'Bryant, H.S., Johnson, R.L., et al. (2000). Short term performance effects of weight training with multiple sets not to failure vs a single set to failure in women. *Journal of Strength and Conditioning Research, 14,* 328-331.

Sapega, A.A., Nicholas, J.A., Sokolow, D., & Saraniti, A. (1982). The nature of torque overshoot in Cybex isokinetic dynamometry. *Medicine and Science in Sports and Exercise, 14,* 368-375.

Schlicht, J., Camaione, D.N., & Owens, S.V. (2001). Effect of intense strength training on standing, balance, walking speed, and sit-to-stand performance in older adults. *Journal of Gerontology: A Biological Science and Medical Science, 56,* M281-M286.

Schoffstall, J.E., Branch, J.D., Leutholtz, B.C., & Swain, D.E. (2001). Effects of dehydration and rehydration on the one-repetition maximum bench press of weight-trained males. *Journal of Strength and Conditioning Research, 15,* 102-108.

Seger, J.Y., Westing, S.H., Hanson, M., Karlson, E., & Ekblom, B. (1988). A new dynamometer measuring concentric and eccentric muscle strength in accelerated, decelerated, or isokinetic movements. *Journal of Applied Physiology, 57,* 526-550.

Sewall, L.P., & Lander, J.E. (1991). The effects of rest on maximal efforts in the squat and bench press. *Journal of Applied Sport Science Research, 5,* 96-99.

Shankman, G.A. (1984). Training related injuries in progressive resistance exercise programs. *National Strength and Conditioning Association Journal, 6,* 36-37.

Sharp, M.A., Harman, E., Vogel, J.A., Knapik, J.J., & Legg, S.J. (1988). Maximal aerobic capacity for repetitive lifting: Comparison with three standard exercise testing modes. *European Journal of Applied Physiology, 57,* 753-760.

Shaw, C.E., McCully, K.K., & Posner, J.D. (1995). Injuries during the one repetition maximum assessment in the elderly. *Journal of Cardiopulmonary Rehabilitation, 15,* 283-287.

Smidt, G.L., Albright, J.P., & Densingerm, R.H. (1984). Pre- and postoperative functional changes in total knee patients. *Journal of Orthopedic Sport, Medicine and Physical Therapy, 6,* 25-29.

Smith, T.K. (1984). Preadolescent strength training: Some considerations. *Journal of Physical Education, Recreation and Dance, 80,* 43-44.

Stone, M., & O'Bryant, H. (1987). *Weight training: A scientific approach.* Minneapolis: Bellwether Press.

Stumbo, T.A., Merriam, S., Nies, K., Smith, A., Spurgeon, D., & Weir, J.P. (2001). The effect of hand-grip stabilization on isokinetic torque at the knee. *Journal of Strength and Conditioning Research, 15,* 372-377.

Thomas, L. (1984). Isokinetic torque levels for adult females: Effects of age and body size. *Journal of Orthopedic Sport, Medicine and Physical Therapy, 6,* 21-24.

Thomee, R., Renstrom, P., Grimby, G., & Peterson, L. (1987). Slow or fast isokinetic training after knee ligament surgery. *Journal of Orthopedic Sport, Medicine and Physical Therapy, 8,* 495-499.

Thomson, D.B., & Chapman, A.E. (1988). The mechanical response of active human muscle during and after stretch. *European Journal of Applied Physiology, 57,* 691-697.

Thomson, R., Fix, B., White, P., Moran, R., Longo, P., Van Haianger, D., et al. (1989). Roundtable: Safe facility design and standards for safe equipment. *National Strength and Conditioning Association Journal, 11,* 14-22.

Thorstensson, A., Grimby, G., & Karlsson, J. (1976). Force-velocity relations and fiber composition in human knee extensor muscles. *Journal of Applied Physiology, 40,* 12-16.

Tomberline, J.P., Basford, J.R., Schwen, E.E., Orte, P.A., Scott, S.G., Laughman, R.K., et al. (1991). Comparative study of isokinetic eccentric and concentric quadriceps training. *Journal of Orthopedics and Sports Physical Therapy, 14,* 31-36.

Van der Wall, H., McLaughlin, A., Bruce, W., Orth, F.A., Frater, C.J., Kannangara, S., et al. (1999). Scintigraphic patterns of injury in amateur weight lifters. *Clinical Nuclear Medicine, 24,* 915-920.

Verdera, F., Champavier, L., Schmidt, C., Bermon, S., & Marconnet, P. (1999). Reliability and validity of a new device to measure isometric strength in polyarticular exercises. *Journal of Sports Medicine and Physical Fitness, 39,* 113-119.

Viitasalo, J.T., Era, P., Leskinen, A.L., & Häkkinen, K. (1985). Muscular strength profiles and anthropometry in random samples of men aged 31-35, 51-55 and 71-75 years. *Ergonomics, 28,* 1563-1574.

Walmsley, R.P., & Szybbo, C. (1987). A comparative study of the torque generated by the shoulder internal and external rotator muscles in different positions and by varying speeds. *Journal of Orthopedic Sport, Medicine and Physical Therapy, 9,* 217-222.

Ware, J.S., Clemens, C.T., Mayhew, J.L., & Johnston, T.J. (1995). Muscular endurance repetitions to predict bench press and squat strength in college football players. *Journal of Strength and Conditioning Research, 9,* 99-103.

Wathen, D., Borden, R., Dunn, B., Everson, J., Gieck, J., Hill, B., et al. (1983). Prevention of athletic injuries through strength training and conditioning. *National Strength and Conditioning Association Journal, 5,* 14-19.

Wathen, D. (1994). Load assignment. In T. Baechle (Ed.), *Essentials of strength training and conditioning* (pp. 435-446). Champaign, IL: Human Kinetics.

Weir, J.P., Wagner, L.L., & Housh, T.J. (1994). The effect of rest interval length on repeated maximal bench presses. *Journal of Strength and Conditioning Research, 8,* 58-60.

Weiss, L.W., Fry, A.C., Wood, L.E., & Melton, C. (2000). Comparative effects of deep versus shallow squat and leg-press training on vertical jumping ability and related factors. *Journal of Strength and Conditioning Research, 14,* 241–247.

Weltman, R., Tippett, S., Janney, C., Strand, K., Rians, C., Cahill, B.R., et al. (1988). Measurement of isokinetic strength in prepubertal males. *Journal of Orthopedic Sport, Medicine and Physical Therapy, 9*, 345-351.

Williams, P.A., & Cash, T.F. (2001). Effects of a circuit weight training program on the body images of college students. *International Journal of Eating Disorders, 30*, 75-82.

Wilson, G.J., Elliott, B.C., & Wood, G.A. (1991). The effect on performance of imposing a delay during a stretch-shorten cycle movement. *Medicine and Science in Sports and Exercise, 23*, 364-370.

Wilson, G.J., Murphy, A.J., & Pryor, J.F. (1994). Musculotendinous stiffness: Its relationship to eccentric, isometric, and concentric performance. *Journal of Applied Physiology, 76*, 2714-2719.

Wilson, G.J., & Murphy, A.J. (1996). Strength diagnosis: The use of test data to determine specific strength training. *Journal of Sports Sciences, 14*, 167-173.

Wilson, G.J., Murphy, A.J., & Walshe, A. (1996). The specificity of strength training: The effect of posture. *European Journal of Applied Physiology, 73*, 346-352.

Wilson, G.J., Walshe, A.D., & Fisher, M.R. (1997). The development of an isokinetic squat device: Reliability and relationship to functional performance. *European Journal of Applied Physiology, 75*, 455-461.

Wrigley, T., & Strauss, G. (2000). Strength assessment by isokinetic dynamometry. In C.J. Gore (Ed.), *Physiological tests for elite athletes* (pp. 155-199). Champaign, IL: Human Kinetics.

Wyse, J.P., Mercer, T.H., & Gleeson, N.P. (1994). Time of day dependence of isokinetic leg strength and associated interday variability. *British Journal of Sports Medicine, 28*, 167-170.

Zemper, E.D. (1990). Weightroom safety: Four year study of weightroom injuries in a national sample of college football teams. *National Strength and Conditioning Association Journal, 12*, 32-34.

# Chapter 9

Alway, S.E., Grumbt, W.H., Gonyea, W.J., & Stray-Gundersen, J. (1989). Contrasts in muscle and myofibers of elite male and female bodybuilders. *Journal of Applied Physiology, 67*, 24-31.

Andersen, P. (1975). Capillary density in skeletal muscle of man. *Acta Physiologica Scandinavica, 95*, 203-205.

Andersen, P., & Henriksson, J. (1977). Capillary supply to the quadriceps femoris muscle of man: Adaptive response to exercise. *Journal of Physiology, 270*, 677-690.

Ausubel, F.M. (Ed.). (1994). *Current protocols in molecular biology.* New York: Wiley.

Bamman, M.M., Clarke, M.S.F., Talmadge, R.J., & Feeback, D.L. (1999). Enhanced protein electrophoresis technique for separating human skeletal muscle myosin heavy chain isoforms. *Electrophoresis, 20*, 466-468.

Barany, M. (1967). ATPase activity of myosin correlated with speed of muscle shortening. *Journal of General Physiology, 50*, 197-218.

Barany, M., Barany, K., & Giometti, C.S. (1998). Gel electrophoresis for studying biological function. *Analytica Chimica Acta, 372*, 33-66.

Berchtold, M.W., Brinkmeier, H., & Muntener, M. (2000). Calcium ion in skeletal muscle: Its crucial role for muscle function, plasticity, and disease. *Physiology Reviews, 80*, 1215-65.

Bergstrom, J. (1992). Muscle electrolytes in man. *Scandinavian Journal of Clinical and Laboratory Investigation, Supplement, 68*, 1-110.

Bergstom, J., Hermansen, L., Hultman, E., & Saltin, B. (1967). Diet, muscle glycogen and physical performance. *Acta Physiologica Scandinavica, 71*, 140-150.

Bradford, M.M. (1976). A rapid and sensitive method for the quantification of microgram quantities of protein utilizing the principle of protein-dye binding. *Annals of Biochemistry, 72*, 248-54.

Brooke, M.H., & Kaiser, K.K. (1970). Muscle fiber types: How many and what kind? *Archives of Neurology, 23*, 369-379.

Dubowitz, V., & Brooke, M.H. (1973). *Muscle biopsy: A modern approach.* London: Saunders.

Dwyer, D., Browning, J., & Weinstein, S. (1999). The reliability of muscle biopsies taken from vastus lateralis. *Journal of Science and Medicine in Sports, 2*, 333-340.

Epstein, E. (1986). Analytical procedures and instrumentation. In R.W. Tietz (Ed.), *Textbook of clinical chemistry* (pp. 98-109). Philadelphia: Saunders.

Evans, W.J., Phinney, S.D., & Young, V.R. (1982). Suction applied to a muscle biopsy maximises muscle sample size. *Medicine and Science in Sports, 14*, 101-102.

Fry, A.C., Allemeir, C.A., & Staron, R.S. (1994). Correlation between percentage fibre type area and myosin heavy chain content in human skeletal muscle. *European Journal of Applied Physiology, 68*, 246-251.

Halkjaer-Kristensen, J., & Ingemann-Hansen, T. (1979). Microphotometric analysis of NADH-tetrazolium reductase and α-glycerophosphate dehydrogenase in human quadriceps muscle. *Histochemistry Journal, 11*, 127-136.

Hartree, E.F. (1972). Determination of protein: A modification of the Lowry method that gives a linear photometric response. *Annals of Biochemistry, 48*, 422-427.

Hennessey, J.V., Chromiak, J.A., Della Ventura, S., Guertin, J., & MacLean, D.B. (1997). Increase in percutaneous muscle biopsy yield with a suction-enhancement technique. *Journal of Applied Physiology, 82*, 1739-1742.

Hepple, R.T. (1997). A new measurement of tissue capillarity: The capillary-to-fibre perimeter exchange index. *Canadian Journal of Applied Physiology, 22*, 11-22.

Hikida, R.S., Staron, R.S., Hagerman, F.C., Walsh, S., Kaiser, E., Shell, S., et al. (2000). Effects of high-intensity resistance training on untrained older men: II. Muscle fiber characteristics and nucleo-cytoplasmic relationships. *Journal of Gerontology: Biological Sciences, 55A,* B347-B354.

Humphries, B.J., Newton, R.U., Abernethy, P.J., & Blake, K.D. (1997). Reliability of an electrophoretic and image processing analysis of human skeletal muscle taken from m. vastus lateralis. *European Journal of Applied Physiology, 75,* 532-536.

Kraemer, W.J., Patton, J.F., Gordon, S.E., et al. (1995). Compatibility of high-intensity strength and endurance training in hormonal and skeletal muscle adaptations. *Journal of Applied Physiology, 78,* 976-989.

Laemmli, U.K. (1970). Cleavage of structural proteins during the assembly of the head of bacteriophage T4. *Nature, 227,* 680-685.

Lexell, J., & Taylor, C.C. (1989). Variability in muscle fibre areas in whole human quadriceps muscle. How much and why? *Acta Physiologica Scandinavica, 136,* 561-568.

Lowry, O.H. (1951). Protein measurement with the folin phenol reagent. *Journal of Biological Chemistry, 193,* 265.

McCall, G.E., Byrnes, W.C., Dickinson, A., Pattany, P.M., & Fleck, S.J. (1996). Muscle fiber hypertrophy, hyperplasia, and capillary density in college men after resistance training. *Journal of Applied Physiology, 81,* 2004-2012.

McCall, G.E., Byrnes, W.C., Dickinson, A.L., & Fleck, S.J. (1998). Sample size required for the accurate determination of fiber area and capillarity of human skeletal muscle. *Canadian Journal of Applied Physiology, 23,* 594-599.

McGuigan, M.R., Kraemer, W.J., Deschenes, M.R., et al. (2002). Statistical analysis of fiber area in human skeletal muscle. *Canadian Journal of Applied Physiology.*

Padykula, H.A., & Herman, E. (1955). The specificity of the histochemical method for adenosine triphosphatase. *Journal of Histochemistry and Cytochemistry, 3,* 170-195.

Pette, D. (1998). Training effects on the contractile apparatus. *Acta Physiologica Scandinavica, 162,* 367-376.

Pette, D., & Staron, R.S. (1990). Cellular and molecular diversities of mammalian skeletal muscle fibers. *Reviews of Physiology, Biochemistry and Pharmacology, 116,* 1-76.

Pette, D., & Vrbova, G. (1992). Adaptation of mammalian skeletal muscle fibers to chronic electrical stimulation. *Reviews of Physiology, Biochemistry and Pharmacology, 120,* 115-202.

Qu, Z., Andersen, J.L., & Zhou, S. (1997). Visualisation of capillaries in human skeletal muscle. *Histochemistry and Cell Biology, 107,* 169-174.

Schiaffino, A., & Reggiani, C. (1996). Molecular diversity of myofibrillar proteins: Gene regulation and functional significance. *Physiology Reviews, 76,* 371-423.

Smerdu, V., Karsch-Mizrachi, I., Campione, M., Leinwand, L., & Schiaffino, S. (1994). Type IIx myosin heavy chain transcripts are expressed in type IIb fibers of human skeletal muscle. *American Journal of Physiology, 267,* C1723-1728.

Staron, R.S., & Hikida, R.S. (1992a). Histochemical, biochemical, and ultrastructural analyses of single human muscle fibers, with special reference to the C-fiber population. *Journal of Histochemistry and Cytochemistry, 40,* 563-568.

Staron, R.S., & Hikida, R.S. (1992b). Progressive resistance training reduces myosin heavy chain coexpression in single muscle fibers from older men. *Journal of Applied Physiology, 88,* 627-633.

Staron, R.S., Hikida, R.S., & Hagerman, F.C. (1983). Myofibrillar ATPase activity in human muscle fast-twitch subtypes. *Histochemistry, 78,* 405-408.

Staron, R.S., Karapondo, D.L., Kraemer, W.J., et al. (1994). Skeletal muscle adaptations during early phase of heavy-resistance training in men and women. *Journal of Applied Physiology, 76,* 1247-1255.

Stoscheck, C.M. (1990). Quantitation of protein. *Methods of Enzymology, 182,* 50-68.

Talmadge, R.J., & Roy, R.R. (1993). Electrophoretic separation of rat skeletal muscle myosin heavy-chain isoforms. *Journal of Applied Physiology, 75,* 2337-2340.

Tesch, P.A., Thorsson, A., & Kaiser, P. (1984). Muscle capillary supply and fiber type characteristics in weight and power lifters. *Journal of Applied Physiology, 56,* 35-38.

Thompson, L.V. (1994). Effects of age and training on skeletal muscle physiology and performance. *Physical Therapy, 74,* 71-81.

Wagner, P.D. (1981). Formation and characterisation of myosin hybrids containing essential light chains and heavy chains from different muscle myosins. *Journal of Biological Chemistry, 256,* 2493-2498.

Wagner, P.D., & Giniger, W. (1981). Hydrolysis of ATP and reversible binding of F-actin by myosin heavy chains free of all light chains. *Nature, 292,* 560-562.

Weber, K., & Osborn, M. (1969). The reliability of molecular weight determinations by dodecyl sulfate-polyacrylamide gel electrophoresis. *Journal of Biological Chemistry, 244,* 4406-4412.

# Chapter 10

Alfonsi, E., Pavesi, R., Merlo, I.M., Gelmetti, A., et al. (1999). Hemoglobin near-infrared spectroscopy and surface EMG study in muscle ischaemia and fatiguing isometric contraction. *Journal of Sports Medicine and Physical Fitness, 39,* 83-92.

Angus, C., Welford, D., Sellens, M., Thompson, S., & Cooper, C.E. (1999). Estimation of lactate threshold by near infrared spectroscopy. *Advances in Experimental Medicine and Biology, 471,* 283-8.

Asanoi, H., Wada, O., Miyagi, K., Ishizaka, S., Kameyama, T., Seto, H., et al. (1992). New redistribution index of nutritive blood flow to skeletal muscle during dynamic exercise. *Circulation, 85*(4),1457-63.

Aunola, S., & Rusko, H. (1986). Aerobic and anaerobic threshold determined from venous lactate or from ventilation and gas exchange in relation to muscle fiber composition. *International Journal of Sports Medicine, 7,* 161-166.

Azuma, K., Homma, S., & Kagaya, A. (2000). Oxygen supply-consumption balance in the thigh muscles during exhausting knee-extension exercise. *Journal of Biomedical Optics, 5,* 97-101.

Bae, S.Y., Hamaoka, T., Katsumura, T., Shiga, T., Ohno, H., & Haga, S. (2000). Comparison of muscle oxygen consumption measured by near infrared continuous wave spectroscopy during supramaximal and intermittent pedalling exercise. *International Journal of Sports Medicine, 21,* 168-74.

Belardinelli, R., Barstow, T.J., Porszasz, J., & Wasserman, K. (1995a). Changes in skeletal muscle oxygenation during incremental exercise measured with near infrared spectroscopy. *European Journal of Applied Physiology, 70,* 487-92.

Belardinelli, R., Barstow, T.J., Porszasz, J., & Wasserman, K. (1995b). Skeletal muscle oxygenation during constant work rate exercise. *Medicine and Science in Sports and Exercise, 27,* 512-9.

Bhambhani, Y., Buckley, S., & Susaki, T. (1999). Muscle oxygenation trends during constant work rate cycle exercise in men and women. *Medicine and Science in Sports and Exercise, 31,* 90-8.

Bhambhani, Y., Maikala, R., & Buckley, S. (1998). Muscle oxygenation during incremental arm and leg exercise in men and women. *European Journal of Applied Physiology, 78,* 422-31.

Bhambhani, Y., Maikala, R., & Esmail, S. (2001). Oxygenation trends in vastus lateralis muscle during incremental and intense anaerobic cycle exercise in young men and women. *European Journal of Applied Physiology, 84,* 547-56.

Binzoni, T., Quaresima, V., Barattelli, G., et al. (1998). Energy metabolism and interstitial fluid displacement in human gastrocnemius during short ischemic cycles. *Journal of Applied Physiology, 85,* 1244-51.

Blei, M.L., Conley, K.E., Odderson, I.B., Esselman, P.C., & Kushmerick, M.J. (1993). Individual variation in contractile cost and recovery in a human skeletal muscle. *Proceedings of the National Academy of Sciences of the United States of America, 90*(15), 7396-400.

Boushel, R., Langberg, H., Green, S., Skovgaard, D., Bulow, J., & Kjaer, M. (2000). Blood flow and oxygenation in peritendinous tissue and calf muscle during dynamic exercise in humans. *Journal of Physiology, 524,* 305-13.

Boushel, R., Langberg, H., Olesen, J., et al. (2000). Regional blood flow during exercise in humans measured by near-infrared spectroscopy and indocyanine green. *Journal of Applied Physiology, 89,* 1868-78.

Boushel, R., & Piantadosi, C.A. (2000). Near-infrared spectroscopy for monitoring muscle oxygenation. *Acta Physiologica Scandinavica, 168,* 615-22.

Cerretelli, P., & Binzoni, T. (1997). The contribution of NMR, NIRS and their combination to the functional assessment of human muscle. *International Journal of Sports Medicine, 18,* S270-9.

Chance, B., Borer, E., Evans, A., Holtom, G., et al. (1989). Optical and nuclear magnetic resonance studies of hypoxia in human tissue and tumors. *Annals of the New York Academy of Sciences, 551,* 1-16.

Chance, B., Dait, M.T., Zhang, C., Hamaoka, T., & Hagerman, F. (1992). Recovery from exercise-induced desaturation in the quadriceps muscles of elite competitive rowers. *American Journal of Physiology, 262,* C766-75.

Chance, B., Leigh, J.S., Jr., Clark, B.J., Maris, J., Kent, J., Nioka, S., et al. (1985). Control of oxidative metabolism and oxygen delivery in human skeletal muscle: A steady state analysis of the work/energy cost transfer function. *Proceedings of the National Academy of Sciences of the United States of America, 82,* 8384-8388.

Chance, B., Nioka, S., Kent, J., McCully, F., et al. (1988). Time-resolved spectroscopy of hemoglobin and myoglobin in resting and ischemic muscle. *Analytical Biochemistry, 174,* 698-707.

Cheatle, T.R., Potter, L.A., Cope, M., et al. (1991). Near-infrared spectroscopy in peripheral vascular disease. *British Journal of Surgery, 78,* 405-8.

Colier, W.N., Meeuwsen, I.B., Degens, H., & Oeseburg, B. (1995). Determination of oxygen consumption in muscle during exercise using near infrared spectroscopy. *Acta Anaesthesiologica Scandinavica Supplementum, 107,* 151-5.

Conley, K.E., Ordway, G.A., & Richardson, R.S. (2000). Deciphering the mysteries of myoglobin in striated muscle. *Acta Physiologica Scandinavica, 168,* 623-34.

Costes, F., Barthelemy, J.C., Feasson, L., Busso, T., Geyssant, A., & Denis, C. (1996). Comparison of muscle near-infrared spectroscopy and femoral blood gases during steady-state exercise in humans. *Journal of Applied Physiology, 80,* 1345-50.

Costes, F., Denis, C., Roche, F., Prieur, F., Enjolras, F., & Barthelemy, J.C. (1999). Age-associated alteration of muscle oxygenation measured by near infrared spectroscopy during exercise. *Archives of Physiology and Biochemistry, 107,* 159-67.

Costes, F., Prieur, F., Feasson, L., Geyssant, A., Barthelemy, J.C., & Denis, C. (2001). Influence of training on NIRS muscle oxygen saturation during submaximal exercise. *Medicine and Science in Sports and Exercise, 33,* 1484-9.

Coyle, E. (1995). Integration of the physiological factors determining endurance performance ability. *Exercise and Sport Sciences Reviews, 23,* 25-63.

De Blasi, R.A., Almenräder, N., & Ferrari, M. (1997). Comparison of 2 methods of measurement of forearm oxygen consumption by near infrared spectroscopy. *Journal of Biomedical Optics, 2,* 171-5.

De Blasi, R.A., Cope, M., Elwell, C., Safoue, F., & Ferrari, M. (1993). Noninvasive measurement of human forearm oxygen consumption by near infrared spectroscopy. *European Journal of Applied Physiology, 67*, 20-5.

De Blasi, R.A., Fantini, S., Franceschini, M.A., Ferrari, M., & Gratton, E. (1995). Cerebral and muscle oxygen saturation measurement by frequency-domain near-infrared spectrometer. *Medical and Biological Engineering and Computing, 33*, 228-30.

De Blasi, R.A., Ferrari, M., Natali, A., Conti, G., Mega, A., & Gasparetto, A. (1994). Noninvasive measurement of forearm blood flow and oxygen consumption by near-infrared spectroscopy. *Journal of Applied Physiology, 76*, 1388-93.

Delpy, D.T., & Cope M. (1997). Quantification in tissue near-infrared spectroscopy. *Philosophical Transactions of the Royal Society of London, Series B, Biological Sciences, 352*, 649-59.

Demarie, S., Quaresima, V., Ferrari, M., Sardella, F., Billat, V., & Faina, M. (2001). $\dot{V}O_2$ slow component correlates with vastus lateralis de-oxygenation and blood lactate accumulation during running. *Journal of Sports Medicine and Physical Fitness, 41*, 448-55.

Ding, H., Wang, G., Lei, W., et al. (2001). Non-invasive quantitative assessment of oxidative metabolism in quadriceps muscles by near infrared spectroscopy. *British Journal of Sports Medicine, 35*, 441-4.

Dulieu, V., Casillas, J.M., Maillefert, J.F., et al. (1997). Muscle metabolism changes with training in the non-amputated limb after vascular amputation: Interest of phosphorus 31 NMR spectroscopy. *Archives of Physical Medicine and Rehabilitation, 78*(8), 867-71.

Ferrari, M., Binzoni, T., & Quaresima, V. (1997). Oxidative metabolism in muscle. *Philosophical Transactions of the Royal Society of London, Series B, Biological Sciences, 352*, 677-83.

Ferrari, M., Wei, Q., Carraresi, L., De Blasi, R.A., & Zaccanti, G. (1992). Time resolved spectroscopy of the human forearm. *Journal of Photochemistry and Photobiology B, 16*, 141-53.

Foster, C., Rundell, K.W., Snyder, A.C., et al. (1999). Evidence for restricted muscle blood flow during speed skating. *Medicine and Science in Sports and Exercise, 31*, 1433-40.

Franceschini, M.A., Boas, D.A., Zourabian, A., et al. (2002). Near-infrared spiroximetry: Noninvasive measurements of venous saturation in piglets and human subjects. *Journal of Applied Physiology, 92*, 372-84.

Gaesser, G.A., & Poole, D.C. (1986). Lactate and ventilatory thresholds: Disparity in time course of adaptations to training. *Journal of Applied Physiology, 61*(3), 999-1004.

Grassi, B., Quaresima, V., Marconi, C., Ferrari, M., & Cerretelli, P. (1999). Blood lactate accumulation and muscle deoxygenation during incremental exercise. *Journal of Applied Physiology, 87*, 348-55.

Green, H.J., Hughson, R.L., Orr, G.W., & Ranney, D.A. (1983). Anaerobic threshold, blood lactate, and muscle metabolites in progressive exercise. *Journal of Applied Physiology, 54*(3), 1032-1038.

Haga, S., Bae, S.Y., Hamaoka, T., et al. (1998). Oxidative metabolism in skeletal muscle measured during supramaximal exercise in sprinter and active control groups by near infrared continuous wave spectroscopy. *Advances in Exercise and Sports Physiology, 4*, 57-64.

Hamaoka, T., McCully, K.K., Katsumura, T., Teruiki, S., & Chance, B. (2000). Non-invasive measures of muscle metabolism. In C.K. Sen, L. Packer, & O.O.P. Hanninen (Eds.). *Handbook of oxidants and antioxidants in exercise* (pp. 485-509). Amsterdam: Elsevier.

Hamaoka, T., Mizuno, M., Katsumura, T., Osada, T., Shimomitsu, T., & Quistorff, B. (1998). Correlation between indicators determined by near infrared spectroscopy and muscle fiber types in humans. *Japanese Journal of Applied Physiology, 28*(5), 243-248.

Hamaoka, T., Iwane, H., Shimomitsu, T., et al. (1996). Non-invasive measures of oxidative metabolism on working human muscles by near-infrared spectroscopy. *Journal of Applied Physiology, 81*, 1410-7.

Hampson, N.B., & Piantadosi, C.A. (1988). Near infrared monitoring of human skeletal muscle oxygenation during forearm ischemia. *Journal of Applied Physiology, 64*, 2449-57.

Hicks, A., McGill, S., & Hughson, R.L. (1999). Tissue oxygenation by near-infrared spectroscopy and muscle blood flow during isometric contractions of the forearm. *Canadian Journal of Applied Physiology, 24*, 216-30.

Homma, S., Eda, H., Ogasawara, S., & Kagaya, A. (1996). Near-infrared estimation of $O_2$ supply and consumption in forearm muscles working at varying intensity. *Journal of Applied Physiology, 80*, 1279-84.

Im, J., Nioka, S., Chance, B., Kime, R., & Rundell, K.W. (In review). Blood volume change and deoxygenation rate determined by NIRS is related to the anaerobic threshold in cross-country skiers. *European Journal of Applied Physiology.*

Im, J., Nioka, S., Chance, B., & Rundell, K.W. (2000). Muscle oxygen desaturation is related to whole body $\dot{V}O_2$ during cross-country ski skating. *International Journal of Sports Medicine, 22*, 356-60.

Jensen-Urstad, M., Hallback, I., & Sahlin, K. (1995). Effect of hypoxia on muscle oxygenation and metabolism during arm exercise in humans. *Clinical Physiology, 15*, 27-37.

Kagaya, A. (1990). Levelling-off of calf blood flow during walking and running, and its relationship to anaerobic threshold. *Annals of Physiological Anthropology, 9*(2), 219-24.

Kemp, G.J., Hands, L.J., Ramaswami, G., et al. (1995). Calf muscle mitochondrial and glycogenolytic ATP synthesis in patients with claudication due to peripheral vascular disease analysed using 31P magnetic resonance spectroscopy. *Clinical Science (London), 89*(6), 581-90.

Kemp, G.J., Roberts, N., Bimson, W.E., et al. (2001). Mitochondrial function and oxygen supply in normal and in chronically ischemic muscle: A combined 31P magnetic resonance spectroscopy and near infrared spectroscopy study in vivo. *Journal of Vascular Surgery, 34*(6), 1103-10.

Kouzaki, M., Shinohara, M., Ikeda, H., Watarai, K., & Fukunaga, T. (2001). Reduced slow component of oxygen uptake by endurance training is associated with reduced muscle deoxygenation [Abstract]. *Medicine and Science in Sports and Exercise, 33*, S201.

MacDonald, M.J., Tarnopolsky, M.A., Green, H.J., & Hughson, R.L. (1999). Comparison of femoral blood gases and muscle near-infrared spectroscopy at exercise onset in humans. *Journal of Applied Physiology, 86*, 687-93.

Mancini, D.M., Bolinger, L., Li, H., Kendrick, K., Chance, B., & Wilson, J.R. (1994). Validation of near-infrared spectroscopy in humans. *Journal of Applied Physiology, 77*, 2740-7.

Matsushita, K., Homma, S., & Okada, E. (1998). Influence of adipose tissue on muscle oxygenation measurement with NIRS instrument. *Proceedings of the Society of Photo-Optical Instrumentation Engineers, 3194*, 151-65.

McCully, K.K. (1993). 31P-MRS of quadriceps reveals quantitative differences between sprinters and long-distance runners. *Medicine and Science in Sports and Exercise, 25*(11), 1299-300.

McCully, K.K., & Hamaoka, T. (2000). Near-infrared spectroscopy: What can it tell us about oxygen saturation in skeletal muscle? *Exercise and Sport Sciences Reviews, 28*, 123-7.

McCully, K.K., Iotti, S., Kendrick, K., et al. (1994). Simultaneous *in vivo* measurements of $HbO_2$ saturation and PCr kinetics after exercise in normal humans. *Journal of Applied Physiology, 77*, 5-10.

Meyer, R. (1988). A linear model of muscle respiration explains monoexponential phosphocreatine changes. *American Journal of Physiology, 254 (Cell Physiology 23)*, C548-C553.

Millikan, G.A. (1942). The oximeter, an instrument for measuring continuously the oxygen saturation of arterial blood in man. *Review of Scientific Instruments, 13*, 434-444.

Miura, H., McCully, K., Hong, L., Nioka, S., & Chance, B. (2001). Regional difference of muscle oxygen saturation and blood volume during exercise determined by near infrared imaging device. *Japanese Journal of Physiology, 51*, 599-606.

Miura, H., Araki, H., Matoba, H., & Kitagawa, K. (2000). Relationship among oxygenation, myoelectric activity, and lactic acid accumulation in vastus lateralis muscle during exercise with constant work rate. *International Journal of Sports Medicine, 21*, 180-4.

Miura, T., Takeuchi, T., Sato, H., et al. (1998). Skeletal muscle deoxygenation during exercise assessed by nearinfrared spectroscopy and its relation to expired gas analysis parameters. *Japanese Circulation Journal, 62*, 649-57.

Moraine, J.J., Lamotte, M., Berre, J., Niset, G., Leduc, A., & Naeije, R. (1993). Relationship of middle cerebral artery blood flow velocity to intensity during dynamic exercise in normal subjects. *European Journal of Applied Physiology and Occupational Physiology, 67*(1), 35-8.

Moritani, T., Berry, M.J., Bacharach, D.W., & Nakamura, E. (1987). Gas exchange parameters, muscle blood flow and electromechanical properties of the plantar flexors. *European Journal of Applied Physiology and Occupational Physiology, 56*(1), 30-7.

Neary, J.P., Hall, K., & Bhambhani, Y.N. (2001). Vastus medialis muscle oxygenation trends during a simulated 20-km cycle time trial. *European Journal of Applied Physiology, 85*, 427-33.

Neary, J.P., McKenzie, D.C., & Bhambhani, Y.N. (2002). Effects of short-term endurance training on muscle deoxygenation trends using NIRS. *Medicine and Science in Sports and Exercise, 34*(11), 725-32.

Nioka, S., Miura, H., Long, H., Peery, A., Moser, D., & Chance, B. (1999). Functional quadriceps and gastrocnemius imaging in elite and untrained subjects. *Proceedings of the Society of Photo-Optical Instrumentation Engineers, 3597*, 282-90.

Nioka, S., Moser, D., Lech, G., et al. (1998). Muscle deoxygenation in aerobic and anaerobic exercise. *Advances in Experimental Medicine and Biology, 454*, 63-70.

Niwayama, M., Hamaoka, T., Lin, L., et al. (2000). Quantitative muscle oxygenation measurement using NIRS with correction for the influence of a fat layer: Comparison of oxygen consumption rates with measurements by other techniques. *Proceedings of the Society of Photo-Optical Instrumentation Engineers, 3911*, 256-65.

Niwayama, M., Lin, L., Shao, J., Kudo, N., & Yamamoto, K. (2000). Quantitative measurement of muscle hemoglobin oxygenation using near-infrared spectroscopy with correction for the influence of a subcutaneous fat layer. *Review of Scientific Instruments, 71*, 4571-5.

Oda, M., Yamashita, Y., Nakano, T., et al. (2000). Near infrared time-resolved spectroscopy system for tissue oxygenation monitor. *Proceedings of the Society of Photo-Optical Instrumentation Engineers, 4160*, 204-10.

Piantadosi, C.A., Hemstreet, T.M., & Jobsis-Vandervliet, F.F. (1986). Near-infrared spectrophotometric monitoring of oxygen distribution to intact brain and skeletal muscle tissues. *Critical Care Medicine, 14*(8), 698-706.

Poole, D., & Gaesser, G.A. (1985). Response of ventilatory and lactate thresholds to continuous and interval training. *Journal of Applied Physiology, 58*(4), 1115-1121.

Quaresima, V., Colier, W.N., van der Sluijs, M., & Ferrari, M. (2001). Nonuniform quadriceps $O_2$ consumption revealed by near infrared multipoint measurements. *Biochemical and Biophysical Research Communications, 285*, 1034-9.

Quaresima, V., Ferrari, M., Ciabattoni, M., Cantò, U., & Colonna, R. (1999). Oxygenation kinetics of different leg muscle groups measured during a 100-m sprint run by a portable near-infrared photometer. *Italian Journal of Sport Science, 6*, 20-3.

Quaresima, V., Homma, S., Azuma, K., et al. (2001). Calf and shin muscle oxygenation patterns and femoral artery blood flow during dynamic plantar flexion exercise in humans. *European Journal of Applied Physiology, 84*, 387-94.

Quaresima, V., Komiyama, T., & Ferrari, M. (2002). Differences in oxygen re-saturation of thigh and calf muscles after 2 treadmill stress tests. *Comparative Biochemistry and Physiology A, 132,* 67-73.

Quaresima, V., Pizzi, A., De Blasi, R.A., Ferrari, A., & Ferrari, M. (1996). Influence of the treadmill speed/slope on quadriceps oxygenation during dynamic exercise. *Advances in Experimental Medicine and Biology, 388,* 231-5.

Richardson, R.S., Knight, D.R., Poole, D.C., et al. (1995). Determinants of maximal exercise $\dot{V}O_2$ during single leg knee-extensor exercise in humans. *American Journal of Physiology, 268*(4 Pt. 2), H1453-61.

Rundell, K.W. (1996). Compromised oxygen uptake in speed skaters during treadmill in-line skating. *Medicine and Science in Sports and Exercise, 28*(1), 120-7.

Rundell, K.W., Nioka, S., & Chance, B. (1997). Hemoglobin/myoglobin desaturation during speed skating. *Medicine and Science in Sports and Exercise, 29,* 248-58.

Rundell, K.W., Szmedra, L., Im, J., Nioka, S., & Ploetz, J.A. (1998). Subcutaneous fat and optical density measurements using near-infrared spectrophotometry [Abstract]. *Medicine and Science in Sports and Exercise, 30*(Suppl.), 5.

Sahlin K. (1992). Non-invasive measurements of $O_2$ availability in human skeletal muscle with near-infrared spectroscopy. *International Journal of Sports Medicine, 13,* S157-60.

Sako, T., Hamaoka, T., Higuchi, H., Kurosawa, Y., & Katsumura, T. (2001). Validity of NIR spectroscopy for quantitatively measuring muscle oxidative metabolic rate in exercise. *Journal of Applied Physiology, 90,* 338-44.

Seiyama, A., Hazeki, O., & Tamura, M. (1988). Noninvasive quantitative analysis of blood oxygenation in rat skeletal muscle. *Journal of Biochemistry (Tokyo), 103*(3), 419-24.

Shiga, T., Tanabe, K., Nakase, Y., Shida, T., & Chance, B. (1995). Development of a portable tissue oximeter using near infrared spectroscopy. *Medical and Biological Engineering and Computing, 33,* 622-6.

Simon, J., Young, J.L., Blood, D.K., Segal, K.R., Case, R.B., & Butin, B. (1986). Plasma lactate and ventilation thresholds in trained and untrained cyclists. *Journal of Applied Physiology, 60*(3), 777-781.

Simonson, S.G., & Piantadosi, C.A. (1996). Near-infrared spectroscopy. Clinical applications. *Critical Care Clinics, 12*(4), 1019-29.

Szmedra, L., Im, J., Nioka, S., Chance, B., & Rundell, K.W. (2001). Hemoglobin/myoglobin oxygen desaturation during Alpine skiing. *Medicine and Science in Sports and Exercise, 33,* 232-6.

Takaishi, T., Sugiura, T., Katayama, K., et al. (2002). Changes in blood volume in a working muscle during crank cycle. *Medicine and Science in Sports and Exercise, 33,* 520-8.

Tamaki, T., Uchiyama, S., Tamura, T., & Nakano, S. (1994). Changes in muscle oxygenation during weight-lifting exercise. *European Journal of Applied Physiology, 68,* 465-9.

Thompson, C.H., Kemp, G.J., Sanderson, A.L., et al. (1996). Effect of creatine on aerobic and anaerobic metabolism in skeletal muscle in swimmers. *British Journal of Sports Medicine, 30,* 222-5.

Tran, T.K., Sailasuta, N., Kreutzer, U., et al. (1999). Comparative analysis of NMR and NIRS measurements of intracellular $PO_2$ in human skeletal muscle. *American Journal of Physiology, 276,* R1682-90.

van Beekvelt, M.C., Borghuis, M.S., van Engelen, B.G., Wevers, R.A., & Colier, W.N. (2001). Adipose tissue thickness affects *in vivo* quantitative near-IR spectroscopy in human skeletal muscle. *Clinical Science, 101,* 21-8.

van Beekvelt, M.C., Colier, W.N., Wevers, R.A., & Van Engelen, B.G. (2001). Performance of near-infrared spectroscopy in measuring local $O_2$ consumption and blood flow in skeletal muscle. *Journal of Applied Physiology, 90,* 511-9.

Wang, Z., Noyszewski, E.A., & Leigh, J.R. (1990). In vivo MRS measurement of deoxymyoglobin in human forearms. *Magnetic Resonance in Medicine, 14,* 562-567.

Webster, A.L., Syrotuik, D.G., Bell, G.J., Jones, R.L., Bhambhani, Y., & Young, M. (1998). Exercise after acute hyperbaric oxygenation: Is there an ergogenic effect? *Undersea and Hyperbaric Medicine, 25,* 153-9.

Wilson, J.R., Mancini, D.M., McCully, K., Ferraro, N., Lanoce, V., & Chance, B. (1989). Noninvasive detection of skeletal muscle underperfusion with near-infrared spectroscopy in patients with heart failure. *Circulation, 80*(6), 1668-74.

Wittenberg, J.B. (1970). Myoglobin-facilitated oxygen diffusion: Role of myoglobin in oxygen entry into muscle. *Physiological Reviews, 50*(4), 559-636.

Yoxall, C.W., & Weindling, A.M. (1997). Measurement of venous oxyhaemoglobin saturation in the adult human forearm by near infrared spectroscopy with venous occlusion. *Medical and Biological Engineering and Computing, 35,* 331-6.

Zatina, M.A., Berkowitz, H.D., Gross, G.M., Maris, J.M., & Chance, B. (1986). 31P nuclear magnetic resonance spectroscopy: Noninvasive biochemical analysis of the ischemic extremity. *Journal of Vascular Surgery, 3*(3), 411-20.

# Chapter 11

Abate, N., Burns, D., Peshock, R., Garg, A., & Grundy, D. (1994). Estimation of adipose tissue mass by magnetic resonance imaging: Validation against dissection in human cadavers. *Journal of Lipid Research, 35,* 1490-1496.

Abate, N., Garg, A., Coleman, R., Grundy, S., & Peshock, R. (1997). Prediction of total subcutaneous abdominal, intraperitoneal and retroperitoneal adipose tissue masses in men by a single axial magnetic resonance imaging slice. *American Journal of Clinical Nutrition, 65,* 403-408.

Baker, L. (1989). Principles of the impedance technique. *Institute of Electrical and Electronic Engineers, Engineering in Medicine, 3*, 11-15.

Baumgartner, R., Chumlea, W., & Roche, A. (1990). Bioelectrical impedance for body composition. In K. Pandolf & J. Holloszy (eds.), *Exercise and sport sciences reviews* (pp. 193-224). Baltimore: Williams & Wilkins.

Behnke, A., & Wilmore, J. (1974). *Evaluation and regulation of body build and composition.* Englewood Cliffs, NJ: Prentice Hall.

Bertin, E., Marcus, C., Ruiz, J., Eschard, J., & Leutenegger, M. (2000). Measurement of visceral adipose tissue by DXA combined with anthropometry in obese humans. *International Journal of Obesity, 24*, 263-270.

Booth, R., Goddard, B., & Paton, A. (1966). Measurement of ultrasound, calipers, and electrical conductivity. *British Journal of Nutrition, 20*, 719-727.

Borkan, C., Halts, D., Cardarelli, J., & Burrows, B. (1982). Comparison of ultrasound and skinfold measurements in assessments of subcutaneous and total fatness. *American Journal of Physical Anthropology, 58*, 307-313.

Braggi, R., Vollman, M., Nies, M., Brener, C., Flakoll, P., Levenhagen, D., et al. (1999). Comparison of air-displacement plethysmography with hydrostatic weighing and bioelectrical impedance analysis for the assessment of body composition in healthy adults. *American Journal of Clinical Nutrition, 69*, 898-903.

Brozek, J., Grande, J., Anderson, T., & Keys, A. (1963). Densitometric analysis of body composition: A review of some quantitative assumptions. *Annals of the New York Academy of Sciences, 110*, 113-140.

Brozek, J., & Keys, A. (1951). The evaluation of leanness-fatness in man: Norms and intercorrelations. *British Journal of Nutrition, 5*, 194-206.

Bulbulian, R. (1984). The influence of somatotype on anthropometric prediction of body composition in young women. *Medicine and Science in Sports and Exercise, 16*, 389-397.

Bullen, B., Quaade, F., Oleson, E., & Lund, S. (1965). Ultrasonic reflections used for measuring subcutaneous fat in humans. *Human Biology, 37*, 375-384.

Bushberg, J., Seibert, A., Leidholdt, E., & Boone, J. (1994). X-ray computed tomography. In W. Passano (ed.), *The essentials of medical imaging* (pp. 239-289). Baltimore: Williams & Wilkins.

Buskirk, E. (1961). Underwater weighing and body density: A review of procedures. In J. Brozek & A. Henschel (eds.), *Techniques for measuring body composition* (pp. 90-105). Washington, DC: National Academy of Science.

Campos, A., Chen, M., & Meguid, M. (1989). Comparisons of body composition derived from anthropometric and bioelectrical impedance methods. *Journal of the American College of Nutrition, 8*, 484-489.

Cataldo, D., & Heyward, V. (2000). Pinch an inch: A comparison of several high-quality and plastic skinfold calipers. *ACSM's Health & Fitness Journal, 4*, 12-16.

Caton, J., Mole, P., Adams, W., & Heustis, D. (1988). Body composition analysis by bioelectrical impedance: Effect of skin temperature. *Medicine and Science in Sports and Exercise, 20*, 489-491.

Chumlea, W., & Baumgartner, R. (1990). Bioelectrical impedance methods for the estimation of body composition. *Canadian Journal of Sport Sciences, 15*, 172-179.

Chumlea, W., Baumgartner, R., & Roche, A. (1988). The use of specific resistivity to estimate fat-free mass from segmental body measures of bioelectric impedance. *American Journal of Clinical Nutrition, 48*, 7-15.

Chumlea, W., Roche, A., Guo, S., & Woynarowska, B. (1987). The influence of physiological variables and oral contraceptives on bioelectric impedance. *Human Biology, 59*, 257-270.

Clasey, J., Hartman, M., Kanaley, J., Wideman, L., Teates, C., Bouchard, C., et al. (1997). Body composition by DEXA in older adults: Accuracy and influence of scan mode. *Medicine and Science in Sports and Exercise, 29*, 560-567.

Clasey, J., Kanaley, J., Wideman, L., Heymsfield, S., Teates, C., Gutgesell, M., et al. (1999). Validity of methods of body-composition assessment in young and older men and women. *Journal of Applied Physiology, 86*, 1728-1738.

Claus, A. (1957). The measurement of natural interfaces in the pig's body with ultrasound. *Fleischwirtschaft, 9*, 552-554.

Cordain, L., Whiker, R., & Johnson, J. (1988). Body composition determination in children using bioelectrical impedance. *Growth, Development and Aging, 52*, 37-40.

Cromwell, L., Weibell, F., & Pfeiffer, E. (1980). *Biomedical instrumentation and measurement* (2nd ed.). Englewood Cliffs, NJ: Prentice Hall.

Cugini, P., Salandri, A., Petrangeli, C., Capodaglio, P., & Giovannini, C. (1996). Circadian rhythms in human body composition. *Chronobiology International, 13*, 359-371.

Demerath, E., Guo, S., Chumlea, W., Towne, B., Roche, A., & Siervogel, R. (2002). Comparison of percent body fat estimates using air displacement plethysmography and hydrodensitometry in adults and children. *International Journal of Obesity, 26*, 389-397.

Dempster, P., & Aitkens, S. (1995). A new air displacement method for the determination of human body composition. *Medicine and Science in Sports and Exercise, 27*, 1692-1697.

Deurenberg, P., Kusters, C., & Smit, H. (1990). Assessment of body composition by bioelectrical impedance in children and young adults is strongly age-dependent. *European Journal of Clinical Nutrition, 44*, 261-268.

Deurenberg, P., Weststrate, J., & van der Kooy, K. (1989). Body composition changes assessed by bioelectrical impedance measurements. *American Journal of Clinical Nutrition, 49*, 401-403.

Dumont, B.L. (1957, July). *New methods of estimation of carcass quality on live pigs.* Paper presented at the joint

FAO/EAAP Meeting on Pig Progeny Testing, Copenhagen, Denmark.

Dumont, B.L. (1959). Measure of the fatness of hogs by the method of ultrasonic echoes. *C.R. Academy of Agriculture, 45*, 628.

Durnin, J., & Womersley, J. (1974). Body fat assessed from total body density and its estimation from skinfold thickness: Measurements on 481 men and women aged from 16 to 72 years. *British Journal of Nutrition, 32*, 77-92.

Edelman, I., Olney, J., & James, A. (1952). Body composition: Studies in the human being by the dilution principle. *Science, 115*, 447-454.

Edwards, D., Hammond, W., Healy, M., Tanner, J., & Whitehouse, R. (1955). Design and accuracy of calipers for measuring subcutaneous fat thickness. *British Journal of Nutrition, 9*, 133-143.

Ellis, K. (2000). Human body composition: In vivo methods. *Physiological Reviews, 80*, 649-680.

Elsen, R., Siu, M.-L., Pineda, O., & Solomons, N. (1987). Sources of variability in bioelectrical impedance determinations in adults. In K. Ellis, S. Yasumura, & W. Morgan (eds.), *In vivo body composition studies* (pp. 184-188), New York: Plenum Press.

Evans, E., Arngrimsson, S., & Cureton, K. (2001). Body composition estimates from multicomponent models using BIA to determine body water. *Medicine and Science in Sports and Exercise, 33*, 839-845.

Fanelli, M., & Kuczmarski, R. (1984). Ultrasound as an approach to assessing body composition. *American Journal of Clinical Nutrition, 39*, 703-709.

Fowler, P., Fuller, M., Glasbey, C., Cameron, G., & Foster, M. (1992). Validation of the in-vivo measurement of adipose tissue by magnetic resonance imaging of lean and obese pigs. *American Journal of Clinical Nutrition, 56*, 7-13.

Fukunaga, T., Matsuo, A., Ishida, Y., Tsunoda, N., Uchino, S., & Ohkubo, M. (1989). Study for measurement of muscle and subcutaneous fat thickness by means of ultrasonic B-mode method. *Japanese Journal of Medical Ultrasonics, 16,* 50-57.

Fuller, N., & Elia, M. (1989). Potential use of bioelectrical impedance of the "whole body" and of body segments for the assessment of body composition: Comparison with densitometry and anthropometry. *European Journal of Clinical Nutrition, 43*, 779-791.

Garzarella, L., Ishida, Y., Graves, J., Leggett, S., Pollock, M., Carroll, J., et al. (1991). The development of prediction equations for estimating body composition in females by B-mode ultrasound. *Medicine and Science in Sports and Exercise, 23*, S90.

Girandola, R., Wiswell, R., & Romero, G. (1977). Body composition changes resulting from fluid ingestion and dehydration. *Research Quarterly, 48*, 299-303.

Goldman, H., & Becklace, M. (1959). Respiratory function tests: Normal values of medium altitudes and the prediction of normal results. *American Review of Tuberculosis and Respiratory Diseases, 79*, 457-467.

Goldman, R., & Buskirk, E. (1961). Body volume measurement by underwater weighing: Description of a technique. In J. Brozek & A. Henschel (eds.), *Techniques for measuring body composition* (pp. 78-89). Washington, DC: National Academy of Science.

Goodpaster, B., Thaete, F., & Kelley, D. (2000). Composition of skeletal muscle evaluated with computed tomography. *Annals of the New York Academy of Sciences, 904*, 19-24.

Goodpaster, B., Thaete, F., Simoneau, J.-A., & Kelley, D. (1997). Subcutaneous abdominal fat and thigh muscle composition predict insulin sensitivity independently of visceral fat. *Diabetes, 46*, 1579-1585.

Graves, J., Pollock, M., Colvin, A., Van Loan, M., & Lohman, T. (1989). Comparison of different bioelectrical impedance analyzers in the prediction of body composition. *American Journal of Human Biology, 1*, 603-611.

Gruber, J., Pollock, M., Graves, J., Colvin, A., & Braith, R. (1990). Comparison of Harpenden and Lange calipers in predicting body composition. *Research Quarterly for Exercise and Sport, 61*, 184-190.

Guo, S., Roche, A., Chumlea, W., Miles, D., & Pohlman, R. (1987). Body composition predictions from bioelectrical impedance. *Human Biology, 59*, 221-223.

Guo, S., Roche, A., & Houtkooper, L. (1989). Fat-free mass in children and young adults predicted from bioelectrical impedance and anthropometric variables. *American Journal of Clinical Nutrition, 50*, 435-443.

Haarbo, J., Gotfredsen, C., Hassager, C., & Christiansen, C. (1991). Validation of body composition by dual energy x-ray absorptiometry (DEXA). *Clinical Physiology, 11*, 331-341.

Hannan, W., Cowen, S., Freeman, C., & Shapiro, C. (1990). Evaluation of bioelectrical impedance analysis for body composition measurements in anorexia nervosa. *Clinics in Physical and Physiological Measurement, 11*, 209-216.

Harrison, G.G. (1985). Height-weight tables. *Annals of Internal Medicine, 103*.

Haymes, E., Lundegren, H., Loomis, J., & Buskirk, E. (1976). Validity of the ultrasonic technique as a method of measuring subcutaneous adipose tissue. *Annals of Human Biology, 3*, 245-251.

Heymsfield, S., Lichtman, S., Baumgartner, R., Wang, J., Kamen, Y., Aliprantis, A., et al. (1990). Body composition of humans: Comparison of two improved four-compartment models that differ in expense, technical complexity, and radiation exposure. *American Journal of Clinical Nutrition, 52*, 52-58.

Heymsfield, S., Ross, R., Wang, A., & Frager, D. (1997). Imaging techniques of body composition: Advantages of measurement and new uses. In *Emerging technologies for nutrition research* (pp. 127-150). Washington, DC: National Academy Press.

Heymsfield, S., & Waki, M. (1991). Body composition in humans: Advances in the development of multicompartment chemical models. *Nutrition Reviews, 49*, 97-108.

Heymsfield, S., Wang, J., & Aulet, M. (1990). Dual photon absorptiometry: Validation of mineral and fat measurements. In S. Yasumura & J. Harrison (eds.), *In vivo body composition studies* (pp. 327-337). New York: Plenum Press.

Heymsfield, S., Wang, J., Heshka, S., Kehayias, J., & Pierson, R. (1989). Dual-photon absorptiometry: Comparison of bone mineral and soft tissue mass measurements in vivo with established methods. *American Journal of Clinical Nutrition, 49,* 1283-1289.

Heyward, V., & Stolarczyk, L. (1996). *Applied body composition assessment.* Champaign, IL: Human Kinetics.

Hoffer, B., Meador, C., & Simpson, D. (1969). Correlation of whole-body impedance with total body water volume. *Journal of Applied Physiology, 27,* 531-534.

Hubert, H.A., Feinlab, M., McNamara, P.M., & Castelli, W.P. (1983). Obesity as an independent risk factor for cardiovascular disease: A 26-year follow-up of the participants in the Framingham Heart Study. *Circulation, 67,* 968-977.

Ishida, Y., Carroll, J., Pollock, M., Graves, J., & Leggett, S. (1990). Reliability of B-mode ultrasound in the measurement of body fat and muscle thickness. *Medicine and Science in Sports and Exercise, 22,* S111.

Jackson, A., & Pollock, M. (1976). Factor analysis and multivariate scaling of anthropometric variables for the assessment of body composition. *Medicine and Science in Sports and Exercise, 8,* 196-203.

Jackson, A., & Pollock, M. (1982). Steps toward the development of generalized equations for predicting body composition of adults. *Canadian Journal of Applied Sports Science, 7,* 189-196.

Jackson, A., Pollock, M., & Gettman, L. (1978). Intertester reliability of selected skinfold and circumference measurements and percent fat estimates. *Research Quarterly, 49,* 546-551.

Jackson, A., Pollock, M., Graves, J., & Mahar, M. (1988). Reliability and validity of bioelectrical impedance in determining body composition. *Journal of Applied Physiology, 64,* 529-534.

Jackson, A., Pollock, M., & Ward, A. (1978). Generalized equations for predicting body density of men. *British Journal of Nutrition, 40,* 497-504.

Jackson, A., Pollock, M., & Ward, A. (1980). Generalized equations for predicting body density of women. *Medicine and Science in Sports and Exercise, 16,* 606-613.

Janssen, I., Katzmarzyk, P., & Ross, R. (2002). Body mass index, waist circumference, and health risk. *Archives of Internal Medicine, 162,* 2074-2079.

Jensen, M., Kanaley, J., Reed, J., & Sheedy, P. (1995). Measurement of abdominal and visceral fat with computed tomography and dual-energy x-ray absorptiometry. *American Journal of Clinical Nutrition, 61,* 274-278.

Katch, F. (1968). Apparent body density and variability during underwater weighing. *Research Quarterly, 39,* 993-999.

Katch, F., & Katch, V. (1980). Measurement and prediction errors in body composition assessment and search for the perfect equation. *Research Quarterly for Exercise and Sport, 51,* 249-260.

Katch, F., & McArdle, W. (1973). Prediction of body density from simple anthropometric measurements in college-age men and women. *Human Biology, 45,* 445-454.

Kekes-Szabo, T., Hunter, G., Nyikos, I., Nicholson, C., Snyder, S., & Lincoln, B. (1994). Development and validation of computed tomography derived anthropometric regression equations for estimating abdominal adipose tissue distribution. *Obesity Research, 2,* 450-457.

Kelsey, C. (1993). Introduction. In J. Juhl & A. Crummy (eds.), *Essentials of radiologic imaging* (6th ed.) (pp. 1-18). New York: Lippincott.

Keys, A., & Brozek, J. (1953). Body fat in adult men. *Physiological Reviews, 33,* 245-325.

Khaled, M., Lukaski, H., & Watkins, C. (1987). Determination of total body water by deuterium NMR. *American Journal of Clinical Nutrition, 45,* 1-6.

Kohrt, W. (1998). Preliminary evidence that DEXA provides an accurate assessment of body composition. *Journal of Applied Physiology, 84,* 372-377.

Kuczmarski, R., Fanelli, M., & Koch, G. (1987). Ultrasonic assessment of body composition in obese adults: Overcoming the limitation of the skinfold caliper. *American Journal of Clinical Nutrition, 45,* 717-724.

Kushner, R., & Haas, A. (1988). Estimation of lean body mass by bioelectrical impedance analysis compared to skinfold anthropometry. *European Journal of Clinical Nutrition, 41,* 101-106.

Kushner, R., & Schoeller, D. (1986). Estimation of total body water by bioelectrical impedance analysis. *American Journal of Clinical Nutrition, 44,* 417-424.

Lantz, H., Samuelson, G., Bratteby, L., Mallmin, H., & Sjostrom, L. (1999). Differences in whole body measurements by DXA-scanning using two Lunar DPX-L machines. *International Journal of Obesity, 23,* 764-770.

Latin, R. (1987). Percent body fat determinations by body impedance analysis and skinfold measurements. *Fitness in Business, 2,* 24-27.

Lemieux, S., Prud'homme, D., Bouchard, C., Tremblay, A., & Despres, J. (1996). A single threshold value of waist girth identifies normal-weight and overweight subjects with excess visceral adipose tissue. *American Journal of Clinical Nutrition, 64,* 685-693.

Lichtman, S., Heymsfield, S., & Kehayias, J. (1990). Elemental reconstruction of human composition in vivo. *Federation of American Societies for Experimental Biology, 4,* A2261.

Lockner, D., Heyward, V., Baumgartner, R., & Jenkins, K. (2000). Comparison of air displacement plethysmography, hydrodensitometry and dual x-ray absorptiometry for assessing body composition of children 10 to 18 years of age. *Annals of the New York Academy of Science, 9004,* 72-77.

Lohman, T. (1981). Skinfolds and body density and their relation to body fatness: A review. *Human Biology, 53,* 181-225.

Lohman, T. (1986). Applicability of body composition techniques and constants for children and youths. *Exercise and Sports Science Reviews, 14,* 325-357.

Lohman, T. (1992). Advances in body composition assessment. In *Current issues in exercise science series* (p. 12). Champaign, IL: Human Kinetics.

Lohman, T., Pollock, M., Slaughter, M., Brandon, L., & Boileau, R. (1984). Methodological factors and the prediction of body fat in female athletes. *Medicine and Science in Sports and Exercise, 16,* 92-96.

Lohman, T., Roche, A., & Martorell, R. (1988). *Anthropometric standardization reference manual.* Champaign, IL: Human Kinetics.

Lukaski, H. (1987). Methods for the assessment of human body composition: Traditional and new. *American Journal of Clinical Nutrition, 46,* 537-556.

Lukaski, H., Bolonchuk, W., Hall, C., & Siders, W. (1986). Validation of tetrapolar bioelectrical impedance method to assess human body composition. *Journal of Applied Physiology, 60,* 1327-1332.

Lukaski, H., Bolonchuk, W., Johnson, P., Lykken, G., & Sandstead, H. (1984). Assessment of fat-free mass using bioelectrical impedance measurements of the human body. *American Journal of Clinical Nutrition, 41,* 657-658.

Lukaski, H., Bolonchuk, W., Siders, W., & Hall, C. (1990). Body composition assessment of athletes using bioelectrical impedance measurements. *Journal of Sports Medicine and Physical Fitness, 30,* 434-440.

Lukaski, H., Johnson, P., Bolonchuk, W., & Lykken, G. (1985). Assessment of fat-free mass using bioelectrical impedance measurements of the human body. *American Journal of Clinical Nutrition, 41,* 810-817.

Mazess, R., Barden, H., Bisek, J., & Hanson, J. (1990). Dual-energy x-ray absorptiometry for total-body and regional bone-mineral and soft-tissue composition. *American Journal of Clinical Nutrition, 51,* 1106-1112.

McArdle, W., Katch, F., & Katch, V. (2001). *Exercise physiology: Energy, nutrition, and human performance.* Baltimore: Lippincott, Williams & Wilkins.

McCrory, M., Gomex, T., Bernauer, E., & Mole, P.A. (1995). Evaluation of a new air displacement plethysmograph for measuring human body composition. *Medicine and Science in Sports and Exercise, 27,* 1686-1691.

McCrory, M., Mole, P., Gomez, T., Dewey, K., & Bernauer, E. (1998). Body composition by air-displacement plethysmography by using predicted and measured thoracic gas volumes. *Journal of Applied Physiology, 84,* 1475-1479.

Mendez, J., & Lukaski, H. (1981). Variability of body density in ambulatory subjects measured at different days. *American Journal of Clinical Nutrition, 34,* 78-81.

Millard-Stafford, M., Collins, M., Evans, E., Snow, T., Cureton, K., & Rosskopf, L. (2001). Use of air displacement plethysmography for estimating body fat in a four-component model. *Medicine and Science in Sports and Exercise, 33,* 1311-1317.

Milliken, L., Going, S., & Lohman, T. (1996). Effects of variations in regional composition on soft tissue measurements by dual X-ray absorptiometry. *International Journal of Obesity, 20,* 677-682.

Mitsiopoulos, N., Baumgartner, R., Heymsfield, S., Lyons, W., Gallager, D., & Ross, R. (1998). Cadaver validation of skeletal muscle measurement by magnetic resonance imaging and computerized tomography. *Journal of Applied Physiology, 85,* 115-122.

Munro, A., Joffe, A., Ward, J., Syndham, C., & Fleming, P. (1966). An analysis of the errors in certain anthropometric measurements. *Internationale Zeitschrift Fur Angewandte Physiologic Einschliesslish Arbeitphysiologic, 23,* 93-106.

National Heart, Lung, and Blood Institute (1998). HLBI Obesity Education Initiative expert panel. Clinical Guidelines on the Identification, Evaluation, and Treatment of Overweight and Obesity in Adults. www.nhibi.nih.gov/nhlbi/.

Newell, J. (1961). Ultrasonic localisation. *British Journal of Radiology, 34,* 539-546.

Oppliger, R., Looney, M., & Tipton, C. (1987). Reliability of hydrostatic weighing and skinfold measurements of body composition using generalizability study. *Human Biology, 59,* 77-96.

Organ, L., Eklund, A., & Ledbetter, J. (1994). An automated real time underwater weighing system. *Medicine and Science in Sports and Exercise, 26,* 383-391.

Pace, N., & Rathbun, E. (1945). Studies on body composition. III. The body water and chemically combined nitrogen content in relation to fat content. *Journal of Biological Chemistry, 158,* 685-691.

Park, Y.-W., Heymsfield, S., & Gallagher, D. (2002). Are dual-x-ray absorptiometry regional estimates associated with visceral adipose tissue mass? *International Journal of Obesity, 26,* 978-983.

Pascale, L., Grossman, M., Sloan, H., & Frankel, T. (1956). Correlations between thickness of skinfolds and body density in 88 soldiers. *Human Biology, 28,* 165-176.

Peppler, W., & Mazess, R. (1981). Total body bone mineral and lean body mass by dual photon absorptiometry. I. Theory and measurement procedure. *Calcified Tissue International, 33,* 353-359.

Plourde, G. (1997). The role of radiologic methods in assessing body composition and related metabolic parameters. *Nutrition Reviews, 55,* 280-296.

Ploutz-Snyder, L., Foley, J., Ploutz-Snyder, R., Kanaley, J., Sagendorf, K., & Meyer, R. (1999). Gastric gas and fluid emptying assessed by magnetic resonance imaging. *European Journal of Applied Physiology and Occupational Physiology, 79,* 212-220.

Pollock, M., Hickman, T., Kendrick, Z., Jackson, A., Linnerud, A., & Dawson, G. (1976). Prediction of body density in young and middle-aged men. *Journal of Applied Physiology, 40*, 300-304.

Pollock, M., Jackson, A., & Graves, J. (1986). Analysis of measurement error related to skinfold site, quantity of skinfold fat, and sex. *Medicine and Science in Sports and Exercise, 18*, S32.

Pollock, M., Laughbridge, E., Coleman, B., Linnerud, A., & Jackson, A.E. (1975). Prediction of body density in young and middle-aged women. *Journal of Applied Physiology, 38*, 745-749.

Pollock, M., Schmidt, D., & Jackson, A. (1980). Measurement of cardiorespiratory fitness and body composition in the clinical setting. *Comprehensive Therapy, 6*, 12-27.

Pollock, M., & Wilmore, J. (1990). *Exercise in health and disease: Evaluation and prescription for prevention and rehabilitation.*

Roche, A., Heymsfield, S., & Lohman, T. (1996). *Current issues in exercise science* (Monograph Number 3). Champaign, IL: Human Kinetics.

Ross, R. (1996). Magnetic resonance imaging provides new insights into the characterization of adipose and lean tissue distribution. *Canadian Journal of Physiology and Pharmacology, 74*, 778-785.

Ross, R., Léger, L., Guardo, R., DeGuise, J., & Pike, B. (1991). Adipose tissue volume measured by magnetic resonance imaging and computerized tomography in rats. *Journal of Applied Physiology, 70*, 2164-2172.

Ross, R., Léger, L., Morris, D., DeGuise, J., & Guardo, R. (1992). Quantification of adipose tissue by MRI: Relationship with anthropometric variables. *Journal of Applied Physiology, 72*, 787-795.

Rush, S., Abidskoo, J., & Fee, R. (1963). Resistivity of body tissues at low frequencies. *Circulation Research, 12*, 40-50.

Salamone, L., Fuerst, T., Visser, M., Kern, M., Lang, T., Dockrell, M., et al. (2000). Measurement of fat mass using DEXA: A validation study in elderly adults. *Journal of Applied Physiology, 89*, 345-352.

Sardinha, L., Lohman, T., Teixeira, P., Guedes, D., & Going, S. (1998). Comparison of air displacement plethysmography with dual-energy X-ray absorptiometry and 3 field methods for estimating body composition in middle-aged men. *American Journal of Clinical Nutrition, 68*, 786-793.

Schell, B., & Gross, R. (1987). The reliability of bioelectrical impedance measurements in the assessment of body composition in healthy adults. *Nutrition Report International, 36*, 449-459.

Schoeller, D. (1991). Isotope dilution methods. In P. Bjorntorp & B. Brodoff (eds.), *Obesity* (pp. 80-88). New York: Lippincott.

Schoeller, D. (1996). Hydrometry. In A. Roche, S. Heymsfield, & T. Lohman (eds.), *Human body composition* (pp. 25-43). Champaign, IL: Human Kinetics.

Segal, K., Gutin, B., Presta, E., Wang, J., & Van Itallie, T. (1985). Estimation of human body composition by electrical impedance methods: A comparative study. *Journal of Applied Physiology, 58*, 1565-1571.

Segal, K., Van Loan, M., Fitzgerald, P., Hogdon, J., & Van Itallie, T. (1988). Lean body mass estimation by bioelectrical impedance analysis: A four-site cross-validation study. *American Journal of Clinical Nutrition, 47*, 7-14.

Sheng, H., & Huggins, R. (1979). A review of body composition studies with emphasis on total body water and fat. *American Journal of Clinical Nutrition, 32*, 630-647.

Shols, A., Wouters, E., Soeters, P., & Westerterp, K. (1991). Body composition by bioelectrical-impedance analysis compared with deuterium dilution and skinfold anthropometry in patients with chronic obstructive pulmonary disease. *American Journal of Clinical Nutrition, 53*, 421-424.

Sinning, W. (1978). Anthropometric estimation of $D_b$, fat, and LBW in women gymnasts. *Medicine and Science in Sports and Exercise, 10*, 243-249.

Sinning, W., & Wilson, J. (1984). Validity of "generalized" equations for body composition, *Research Quarterly, 55*, 153-160.

Siri, W. (1956). The gross composition of the body. In C. Tobias & J. Lawrence (eds.), *Advances in biological and medical physics* (pp. 239-240). New York: Academic Press.

Siri, W. (1961). Body composition from fluid spaces and density: Analysis of methods. In J. Brozek & A. Henschel (Eds.), *Techniques for measuring body composition* (pp. 223-244). Washington, DC: National Academy of Science.

Siu, M.-L., Elsen, R., Mazariegos, M., Solomons, N., & Pineda, O. (1987). Evaluation through sequential determination of the stability of bioelectrical impedance measurements for body composition analysis. In K. Ellis, S. Yasumura, & W. Morgan (ed.), *In vivo body composition studies* (pp. 189-194). New York: Plenum Press.

Sloan, A. (1967). Estimation of body fat in young men. *Journal of Applied Physiology, 23*, 311-315.

Sloan, A., Burt, J., & Blyth, C. (1962). Estimation of body fat in young women. *Journal of Applied Physiology, 17*, 967-970.

Sloan, A., & Shapiro, A. (1972). A comparison of skinfold measurements in three standard calipers. *Human Biology, 44*, 29-36.

Smith, J., & Mansfield, E. (1984). Body composition prediction in university football players. *Medicine and Science in Sports and Exercise, 16*, 398-405.

Snead, D., Birge, S., & Kohrt, W. (1993). Age-related differences in body composition by hydrodensitometry and dual energy X-ray absorptiometry. *Journal of Applied Physiology, 74*, 770-775.

Snijder, M., Visser, M., Dekker, J., Seidell, J., Fuerst, T., Tylavsky, F., et al. (2002). The prediction of visceral fat by dual-energy X-ray absorptiometry in the elderly: A comparison with computed tomography and anthropometry. *International Journal of Obesity, 26*, 984-993.

Svendsen, O., Hassage, C., Bergmann, I., & Christiansen, C. (1993). Measurement of abdominal and intra-abdominal fat in postmenopausal women by dual energy x-ray absorptiometry and anthropometry: Comparison with computerized tomography. *International Journal of Obesity, 17*, 45-51.

Tang, H., Vasselli, J., Wu, E., Boozer, C., & Gallagher, D. (2002). High-resolution magnetic resonance imaging tracks changes in organ tissue mass in obese and aging rats. *American Journal of Physiology, 282*, R890-R899.

Thomas, E., Saeed, N., Hajnal, J., Brynes, A., Goldstone, A., Fros, T.G., et al. (1998). Magnetic resonance imaging of total body fat. *Journal of Applied Physiology, 85*, 1778-1785.

Thorland, W., Johnson, G., Tharp, G., Fagot, T., & Hammer, R. (1984). Validity of anthropometric equations for the estimation of body density in adolescent athletes. *Medicine and Science in Sports and Exercise, 16*, 77-81.

Tothill, P., Han, T., Avenell, A., McNeill, G., & Reid, D. (1996). Comparisons between fat measurements by dual-energy x-ray absorptiometry, underwater weighing and magnetic resonance imaging in healthy women. *European Journal of Clinical Nutrition, 50*, 747-752.

Tucci, J., Carpenter, D., Graves, J., Pollock, M., Felheim, R., & Mananquil, R. (1991). Interday reliability of bone mineral density measurements using dual energy x-ray absorptiometry. *Medicine and Science in Sports and Exercise, 23*, S115.

Van der Kooy, K., & Siedell, J. (1993). Technique for the measurement of visceral fat: A practical guide. *International Journal of Obesity, 17*, 187-196.

Van Loan, M. (1990). Bioelectrical impedance analysis to determine fat-free mass, total body water and body fat. *Sports Medicine, 10*, 205-217.

Verlooy, H., Dequeker, J., Geusens, P., Nijs, J., & Goris, M. (1991). Body composition by intercomparison of underwater weighing, skinfold measurements, and dual-photon absorptiometry. *British Journal of Radiology, 64*, 765-767.

Vescovi, J., Zimmerman, S., Miller, W., & Fernhall, B. (2002). Effects of clothing on accuracy and reliability of air displacement plethysmography. *Medicine and Science in Sports and Exercise, 34*, 282-285.

Vettorazzi, C., Barillas, C., Pineda, O., & Solomons, N. (1987). A model for assessing body composition in amputees using bioelectrical impedance analysis. *Federation Proceedings, 46*, 1186.

Volz, P., & Ostrove, S. (1984). Evaluation of a portable ultrasonoscope in assessing the body composition of college-age women. *Medicine and Science in Sports and Exercise, 16*, 97-102.

Wagner, D., & Heyward, V. (1999). Techniques of body composition assessment: A review of laboratory and field methods. *Research Quarterly for Exercise and Sport, 70*, 135-149.

Wagner, D., Heyward, V., & Gibson, A. (2000). Validation of air displacement plethysmography for assessing body composition. *Medicine and Science in Sports and Exercise, 32*, 1339-1344.

Wang, A., Deurenberg, P., Guo, S., Pietrobelli, A., Wang, J., Pierson, R., et al. (1998). Six-compartment body composition model: Inter-method comparison of total body fat measurement. *International Journal of Obesity, 22*, 329-337.

Wang, J., Dilmanian, F., Thornton, J., Russell, M., Burastero, S., Mazariegos, M., et al. (1998). In vivo neutron activation analysis for body fat: Comparisons by seven methods. *Basic Life Sciences, 60*, 31-34.

Wang, J., Heymsfield, S., Aulet, M., Thornton, J., & Pierson, R. (1989). Body fat from body density: Underwater weighing vs. dual-photon absorptiometry. *American Journal of Physiology, 256*, E8329-E8334.

Wang, S., Pierson, R., Jr., & Heymsfield, S. (1992). The five level model: A new approach to organizing body composition research. *American Journal of Clinical Nutrition, 56*, 1028.

Weltman, A., Levine, S., Seip, R., & Tran, Z. (1988). Accurate assessment of body composition in obese females. *American Journal of Clinical Nutrition, 48*, 1179-1183.

Weltman, A., Seip, R., & Tran, Z. (1987). Practical assessment of body composition in obese males. *Human Biology, 59*, 523-536.

Werdein, E., & Kyle, L. (1960). Estimation of the constancy of density of the fat-free body. *Journal of Clinical Investigation, 39*, 626-629.

Weyers, A., Mazzetti, S., Love, D., Gomez, A., Kraemer, W., & Volek, J. (2002). Comparison of methods for assessing body composition changes during weight loss. *Medicine and Science in Sports and Exercise, 34*, 497-502.

Wilmore, J., & Costill, D. (1988). *Training for sport and activity: The physiological basis of the conditioning process.* Boston: Allyn and Bacon.

Wilmore, J.H. (1969). The use of actual, predicted, and constant residual volumes in the assessment of body composition by underwater weighing. *Medicine and Science in Sports, 1*, 87-90.

Wilmore, J.H., & Behnke, A. (1970). An anthropometric estimation of body density and lean body weight in young women. *American Journal of Clinical Nutrition, 23*, 267-274.

Witt, R., & Mazess, R. (1978). Photon absorptiometry of soft-tissue and fluid content: The method and its precision and accuracy. *Physics in Medicine and Biology, 23*, 620-629.

Yee, A., Fuerst, T., Salamone, L., Visser, M., Dockrell, M., Van Loan, M., et al. (2001). Calibration and validation of an air-displacement plethysmography method for estimating percentage body fat in an elderly population: A comparison among compartmental models. *American Journal of Clinical Nutrition, 74*, 637-642.

Young, C. (1964). Prediction of specific gravity and body fatness in older women. *Journal of the American Dietetic Association, 45,* 333-338.

Young, C., Martin, M., Tensuan, R., & Blondin, J. (1962). Predicting specific gravity and body fatness in young women. *Journal of the American Dietetic Association, 40,* 102-107.

Zebatakis, P., Gleim, G., Vitting, K., Gardemswartz, M., Agrawal, M., Michelos, M., et al. (1987). Volume changes affect electrical impedance measurement of body composition. *Medicine and Science in Sports and Exercise, 19,* S40.

# Chapter 12

Alter, M.J. (1996). *Science of flexibility* (2nd ed.). Champaign, IL: Human Kinetics.

American Academy of Orthopaedic Surgeons. (1994). *The clinical measurement of joint motion.* Chicago: Author.

American Alliance for Health, Physical Education, Recreation and Dance. (1984). *AAHPERD technical manual: Health related physical fitness.* Reston, VA: Author.

American Medical Association. (1984). *Guides to the evaluation of permanent impairment* (2nd ed.). Chicago: Author.

American Medical Association. (1988). *Guides to the evaluation of permanent impairment* (3rd ed.). Chicago: Author.

Blair, S.N., Falls, H.B., & Pate, R.R. (1983). A new physical fitness test. *Physician and Sports Medicine, 11,* 87-95.

Broer, M.R., & Galles, N.R.G. (1958). Importance of relationship between various body measurements in performance of the toe-touch test. *Research Quarterly, 29,* 253-263.

Clarkson, H.M., & Gilewich, G.B. (1989). *Musculoskeletal assessment. Joint range of motion and manual muscle strength.* Baltimore: Williams and Wilkins.

Corbin, C.B., & Noble, L. (1980). Flexibility: A major component of physical fitness. *Journal of Physical Education and Recreation, 51,* 23-24, 57-60.

Cowan, D., Jones, B., Tomlinson, P., Robinson, J., & Polly, D. (1988). *The epidemiology of physical training injuries in U.S. Army infantry trainees: Methodology, population, and risk factors* (Tech. Rep. No. T4-89). U.S. Army Research Institute of Environmental Medicine Technical Report No. T4-89.

Dickinson, R.V. (1968). The specificity of flexibility. *Research Quarterly, 39,* 729-794.

Elveru, R.A., Rothstein, J.M., & Lamb, R.L. (1988). Goniometric reliability in a clinical setting. *Physical Therapy, 68*(5), 672-677.

Ekstrand, J., & Gillquist, J. (1983). The avoidability of soccer injuries. *International Journal of Sports Medicine, 4,* 124-128.

Galley, P.M., & Forster, A.L. (1982). *Human movement. An introductory text for physiotherapy students.* Edinburgh: Churchill Livingstone.

Gogia, P.B., Braatz, J.H., Rose, S.J., & Norton, B.J. (1987). Reliability and validity of goniometric measurements at the knee. *Physical Therapy, 67*(2), 192-195.

Golding, L.A., Myers, C.R., & Sinning, W.E. (1989). *The Y way to physical fitness.* Champaign, IL: Human Kinetics.

Hoeger, W.K. and Hoeger, S.A. (2005). *Lifetime physical fitness and wellness: A personalized program.* Belmont, CA: Thomas Wadsworth.

Hoeger, W.K. (1987). *The complete guide for the development and implementation of health promotion programs.* Englewood, CO: Morton.

Hoeger, W.K. (1991). *Principles and labs for physical fitness and wellness.* Englewood, CO: Morton.

Hoeger, W.K., & Hopkins, D.R. (1990). Comparisons of the sit and reach and the modified sit and reach flexibility tests. *Medicine and Science in Sports and Exercise, 22*(Suppl. 2), S10.

Hoppenfield, S. (1976). *Physical examination of the spine and extremities.* Norwalk, CT: Appleton-Century-Crofts.

Hubley-Kozey, C.L. (1990). Testing flexibility. In J.D. MacDougall, H.A. Wenger, & H.J. Green (Eds.), *Physiological testing of the high-performance athlete* (pp. 309-359). Champaign, IL: Human Kinetics.

Jackson, A., & Langford, N.J. (1989). The criterion-related validity of the sit and reach test: Replication and extension of previous findings. *Research Quarterly for Exercise and Sport, 60,* 384-387.

Jackson, A.W., & Baker, A.A. (1986). The relationship of the sit and reach test to criterion measures of hamstring and back flexibility in young females. *Research Quarterly for Exercise and Sport, 57,* 183-186.

Johnson, B.L. (1972). *Flexomeasure instructional manual.* Portland, TX: Littleman Books.

Kendall, F.P., McCreary, E.K., & Provance, P.G. (1993). *Muscles. Testing and function* (4th ed.) Philadelphia: Lippincott, Williams and Wilkins.

Kippers, V., & Parker, A. (1987). Toe touch test: A measure of its validity. *Physical Therapy, 67,* 1680-1684.

Knapik, J.J., Bauman, C.L., Jones, B.H., Harris, J.M., & Vaughan, L. (1991). Preseason strength and flexibility imbalances associated with athletic injuries in female collegiate athletes. *American Journal of Sports Medicine, 19,* 76-81.

Knapik, J.J., Jones, B.H., Bauman, C.L., & Harris, J.M. (1992). Strength, flexibility and athletic injuries. *Sports Medicine, 14*(5), 277-288.

Leighton, J.R. (1964a). Flexibility characteristics of males six to ten years of age. *Journal of the Association for Physical and Mental Rehabilitation, 18,* 19-25.

Leighton, J.R. (1964b). A study of the effect of progressive weight training on flexibility. *Journal of the Association for Physical and Mental Rehabilitation, 18,* 101-105.

Leighton, J.R. (1966). The Leighton flexometer and flexibility test. *Journal of the Association for Physical and Mental Rehabilitation, 20,* 86-93.

Macrea, I.F., & Wright, V. (1969). Measurement of back movement. *Annals of Rheumatological Disease, 28,* 584-589.

Mayer, T.G. (1990). Discussion: Exercise, fitness and back pain. In C. Bouchard, R.J. Shephard, T. Stephens, J.R. Sutton, & B.D. McPherson (Eds.), *Exercise, fitness, and health: A consensus of current knowledge* (pp. 541-546). Champaign, IL: Human Kinetics.

Mellin, G.P. (1986). Accuracy of measuring lateral flexion of the spine with tape. *Clinical Biomechanics, 1,* 85-89.

Minister of State, Fitness and Amateur Sports. (1987). *Canadian standardized test of fitness (CSTF) operations manual (for 15 to 69 years-of-age).* Ottawa, ON: Author.

Mitchell, W.S., Millar, J., & Sturrock, R.D. (1975). An evaluation of goniometry as an objective parameter for measuring joint motion. *Scottish Medical Journal, 20,* 57-59.

Nachemson, A.L. (1990). Exercise, fitness and back pain. In C. Bouchard, R.J. Shephard, T. Stephens, J.R. Sutton, & B.D. McPherson (Eds.), *Exercise, fitness, and health: A consensus of current knowledge* (pp. 533-540). Champaign, IL: Human Kinetics.

Newton, M., & Waddell, G. (1991). Reliability and validity of clinical measurement of the lumbar spine in patients with low back pain. *Physiotherapy, 77,* 12, 796-800.

Norkin, C.C., & White, D.J. (1995). *Measurement of joint motion: A guide to goniometry.* (2nd ed). Philadelphia: Davis.

Rothstein, J.M. (Ed). (1985). *Measurement in physical therapy.* Edinburgh: Churchill Livingstone.

Schober, P. (1937). The lumbar vertebral column in backache. *Munchener Medizinisch Wochenschrift, 84,* 336-338.

Smith, J.F., & Miller, C.V. (1985). The effect of head position in sit and reach performance. *Research Quarterly for Exercise and Sport, 56,* 84-85.

Summary of findings from national children and youth fitness study. (1985, January). *Journal of Physical Education, Recreation and Dance,* 44-90.

Van Adrichem, J.A.M., & van der Korst, J.K. (1973). Assessment of the flexibility of the lumbar spine: A pilot study in children and adolescents. *Scandinavian Journal of Rheumatology, 2,* 87-91.

Watkins, M.A., Riddle, D.L., Lamb, R.L., & Personius, W.J. (1991). Reliability of goniometric measurements and visual estimates of knee range of motion obtained in a clinical setting. *Physical Therapy, 71,* 2, 90/15-96/21.

Wear, C.L. (1963). Relationships of flexibility measurements to length of body segments. *Research Quarterly, 34,* 234-238.

Williams, R., Binkley, J., Bloch, R., Goldsmith, C.H., & Minuk, T. (1993). Reliability of the modified-modified Schober and double inclinometer methods for measuring lumbar flexion and extension. *Physical Therapy, 73,* 26-37.

Youth fitness in 1985. The President's Council on Physical Fitness and Sports school population fitness survey. Washington, DC: President's Council on Physical Fitness and Sports, 1985.

# index

Note: The italicized *t* and *f* following page numbers refer to tables and figures, respectively.

# about the editors

**Peter J. Maud, PhD** is a professor and currently serves as academic department head at New Mexico State University. His primary academic interests have been in the areas of exercise physiology and, more recently, in wellness or health fitness. He has taught in nine different colleges and universities with administrative responsibilities as either department head or chair at four of these. His article entitled *The Grass is Always Greener: The Rambling Reminiscences of a Wandering Professor* perhaps explains his philosophy for these numerous moves. Born and raised in England, Peter attended Loughborough and Chester Colleges before teaching in the public schools in Canada, England, and Australia. He subsequently received his bachelor's degree from the University of Oregon and his master's and doctoral degrees from the University of New Mexico. An avid believer in "practicing what one preaches," he played rugby for thirty-five years and enjoyed badminton, fives, squash, handball, and racquetball before moving to endurance sports. He completed a rather slow Hawaii Iron Man triathlon in 1986 and, despite several operations, is still active as a racquetball player, and a cyclist. He is a lousy golfer but still insists on following that little white ball around the course.

**Carl Foster, PhD** is a Professor in the Department of Exercise and Sport Science and Director of the Human Performance Laboratory at the University of Wisconsin-La Crosse. He did his doctoral work at the University of Texas at Austin and did postdoctoral work at Ball State University. His research interests range from clinical exercise physiology to high performance sports physiology. He is the Chair of the Sports Medicine/Sports Science committee of U.S. Speedskating. Dr. Foster is a fellow of the American College of Sports Medicine and of the American Association for Cardiovascular and Pulmonary Rehabilitation, a former Associate Editor in Chief of *Medicine and Science in Sports and Exercise*, and is the 2005-2006 President of the American College of Sports Medicine.

# about the contributors

**Joseph M. Berning, PhD,** is an assistant professor in the physical education, recreation and dance department and director of the Exercise Physiology Laboratory at New Mexico State University (NMSU). He is an NSCA-certified strength and conditioning specialist and NSCA academic program director for NMSU. He currently teaches six individual kinesiology-based courses and conducts research in the areas of anabolic steroids and anaerobic exercise. In addition, he examines physiological responses of resistance training with various populations.

**Veronique Billat, PhD,** is a professor in the department of sport and exercise science at the University of Evry in France and director of the Laboratory of Exercise Physiology (LEPHE). She graduated from the University of Grenoble in France in 1988 and held positions of associate professor and professor at the University of Creteil and University of Lille. Her research interests include the optimization of training in middle- and long-distance running. She is also a fellow of the American College of Sports Medicine and a member of the American Physiology Society and the European College of Sport Science.

**Holly M. Cotter, MS,** is a student in the doctorate of physical therapy program at Concordia University in Mequon, Wisconsin.

**Jack T. Daniels, PhD,** is director of the Running Research Center at the University of Northern Arizona. A two-time Olympian in the modern pentathlon (silver team medal, Melbourne 1956; bronze team medal, Rome 1960), he received his PhD from the University of Wisconsin at Madison. His early research with adaptation to altitude, conducted before the 1968 Mexico City Olympics, set the stage for our understanding of the use of altitude training for both altitude-level and sea-level competitions. He is well known as both a running coach (*Runner's World* featured him on the cover as the World's Best Coach, and he has been designated a master coach by USA Track & Field) and applied physiologist, specializing in studies of the economy of movement. He is the author of *Daniels' Running Formula, Second Edition* (published by Human Kinetics in 2005).

**James A. Davis, PhD,** is a professor in the department of kinesiology and director of the Laboratory of Applied Physiology at California State University at Long Beach. He received his PhD in physiology from the University of California at Davis in 1975. His research interests include acid/base balance and pulmonary gas exchange during exercise. He has been a member of the board of trustees for the American College of Sports Medicine. Additionally, he was a member of the editorial board for the journal *Medicine and Science in Sports and Exercise.*

**Jos J. deKoning, PhD,** is a Universiteit Docent in the faculty of human movement studies at the Vrije Universiteit Amsterdam in the Netherlands, as well as an adjunct professor in the department of exercise and sport science at the University of Wisconsin at La Crosse. His research interests include modeling cyclic exercise performance. He is regarded as one of the world's foremost experts on the physiology and biomechanics of speed skating.

**Christopher Dodge, MS,** is the manager of the Human Performance Laboratory at the University of Wisconsin at La Crosse.

**Duncan N. French, PhD, CSCS,** is the performance sport manager at Northumbria University, UK. He also works as a lecturer within the Division of Sports Sciences at Northumbria University and is a consultant strength and conditioning coach to the English Institute of Sport. Duncan is the strength and conditioning coordinator for the England Basketball national programmes. His research interests focus on the hormonal responses to resistance exercise, athletic development in elite performers, and neuromuscular adaptation to resistance training. Duncan is a graduate of the University of Connecticut.

**Andrew C. Fry, PhD, CSCS,** is a full professor in the department of health and sport sciences at the University of Memphis and the director of the Exercise Biochemistry Laboratory. He is a former competitive wrestler and lifter. His recent research has focused on resistance exercise physiology and performance. Dr. Fry has extensively studied the phenomenon of overtraining, adaptations to resistance exercise, and the relationship between cellular and molecular characteristics of muscle and voluntary muscle performance. He is a past vice president of the National Strength and Conditioning Association.

**Linda Garzarella, MS,** is a coordinator of statistical research in the department of biostatistics in the

College of Medicine at the University of Florida at Gainesville. She received her BS in nutrition science from Drexel University in Philadelphia and MS in exercise physiology from the University of Florida. Her thesis project, "Predicting Body Composition of Healthy Females by B-Mode Ultrasound: Comparison with Anthropometry and Bioelectrical Impedance," was conducted under the guidance of Drs. M.L. Pollock and J.E. Graves. Her other interests include weight training and studying Hebrew.

**James E. Graves, PhD,** is a professor of exercise and sport science and dean of the College of Health at the University of Utah. He previously served as professor of exercise science and associate dean for graduate studies, budget and research in the School of Education at Syracuse University, where his accomplishments include the establishment of the department of exercise science and foundation of the Musculoskeletal Research Laboratory. He is a former chair of the exercise science department and held an adjunct appointment in the department of physical medicine and rehabilitation at the State University of New York Upstate Medical University. He also assisted in the development of the Center for Exercise Science, located in the Colleges of Medicine and Health and Human Performance, at the University of Florida at Gainesville. Dr. Graves is coauthor of *The Lumbar Spine* (Sequoia Communications 1988). His research interests and experience include the influence of physical activity on aging and the prescription of resistance training for the prevention and rehabilitation of orthopedic problems including low back pain. His most recent publication is an edited text titled *Resistance Training for Health and Rehabilitation* (Human Kinetics 2002). In 1999 Dr. Graves was named Syracuse University's United Methodist Scholar Teacher of the Year. He earned his PhD in exercise science from the University of Massachusetts at Amherst.

**Everett Harman, PhD,** completed his doctoral degree at the University of Massachusetts at Amherst in 1984. He currently works as a research physiologist and head of the biomechanics laboratory at the U.S. Army Research Institute of Environmental Medicine in Natick, Massachusetts, where he has served for more than 20 years. His research has focused on the biomechanics and physiology of physically demanding activities. He served as vice president for research of the National Strength and Conditioning Association and currently serves on the development committee for the organization's certified personal trainer examination. Dr. Harman has conducted major physical training studies and has developed several unique tests to measure activity-specific physical abilities.

**Floor J. Hettinga, MS,** is a PhD student in the faculty of human movement Studies at the Vrije Universiteit Amsterdam. Her research interests include modeling cyclic exercise performance.

**Joohee Im** is completing her PhD at Temple University while working in biophysics at the University of Pennsylvania. She worked as a physiologist for the U.S. Olympic Committee before beginning her PhD work. Her research interests include skeletal muscle energetics and the study of mitochondrial and hemodynamic function in diabetes, using NIRS and MRS.

**Jill Kanaley, PhD,** is an associate professor in the department of exercise science at Syracuse University and director of the Human Performance Laboratory. She graduated from the University of Illinois at Urbana-Champaign and completed postdoctoral fellowships at the Mayo Clinic and the University of Virginia. Her research interests include exercise endocrinology and metabolism with a focus on type 2 diabetes and obesity. She has also conducted additional research on issues related to body composition and growth hormone. Dr. Kanaley is a fellow in the American College of Sports Medicine and is currently the president of the Mid-Atlantic Regional Chapter of the American College of Sports Medicine.

**Kate Kerr, PhD,** trained as a physiotherapist in Belfast, Northern Ireland, and worked as a clinical physiotherapist in Northern Ireland and the United States. She moved into academia as a lecturer in physiotherapy at the University of Ulster and recently as senior lecturer at the University of Nottingham, where her main interests include kinesiology, sports medicine, and health education. She has been extensively involved in professional activities with the Chartered Society of Physiotherapy (UK) and has published widely in books and journals.

**William J. Kraemer, PhD, FACSM, CSCS,** is a full professor in the department of kinesiology in the Neag School of Education, working in the Human Performance Laboratory at the University of Connecticut at Storrs. He holds an appointment as a full professor in the department of physiology and neurobiology along with an appointment as a full professor of medicine at the UCONN Health Center School of Medicine. Dr. Kraemer is a fellow of the American College of Sports Medicine and a past president of the National Strength and Conditioning Association.

**Joann Lampen, MS,** is a graduate student in the faculty of human movement studies at the Vrije Universiteit Amsterdam.

Philippe Lopes, PhD, is an associate professor in the department of sport and exercise science at the University of Evry in France. He graduated from the University of Ulster in exercise physiology in 1998 and became senior lecturer in exercise physiology at the University of Brighton (Chelsea School of Sport and Exercise Science, Physical Education and Dance) and subsequently at the University of East London. His research interests include heart rate variability at rest and during exercise in sedentary, recreational, and elite populations and cardiorespiratory interactions during exercise.

Michael R. McGuigan, PhD, is a lecturer in exercise physiology in the School of Exercise, Biomedical and Health Sciences at Edith Cowan University, Australia. He completed his BPhEd (Hons) at the University of Otago and a PhD at Southern Cross University, Australia. After completing his postdoctoral work with Dr. William Kraemer at Ball State University, he was an assistant professor at the University of Wisconsin at La Crosse for three years. His research focuses on the physiological responses to resistance exercise and the use of resistance training as a health intervention for various populations.

At the time of his death in 1998, Michael L. Pollock, PhD, was a professor of medicine, physiology, and health and human performance and founding director of the Center for Exercise Science at the University of Florida at Gainesville. Born in 1936 in Los Angeles, Dr. Pollock earned a bachelor's degree from the University of Arizona (1958) and master's (1961) and doctoral (1967) degrees from the University of Illinois. A prolific writer and researcher, Dr. Pollock was the author of three books and more than 300 journal articles related to exercise and the influence of exercise health maintenance and disease prevention. Dr. Pollock's work involved the prescription of various forms of physical activity, cardiopulmonary rehabilitation, and body composition analysis. Dr. Pollock was the lead author on the initial position stand of the American College of Sports Medicine on the recommended quantity and quality of exercise for developing and maintaining fitness in healthy adults. He served as president of the American College of Sports Medicine from 1982 to 1983.

Nicholas A. Ratamess, PhD, CSCS, is an assistant professor in the department of health and exercise science at the College of New Jersey in Ewing, New Jersey. He received his PhD in kinesiology and exercise science from the University of Connecticut. His major research interest is examining the physiological adaptations to resistance training. He has authored or coauthored more than 50 scientific and educational publications in the strength and conditioning field.

Kenneth W. Rundell, PhD, completed his dissertation and postdoctoral work at SUNY Health Science Center at Syracuse. He worked as a senior physiologist for the U.S. Olympic Committee for 10 years. He is currently director of the Human Performance Laboratory and professor of health science at Marywood University in Scranton, Pennsylvania. His research interests include the effects of inhalation of ultrafine and fine particles on airway and vascular function.

Matthew J. Sharman, PhD, is a postdoctoral research fellow in the School of Exercise, Biomedical and Health Sciences at Edith Cowan University, Australia. He completed a BHMS and BA (Hons) through Southern Cross University and University of Technology Sydney in Australia and MA and PhD at the University of Connecticut. His research focuses on the role of the immune response in atherosclerosis and the effect of diet and exercise interventions on this disease.

John White, PhD, is a University of Nottingham emeritus senior lecturer in sports and exercise medicine, having spent a career in university education in England after completing graduate studies as a Fulbright scholar in the United States. He has broad experience in sport and exercise sciences, health screening, and lifestyle evaluation. He has published extensively in journals and books as an author, editor, and reviewer and has widespread exposure in media and communications. He is currently a freelance consultant.

*You'll find
other outstanding
exercise physiology
resources at*

# www.HumanKinetics.com

*In the U.S. call*

# 1-800-747-4457

Australia.................................................. 08 8277 1555
Canada .................................................. 1-800-465-7301
Europe........................................+44 (0) 113 255 5665
New Zealand.................................... 0064 9 448 1207

**HUMAN KINETICS**
*The Information Leader in Physical Activity*
P.O. Box 5076 • Champaign, IL  61825-5076 USA